Petrology (Igneous and Metamorphic Rocks)

Proceedings of the 27th International Geological Congress

Stratigraphy	Volume 1
Palaeontology	Volume 2
Quaternary Geology and Geomorphology	Volume 3
Sedimentology	Volume 4
Precambrian Geology	Volume 5
Geology of Ocean Basins	Volume 6
Tectonics	Volume 7
Geophysics	Volume 8
Petrology (Igneous and Metamorphic Rocks)	Volume 9
Mineralogy	Volume 10
Geochemistry and Cosmochemistry	Volume 11
Metallogenesis and Mineral Ore Deposits	Volume 12
Oil and Gas Fields	Volume 13
Solid Fuel Mineral Deposits	Volume 14
Non-metallic Mineral Ores	Volume 15
Hydrogeology	Volume 16
Engineering Geology	Volume 17
Remote Sensing	Volume 18
Comparative Planetology	Volume 19
Mathematical Geology and Geological Information	Volume 20
History of Geology	Volume 21
Geological Education	Volume 22
Origin and History of Marginal and Inland Seas	Volume 23

Proceedings of the
27th International Geological Congress
Moscow 4–14 August 1984

Volume 9

PETROLOGY (IGNEOUS AND METAMORPHIC ROCKS)

VNU SCIENCE PRESS
Utrecht, The Netherlands
1984

VNU Science Press BV
P.O. Box 2073
3500 GB Utrecht
The Netherlands

First published 1984

ISBN 90–6764–018–2 Volume 9
ISBN 90–6764–009–3 set of 23 volumes

Printed in Great Britain by J. W. Arrowsmith Ltd, Bristol

PREFACE

The leading scientists in the many different fields of geology were invited by the Organizing Committee to present a paper on a specific problem in present-day geological science at the 27th International Geological Congress. The published proceedings of the Congress consist of twenty-three volumes. Each volume is dedicated to a particular aspect of geology. Together the volumes contain all of the contributions presented at the Congress.

The Organizing Committee is pleased to acknowledge the efforts of all of the participating scientists in helping to produce these proceedings.

Professor N. A. BOGDANOV
General-Secretary of the
Organizing Committee

CONTENTS

MAGMATISM AND GEODYNAMICS

O.A. BOGATIKOV, L.P. ZONENSHAIN, V.I. KOVALENKO
Institute of Geology of Ore Deposits, Petrography, Mineralogy, and Geochemistry, Acad. Sci. USSR;
Institute of Oceanology, Acad. Sci. USSR, Moscow.

ABSTRACT

The history of the Earth is classified into the early (gefore 2,500 my B.P.) and late (after that date) stages. The former stage resulted in the formation of continental nuclei, greenstone belts, the world ocean. The occurrence of magmatism (M) is controlled by geodynamical environments specific for the early Precambrian alone. The later stage M is believed to be associated with the dynamics of lithosphere plates. We distinguish the elementary (simple) - oceanic and continental rifts, intraoceanic island arcs, active continental margins, collision zones of continents, intraplate; and mixed (combinations of several simple ones) geodynamical environments with mixed magmatism. Using Central Asian, Uralian, Mediterranean fold belts as examples, we demonstrate different successions of gedynamical environments in time and space. However, all these belts arose in the place of oceanic basins of different types following the convergence and subsequent collision of continents.

INTRODUCTION

Geodynamics studies deep-seated forces and processes caused by the evolution of the planet Earth, and which in turn control the movement of material and energy within the Earth and its upper solid shells. In this sense the relationschips between magmatism and geodynamics are obvious. It is also clear that magmatism can serve as an indicator of those geodynamical environments in which it arises.

Proceedings of the 27th International
Geological Congress, Volume 9, pp. 1–25
PETROLOGY (IGNEOUS AND METAMORPHIC ROCKS)

Changes in time

Both magmatism and those geodynamical conditions in which it appeared have not been uniform through the Earth's geologic history. Isotope studies of Sr, Nd, Nf, Pb, He, Ar, Xe in the basalts of the mid-oceanic ridges (MOR basalts), in intraplate basalts, and continental magmatic rocks have revealed an important mark in the history of the Earth about 2,000 m.y. ago (6, 20, 21, 26), when a layer appeared in the Earth's mantle that gave rise to MOR basalts depleted in radiogenic strontium and lithophilous elements with large ionic radii (LIL elements). Geologic eveidence tells us that at least 80% of the continental crust had been created before that time /8, 3, 5/ and, since this is rich in LIL elements, and is complementary to MOR basalts in this respect, it can be hypothesized that the upper crust had before 2,000 m.y. ago differentiated into a continental lithosphere enriched by LIR elements and the depleted mantle layer. There was also the undifferentiated lower mantle, or the undepleted mantle. It has left its imprints in the isotope characteristics of intraplate volcanism both on the continents and in the oceans. We agree with many other investigators in accepting the depleted-undepleted mantle boundary to lie at a depth of 650 km, where there is an interface between the lower and upper mantles (Fig.1). It is at this boundary that phase transitions into high-pressure minerals occur, as can be inferred from experimental evidence. Below it there are no Zavaritsky-Benioff seismic focal zones. The upper mantle was subjected to metasomatic alteration by fluids probably coming from the undepleted lower mantle /24, 2/. This has led to some portions of the depleted upper mantle being enriched by LIL elements.

The history of the Earth is thus divided into two distinct stages; relationships between magmatism and geodynamics should be discussed separately for each.

The early stage

The Earth's geologic history during the first

2,500 or 3,000 m.y. can be divided into three periods /3, 4/: 1) The lunar period, older than 3,500 m.y.; 2) the nuclear one (3,500 to 2,500 m.y. ago); 3) the craton one (2,500 to 1,500 m.y. ago). The lunar period corresponds to the formation of the protocrust. The oldest known rocks, of age 3,800-3,850 m.y., are "grey gneisses" /19/ which are like andesites compositionally, i.e., have the mean composition of the continental crust. If the "grey gneisses" were actually the protocrustal rocks, then they were probably formed by coming directly from the mantle owing to partial melting of high water content mantle material prior to the separation of the hydrosphere water.

The main event of the nuclear period was the formation of older continental nuclei. Other events include the formation of greenstone belts where magmatic rocks of different compositions are present /18/: tholeiitic basalts, calc-alkaline andesites, as well as highly typical komatiite lavas whose formation requires temperatures above 1500^{0}C. Widespread komatiites are evidence for higher temperatures in the older mantle as compared with the present-day mantle due to a higher geothermal gradient. There is some evidence /9/ that komatiites come from the primary undepleted mantle. There are vertually no real ophiolites within the greenstone belts. Greenstone belts have been compared with fold belts that have experienced the entire cycle of evolution, from opening to closure of the oceanic basin, with volcanic island arcs and marginal basins behind them, have been identified as old rift zones, been regarded as a result of interaction between a great number of thin lithosphere plates, and even of impact processes. The magmatism and structure of greenstone belts is so peculiar, however, that not one of these hypotheses can be regarded as definitively proven. Plate tectonics was apparently not operative in the Archean, at least in its present form. Numerous tension zones might exist (not a single worldwide system as they are now) with plenty of basaltic melt coming from a heated mantle. This widely dispersed spreading of crustal blocks might be compensated by crustal compression and thickening between greenstone belts with simultaneous high-temperature metamorphism, granite melting, and growth of granite-greiss domes, thus increasing continental nuclei.

The formation of continents was completed during the craton period, but many formations of the nuclear period continued their evolution. It is during this time that first large-scale effusions of acid lavas occurred forming extended volcanic belts like the Akitkan one near Lake Baikal, as well as rapakiwi-granite massifs. These belts were located in continental margins and so were close to Andean type belts.

Late stage

The past 2,000 m.y. of the Earth's history are largely characterized by the fact that the chief events in its upper shells have been controlled by the movement of lithosphere plates.

The late stage has four sources of magmatism: 1) continental crust enriched by LIL elements (crustal magmatism), 2) a mantle depleted in LIL elements (oceanic depleted mantle magmatism), 3) undifferentiated primary mantle (undepleted mantle magmatism), 4) a metasomatically altered depleted mantle enriched by LIL elements (undepleted mantle magmatism).

Present and past manifestations of magmatism fall into two large groups: those occurring at the boundaries of lithosphere plates and intraplate ones. The source of the former group magmatism lies within the depleted mantle, enriched continental and oceanic crust, or a mixture of both. The latter group magmatism requires material from the undepleted mantle. The geodynamical environments, their magmatism and its sources are shown in outline in Figures 1 and 2. These environments can be regarded as simple, since there are also various complicated combinations of these that lead to the appearance of composite magmatism (Figures 3 through 6).

Plate boundary magmatism

Plate boundary magmatism is by far the strongest,

both for extent and intensity (Figure 2).

Magmatism of constructive spreading boundaries

Plates may be spreading both in oceanic and continental crust. Oceanic and continental rift zones are a structural expression of this process.

Oceanic rift magmatism

The oceanic crust created at the axes of mid-oceanic ridges is formed by magmatic rocks alone, ranging from the layered ultrabasic-basic complex to gabbro to MOR basalts. Ophiolites correspond to the past oceanic crust in fold belts. MOR basalts represented by low-potassium tholeiites are formed during partial melting of the depleted mantle under "dry" conditions only. Geophysical evidence from mid-oceanic ridges and studies in ophiolite massifs /7, 22/ indicate that basalts are coming from a funnel-like chamber at depths 5-10 km with a width of 30-40- km which is continually supplied from below. Here a steady-state process goes on creating the oceanic crust with cumulative precipitation of heavy minerals on the chamber bottom, downward building of the chamber roof and magma upwelling, forming a series of dykes which feed basalt effusions at the sea floor. Most of the magma is differentiated within the chamber and during the travel towards the surface. Volcanic features are localized within narrow (1 to 3 km) axial extrusive zones above the centre of the magma chamber. It is here that the bulk of crustal igneous rocks is created: each year 2 to 4 km^2 area of new oceanic crust has been formed at spreading axes during the Cenozoic, i.e., taking a minimum thickness of oceanic crust to be 5 km we get 10 to 20 km^3 annually. This is an order of magnitude greater than the volume of volcanism above subduction zones.

Continental rift magmatism

A bimodal contrast volcanism of basalt-rhyolitic composition is typical for intracontinental rifts. It includes occasionally comendites, ongonites and topaz-bearing rhyolites. The basalts are of the so-called transitional type, somewhat enriched by LIL elements compared with MOR basalts, apparently owing to the contamination with crustal material. Rhyolites are generated due to melting of the continental crust above a rising heated mantle diapir. Data from the East African rift system show /12,16/ that felsic lava cruptous are dominant in slightly opened rifts opening, basalts appearing only later. Alkali basalts are known to occur in many continental rifts, as well as other alkaline and alkali-ultrabasic rocks.

Magmatism of destructive boundaries (plate collision)

With lithosphere plates converge a great variety of steructures result above subduction zones, depending on the type of plate (continental or oceanic ones), their age, and on how they more relative to subduction zones themselves. An important factor, one that has been nearly overlooked until now, is the degree of coupling between converging plates. This degree of coupling may be very small, the plates are then torn off ("decoupling") one from another /11,32/ with one of the plates being freely downwelled into the mantle. This is the picture presented by the Mariana subduction zone, expressed as absence of large earthquakes with magnitudes >5-6. When coupling is weak the usual result is the appearance of extension and the opening of interarc basins. Again, coupling between plates may be strong, resulting in predominant compression. An example is the Andean margin of South America. Strong coupling is shown in numerous earthquakes with magnitudes 6. The conditions of strong coupling may continue for a long time.

Structures arising at the convergent boundaries can be classified as follows /31,4/ : 1) intraoceanic island arcs; 2) active continental margins (with island arcs of the Kamchatka type and margins of the Andean type); 3) zones of continental collision.

Magmatism of intraoceanic island arcs and active continental margins with island arcs.

Strong coupling if only episodically is necessary for the island arcs. An oceanic plate or sometimes (as exemplified by the Banda arc) a very thin continental plate is underthrusted into subduction zone. Basically we have very different types of volcanism, from island-arc tholeiites to the commonest lavas (andesites, andesite-basalts) of the calc-alkaline series to subalkaline (Shoshonites) and alkaline basalts. The lavas are frequently accompanied by intrusions of quartz diorite and plagio-granite.

Composition of arc volcanic products depends on a number of local conditions: 1) coupling between plates: when coupling is weak, extention arises, even in the frontal part of arcs, resulting in the appearance of island-arc type tholeiitic basalts (the Mariana arc), while strong coupling produces dominant compression with a greater proportion of calc-alkaline andesites (the Kuril arc); 2) the rate of subduction; the greater the rate the larger is the volume of volcanic products and the smaller the bulk alkali content (Miyashiro); 3) the time of subduction; the longer the time the more matured becomes the arc, the thicker its crust, the greater the opportunities for the production of calc-alkaline magma (the New Hebrides arc); 4) the crust and mantle composition of the overriding plate zone on which an island arc is being formed; if it is and oceanic plate, the dominant rocks are basalts and such characteristic rocks as boninites and marianites, while if the plate includes continental crustal blocks, then acid magmas are highly probable etc. Features common to the magmatism of any isalnd arc remain: 1) a voluminous fluid phase due to dehydration of the undergoing ocean plate and subducted sediments /23/, this being the essential difference between relatively water-rich island-arc magmatism and the "dry" magmatism incident to the divergent plute boundaries it may be that the high water content of island arc magmas is in some degree related to Korzhinsky's fluid flows /14/; 2) the presence of calc-alkaline, strongly differentiated series; 3) lateral zoning of polarity in the increase in alkali content, in particular in potassium content with constant concentration of silica as we go from the fromtal to back part of an island arc /10,30/.

The source of island-arc magmatism lies in the material sinking into the subduction zone and in a wedge of crust of upper mantle overriding it, i.e. mostly in the depleted mantle (possibly with some parts of metasomatically altered mantle nock) and in the continental (enriched) crust if it is present.

Magmatism of active continental margins of the Andean type

Ussually, a young oceanic plate, is subducted directly a continent along active continental margins. The conditions for the magma generation are those prevailing under intraoceanic island arcs, but here a thick continental crust rich in lithophile elements takes part in magma formation, and, as a result, aboundant felsire of the calc-alkaline series appear. Many investigators, relate latites and shoshonites which are typically for the Andean type margins with the continental crust influence though they may have been produced by melting of a metasomatically altered and, possibly, undepleted mantle. Palingenic granitic batholiths arise. A comparatively well-defined lateral magmatic zonal pattern emerges, calc-alkaline volcanism being replaced as we move inland by latites and shoshonites, and granite-granodiorite batholiths by diorite-monzonites, "standard" and lithium-fluorine rare-metal granites and alkaline rocks /15,30/.

Magmatism of plate collision zones.

There may be different cases of collision, the colliding objects being island arcs (Sulawesi island), an island arc and passive continental margin (Timor and Australia), two continents (India and Eurasia in the Himalayas) etc. In all cases the subduction zone is blocked and consumption stops owing to one of the colliding lithospheric segments being buoyant. There is empirical evidence indicating that the continental lithosphere, if thinned within a passive margin, can be subducted through 100 to 200 km (Zagros), rarely a greater distance (Tibet). Collision creates conditions favourable for obduction of ophiolites, i.e., for the oceanic basement of island arcs being thrusted onto a continental adge. When subduction has ceased the subducted continental lithosphere experiences high temperature metamorphism (proceeding as far as granulites) /28/, is mobilized, and produces granite-gneiss domes. Collision causes piling up and thickening of the crust accompanied by generation of the magmas that have produced granite batholiths.

Intraplate magmatism

All extrusive and intrusive activities that are not controlled by boundaries of lithospheric plates can be considered as intraplate magmatism (Fig.2): examples are volcanic oceanic islands (Hawaii), isolated volcanoes (Tibesti) and extensive flood basalts (traps) on the continents /33/. The most typical rocks are basalts, both of tholeiitic and alkaline type; alkaline and alkali-ultramafic rocks, kimberlite included, are also common. Acid varieties sometimes appear, granite included (Jos plateau in Nigeria, White Mountains massif in the Appalachians). An obligatory component is material from the undepleted mantle or metasomatically altered mantle, resulting in the magmatic products being enriched by radiogenic strontium and LIL elements. For example, tholeiitic basalts of the oceanic islands are at least twice as rich in LIL elements as are (also tholeiitic) MOR basalts. Traps seem to be comparatively "clean" products derived from the undepleted mantle, though some investigators quote evidence for their contamination by continental crustal material. In the other cases it is argued either in favour of contamination by enriched continental crust (volcanous in the continents) or dilution by the depleted mantle (volcanic oceanic islands, Iceland etc.) /1,27/. Intraplate volcanism, traps, for instance, can initiate rifting of the continents, i.e. formation of boundaries where divergence of the lithosphere plates will take place. There seems to be some significance in the fact that it is frequently difficult to draw a districtuion between the magmatism of continental rifts and intrap plate magmatism. It should be stressed that manifestations of intraplate magmatism vary widely from place to place in composition: it may be trapps, then kimberlites or alkali-ultrabasic massifs, or again granite, with combinations of the above rocks in some cases.

Composite magmatism

We have discussed relationships of magmatism has to elementary (simple) geodynamical features (Fig.1). Situations are fairly common, however, in which either magmatic products proper to boundaries between lithoshere plates of different types coicide spatially or intraplate magmatism is superposed upon the plate boundary magmatism (Figures 2 through 5). Possible combinations of actual environments for mixed magmatism are shown in Fig.6. Some modern examples are: 1) Iceland (Fig.3), where intraplate volcanism is superposed on the spreading boundary, and wher MOR basalts are being mixed with undepleted mantle material /25,27,20/; 2) the Mediterranium belt, where intraplate magmatism is superposed upon the boundary where the African and Eurasian plates converge; 3) western North America (Fig.4), where there is spatial coincidence of Cascade Range calc-alkaline volcanism associated with subduction of the minor Juan-de-Fuca plate under North America, rift magmatism of the Basin and Range Province, and Columbian plateau basalts associated with intraplate volcanism; 4) interarc basins which are opening back of island arcs (Fig.5), where typical MOR basalts are effused in basin centres, at spreading axes; while at the edges, sometimes in basin centres, lavas associated with subduction zones appear; i.e., products of "watered" magmas are here side by side with "dry" basalts. There may be also other cases in which composite magmatism occurs in different elementary geodynamical features.

Evolution of magmatism and paleogeodynamical environments of the geoligic past in fold belts

Terrestrial fold belts contain magmatic formations that are paleoanalogs of present-day magmatic manifestations being formed in different geodynamical environments. The entire spectrum of types of magmatism we encounter in fold zones can be explained by various combinations of these environments /31,4/. Illustrations follow (Fig.7).

North Mongolia (Fig.7a). Caledonian structures occupy the northern part of the Central Asia belt wihtin Mongolia. They consist of late Precambrian to Cambrian, at places (Mongolian Altai) Ordovician sequences that experienced folding activity in the middle of the Cambrian and Silurian. The area involved in Cambrian (Salairian) folding shows the following succession of magmatism (upwards):

1) Upper Riphean - bimodel basalt-rhyolite series (Dzabhan suite, southwestern Khangai Highland; rocks of the Sarkhoi suite (East Sayan mountains, Prihubsugugulye); overlies strongly unconformably the old sialic basement, reflecting the conditions of continental riftogenesis;

2) The Vendian (at places with lower Cambrian) - ophiolites represented by a complete section (Khan-Taishiri Range), form metamorphosed peridotites to a layered complex to a series of parrallel dykes to pillow lavas; correspond to the floor of ocean basins;

3) Lower to Middle Cambrian (Vendian at places) - calc-alkaline volcanites with shoshonites (Lake and Ider zones) corresponding to island arcs, at places (Khan-Kluhei range) basalts closely resembling oceanic tholeiites; may correspond to the floor of a marginal sea;

4) Middle to Upper Cambrian - tonalite-granodiorite batholiths of the Telmin complex; correspond to collision of the North Mongolian microcontinent and Siberian continent;

5) Lower to middle Ordovician - calc-alkaline island-arc volcanites (eastern Mongolian Altai); corresponds to the island arc along the edge of the Siberian continent outside the collision zone;

6) Lower to Middle Cevonian - volcanic series of the bimodal basalt-rhyolite association, subalkalic volcanites and granites with granosyenite and alkalic granites, alkalinc gabbro, ijolite - urtite, syenite (Great Lakes basin, southwestern East Sayan Mountains, northwestern Khangai); calc-alkaline volcanities of Mongolian Altai; corresponds to composite magmatism as a result of different geodynamical envirionments superposed (Fig.6): an island arc, a continental rift

and, possibly, intraplate magmatism;

7) Permian - Triassic-subalkalic and alkalic volcano-plutonic series: trachybasalts, trachyandesites, trachyrhyolites, comendites, alkalic granites; corresponds to composite magmatism as does the Devonian.

To sum up, the evolution of magmatism in North Mongolia presents evidence in favour of a continent in the Riphean, opening of the ocean, and the appearance of an island arc system in the Vendian closure of the oceanic basin in the middle Cambrian due to collision of the continents and accretion of the Siberian continent at the expense of this area. The oceanic basin existed at that time (until the Carboniferous) in the place of South Mongolian Hercynides. Devonian and Permian-Triassic composite magmatism may be associated with the edge of the Siberian continent approaching the spreading axes of the South Mongolian oceanic basin, which determines the association of island-arc and riftogenic magmatic series. Intraplate magmatism may also have occured at that time.

South Mongolia (Fig.7b). This region belongs to typical Hercynides with widely occurring Middle Paleozoic deposits, from Upper Ordovician to Lower Carboniferous, which had been deformed during the period from latest Devonian until the middle of the Carboniferous. The evolution of magmatism here can be tentatively outlined as follows:

1) Late Ordovician-Silurian - fragmentarily disjuncted ophiolites consisting of serpentinites, gabbro, pillow lavas, included as blocks in strips of serpentine melange (Dzolen Range, Gurvan-Saikhan, Manlai etc.); mark the floor of the Middle Paleozoic Paleothetys ocean;

2) The Devonian - widely occuring calc-alkaline andesite basalt sequences corresponding to an island arc that arose on the Paleothetys ocean floor; occur together with thick silica-gray-wacke sequences that in all probability indicate the existence of marginal basins back of island arcs; first tectonic sheets and olistostromes aros in the latest Devonian;

3) Early Carboniferous - an extended strip of calc-alkaline basalt-andesite-rhyolite volcanites in the

south and shoshonite ones in the north which controlled the Zavaritsky-Benioff zone inclining northwards under the Siberian continent /29/; it southern part unconformably overlies a folded basement, the northern part overlies Devonian sequences, at places conformably;

4) Middle Carboniferous - granodiorite-granitic intrusions in the north of the South Mongolian zone which temporally coincide with folding in this part of the zone (collision of the South Gobi microcontinent and Siberia);

5) Late Carboniferous-Permian - calc-alkaline plutonic and volcanic magmatism in combination with alkalic and bimodal magmatism; calc-alkaline magmatism occurred in a continuous belt along the southern margin of Mongolia and was controlled by a subduction zone inclining northwards of the late Paleozoic Paleothetys (Inner Mongolian zone); alkalic (pseudoleucite syenites of the Lutingol massif) and bimodal basalt-trachyrhyolite-comendite-alkali-granitic associations occur as sublatitudinal intermittent strips (South Gobi belt of alkalic rocks); reflect a magmatism close to the rift one with possibly intraplate magmatism superposed upon the subduction magmatism;

6) Triassic-Early Jurassic - dyke belts of subalkalic basalts, trachyrhyolites, alkali granites, which are probably indicators of continental riftogenesis; possibly with intraplate magmatism;

7) Late Jurassic - early Cretaceous - alkali complexes (melanephelinites, trachytes, syenites, latites) and bimodal alkali-basalt-rhyolite-ongonite associations, indicating breakup (rifto-genesis) of this area, possibly in relation to passage over a "hotspot".

To sum up, South Mongolia experienced esland-arc environments that were formed on the Paleo-Tethys ocean floor replaced by mixed environments in which there was spatial coincidence of conditions for Andean type active margins and continental rifting, initiated by passage above abnormally heated mantle regions.

The alternation of magmatism and geodynamical environments in the South Urals is as follows (Fig.7c):

1) Early Ordovician - alkalic-basalt and bimodal (basalt-rhyolite) volcanism (Sakmar zone, Trans-Ural rise), reflecting a continental rifting and breakup of the East European continental margin.

2) Lower Silurian oceanic complex making up allochthone nappes consisting of large ultramafic bodies and rocks of a layered series below (Kempirsai and Khabarny massifs and others) and low potassium content basalts above (Polyakov and Sugralin suites and others); corresponds to the opening time for the Ural paleoocean;

3) Silurian calc-alkaline andesite-dacite-rhyolite series in the Sakmar zone; they mark the formation of an island arc on the sea floor and separation of a marginal basin near the edge of the East European continent;

4) Devonian and Early Carboniferous (until Early Visean) calc-alkaline andesite-basalt and andesite-dacite-rhyolite series in the inner zones of the Urals with volcanism migrating eastwards from the Early-Middle Devonian (Irendyk suite) to Middle-Late Devonian (Ulutaus and Koltuban suites) and Early Caboniferous (Berezov suite); characterize a long-continued existence of island arcs with an eastward displacement towards the Ural paleoocean; tholeiitic basalts and bimodal basalt-rhyolite sequences (Lower Devonian Baimak-Burebai suite, Middle Devonian Karamalytash suite), appears behind the arcs nearest to the East European continent, indicating extension and back-arc spreading with the formation of marginal basins; Givetian basalts are widely spread in the southern most Urals (Mugodjary Mts); they include series of shected dykes and pillow lavas having mixed oceanic and island-arc characteristics; they are a remnant of a back-arc ocean floor behind the Devonian island arc; alkalic lavas (chancharites) found in the Saumara zone and other localities seem to be related toa "hotspot".

5) Granites and granite-gneiss domes of the Mugodjary anticlinorium with an age of 370-310 m.y.; the

formation of the domes was associated with collision between the Mugodjary micro-continent and the island arc.

6) Calc-alkaline, andesite-basalt volcanic series of the upper Visean, Namurian, and partly of the Bashkirian stage that makes up the extended Valeryan volcanic belt along the Ural-Kazakhstan boundary corresponds to an island arc that arose near the edge of the Kazakhstan continent; the floor of the Ural paleoocean has been consumed under it.

7) Late Carboniferous and Permian granites and granite-gneiss domes whose formation was related with collision between the East European and Kazakhstan continents, which began in the middle of the Caboniferous and had continued 90 m.y. until the earliest Triassic; this extremely long collision period has created the present-day nappe-fold structure of the Ural mountains, thickened its crust, and produced a tectonic layering of the lithosphere as a series of superposed nappes; the massifs miaskite in ilmens and the Vishnevye mountains were generated just at that time. As indicated by recent data obtained by V.A. Kononova /13/, metamorphism and remobilization of the miaskites occurred during late Carboniferous-early Permian time, while their formation must have occurred at an earlier date, possibly that corresponding to continental riftogenesis.

Typical features of the Uralian evolution are thus (1) development of a system of island arcs and marginal seas, producing copions volcanic rocks and chalkophilous metallogeny, and (2) a very long period of continental collision, causing the present linear structure of the Urals.

The Mesozoic-Cenozoic evolution of the Mediterranean belt bears the history of the thys oceanic closure recorded in it. The succession of events reflected in the magmatism can be represented as follows (Fig.7d):

1) Late Triassic (at places Early Jurassic) calc-alkaline volcanism in northern Predkavkazye, or Antecaucasus (Nogai suite) and in the Crimea marks the northern active Eurasian margin of the Thetys, in contrast to the southern passive Gondwana margin;

the volcanism contains evidence for crustal consumption of the Late Paleozoic Tethys; at the latest Triassic, the Iranian continent collided against an Eurasian margin, causing Cimmerian folding;

2) The first half of the Jurassic saw the formation of Sevan-Akkerin ophiolites, which indicates continues spreading in the Tethys ocean; the formation of the lesser Caucasus-East-Pontian volcanic belt, made up of volcanic rocks (andesites and andesite-basalts) of the calc-alkaline series, took place from the Middle, at places Early, Jurassic until the end of Neocomian. The belt corresponds to an island arc of the relevant age; during Early Jurassic time, back of the arc within Greater Caucasus, alkali and tholeiitic basalts were extended probably owing to wpreading, which had created a deep-sea marginal basin extending from the eastern Black Sea to Kopet-Cag towards the middle of the Jurassic;

3) Africa and Arabia converging towards Eurasia in the middle of the Cretaceous caused first major collisions between the southern passive margin and continental fragments of the Gondwana and the island arc, resulting in closure of large portion of the Tethys;

4) During Late Cretaceous time (from the Turonian, at places from the Campanian) and in the Paleogene, a new volcanic belt was formed, extending from the southwestern edge of Iran through Trans Caucasus, i.e. area south of the Caucasus range (Talysh, Ajaro-Trialet zone) and the Pontos into Bulgaria (Burgas synclinorium); the belt consists of calc-alkaline (in the south) and shoshonite (in the north) lavas, marking the existence of an active margin under which northward consumption of the residual Tethys ocean took place; the basin of the western Black Sea opened in latest Cretaceous time back of the arc, the evidence for this being riftogenetic alkali basalts in the axial part of the Ajaro-Trialet zone and the Burgas synclinorium; granite-granodiorite massifs in Trans Caucasus and Iran are incident to the volcanic belt; towards the end of the Paleogene, continued convergence of Africa and Arabia towards Eurasia caused (1) a complete consumption of the

oceanic Tethys crust under the margin Eurasia, (2) a substantial decrease in the area of the Black Sea basine, and (3) initiation of a new phase of continental collision.

5) The neotectonic Caucasian volcanism since 10 m.y. B.P. until now has produced calc-alkaline series in Armenia and the Greater Caucasus on the one hand and, on the other, alkali basalts; the former can be successfully correlated to the increased rate of the convergence of Arabia and Eurasia due to the opening of the Red Sea, while the latter seem to be associated with intraplate volcanism.

It follows that the evolution of the Mediterranean belt within Caucasus is characterized by the existence of (at least three) volcanic arcs and by several collision episodes involving the southern continental massifs and the northern active margin. The collision still continues.

The Verkhoyansk-Kolyma fold belt shows a significantly different magmatic history (Fig.7e). During a long time, since the Riphean until the Late Jurassic, i.e., during 500 m.y., its area had been occupied by an amagmatic passive margin with only two short episodes of intraplate basaltic volcanism, in the Devonian and Early Triassic. Approach towards the passive margin of the Kolyma-Omolonsk massif during Late Jurassic produced a volcanic island arc (Uyandin-Yasachnen or Ilinotass volcanic belt of the Chersky ridge). After a short period of volcanism, the massif collided with the Siberian continent, producing gold-bearing granit-granodiorite batholiths of the Kolyma belt in the earliest Cretaceous. Since the middle of the Cretaceous, when the continental collision came to an end, the Okhotsk-Chukotka volcano-plutonic belt was superposed upon the fold belt, indicating that North-East Asia belonged at that time to the Andean type active continental margin. The Verkhoyansk-Kolyma fold belt is thus characterized by an extremely short (30 to 40 m.y.) span of existence, of both the island arcs and continental collision process.

Conclusions

The history of the Earth during the past 2.000 m.y. contains repeated similar magmatic assemblages characterizing comparatively few geodynamical environments of their formation. However, their succession in time and space vary widely from place to place. This is clearly seen from the examples considered, showing the evolution of magmatism in different Eurasian fold belts. The feature all fold belts have in common is that they were born from oceanic basins of continents. Because of this, all fold belts nearly always contain remnants of oceanic crust, presented by ophiolites and complexes formed in the environments of converging lithosphere plates.

The Ural mountains were mainly formed in the place of island arcs and marginal basins, typically with a very long period of continental collision. The Moditerranean belt has a similar history: there have been several collision episodes within it. The Verkhoyansk-Kolyma belt is remarkable for a very short period of convergence of the continents and the associated magmatism. The Mongolian mountains both northern and southern, had the initial stages of their history similar to that of the Urals, evolving in the place of island arcs and marginal basins. However, the later stages within Mongolia involved mixed magmatism reflecting coexistence of contrasting geodynamical conditions including active continental margins of the Andean type and continental rifts, possibly with some influence of intraplate magmatism. This kind of complex environment, denoted as the Mongol-Okhotsk one, seems most probably to arise when a continental margin overthrusts an oceanic spreading ridge. An example of such an environment during Late Cenozoic time is the North American margin.

REFERENCES

1. Allegre C.J. Chemical geodynamics.-Tectonophysics, v.81, N 3/4, 1982, p.109-132.

2. Beottcher A.L., O'Neil J.R. Stable isotope, chemical and petrographic studies of high-pressure amphi-

boles and micas: evidence for metasomatism in the mantle source regions of alcali basalts and kimberlites.-Amer.J.Sci., 1980, v.280A, p.594.
3. Bogatikov O.A., Borsuk A.M., Dmitriev Yu.I., Kovalenko V.I., Ryabchikov I.D. The evolution of magmatism in the Earth's history. In: "Petrology", Moscow, "Nauka", 1980, p.3-13 (in Russian).
4. Bogatikov O.A., Borsuk A.M., Dmitriev Yu.I., Kovalenko V.I., Ryabchikov I.D. Magmatic formations in the evolution of the lithosphere. Izv., Acsd. Sci USSR, geol. ser., 1983, No.1, p.3-16 (in Russian).
5. Borukaev Ch.B. Principal structural elements of the Precambrian. Abstract of doct. diss. Institute of Geology and Geophysics, Novosibirsk, 1983, 45 pp. (in Russian).
6. Brocks C., Hart S.R., Horman A., James D.E. Rd-Sr-mantle isochrons from oceanic regions.-Earth Planet. Sci.Lett., 1976, v.32, p.51-61.
7. East Pacific Rise Study Group. Crustal processes of the Mid-Ocean Ridge.-Science, 1981, v.213, N 4503, p.31-40.
8. Engel A.E.J. et al. Crustal evolution and global tectonics: a petrogenic view.-Bull. Geol. Amer., 1974, v.85, p.6.
9. Gruau G., Glikson A.Y. Komatiites of the Onwerwacht Group, S.Africa: REE Geochemistry, Sm/Nd age and mantle evolution. Contib. Miner. Petrol., 1982, v. 80, N 4, p.197.
10. Hatherton T., Dickinson W.R. The relationship between andesitic volcanism and seismicity in Indonesia, the Lesser Antilles, and other island arcs.-J.Geophys.Res., 1969, v.74, N 22, p.5301-5310.
11. Kanamori H. Seismic and aseismic slip along subduction zones and their tectonic implications. In: M.Talwani and N.C.Pitman 111(Eds), Island Arcs, Deep-Sea Trenches and Bauc-arc Basins Am. Geoth., Union, Washingtion, D.S., 1977, p.162-174.
12. Kamin V.G. The evolution of continental rifts. Abstract of doct. diss. Geological Institute, Acad. Sci. USSR, Moscow, 1982, 47 pp. (in Russian).
13. Kononova V.A. Alkalic magmatic rock series and sources of their material (from isotope ratios for

Sr, O, C). Inâ "Petrology", Moscow, "Nauka", 1980, p.30-40 (in Russian).
14. Korzhinsky D.S. Currents of transmagmatic solutions and granitization. In "Magmatism, formations of basement rocks, and the deeper earth", vol. 1. Moscow, "Nauka", 1972, p.144-152 (in Russian).
15. Kovalenko V.I. Petrology and geochemistry of rare-mentral granitoids. Novosibirsk, "Nauka", 1977, 207 pp. (in Russian).
16. Milanovsky E.E. Rift zones in the continents. Moscow, "Nedra", 1976, 278 pp. (in Russian).
17. Miyashiro A. Volcanic rock series and tectonic setting.-Amer.Rev.Earth and Planet. Sci., 1975, v.3, p.251-269.
18. Moralev V.M. Tectonic conditions of the formation of early Precambrian ultrabasic-basic associations exemplified by the Aldan and Hindustan shields. Abstract doct. diss., Moscow, 1982, 35 pp. (in Russian).
19. Moorbath S. Age and isotope constraints for the evolution of Archaean crust. In: The Early History of the Earth. Wiley, London, pp.351-360, 1976.
20. ¾'Nions R.K. Isotopic abundance relevant to identification of magma sources. Abstracts of a meeting for discussion "The relative contributions of mantle, oceanic crust and continental crust to magma genesis. "The Royal Society, London, 1983, p.5-6.
21. O'Nions R.K., Hamilton P.J., Evensen N.M. The chemical evolution of the Earth's mantle.-Sci. Amer., 1980, v.242, N 5, p.91-101.
22. Pallister J.S., Hopson C.A. Samail ophiolite plutonic suite: Field relations, phase variation, eruptive variation and layering, and a model of a spreading ridge magma chamber.-J.Geoph. Res., 1981, v.86, N.B4, 1981, p.2593-2661.
23. Ryabchikov I.D., Bogatikov O.A., Babansky A.D. Physical and chemical problems in the origin of calc-alkaline magmas. Izv., Acad. Sci. USSR, geol. ser., 1978, N.8, p.5 (in Russian).
24. Ryabchikov I.D., Kovalenko V.I., Dikov Yu.P., Vladykin N.V. Titanium-containing mantle micas: composition, conditions of formation, possiblerole in

the genesis of potassium alkalic magmas. Geokhimia, 1981. N.6, p.873-889 (in Russian).
25. Schilling J.G. Iceland mantle plume: geochemical study of Reykjanes Ridge.-Nature, 1973, v.242, p.565-571.
26. Sun S.S., Hakson J. Evolution of the mantle: geochemical evodence from alkali basalts. Geology, 1975, v.3, p.297-302.
27. Wasserburg G.J., De Paolo D.J. Models of Earth structure inferred from neodymium and strontium isotopic abundances.-Proc.Nation.Acad.Sci., USA, 1979, v.76, p.3594-3596.
28. Windley B.F. Phanerozoic granylites.-J.Geol.Soc. London, v.138, 1981, p.745-571.
29. Yarmolyuk V.V., Kovalenko V.I. Carboniferous volcanism of Mongolia and the geodynamics of its formation. DAN SSSR, 1982, v.263, N.1, p.172-174 (in Russian).
30. Zonenshain L.P., Kuzmin M.I., Kovalenko V.I., Saltykowsky A.S. Mwzozoic structural-magmatic pattern and metallogeny of the western part of the Pacific Belt.-Earth and Planet.Sci.Let., 1974, v.22, p.96-109.
31. Zonenshain L.P., Kuzmin M.I., Moralev V.M. Global tectonics, magmatism, and metallogeny. Moscow, "Nedra", 1976, 223 pp. (in Russian).
32. Zonenshain L.P., Savostin L.A., Movements of lithosphere plates relative to subduction zones: formation of marginal seas and active continental margins.-Tectonophysics, 1981, v.74, p.57-87.
33. Zonenshain L.P., Kuzmin M.I. Intraplate magmatism and its significance for understanding mantle processes. Geotectonica, 1983, N.1, p.28-45 (in Russian).

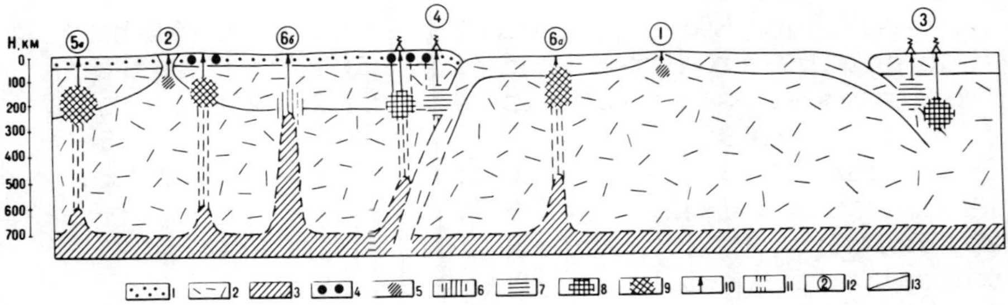

FIGURE 1. Correlation of magmatism and simple geodynamical environments. Magma sources (1-3): 1 - continental crust; 2 - depleted upper mantle; 3 - undepleted lower mantle; magma types (4-9); 4 - palingenetic sialic (continental); 5 - tholeiitic basaltic (rift type); 6 - tholeiitic basaltic intracontinental (trap type); 7 - calc-alkaline, with normal alkali content (andesitic type); 8 - calc-alkaline, with increased alkali content (shoshonite type); 9 - alkalic, kimberlite, and carbonatite type; 10 - direction of magma motion; 11 - zone of metasomatically altered mantle (enriched mantle); 12 - geodynamical environments described in text (encircled numbers refer to this and following figures): 1 - oceanic rift; 2 - continental rift= 3 - intraoceanic island arc; 4 - active continental margin of the Andean type; 5 - collision of continents (not shown); 6 - intraplate volcanism: a) oceanic; b,c) continental; 13 - - lower boundary of lithosphere plates.

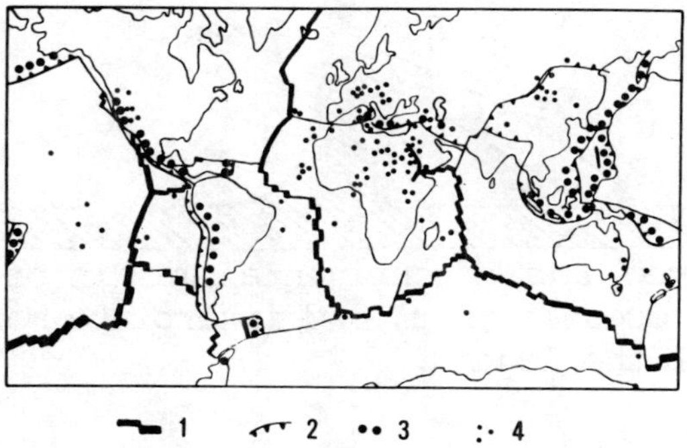

FIGURE 2. Spatial distribution of magmatic rocks younger than 40 m.y.: 1 - divergent plate boundary with tholeiitic basalts; 2 - convergent plate boundary; 3 - calc-alkaline series; 4 - intraplate magmatism.

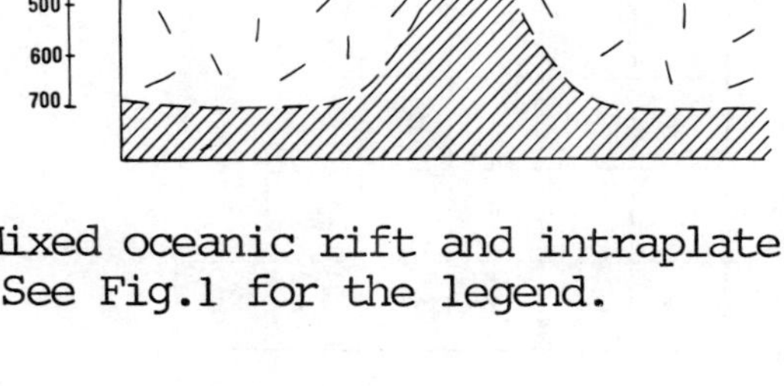

FIGURE 3. Mixed oceanic rift and intraplate magmatism (Iceland). See Fig.1 for the legend.

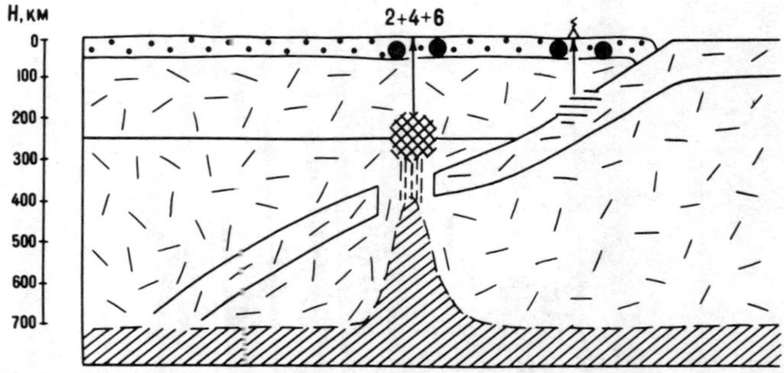

FIGURE 4. Mixed active continental margin of the Andean type, oceanic rift, and interplate magmatism. See Fig.1 for the legend.

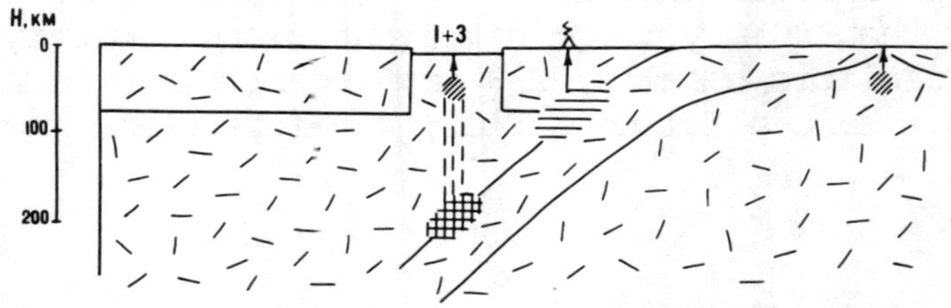

FIGURE 5. Mixed, island arc and an oceanic rift (interarc basin) magmatism. See Fig.1 for the legend.

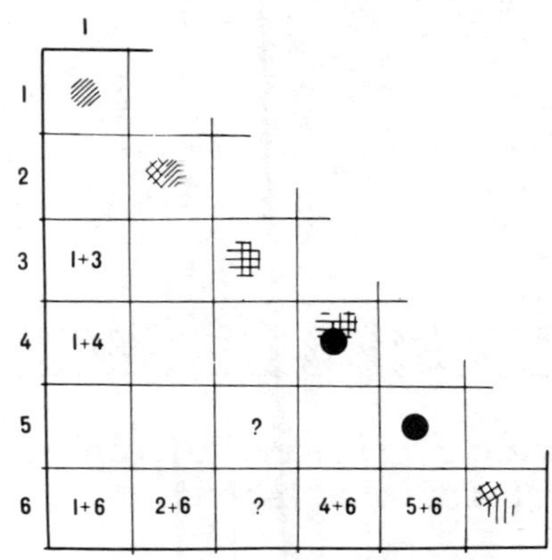

FIGURE 6. Correlation of elementary and known mixed magmatism with geodynamical environment of their formation. See Fig.1 for the legend.

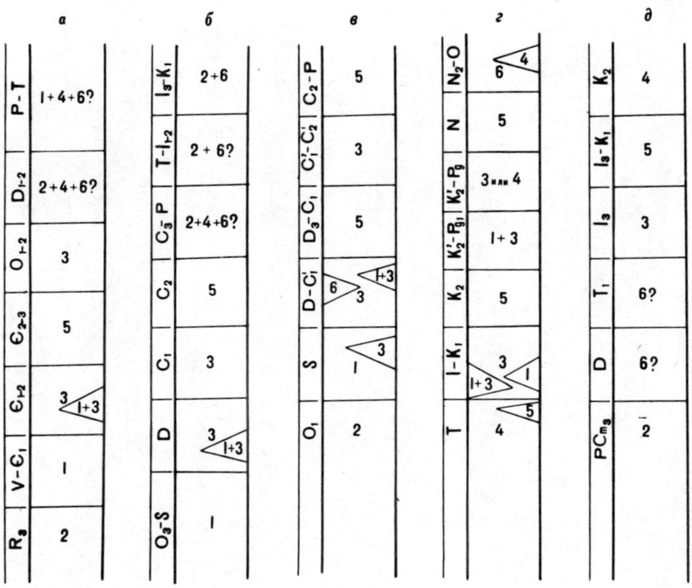

FIGURE 7. Evolution of magmatism and geodynamical environments (for legend see Fig.1 and 6) in some foldbelts: a - North Mongolia; b - South Mongolia; c - Urals; d - Mediterranian belt; e - Verkhoyansk area (Chersky range).

RELATION OF LITHOPHILE ELEMENT MINERALIZATION TO ACID MAGMATISM, WESTERN USA

BURT, D.M.
Department of Geology, Arizona State University, Tempe, Arizona 85287, USA.

ABSTRACT

In the western United States, deposits of the lithophile metals Li, Be, W, Mo, Ta, Nb and U range in level of emplacement from Precambrian rare metal pegmatites, through greisenized W-Be veins, to Tertiary subvolcanic or porphyry Mo-W deposits (Climax-type), and finally to volcanic - surficial deposits such as tuff-hosted Be deposits at Spor Mountain, Utah. Deposits of these four major depth facies appear to be related to essentially the same fluorine-rich granitic magma type, whose fractionation and devolatilization was directly or indirectly responsible for the mineralization. The crustal level at which the devolatilization occurred was mainly a function of the H_2O-content of the magma, with more H_2O-rich magmas being emplaced at greater depths. Volatile saturation in parent granitic magma bodies is a useful criterion for prospecting; it is normally indicated by striking textural variations in rocks emplaced at all levels. The presence or absence of reactive carbonate country rocks is another useful criterion in prospecting.

INTRODUCTION

In the western United States occur a great variety of mineral deposits of the lithophile (or "fluorophile") elements Li, Be, W, Mo, Sn, Ta, Nb and U. In age and environment of formation they range from Precambrian rare metal pegmatites to late Tertiary volcanic and surficial deposits. Despite this variety, all deposits share many geo-

Proceedings of the 27th International
Geological Congress, Volume 9, pp. 27–39
PETROLOGY (IGNEOUS AND METAMORPHIC ROCKS)

chemical and other features in common. For example, all are either directly or indirectly related to acid magmatic rocks (granites and rhyolites). The main purpose of this paper is to consider the magmatic setting of the various deposit types in the context of their depth of formation. This concept was outlined for tin deposits by Varlamoff (1978) and was elaborated to include volcanic and surficial deposits by Burt et al. (1982).

The acid magmas involved in the mineralization are all believed to represent partial melts of older continental crust (perhaps formed in the presence of a high heat flow) All have in common a relatively high content of SiO_2,K_2O, and Na_2O and a relatively low content of CaO and MgO (also SrO and BaO). All tend to be metaluminous to slightly peraluminous (as shown by the common presence of accessory Mn-Fe garnet) and all tend to be enriched in the same suite of trace elements (Cs, Rb, Li, Be, Sn, Ta, Hf, etc.). Those that have been studied have in common a chondrite-normalized rare earth pattern that is distinctly flat,with a large negative europium anomaly. Fluorine is more-or-less enriched in all of them, except in certain phosphate-rich pegmatites.

Mineralization appears to be related to two main magmatic processes: (1) large degrees of fractional crystallization in parent magma chambers, leading to enrichment in fluorine and the ore metals (especially at the tops of magma chambers) and (2) saturation with an aqueous fluid phase, which subsequently escapes along veins (except in the case of pegmatites, which may behave as closed systems).

The main control on the depth of crystallization of a rising body of granitic magma is its H_2O content (Burnham, 1979; Wyllie, 1979). The more H_2O-rich it is, the sooner it becomes saturated with an aqueous fluid phase, which then escapes, leading to rapid crystallization of the residual magma (commonly with porphyritic textures) and mineralization along the path of escape. This process is delayed for relatively H_2O-poor magmas, which can therefore be emplaced at shallower depths, or even be erupted as lavas. A high fluorine content increases the solubility of H_2O in granitic magmas (see review by Bailey, 1977), so that fluorine-rich magmas might be emplaced at relatively shallow depths, despite an appreciable H_2O content.

Fluid saturation in a granitic rock is normally indicated by unusual textural variations. These range from "onion-skin" layering around lithophysal cavities in rhyolitic lavas, to rhythmic banding of euhedral quartz and K-feldspar ("unidirectional solidification textures") in subvolcanic porphyry molybdenum deposits, to pegmatitic or greisenized or miarolitic borders in granites parent to greisen veins, to graphic granite, layered pegmatite-aplite dikes, and similar textures in pegmatitic granites parent to major pegmatite fields. These readily-visible textural variations are useful in prospecting.

The size and frequency of veins is also controlled by depth, with larger but fewer veins typically being formed at greater depths. This factor affects the mining method used. Pegmatites and large greisen veins may thus be mined individually (commonly with hand sorting of ores), but bulk mining and processing methods are required for near-surface, more highly-fractured rocks (stockworks and breccias).

In what follows I shall discuss the various types of lithophile element deposits in the western USA in terms of four main depth facies. From deepest to shallowest, these are (1) pegmatitic (2) greisen vein, (3) subvolcanic (= porphyry = stockwork), and (4) volcanic-surficial. These facies are schematically depicted on Figure 1 (modified from Burt et al., 1982) and 14 deposit types are listed in Table 1. The figure is an over-simplification in that it only considers small individual plutons (not large magma bodies) and does not explicitly consider multiple episodes of intrusion (Multiple episodes are implied by question-marked "ghosts" on the right three sections of the figure). Also, of course, there may be gradations among the various deposit types.

PEGMATITIC FACIES

Aspects of rare metal pegmatite genesis in North America have recently been extensively reviewed (Cerny, 1982). This material need not be repeated here, except to note that rare metal pegmatite fields (deposit type 14) tend to be spatially and geochemically zoned around a parent body of pegmatitic granite. Disseminated heavy minerals in such a granite or in shallower granites (deposit type 13) may become economic if, after weathering, they are concentrated in placer deposits. In the USA, such deposits have been mined for Nb, Ta and rare earth elements in southern Idaho (Dismal Swamp and Bear Valley placers: Barton, 1962).

U.S. granitic pegmatites themselves have been mined for beryl, columbite-tantalite, microlite, lithium minerals, gemstones (tourmaline topaz, and beryl), feldspars, micas, and quartz. The best ores are in subhorizontal bodies, such as the Harding Pegmatite, New Mexico (Jahns and Ewing 1977), but steeply-dipping dikes, such as the Brown Derby Pegmatite in the Quartz Creek district, Colorado (Staatz and Trites, 1955) are more common. The above two pegmatites contain topaz and are rich in Li-Al or Li-Mn-Fe phosphates (montebradite - amblygonite and lithiophilite-triphilite), but are poor in fluorine. Well-known examples occur in the southern Black Hills pegmatite field, south Dakota (Orville, 1960; Redden, 1963) and the White Picacho district, Arizona (London and Burt, 1982).

GREISEN FACIES

The depth of magma crystallization for greisenized veins (deposit type 12a) seems to be only slightly less than for pegmatites. The main difference is that individual pegmatites behave more-or-less as closed systems, whereas a greisenized granite and its related veins are clearly an open system. The pegmatitic borders commonly seen around the granitic plutons parent to greisen veins indicate, nevertheless, that the granite started as a closed system, until the accumulating fluid pressure became high enough to seek release through the country rocks along veins. Also, pegmatitic dikes (called "pegmatoids" in Chinese tungsten deposits and "veindikes" at the Henderson molybdenum mine, Colorado) commonly grade into greisen-bordered quartz veins.

Disseminated tantalum-tin-tungsten mineralization in greisenized albite-lithium mica granites (deposit type 12b) has been described from the USSR ("apogranites" or "rare metal granites"), China, France, and Egypt, but not yet from USA. Similar deposits do occur, however - for example the disseminated scheelite (after wolframite) and beryl mineralization in a greisenized Tertiary (?) pluton at the Boriana Mine (Robinet Ridge), western Arizona (Kerr, 1946). These deposits can probably be thought of as large differentiated bodies of pegmatitic magma that acended high enough in the continental crust to be devolatilized and pressure quenched before the full range of pegmatitic textures could develop.

The greisen vein environment itself is poorly expressed in the western USA, in part because of the unfavorable economics if mining vein-type wolframite and in part because Tertiary magmatic centers, in general, have not yet been eroded deeply enough to expose this type of mineralization. (There may be a real deficiency in elements such as tin in Tertiary acid magmas themselves, of course). Previous producers include the Boriana and Black Pearl mines in western Arizona (Dale, 1961) and the Ima mine, central Ldaho (Kerr, 1946) for wolframite (Tertiary age), and the Boomer mine in central Colorado (Hawley, 1969) for beryl (Precambrian age). Described vein systems are close to vertical (unlike Panasqueira, Portugal).

Where carbonate rocks occur near greisenized granites, they commonly are altered to fluorite-rich or sulfide-rich skarn replacements (deposit type 8), with tin, tungsten, or beryllium mineralization. A familiar example is the Lost River district, Seward Peninsula, Alaska (Dobson, 1982). This type of mineralization is also possible in the shallower subvolcanic environment, as in the beryllium and tungsten-mineralized "ribbon rock" skarns of the Iron Mountain district, southern New Mexico (Jahns, 1944).

SUBVOLCANIC FACIES

The main subvolcanic ore deposits related to fluorine-rich granitic magmatism are Climax (Colorado) - type porphyry molybdenum deposits (deposit type 11). Such deposits are widespread on both sides of the Colorado Plateau and coincide in their distribution with Tertiary topaz-rhyolite lavas (see below), with which they are probably comagmatic.

The rhythmically-banded pegmatitic textures seen in Climax-type deposits (called "unidirectional solidification textures" at the Henderson molybdenum Mine, Colarado: Shannon et al., 1982) and related vein patters are reminiscent of many features seen in deeper greisen-facies granites. The main difference is the much greater intensity of fracturing and veining (making possible bulk mining of molybdenum, and at Climax also tungsten and minor tin) and the much finer grain size of the intrusives and altered rocks. (Some of the intrusive rocks are more like porphyritic rhyolites than granites, and may roughly correspond to rocks that have been called "ongonites" in the USSR). Topaz is a common alteration mineral, and beryllium minerals are also present, if difficult to recognize. The intrusive history of these deposits is known to be exceptionally complex (perhaps because they have been exceptionally well studied), with multiple episodes of intrusion and overlapping zones of mineralization being the rule rather than the exception.

The subvolcanic nature of these deposits is demonstrated by the Pine Grove molybdenum deposit in western Utah, which was discovered by drilling directly down a volcanic neck. Remnants of the garnetiferous tuff erupted from the volcano (now eroded) are found in the surrounding countryside (Keith, 1980). As another example, the Climax deposit itself can be placed directly beneath the nearby Chalk Mountain topaz rhyolite porphyry by reversing the 2000 meter movement along the post-ore Mosquito fault (Climax geological staff, 1982, personal communication). Of course due to erosion the present setting of these deposits may be difficult to determine, and not all of them were necessarily formed beneath volcanic vent areas.

Breccia pipes commonly are spatially related to porphyry molybdenum deposits; some of these, as at Red Mountain, Colorado (Sharp, 1978), may themselves be mineralized (deposit type 10). At greater distances occur base and precious metal veins (deposit type 9). In southern Colorado such veins commonly contain fluorite, huebnerite, helvite, and Mn-silicates or carbonates in addition to Zn, Pb, Ag and minor Au.

VOLCANIC-SURFICIAL FACIES

The volcanic-surficial facies of lithophile element mineralization and its relation to topaz rhyolite volcanism has recently been reviewed by Burt et al.(1982). The widest variety of deposit types occurs in the Spor Mountain district, western Utah. The first deposits to be discovered were fluorite replacements of carbonate host rocks in breccia pipes and volcanic vent areas (deposit type 7). Later it was discovered that the top-most layer of tuff beneath a topaz-bearing rhyolite lava flow was economically mineralized in beryllium (deposit type 5) and was also enriched in tin, tungsten, niobium and other metals (excluding molybdenum). The beryllium ore mineral bertrandite replaces carbonate fragments in the tuff (incorrectly called "water-laid tuff" in earlier literature), together with fine-grained fluorite, manganese oxides, and amorphous silica. The tuffaceous matrix is altered to a lithium-bearing montmorillonite. The presence of the overlying lava flow and the presence of reactive dolomite fragments in the tuff were apparently both necessary for the mineralization. Uranium presumably leached from the same tuff by descending groundwater formed a separate deposit, the Yellow Chief mine, in clastic rocks underneath (deposit type 4).

Similar Tertiary volcanic vent complexes have produced "wood tin" (banded botryoidal cassiterite) that was deposited in topaz rhyolite lavas and breccias (deposit type 6), perhaps due to low temperature dissolution and reprecipitation from earlier fumarolic deposits. This type of deposit is mainly known from northern Mexico, but also from southwestern New Mexico (Taylor Creek district) and northern Nevada (Izenhood Ranch district).

Elsewhere topaz rhyolite volcanic complexes are intermittently mined for gem topaz and red beryl, as well as for "Apache tears" (gemmy obsidian spheroids). The industrial minerals perlite (mainly from northern New Mexico) and pumice are locally important products. Exceptionally young rhyolitic complexes yield geothermal resources (as at Roosevelt Hot Springs in Utah) and recreational opportunities (as at Yellowstone National Park in Wyoming).

At greater distances from the vent areas, and more conjecturally associated with acid magmatism, are hot springs deposits (deposit type 3) mineralized in W,Mn,Ag and Au. An example is the tungsten deposit at Golconda, Nevada (Kerr, 1940). Tuffaceous lake sediments (deposit type 2), such as those at the Anderson uranium mine in Arizona (Sherborne et al., 1979) may be mineralized in uranium, manganese, or lithium. Finally, brines beneath playa lakes (deposit type 1) may contain considerable lithium, as at Silver Peak, Nevada (Kunasz, 1975) or tungsten, as at Searless Lake, California. (The tungsten in Searles Lake probably came from Mesozoic Scheelite skarn deposits in the Sierra Nevada, rather than from Tertiary acid magmatism, and is included here only for the sake of completeness).

SUMMARY AND CONCLUSIONS

The major conclusions of this paper are as follows:

(1) Most of the lithophile metal deposits of the western USA (excluding Sierra Nevada - type scheelite skarns) appear to be associated with essentially the same highly evolved magma type, whose fractionation and devolatilization was directly or indirectly responsible for the mineralization.

(2) Volatile saturation of granitic magmas is evidenced by dramatic textural variations in the resulting rocks. These variations can be seen at every level of emplacement, from pegmatitic to volcanic, and are useful in prospecting.

(3) Differences in the level of emplacement of the magma bodies responsible for mineralization were mainly due to differences in the volatile content of the magma itself. The more volatile-enriched magmas were emplaced at grea-

ter depths, except that fluorine-rich magmas may have been emplaced at shallower levels for a given H_2O content than would have otherwise been possible.

(4) The presence or absence of reactive (i.e. carbonate) wall rocks also affects the type of mineralization, especially at shallower depths.

(5) The same suite of lithophile elements may be concentrated at various levels of emplacement. The elements that are actually recovered during mining may therefore be controlled more by the practicality of various mining and milling methods than by geochemical factors.

ACKNOWLEDGEMENTS

Thanks are due to V.I.Kovalenko and the USSR Academy of Sciences for making available the facilities of the Institute of Geology of Ore Deposits (IGEM) in Moscow during the writing of this paper. I am also grateful to M.F.Sheridan, J.V.Bikun and E.H.Christiansen for their help with the research on topaz rhyolites and to D.London for similar help with lithium pegmatites. Parts of this research were supported by the U.S.Department of Energy (subcontract no 79-270-E on topaz rhyolites) and the US National Science Foundation (Grant no. EAR 7814785 on lithium pegmatites).

REFERENCES

Bailey J.C., 1977, Fluorine in granitic rocks and melts: A review. Chem. Geol., v.19, p.1-42.

Barton W.R., 1962, Columbium and tantalum, a materials survey. U.S.Bur.Mines Inf. Circ. 8120, 110 p.

Burnham C.W., 1979, Magmas and hydrothermal fluids, in: Barnes H.L., Geochemistry of hydrothermal ore deposits, 2nd ed. New York, Wiley, p.71-136.

Burt D.M., Sheridan M.F., Bikun J.V., Christiansen E.H., 1982, Topaz rhyolites - distribution, origin, and significance for exploration. Econ.Geol., v.77, p.1818-1836.

Cerny P., ed., 1982, Granitic pegmatites in science and industry. Mineral.Assoc.Canada, Short Course Handbook, v.8, 555 p.

Dale V.B., 1961, Tungsten deposits of Gila, Yavapai, and Mohave Countries, Arizona U.S.Bur.Mines Inf.Circ. 8078, 104 p.

Dobson D.C., 1982, Geology and alteration of the Lost River tin-tungsten-fluorine deposit, Alaska. Econ. Geol., v.77, p. 1033-1052.

Hawley C.C., 1969, Geology and beryllium deposits of the Lake George (or Badger Flats) beryllium area, Park and Teffusen Counties, Colorado. U.S. Geol. Survey Prof. Paper 608 A, 44p.

Jahns R.H., 1944, "Ribbon rock", an unusual beryllium-bearing tactite. Econ. Geol., v.39, p.173-205.

Jahns R.H., Ewing R.C., 1977, The Harding mine, Taos County, New Mexico. Mineralog. Rec., v.8, p. 115-126.

Keith J.D., 1980, Miocene porphyry intrusions, volcanism and mineralization, southwestern Utah and eastern Nevada. Unpubl. M.S. thesis, Univ.Wisconsin, Madison, 166 p.

Kerr P.F., 1940, Tungsten-bearing manganese deposit at Golconda, Nevada. Geol.Soc.Am.Bull., v.51, p.1359-1390.

Kerr P.F., 1946, Tungsten mineralization in the United States. Geol.Soc.Amer. Mem. 15, 241 p.

Kunasz I.A., 1975, Lithium raw materials, in: Lefond, S.A,, ed., Industrial minerals and rocks, 4th ed. New York, Am. Inst. Min. Met. Petrol. Engineers, p. 791-803.

London D., Burt D.M., 1982, Alteration of spodumene, montebrasite, and lithophilite in pegmatites of the White Picacho district, Arizona. A. Mineralogist, v.67, p.97-113.

Orville P.M., 1960, Petrology of several pegmatites in the Keystone district, Black Hills, South Dakota. Geol.Soc. Am.Bull., v.71, p.1457-1490.

Redden J.A., 1963, Geology and pegmatites of the Four-Mile Quadrangle, Black Hills, South Dakota. U.S.Geol.Survey Prof.Paper 297-D, p.199-287.

Shannon J.R., Walker B.M., Carten R.B., Geraghty E.P., 1982, Unidirectional solidification textures and their significance in determining relative ages of intrusions at the Henderson mine, Colorado. Geology, v.10, p.293-297.

Sharp J.E., 1978, A molybdenum-mineralized breccia pipe complex, Redwell Basin, Colorado. Econ.Geol., v.73, p. 369-382.

Sherborne J.E., Buckovic W.A., Dewitt D.B., Hellinger T.S., Pavlak S.J., 1979, Major uranium discovery in volcanic-clastic sediments, Basin and Range province, Yavapai County, Arizona. Am. Assoc. Petroleum Geologists Bull., v.63, p.621-646.

Staatz M.H., Trites A.F.Jr., 1955, Geology of the Quartz Creek pegmatite district, Gunnison County, Colorado, U.S. Geol. Survey Prof. Paper 265, 111 p.

Varlamoff N., 1978, Classification and spatial-temporal distribution of tin and associated mineral deposits, in: Stemprok M., Burnol L., Tischendorf, eds., Metallization associated with acid magmatism. Prague, Czechoslovakia Geol.Survey, v.3, p.139-156.

Wyllie P.J., 1979, Magmas and volatile components. Am. Mineralogist, v.64, p.469-500.

Table 1. Types of deposits depicted on Figure 1

Number on Figure	Type	Elements
1.	Playa lake brines	Li(W)
2.	Tuffaceous lake sediments	U(Li,Mn)
3.	Hot springs deposits	W,Ag,Mn,Au
4.	Clastic rocks beneath tuffs	U
5.	Mineralized tuffs	Be (F,Li,Cs,U)
6.	Fractured lavas	Wood Sn
7.	Vent and contact breccias in carbonate rocks	F(U)
8.	Fluorite-rich skarn ("ribbon rock")	Sn,W,Be (Fe)
9.	Base and precious metal veins	Ag,Au,Pb,Zn(Sn W)
10.	Mineralized breccia pipes	Mo(ag,Au)
11.	Porphyry (=stockwork) deposits	Mo,W,Sn
12.	a. Greisen-bordered veins	W,Be,Sn
	b. Albitized and greisenized granite	Ta,Sn,W,Be
13.	Disseminated heavy minerals in granite (economic from placers only)	Nb,Ta,Th,REE, U,Hf,Zr,etc.
14.	Rare metal pegmatites	Ta,Be,Li,Cs,Rb etc.

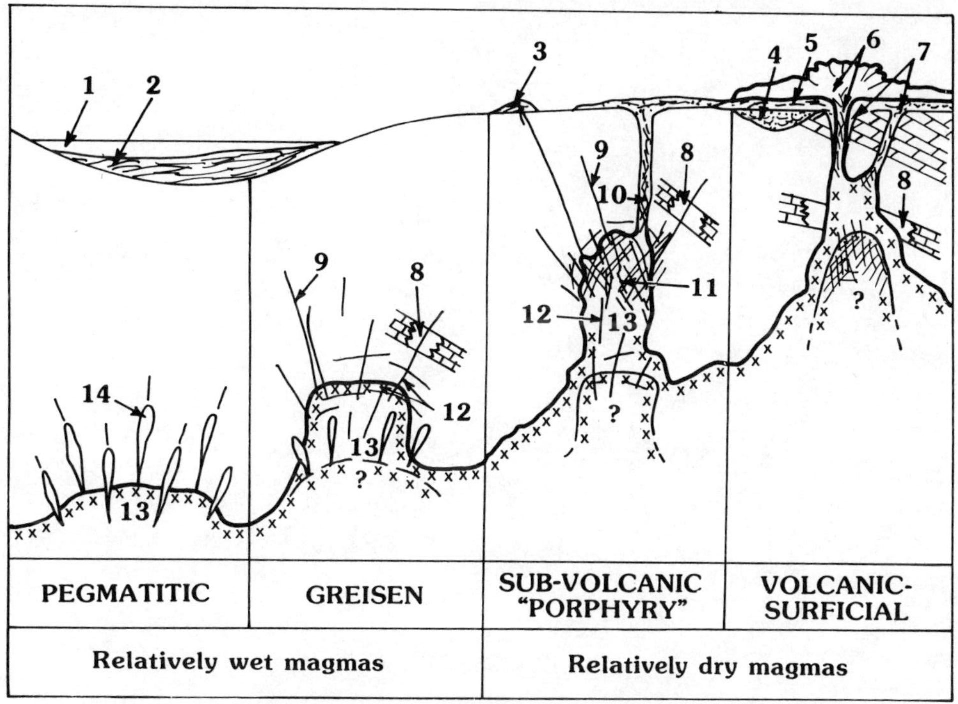

FIGURE 1. Schematic composite cross-section showing types of lithophile element deposits in the western USA and their relation to individual bodies of fluorine-rich magma intruded at various depths. For labels see Table 1 and for discussion, see text.

HIGH -GRADE REGIONAL METAMORPHISM, ANATEXIS, DEGRANITISATION AND THE ORIGIN OF THE GRANITES: AN EXAMPLE FROM THE SOUTH-ALPINE BASEMENT

BORIANI, A. and GIOBBI ORIGONI, E.
Centro di Studio per la Stratigrafia e la Petrografia delle Alpi Centrali, CNR , MILANO and Dipartimento di Scienze della Terra , Università degli Studi di MILANO.

ABSTRACT

The Hercynian belt of the Southern Alps (Italy) consists of Middle Paleozoic sediments (Carnia) and metamorphic rocks ranging from low-grade in the E to high-grade and granulite facies in the W (Ivrea-Verbano zone). The igneous activity is pre-, syn-, post-tectonic. Pre-tectonic volcanism is bimodal with basic extrusives preceding the Caradocian rhyolites; the latter are coeval with widespread granodiorites and granites.

Pre-, syntectonic igneous rocks are the spilite-keratophyre association of Carnia (Carboniferous flysch) and probably also part of the Ivrea-Verbano basic body. The sedimentary sequence contains terrigenous sediments and pelagic limestones as well as pelitic radiolaritic shales (Silurian). Metapelites dominate in the metamorphic rocks but also metagraywackes are important (Strona-Ceneri zone). In the eastern part metarhyolites are intercalated; therefore a pre- and post-Caradocian series can be distinguished. Only pre-Caradocian rocks seem to exist in the W part; accordingly large masses of metaplutonites are present. The Hercynian barrovian-type metamorphism (Rb/Sr mineral ages = 350-325 m.y.) consisted in a relatively low-T synkinematic phase followed by a post-kinematic high-T phase.

Post-tectonic igneous activity is dominantly granitic

Proceedings of the 27th International
Geological Congress, Volume 9, pp. 41–66
PETROLOGY (IGNEOUS AND METAMORPHIC ROCKS)

and rhyolitic-rhyodacitic and postdates of about 50 m.y. the peak of metamorphism. In the Ivrea-Verbano the transition medium-grade - high-grade is marked by limited amounts of anatectic migmatites. Granulites are very probably deeply degranitized; a Rb/Sr w.r. isochron indicates isotopic homogeneisation around 478 m.y. interpreted as peak of metamorphism, implying, if correct, that the Hercynian belt is polymetamorphic (Caledonian + Hercynian). We propose an evolutionary model of ensialic orogen similar to Kröner's (1981) late-proterozoic model but with a more rigid lithospheric behaviour: 1) Continental rift system with initial crustal thinning, subsidence and widespread bimodal volcanism 2) A-type subduction and closure with thrusting and folding followed by high-T post-kinematic metamorphism. 3) Melting of the undergoing crustal slab caused general post-tectonic granitic-rhyolitic magmatism.

INTRODUCTION

The connection between regional metamorphism and the generation of granitic magma was postulated on the basis of geological observation and experimental petrology; this connection has been implied very often in the construction of evolutionary models of orogenic belts. It appears now evident that a straightforward relationship between the two phenomena is no longer sustainable for many good reasons (Pitcher, 1979) but mostly because the intrusion of granites generally postdates the metamorphic climax of a very long time span.

On the other hand it is well documented that high-grade metamorphism is generally connected with the occurrence of migmatites and/or with almost certainly degranitised granulite facies rocks.

These two contrasting evidences are not easily reconciliable in a rational and fully satisfactory theory. The Hercynian belt of the Southern Alps (Northern Italy) offers a good example of a fairly well studied area, in which one can try to propose an evolutionary model, at the present state of the investigations, in which metamorphism and genesis of granitic magmas can be seen as different steps of a single geotectonic process.

THE SOUTH-ALPINE HERCYNIAN BELT

The late-Hercynian cover, of variable age (from Westphalian B, in the western sector, to Westphalian D, or even Permian, in the eastern sector) lies unconformably over a folded, uplifted and eroded belt, which is nearly unmetamorphosed in the Carnic Alps (E) and deeply metamorphosed in the westernmost sector (Ivrea-Verbano zone). It follows that, while the stratigraphic sequence is very well documented in Carnia, in most of the belt only inferred ages and sequences can be put forward.

The Carnic stratigraphic sequence consists (Spalletta et al.,1982), from the bottom upward, of:

1) Middle-Upper Ordovician terrigenous sediments grading into shallow water limestones.

2) Silurian pelagic limestones passing to pelitic, radiolaritic facies with graptolites.

3) Devonian-Dinantian shallow water limestones progressively passing to calcareous pelagic facies, and then to:

4) Lower Silesian flysch with abundant volcanic materials.

In the east-central part (from Comelico to the Giudicarie Line) the Hercynian belt is metamorphosed. It consists of two phyllitic units separated by a complex volcano-sedimentary series with meta-carbonatic layers of unknown age and metarhyolites that are are attributed to the Ordovician (Caradocian) magmatism (Sassi and Zirpoli, 1968). The lower phyllitic unit should be pre-Caradocian and the upper phyllitic unit should be Silurian. Alternatively, Vai et al. (1982) suggest a younger age for the whole sequence: the carbonatic layers could be Devonian and therefore the metarhyolites should be Carboniferous and coeval with the Carnic flysch as well as the upper phyllitic unit.

From the Giudicarie Line to Lake Como the Hercynian belt is largely composed by metapelites (phyllites, micaschists and paragneisses) in which rare metalimestones or carbonate-bearing intercalations and basic-to-intermediate metatuffites are present.

The metamorphic grade increases generally towards W; in detail the gradient was also described S-N or even NE-SW (Liborio and Mottana, 1969) probably on account of the post-Hercynian tectonics. On the whole the paragneisses appear to prevail at higher metamorphic grade and the micaschists (and phyllites) at lower metamorphic grade, as if the stratigraphically deepest parts of the series were more arenaceous.

West of the Val Colla Line the stratigraphic sequence of the "Serie dei Laghi" is known better than that of the Central Alps. It consists of two units ("Scisti dei Laghi" and "Strona-Ceneri"), essentially pelitic the former and arenaceous the latter, with an amphibolite facies metamorphism.

Limestones are absent in both units: metatuffites cluster near the boundary between "Scisti dei Laghi" and "Strona-Ceneri", as well as orthogneisses (from metatonalites to metaleucogranites) with Ordovician intrusion age, that are very abundant.

Only one rock type can be interpreted as a meta-rhyolite (and it has been defined as such by several authors): the "Gneiss chiari" of the Orobic and Lugano region (Boriani and Colombo, 1979). Significantly they occur near the basement-cover boundary and show a low metamorphic grade. Metagranites are rare (Mt Pedena, Mt Legnone in the western Orobic Alps); no radiometric age measurements are available, but, in analogy with the Austridic and "Strona-Ceneri" ortogneisses, they can be referred to the Ordovician magmatic cycle. The sequence seems therefore to be entirely pre-Caradocian and covered by the so-called "Ordovician porphyry plateau".

To the W the "Strona-Ceneri" zone is mostly in tectonic contact with the "Ivrea-Verbano" zone, through the late Hercynian fault Pogallo Line.

The Ivrea-Verbano zone consists of metasediments ranging in metamorphic grade from the sillimanite-muscovite-quartz zone to the granulite facies. The metapelitic rocks prevail by far over the marbles and the metabasites, that are the two other important rock types. The Ivrea-Verbano zone is intruded by a late metamorphic basic igneous body (the Ivrea body with layered sequences) and is characterized by the presence of peridotite lenses mostly along the NW tectonic margin (the Insubric Line).

The last two formations and the problems connected with the transition between them, received particular attention in the last years; they will be described and discussed in detail further on.

IGNEOUS ROCKS (Fig. 1)

Pretectonic magmatism (mostly Ordovician)

The orthogneisses and the metarhyolites mentioned before, represent most of the Ordovician, or supposed Ordovician, magmatism that we know in this region. Recently (AGIP, 1977) in a borehole for oil drilling, near Venezia, it has been found a non-metamorphic granite with a radiometric age of 446 m.y.. The meta-granitoids occur only in the west-central and western sectors of the South Alpine basement, whilst the meta-rhyolites occur within the metasedimentary series in the east-central part and apparently only at the top of the series in the central and western areas.

The afore-mentioned Comelico metarhyolites constitute layers of more than 800 m of thickness (Sassi et al., 1980) with very slight petrographic variations (the lower part seems to be more acidic); the volcanic textures are still recognizable in spite of the meta-morphism (lower temperature part of the greenschist facies). The very narrow compositional range does not allow a chemical characterization of the original magma; nevertheless its calcalcaline affinity is rather clear. A peculiar feature of these rocks is the

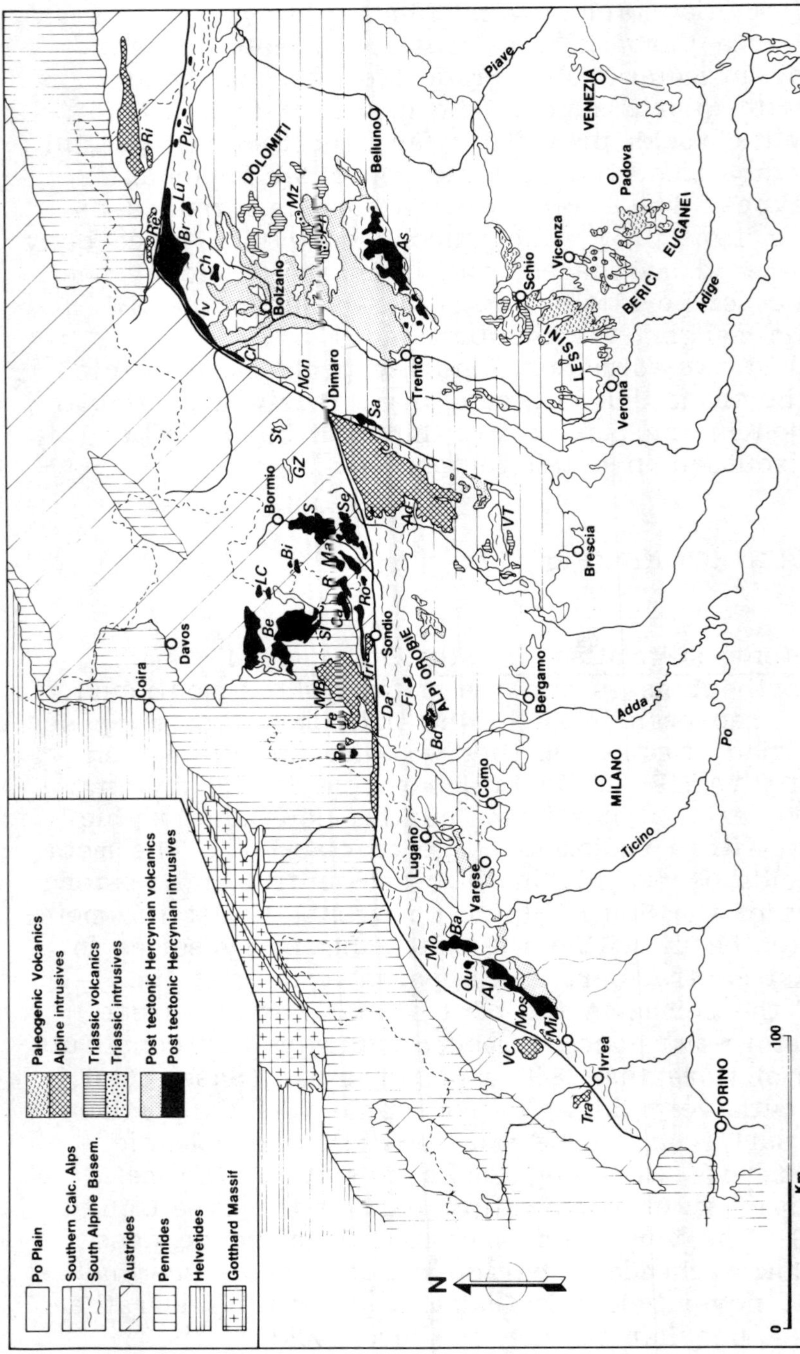
Po Plain
Southern Calc. Alps
South Alpine Basem.
Austrides
Pennides
Helvetides
Gotthard Massif
Paleogenic Volcanics
Alpine intrusives
Triassic volcanics
Triassic intrusives
Post tectonic Hercynian volcanics
Post tectonic Hercynian intrusives
N
0
100
Km
TORINO
Ivrea
Varese
Lugano
Como
MILANO
Bergamo
Brescia
Verona
Vicenza
Schio
Padova
VENEZIA
Belluno
Trento
Bolzano
Dimaro
Bormio
Sondio
Davos
Coira
DOLOMITI
ALPI OROBIE
BERICI
EUGANEI
LESSINI
Po
Ticino
Adda
Adige
Piave

FIGURE 1. Distribution of magmatic rocks in the Italian Alpine area (modified from Schiavinato, 1972). Legend: Ad=Adamello, Al=Alzo, As=Cima d'Asta, B=Brusio, Ba=Baveno, Bd=Val Biandino, Be=Bernina, Bi=Pizzo Bianco, Bo=Val Bodengo, Br=Bressanone, Ca=Pizzo Canciano, Ch=Chiusa, Cr=Monte Croce, Da=Culmine di Dazio, Fe=S.Fedelino, Fi=M.Fioraro, GZ=Gran Zebrù, Iv=Ivigna, LC=Leverone-Cotschen, Lu=Luson, Ma=Masuccio, MB=Masino-Bregaglia, Mi=Miagliano, Mo=Montorfano, Mos=Valle Mosso, Mz=Monzoni, Non=Val di Non, Pr=Predazzo, Pu=Pusteria, Qu=Quarna, Re=Rensenspitze, Ri=Vedrette di Ries, Ro=M.Rolla, Sa=M.Sabion, Se= M.Serottini, Sl=Pizzo Sella, So=Sondalo, St=Cime Sternai, Tra=Traversella, Tri=Triangia, VC=Valle del Cervo, VT=Val Trompia.

FIGURE 2. Geological sketch-map of the "Massiccio dei Laghi". Legend: 1=Pliocene and Quaternary, 2=Volcanic and sedimentary Permo-Mesozoic cover, 3=Late-Hercynian granites, 4=Metapelitic schists and paragneisses of the "Serie dei Laghi" (Scisti dei Laghi), 5=Orthogneisses of the "Serie dei Laghi", 6=Amphibolites of the "Serie dei Laghi", 7=Metapsammitic gneisses of the "Serie dei Laghi" (Strona-Ceneri), 8=Basic rocks, mainly in the granulite facies, including some ultramafites and subordinate metasediments of the "Ivrea-Verbano", 9=Kinzigites (pelitic and semipelitic high-grade metasediments with minor marble and amphibolite intercalations) of the "Ivrea-Verbano", 10=Sesia-Lanzo zone; Rimella and Losone schists; minor intrusives along the Insubric Line, 11=Faults, 12= Overthrusts.

N
Locarno
CENTOVALLI
Brissago
Luino
Finero
M. Zeda
Lago di Mergozzo
M. Mottarone
Lago Maggiore (Verbano)
Ticino
Toce
Ornavasso
M. Cerano
Quarna
Lago d'Orta
Varallo
Roccapietra
Doccio
Borgosesia
Borgomanero
Gattinara
Sesia
Fobello
Valle Mosso
Cossato
Biella
SERIE DEI LAGHI
STRONA CENERI
ZONA IVREA-VERBANO
1
2
3
4
5
6
7
8
9
10
11
12
0
20
Km

variability of the alkalies content, in accordance with all the acidic altered volcanic rocks.

W of the Giudicarie Line, in the Orobic Alps and in the Lugano district, the metarhyolites form a discontinuos horizon at the top or near the top of the metamorphic basement immediately below the post-Hercynian conglomerate, if it is present, or below the Permian volcano-volcanoclastic sequence. They are known as "Gneiss chiari" (Boriani and Colombo, 1979); their mineralogical and textural features are similar to those of the Comelico region, but the recrystallization is more pronounced since their metamorphic grade is higher (lower staurolite zone). Their composition is alkali-rhyolitic. They are generally more silicic than those of the Comelico region; the alkalies are again very variable.

"Serie dei Laghi". (Fig. 2) No metarhyolites occur in this series. The metagranitoids are calcalcaline rocks of mostly granodioritic composition; they are associated with leucogranitic augen gneisses. Boriani et al. (1983), Giobbi Origoni et al. (1983) recently described the stratigraphy of the "Serie dei Laghi" and the relationships between the ortho- and paragneisses. If we subtract the effect of the Hercynian metamorphism and deformation, the inferred stratigraphic sequence consists (from the bottom upwards) (Fig. 3) of rather deep-sea terrigenous sediments (Scisti dei Laghi) of dominating pelitic composition with quartzitic or semipelitic intercalations covered by a volcano-sedimentary horizon with abundant tuffitic material of basaltic composition. The overlying sequence (Strona-Ceneri) consists of coarse grained graywackes. Limestones are absent in the Serie dei Laghi with the exception of widespread carbonate-bearing nodules in the Strona-Ceneri rocks.

The age of intrusion, obtained from a Rb-Sr w.r. isochron (Boriani et al., 1983), is 466 m.y. (± 5) with a $^{87}Sr/^{86}Sr$ initial ratio of 0.7087; a deep source granodioritic crustal magma intruded the sequence producing thermometamorphic effects on the encountered

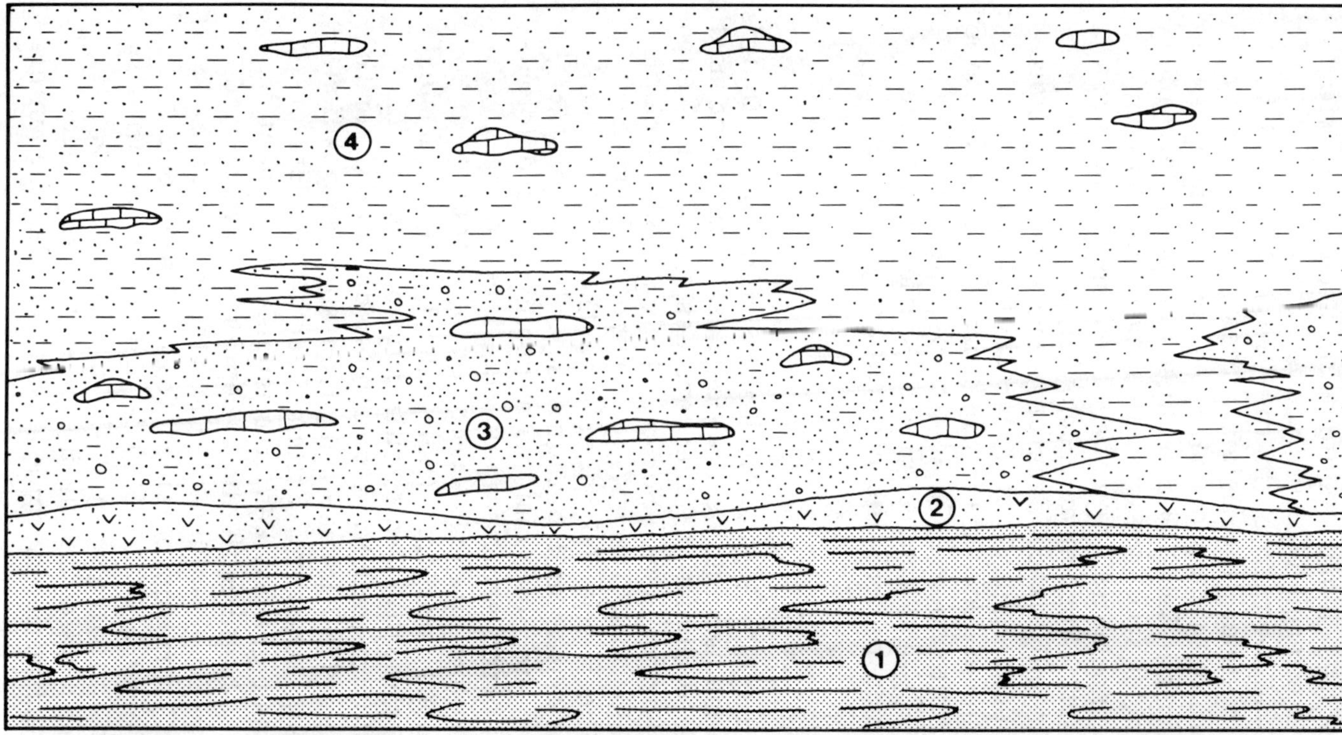

FIGURE 3. The "Serie dei Laghi" stratigraphical sequence: 1=Metapelites (Scisti dei Laghi), 2=Amphibolites (basic metatuffites), 3=Metapsammites-metapsephites (Cenerigneiss), 4=Metapsammites-metasiltites (fine-grained gneisses) with Ca-silicate nodules (also in the Cenerigneiss).

rocks, and inducing also a very important granitisation in the more permeable tuffitic and arenaceous sediments of the higher portions. Parts of the basaltic rocks were feldspathized and now they occur as hornblende augen gneisses; a leucogranitic residuum was emplaced in the highest levels, giving rise to porphyritic granite (the present augen gneisses) and to pegmatitic dykes intruded in the finer grained sediments.

The basaltic tuffites represent an igneous event that certainly preceded the acidic plutonism, but cannot be more precisely dated.

Since the appearance of the basic volcanism marks a well defined lithological change, it looks very likely that its geotectonic meaning was rather important. Whilst the pretectonic plutonic rocks are very well represented in the Serie dei Laghi, metarhyolites are not present; they appear in the adjacent Val Colla zone.

Ivrea-Verbano. The acidic pretectonic plutonites are not represented in this formation, the basic rocks on the contrary are very abundant.

There is not agreement among the authors about the attribution of the single metabasite occurrences to a pre-tectonic, syn-tectonic or late-tectonic magmatism (Boriani and Rivalenti, 1982). It is generally accepted that there are two fundamental types of metabasites:

a) metabasites in thin intercalations within the metapelitic gneisses, representing pre-tectonic marine basalts, flow and/or tuffs (typical presence of Mn mineralization in the associated quartzites)(Bigioggero et al., 1978-79b). The study of these amphibolites is now under way and the few available analythical data do not allow a characterization of this magmatism.

b) syn-to late-tectonic intrusive basic body with cumulitic ultramafites in the lower layered series and pyroxenite, anorthosite rythmic sequences in the upper layered series.

Above the layered series there is a very thick body of gabbro with only a very coarse layering that in the upper part grades into diorites. The series, very well exposed in the talweg of Valsesia,near Varallo, is nearly vertical with the deeper materials on the W side.

The petrochemical characters are very complex and contradictory since the Ivrea body is made up mainly with the products of fractionation of a subalcalic, slightly Na-rich basaltic magma (Bigioggero et al., 1978-79a). The diorites show an alkalic-potassic high-allumina character that can be traced back to a peculiar type of fractionation at high pressure. On the whole it seems likely that emplacement and differentiation took place in an orogenic environment. The syn-tectonic magmatic character is in good agreement with the observed microstructures that range from mainly granoblastic in the lower levels to almost purely hypidiomorphic in the upper parts.

In addition several ultramafic bodies, mainly consisting of lherzolites in the spinel-peridotite facies, occur in the Ivrea-Verbano zone along the Insubric Line. They show features (mostly pyroxenite veins) indicating a state of partial melting, during their emplacement, that could be referred to a nearly solid penetration in the lower crust of subcontinental upper mantle wedges along a deep discontinuity.

Carnia. The Middle Devonian-Lower Silesian flysch of Carnia contains abundant igneous materials consisting of Ti-rich olivine basalts with pillow structures, dolerite sills and volcanoclastic products deeply spilitized and associated with keratophyres (Gentili and Pellizzer, 1964). This volcanism is related (Spalletta et al.,1982) to an extensional tectonic regime as testified also by the sedimentary association.

Post tectonic magmatism.

Although there are igneous rocks that can be considered syn- or post-tectonic when one thinks to the Hercynian magmatism, one generally refers to the widespread granitic plutonism and to the equally widespread volcanism with the characteristic rhyolitic ignimbrites of Permian age.

In the South Alpine domain the highest concentration of granitic plutons is along or near the Insubric Line. Only one important granitic body occurs away from that tectonic lineament: the Cima d'Asta pluton, E of Trento.

Also the much younger post alpine plutonism seems to follow the same pattern (see map of fig.1). The plutons are intruded in a shallow crustal level; they are dominantly granitic, even if more basic types are always present but in extremely subordinate amount. On a broad scale the dominant rock types are granite-granodiorite in the eastern and granite in the western plutons. The measured ages of intrusions (Rb-Sr w.r. isochrons) cluster around the value of 280 m.y. with an initial $^{87}Sr/^{86}Sr$ of about 0.708 -0.709 . The petrochemical, petrographic and isotopic data indicate that at least most of the Permian plutonism is of crustal anatectic origin, whilst the more basic rock types show calcalcaline, probably subcrustal, character, like in the case of the "Appinitic" rocks of the Massiccio dei Laghi (Boriani et al.,1975) that seem to be precursors of the granitic magmatism.

The volcanism is characterized by the very large occurrence of rhyolitic and rhyodacitic ignimbrites with only minor initial andesites, trachiandesites,latites and quartzlatites. The largest outcrop is certainly that of the so-called Atesino plateau (4000 km^2) N of Trento, but their occurrence is almost continuous in the Southern Alps. They were erupted on land in the western and eastern zones (Marinelli,1975) and mostly in lacustrine environment in the Orobic sector (Collio formation containing U deposits) (Cassinis et al., 1975; Origoni Giobbi et al.,1983).

The volcanism is generally postgranitic, because sometime the volcanic deposits lie on eroded granites; the isotopic data do not indicate significant age differences between plutonism and volcanism, thus they can be considered roughly coeval and cogenetic, as supported also by their similar compositional range. Moreover the emplacement of large plutonic bodies and the formation of long eruptive fractures necessary for the outpouring of ignimbritic flows, indicate the same extensional environment.

RELATIONS AMONG DEFORMATION , METAMORPHISM AND IGNEOUS ACTIVITY

The Alpine metamorphism in the Southern Alps is absent or too weak to alter the previous situation. Among the authors that studied the region two parties exist: one party considers the older part of the basement, stratigraphically underlying the metarhyolites, polymetamorphic, i.e. having undergone a first metamorphism coeval with the Caledonian orogeny and a second one of Carboniferous age (Hercynian). The other party considers the whole basement as essentially monometamorphic (Hercynian). Evidences and geological considerations that support polymetamorphism of at least part of the basement are:

a) A Rb/Sr w.r. isochron of 478 m.y. (Hunziker and Zingg, 1980) on the Ivrea-Verbano metasediments, interpreted as corresponding to a high-T reequilibration of the Rb-Sr system and to large scale degranitisation and melting. Small samples give Hercynian ages and are interpreted as due to small scale homogeneization due to slow cooling.

b) U/Pb ages on zircons and monazites from the Strona-Ceneri zone give a "Caledonian" age, while those of the neighbouring Ivrea-Verbano indicate Hercynian reequilibration (Pidgeon et al., 1970; Köppel et al., 1974).

c) The widespread magmatic event around 450 m.y. is seen as late- or post-orogenic by the authors that consider very unlikely a calcalcaline magmatism unrelated to an orogenic event.

d) In many Ordovician granites there are many metasedimentary xenoliths showing a pre-intrusion schistosity.

e) In most of the basement, whatever the reached metamorphic grade, a lower P paragenesis overprints a higher P one; following Zwart's (1967) scheme, the higher pressure phase is attributed to the Caledonian orogenesis and the lower P one to the Hercynian (Bocchio et al., 1980).

f) The orthogneisses apparently show a simpler deformational history in comparison to the metasedimentary country rocks. This observation was considered a proof of an intrusion into an already metamorphosed environment (Boriani et al., 1977).

g) The Austridic basement is very similar to the South-Alpine domain: a Rb/Sr w.r. isochron on amphibolite facies paragneisses, S of the Tauern Window, gives an age of about 497 m.y. (Borsi et al., 1973). Those are covered by paragneissic rocks, interpreted (Sassi and Zanferrari, 1972) as metaconglomerates, to which follow the metarhyolites and a thick phyllitic sequence, that could therefore represent a post-Caradocian series overlying a metamorphic uplifted and eroded basement. This interpretation was extended to the South- Alpine basement.

h) In the Ivrea-Verbano zone there are amphibolites (Boriani, 1965; Boriani and Peyronel Pagliani, 1968) that are clearly derived from eclogites; this could indicate a pre- Hercynian high-P event.

In favour of an essentially monometamorphic (Hercynian) history, the following considerations can be made:

a) The relations between deformation and recrystallisation are fundamentally very simple all over the South-Alpine basement. Most of the deformation occurred at relatively low-T (max. garnet grade); in fact the higher T minerals (oligoclase, staurolite, cordierite, andalusite, kyanite, sillimanite) of the amphibolite facies are essentially post-kinematic. This simple pattern does not appear compatible with the superposition of two important regional metamorphisms, as stated by Gregnanin (1980) for the analogous Austridic basement.

b) A detailed comparative microstructural study of the Ordovician orthogneisses and of the metasedimentary country rocks reveals a common deformational and metamorphic evolution. The metagranitoids were deformed in a relatively low-T phase and then recrystallized statically at higher T.

c) In the Serie dei Laghi the Ordovician granites were intruded in a non metamorphic or at most very weakly metamorphosed sedimentary sequence (see paragraph on the Serie dei Laghi). This statement is supported by the observation of a widespread granitisation, related to the intrusion, conditioned by the relative permeability of the country rocks. In fact the so-called "feldspathized amphibolites" (Boriani and Giobbi Mancini, 1972) turned out to be, at a closer inspection (Giobbi Origoni et al.,1984) more or less coarse grained basic tuffites, granitised before the metamorphism.

d) Another evidence of the non-metamorphic or very low grade metamorphic state of the sedimentary rocks at the moment of the intrusion of the Ordovician granites, is the presence of relics of chiastolite in the contact aureoles of the metapegmatites in the fine grained gneisses of the Strona-Ceneri zone (Bigioggero and Boriani, 1975). Also the "augen gneisses" of this unit were, at least in part, granitised sediments in which the primary permeability played an important role in the granitisation process.

e) The Rb-Sr system equilibration around 480 m.y. could be related to a diffuse metasomatism and granitisation of the sedimentary pile induced by the Ordovician igneous activity.

f) The pre-intrusive schistosity, shown by some xenoliths of the Ordovician granites, could be referred at most to a weak deformational-metamorphic lower Ordovician event (Sardic phase ?) or alternatively to a pre-Hercynian basement not outcropping in the area.

CHARACTERS OF THE METAMORPHISM (Fig.4)

As stated before the metamorphic grade increases from E to W and from S to N, reaching its maximum

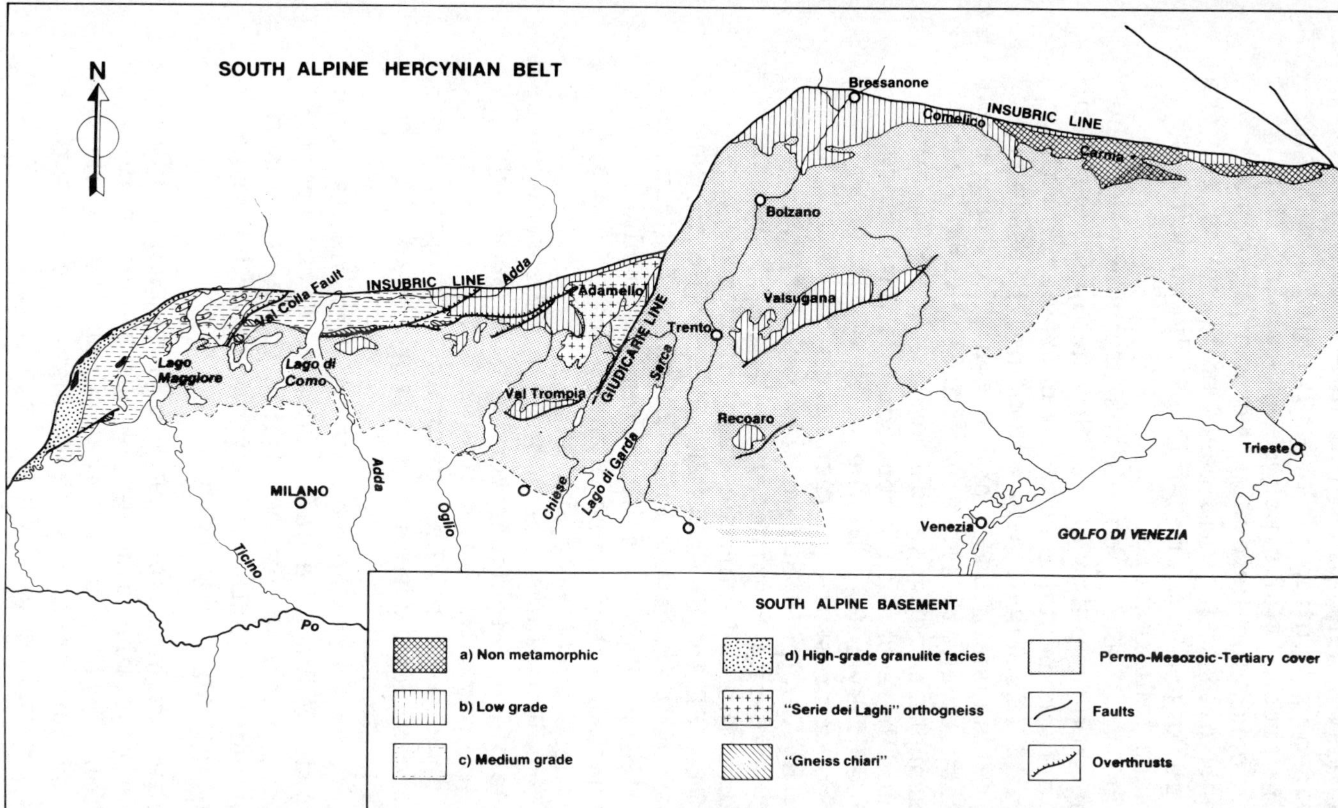

FIGURE 4. Metamorphic sketch-map of the Southern Alps.

in the Ivrea-Verbano zone (granulite facies). The widespread overprinting of lower P parageneses suggested the presence of a polymetamorphism. If this is not the case, the Hercynian metamorphism must have been, at least in some parts of the belt, polyphasic; the geological meaning of this polyphasicity is nevertheless still obscure.

As far as the P and T conditions are concerned, the Hercynian metamorphism of the South-Alpine basement displays a fundamental Barrovian character; in the eastern part (Comelico, Alpi Sarentine, Bressanone, Recoaro, Valsugana) low-T and P conditions are reported (Boriani et al.,1974). W of the Giudicarie Line the metamorphism ranges from low-T, low-P to intermediate and high-T with intermediate P (widespread occurrence of kyanite). In the Ivrea-Verbano the presence of a pressure gradient from SW to NE is proposed by Zingg (1980) on the basis of the occurrence of cordierite + garnet only in the SW sector. In this unit the metamorphic grade increases very rapidly from SE to NW from the amphibolite to the granulite facies (Schmid, 1967; Peyronel Pagliani and Boriani, 1967). This transition from the medium to the high metamorphic grade occurs (Boriani,1983) through these successive steps:

1) In the sillimanite + muscovite + quartz zone towards the sillimanite + K-feldspar zone are present many granitic (or pegmatitic) mobilizates.

2) The breakdown of the muscovite in the presence of quartz coincides with a rapid decrease of the granitic mobilizates.

3) In a very short distance also biotite breaks down reacting with sillimanite and quartz to give garnet + K-feldspar. The appearance of this granulitic assemblage coincides with the appearance of orthopyroxene in the intercalated metabasites.

This sequence indicates that the transition medium-grade – high-grade-granulite is mostly due to the subtraction of water to the system, caused by the hydrous melting of muscovite + plagioclase + quartz + H_2O and the subsequent decrease of X_{H_2O} in the fluid phase. As demonstrated by Kerrick (1972) the reduction of X_{H_2O} causes the breakdown of muscovite and the contemporaneous migration to higher T of the granite solidus. Only if water is continuously supplied from below, one could envisage a conspicuous production of granitic magma during regional metamorphism and a significant degranitisation in the high-grade rocks.

Schmid (1972, 1978-79) affirmed that the Ivrea-Verbano high-grade metapelites are deeply degranitised and Hunziker and Zingg (1980) supported the idea that this degranitisation was connected with the origin of the Ordovician granodiorites and granites of the adjoining Strona-Ceneri zone.

If this is correct we must admit that the Ivrea-Verbano zone played the role of lower crust during the Ordovician magmatic event, when a marked crustal thinning, with conspicuous melting at relatively high temperature, must have occurred.

Moreover the metamorphic sequence of the Ivrea-Verbano, does not seem compatible with a steep thermal gradient (Schmid and Wood, 1976), but it seems mostly due to dehydration.

In addition, if the Ivrea-Verbano was in the lower crust during the Ordovician, its tectonic emplacement shoud have occurred through a sort of "obduction" process during the thrusting and folding of the orogenic belt; then the post-kinematic thermal peak should have cancelled the metamorphic break between the Ivrea-Verbano and the Strona-Ceneri zones.

CONCLUSIONS

The South-Alpine Hercynian belt shows features that are very similar to those of the Late-Proterozoic "mobile belts" (Kröner,1981), maybe with a more rigid behaviour of the lithosphere. The belt seems to be mostly ensialic; only the Ivrea-Verbano rocks (excluding the basic intrusive body) are compatible with an oceanic or a very thinned continental crust, for the abundance of metapelites, Mn-rich quartzites, metabasalts and peridotites. The subcontinental character of the latter indicates that in this part of the ensialic basin the spreading and thinning of the lower crust reached "almost" the oceanisation. We propose that this episode of maximum thinning and spreading occurred in the Ordovician, when the maximum bimodal igneous activity took place with intrusions and extrusions in and above a sedimentary sequence. The origin of these magmas can be found in the hot upwelling astenosphere as well as in thinned lower crust.

After the Ordovician, a progressive cooling down of the lithosphere caused the interruption of the melting; the sedimentary basin passed through alternate subsidences and uprising, probably related also with important transcurrent movements (Vai et al., 1982).

Traces of another crustal thinning can be found in the Lower Carboniferous of Carnia where the presence of the spilite-keratophyre association indicates a beginning of oceanisation.

Meanwhile the cooling of the astenosphere and of the lower crust triggered an A- type subduction with uncoupling of the lower crust from the upper crust.

Thrusting and folding characterized the orogenic phase and the resulting crustal thickening caused a progressive T increase originating a post-kinematic thermal peak. Only very late, in the Permian, when an extensional tectonic regime was set up, essentially acidic magmas could form by melting of the undergoing crustal slab.

In this case it is obvious that absolutely no connection exists between the Hercynian metamorphism, that we can observe in the exposed belt, and the origin of the post-tectonic acidic magmas.

Theoretically the Ivrea-Verbano zone could represent a fragment of the lower crust that during the Ordovician crustal thinning partially melted to give the granodioritic-granitic magmas. The isotopic data (Hunziker and Zingg, 1980), but not the geological evidence (Boriani, 1983), support this hypothesis.

In conclusion, from the study of the South-Alpine Hercynian belt, it does not seem that the transition medium-grade – high-grade in the regional metamorphism could be the seat of the conspicuous anatexis responsible for the formation of crustal anatectic magma. Only water-saturated melts can form at the temperature of regional metamorphism; large scale melting is prevented by the scarcity of water.

The source for the crustal granitic melts must be searched in the lower crust, either during a crustal thinning above a mantle plume, or during an A-type subduction process.

REFERENCES

Bigioggero, B. and Boriani, A. (1975). I noduli a silicati di Al degli gneiss minuti della "Strona-Ceneri". Boll. Soc. Geol. It. 94, 2073-2084.

Bigioggero, B., Boriani, A., Colombo, A. and Gregnanin, A. (1978-79a). The "Diorites" of the Ivrea Basic Complex (Central Alps, Italy). Mem. Sc. Geol. 33, 71-85.

Bigioggero, B., Brigo, L., Ferrario, A., Gregnanin, A., Montrasio, A. and Zuffardi, P. (1978-79b). Strona Valley (Fe-Ni-Cu) and (Fe-Ba) ore deposits. Excursion book. Mem. Sc. Geol. 33, 33-39.

Bocchio, R., Crespi, R., Liborio, G. and Mottana, A. (1980). Variazioni composizionali delle miche chiare nel metamorfismo progrado degli scisti sudalpini dell'alto Lago di Como. Mem. Sc. Geol. 34, 153-176.

Boriani, A. (1965). L'anfibolite a orneblenda e cummingtonite di Ornavasso (Val d'Ossola). Rend. Soc. Min. It.

21, 25-35.

Boriani,A. (1983). The medium-grade - high-grade transition in the regional metamorphism: an example from the Ivrea-Verbano zone (Italy). Rend. Soc. It. Min. Petr. 38(2), 543-548.

Boriani,A., Bigioggero,B. and Origoni Giobbi,E. (1977). Metamorphism, tectonic evolution and tentative stratigraphy of the "Serie dei Laghi" - Geological map of the Verbania area (Northern Italy). Mem. Sc. Geol. 32, 26 pp.

Boriani,A. and Colombo,A. (1979). Gli "Gneiss chiari" tra la Val Sesia e il Lago di Como. Rend. Soc. It. Min. Petr. 35, 299-312.

Boriani,A., Colombo,A., Origoni Giobbi,E. and Peyronel Pagliani,G. (1975). The "Appinite suite" of Massiccio dei Laghi (Northern Italy) and its relationships to the regional metamorphism. Rend. Soc. It. Min. Petr. 30, 893-917.

Boriani,A., Dal Piaz, G.V., Hunziker,J.C.,Von Raumer J. and Sassi,F.P. (1974). Caratteri, distribuzione ed età del metamorfismo prealpino nelle Alpi (Relazione ufficiale). Mem. Soc. Geol. It. 13(suppl.1), 165-225.

Boriani,A. and Giobbi Mancini, E. (1972). The feldspathized amphibolites of the "Strona-Ceneri". Boll. Soc. Geol. It. 91, 655-681.

Boriani,A., Origoni Giobbi,E. and Del Moro,A. (1983). Composition, level of intrusion and age of the "Serie dei Laghi" orthogneisses (Northern Italy, Ticino,

Switzerland). Rend. Soc. It. Min. Petr. 24, 191-205.

Boriani,A. and Peyronel Pagliani,G. (1968). Rapporti fra le plutoniti erciniche e le metamorfiti del "Massiccio dei Laghi" nella zona del M. Cerano (bassa Val d'Ossola). Rend. Soc. It. Min. Petr. 24, 111-142.

Boriani,A. and Rivalenti,G. (in press). Crosta profonda e significato delle rocce basiche e ultrabasiche dell'Ivrea-Verbano in un secolo di studi. Mem. Soc. Geol. It. 24(A)

Borsi,S., Del Moro,A., Sassi,F.P. and Zirpoli,G. (1973). Metamorphic evolution of the Austridic rocks to the South of the Tauern Window (Eastern Alps). Mem. Soc. Geol. It. 12, 549-571.

Cassinis,G., Origoni Giobbi,E. and Peyronel Pagliani,G. (1975). Osservazioni geologiche e petrografiche sul Permiano della bassa Val Caffaro (Lombardia orientale). Atti Ist. Geol. Univ. Pavia 25, 17-71.

Gentili,G. and Pellizzer,R. (1974). Le rocce eruttive del Paleozoico Carnico. Boll. Soc. Geol. It. 83, 151-205.

Giobbi Origoni,E., Testa,B. and Carimati,R. (in press). Contributo alla ricostruzione stratigrafica della "Serie dei Laghi": Litofacies principali della "Strona-Ceneri" a NE del Lago Maggiore (Alpi Meridionali – Italia). Rend. Soc. It. Min. Petr. 38(3).

Gregnanin,A. (1980). Metamorphism and magmatism in the western Italian Tyrol. Rend. Soc. It. Min. Petr. 36, 49-64.

Hunziker,J.C. and Zingg,A. (1980). Lower Paleozoic

amphibolite to granulite facies metamorphism in the Ivrea Zone (Southern Alps, Northern Italy). Schweiz. Min. Petr. Mitt. 60, 181-213.

Kerrick,D.M. (1972). Experimental determination of muscovite + quartz stability with $P_{H_2O} < P_{total}$. Am. Journ. Sc. 272, 946-958.

Köppel,V. (1974). Isotopic U-Pb ages of monazites and zircons from the crust-mantle transition and adjacent units of the Ivrea and Ceneri Zones (Southern Alps, Italy). Contr. Min. Petr. 43, 55-70.

Kröner,A. (1981). Precambrian plate tectonics. In: Precambrian Plate Tectonics, A. Kröner (ed). Elsevier, Amsterdam.

Liborio,G. and Mottana,A. (1969). Lineamenti geologico-petrografici del complesso metamorfico sudalpino nelle Alpi Orobie orientali. Rend. Soc. Min. Petr. 25, 475-519.

Marinelli,G. (1975). Magma evolution in Italy. In: Geology of Italy, C. Squyres (ed). E.S.S.L.A.R., pp. 165-219.

Origoni Giobbi,E., Ravagnani,D. and Bernasconi,A. (1983). Petrologic and metallogenic investigations on the Collio formation of the Novazza U mine. Bergamasc Alps, Italy. Rend. Soc. It. Min. Petr. 38, 293-305.

Peyronel Pagliani,G. and Boriani,A. (1967). Metamorfismo crescente nelle metamorfiti del "Massiccio dei Laghi" nella zona bassa Val d'Ossola - Verbania (Novara). Rend. Soc. Min. It. 23, 351-397.

Pidgeon,R.T., Köppel,V. and Grünenfelder,M. (1970). U-Pb isotopic relationships in zircon suites from a para- and orthogneiss from the Ceneri zone, Southern Switzerland. Contr. Min. Petr. 26, 1-11.

Pitcher,W.S. (1979). The nature, ascent and emplacement of granitic magmas. Journ. Geol. Soc. London 136, 627-662.

Sassi,P.F. and Zanferrari,A. (1972). Il significato geologico del Complesso del Turntaler (Pusteria), con particolare riguardo alla successione di eventi metamorfici prealpini nel basamento austridico delle Alpi Orientali. Boll. Soc. Geol. It. 91, 533-557.

Sassi,P.F., Zanferrari,A. and Zirpoli,G.. (1980). Nuovi dati sulla stratigrafia e i caratteri chimico petrografici dei porfiroidi del Comelico (Alpi orientali). Mem. Soc. Geol. It. 20, 425-434.

Sassi,F.P. and Zirpoli,G. (1968). Il basamento cristallino di Recoaro. Studio petrografico. Mem. Soc. Geol. It. 7, 227-245.

Schiavinato, G. (1972). Plutonismo e vulcanismo sul versante italiano delle Alpi centrali ed orientali. Rend. Acc. Naz. Lincei, Cl. Sc. Fis. Mat. Nat. 52(8), 523-538.

Schmid,R. (1967). Zur Petrographie und Struktur der Zone Ivrea-Verbano zwischen Valle d'Ossola und Val Grande (Prov. Novara, Italien). Schweiz. Min. Petr. Mitt. 47, 935-1117.

Schmid,R. (1972). Substitution von Biotit durch Granat

und, "Degranitisation" in granulitischen Paragneissen der Ivrea-Zone (N-Italien). Habilitationsschrift ETH, Zürich.

Schmid,R. (1978-79). Are the metapelites of the Ivrea-Verbano zone restites? Mem. Sc. Geol. 33, 67-69.

Schmid,R. and Wood,B.J. (1976). Phase relationships in granulitic metapelites from the Ivrea-Verbano zone (Northern Italy). Contr. Min. Petr. 54, 255-279.

Spalletta,C., Vai,G.B. and Venturini,C. (1982). La Catena Paleocarnica. In: Guida alla geologia del sudalpino centro-orientale, A. Castellarin, G.B. Vai (eds). Guide Geol. Reg. Soc. Geol. It., Bologna, pp. 281-292.

Vai,G.B., Boriani,A., Rivalenti,G. and Sassi,P.F. (in press). Catena ercinica e Paleozoico nelle Alpi Meridionali (dagli albori ad oggi). Mem. Soc. Geol. It. 24(A).

Zingg,A. (1980). Regional metamorphism in the Ivrea-zone (Southern Alps, N-Italy): Field and microscopic investigations. Schweiz. Min. Petr. Mitt. 60, 153-178.

Zwart,H.J. (1967). The duality of orogenic belts. Geol. in Mijnb. 46, 283-309.

THE EVOLUTION OF THE ALEUTIAN ISLAND ARC MGMATISM

BORSUK A.M., TSVETKOV A.A., ZHURAVLEV D.Z., CHERNYSHEV I.V.
Institute of Ore Deposits Geology, Petrology, Mineralogy and Geochemistry of the USSR Academy of Sciences, Moscow, USSR.

ABSTRACT

Three different magmatic associations representing the three successive stages of the Aleutian Island arc evolution, have been distinguished and studied on the basis of Komandorkie Islands, which constitute westernmost segment of the arc. These are: 1) basalt-plagiorhyolitic (P_3), 2) basalt-trachydoleritic (N_1^{1-2}), 3) diorite-granitoid (N_2^3). Rocks of andesitic association ($N_2^3-Q_4$) which are abundant within the central and the eastern segments of the Aleutian Island arc, on Komandorskie Islands do not occur. We discuss in this paper some problems of the origin of island-arc magmas, using the data on Nd and Sr isotopes and the on REE element patterns. We believe that the differences in magmatic evolution of the western (komandorskie Islands) and the central plus eastern (Aleutian Islands and Alaska Peninsula) segments of the Aleutian island arc, are determined by the two factors: 1) lateral variations in composition and thickness of the crust - basement of the island arc, caused by its transition from oceanic to continental type, 2) variations in depth of magmatic sources.

In this paper magmatic rocks of Kommandorskie Islands are discussed. For the first time in that region the authers distinguished and investigated three natural magmatic associations representing the three successive

Proceedings of the 27th International
Geological Congress, Volume 9, pp. 67–84
PETROLOGY (IGNEOUS AND METAMORPHIC ROCKS)

stages of the island arc development: 1) basalt-plagiorhyolitic (P_3), 2) basalt-trachy-doleritic (N_1^{1-2}), diorite-granitoid (N_2^{3}-N_2^{1}). (fig.1.).

Chemical compositions of igneous rocks and the results of their raiometric dating are presented in Table 1 and 2 respectively.

Basalt-plagiorhyolitic Association

Period when the Aleutian Trench and the associated submarine ridge was formed is specified by basalt-plagiorhyolitic formation. It is of bimodal type and comprise basalts, plagiorhyolites and plagiorhyodacites. Their spatial distribution and volumes of basic and acidic members siguificantly varies Dredging of the slopes of Komandorskie Islands showed far more extensive distribution of basalts, relative to plagiohyolites, unlike the subaerial parts of the islands, where plagiorhyolites are prevailing.

Basalts make up lava flows, layers of tuffs and tuff-breccias cut by basaltic dykes and occasional stokes (up to 1000M^2), of medium-grained gabbonorites and gabbro. Occuring in basalts are inclusions of ultrabasic rocks: spinel peridotites and pyroxenites /22/.

Basalts are usually porphyric in texture with pilotaxitic or hyalopilitic groundmass. Phenocrysts (30-40 vol.%) are represented by labrador (An_{60-65}), low-Ti augite ($Ca_{40-38}Mg_{45-49}Fe_{12-14}$) and bronzite ($Ca_{2-3}Mg_{65-75}Fe_{27-28}$). Feldspar is often replased by carbonate, chlorite, epidote and zeolites. The groundmass, usually completely chloritized, includes besides plagioclase and clinopyroxene microliths also needles of green hornblende. Magnetite, ilmenite and apatite are common accessories. Cavities in the amygdaloidal basalts and in volcanoclasitic rocks are filled with calcite, quartz, zeolites and chlorite with sometimes found dendrites of native copper. Basaltic dikes and subvolcanic bodies contain phenocrusts of high-Mg hornblend, in addition to plagioclase and clinopyroxene ones.

Major-element data (see Table 2) along with distribution patterns of Co, Ni, Cr, V, Sr, Zr and K/Rb ratios (~800, n=11), show close correspondence of Mednyi Island basic rocks with low-Ti tholeiitic basalts of the earlyest evolutional stages of ensimatic island arcs (Tonga-Kermadek, Mariana etc.), though epimagmatic processes caused considerable variability in consentrations of Fe^{2+}, Fe^{3+}, Mg, Ca, Na /21/.

Among the important geochemical features of Komandorskie Islands basalts is their enrichment in Cu and, to some extend in Ba, which, as it was shown earlier /20/, is due to their primary enrichment in the melt. Postmagmatic processes such as propylitization resulted in leaching of copper from Cu-bearing rock-forming minerals (magnetite and clinopyroxene), and in the formation of native copper deposits, similar in type to the copper-zeolite deposits of Lake Superior in the USA and Canada.

Acid members of the association are represented by plagiorhyolites and plagiohyodacites. They make up thick (up to 50-70 m) and relatively short (1-3 km) lava flows, layers of lava breccias, tuffs, tuff-breccias and of tuff-conglomerates, as well as a number of large hupabyssal stocks, sills and dikes, micro-accoliths and extrusive domes, cutting basic volcanics. Plagiorhyolites and plagiorhyodacites contain as phenocrysts partially or totally albitized andesine (An_{44-48}) (15-25%), resorbed quartz (10-15%) and small quantities of clinopyroxene (1-2 vol.%). The latter varies compositionally from augite ($Ca_{43}Mg_{49}Fe_{8}$) to ferroaugite ($Ca_{51}Mg_{28}Fe_{21}$). Sometimes, plagioclase grains are rimmed by K-Na feldspar $Or_{35-48}Ab_{42-59}An_{5-8}$. Felsitic, occasionally vitrophyric groundmass (60%) contains small impregnations of magmetite, pyroxene, biotite, apatite, sphene and zircon. The biotite is low-Ti (0.16-0.24% TiO_2) and has high Fe-index (f= 56-61%)*. The resorbtion of quartz in rhyolites indicates its non equilibrated state caused by intratelluric crystallization under high water pressure. According to the experimental data /16-19/, the incrase in water saturation of rhyolitic melt lead to abrupt expansion of the near-liquidus quartz crystallization field. Under water content exceeding 5wt%, quartz is the first phase to crystallyze at depths greater than 10-15km. However, with a decrease in water saturation

* f (Fe-index) = $\frac{100\ Fe}{Fe + Mg}$ (at%)

of the melt or with decrease in pressure the system shiffs into the field of joint quartz and feldspar crystallization. This accounts for the nonquilibrity of the intratelluric quartz and the melt within the near surface conditions which lead to the extensive corrosion of phenocrystal quartz.

According to REE distribution patterns, just as to the major and trace-elements, Komandorkie plagorhyolites are close equivalents to the acid rocks formed in continental orogenic belts /21/. The geological and structural relations of basalts and plagioryolites, their occurance within relatively short time-interval along with the above-specified geochemical features, make it possible to include both rocks into the same magmatic association despite the fact that the basic and the acid magmas were probably generated through different mechanisms.

The earliest magmatic manifestations within the central sector of the Aleutian arc corresponding in age, composition and geological setting to basalt-plagiorhyolitic association are known on many islands: Attu and Agattu (basement formation), Amchitka (Krugloi Point, Amchitka, (Krugloi Point, Amchitka, Banjo Point formations), Adak and Ilak (Andrew Lake, Finger Bay formations), Rat (Rat, Ganners Cove formations). At most part they include basaltic pillow lavas which just as Komandorskie basalts have indergone greenstone metamorphism. However the rock-forming clinopyroxenes in these rocks relative to Komandorskie basalts are significantly Fe-enriched and contain more Al_2O_3, Ti_2 and CaO ($Ca_{43-46}Mg_{36-40}Fe_{17-19}$) /5/. The latter circomstance, as we believe, is determined by differences in depths of magmatic sources somewhat greater within the central segment of the Aleutian arc. Petrogenetic differences between Komandorskie and central-Aleutian basalts are also difined through REE patterns /21/.

It should be also noted that besides basic volcanics on central Aleutians are also known some intermediate and acid magmatic bodies, the examples of which are Hidden Bay and Finger Bay layered gabbro-granodiorite plutons on Adak dated 30-35 m.y. /3, 14/. All these features show the continually differnetieted character of basalt plagiorhyolitic association in the central segment of of the Aleutian Island arc.

Within the eastern segment of the Aleutian arc the Unalashka formation is coeval with basalt-plagiorhyolitic association. On Unalashka Island, where the formation is best exposed it is made of coarse and fine-grained pyroclastics, alternating with basaltic, andesitic and dacitic lava flows cut by numerous compositionally similar dikes and sills /7/. Generally, the basalts of the eastern Aleutian segment differ dramatically in major (Mg, Fe, Na, K) and trace-element (Co, Ni, Cr, Sr, Zr) concentrations from tholeiitic basalts of intraoceanic ensimatic island arcs, e.g. Eua Tonga /11/ as well as from Komandorskie basalts /2/.

Nearly sinchronous to Komandorskie-Aleutian basalt-plagiorhyolitic association, numerous intrusive bodies of quartz diorite-granitic association were formed on Kodiak Island and Alaska peninsula. Compositionally they vay from quartz diorites to mozonites and normal granites, while their age scatters whithin 38-26 m.y. /17/.

Basalt-Trachydoleritic Association

On Komandorskie Islands the rocks of this formation are represented by basaltic, trachydoleritic lava flows and sheets, less often - by sills and dikes of trachybasalts, trachydolerites, teschenites and by stocks of subalkaline gabbro /1/.

Typical subalkaline olivine basalts are always massive porphyric rocks with compact fabric. Phenocryst assemblage: zonal labradore (An_{49-65}), clinopyroxene - augite or titan-augite ($Ca_{36-45}Mg_{43-47}Fe_{9-19}$) and high-Mg olivine ($Fo_{88}$). The groundmass with hyalopilitic or pilotaxitic texture consists of plagioclase and clinopyroxene microlites and specs of brown biotite set in a glassy matrix, filled with titanomagnetite micrograins. Vesiculls and inter-mineral spacings are filled with analcime also abundant in the glassy matrix.

Teschenites consist of well shaped titanougite phenocrysts ($Ca_{44-45}Mg_{39-40}Fe_{15-16}$) - more Fe-enriched than clinopyroxene of subalkaline olivine basalts, trachybasalts and trachydolerites, small grains of red-brown amphibole (barkevikite) and biotite in a faintly individualized groundmass of plagioclase (An_{35-49}) anorthoclase, augite, quartz, titanomagnetite and apatite.

Basalt-trachydoleritic association belongs to continuously differentiated type. Its whole-rock chemistry varies gradually from close to average continental flood basalt to average subalkaline gabbro and teschenite /18/. Simultaneously, K_2O content in rocks gradually increases and in some trachydolerites and teschenites exceeds 3 wt.%, being related to appearance of anorthoclase in modal composition. Meanwhile, the bulk Fe-content does not significantly increase, and varies within 8-9% range, whereas Ti-content is exceeds 1 wt.%. Teschenites, the most potassium-rich rocks, contain significantly higher ammounts of Co, Ni, Cr, V, Sr than the typical subalkaline and alkaline oceanic basalts /8/, the variations in these element concentrations, along with potassium, beeng within the evolutional row of subalkaline olivine basalt-teschenite of a continious type /21/. K/Rb ratios fall within 315-340 for subalkaline olivine basalts and gabbro and reach some trachydolerites and teschenites.

Temporal relations and petro-geochemical fearures of magmatic rocks show a progressive rise in alkalinity of a primary basaltic melts. This can be due probably to dry fractionation of 20% Pl, 10% Cpx, 5% 01, 10% Mt, i.e. a mineral association, equal to subalkaline basalt-trachybasalt phenocryst assemblage at moderate depths (within the basaltic minerals stability field /2/.

Of all the Aeutians, subalkaline basalts are known only on Kanaga Island, where they contain inclusions fo dunites and peridotites /4/, whilst true petrographic analogues of Koandorskie trachydoleritic and teschenitic rocks are not yet found either on Aleutian Islands or on Alaska peninsula.

Diorite-Granitoid Association

On Komandorskie Islands, rocks of this mgmatic association make up stocks, small laccolith-type bodies, dikes and lava flows. The largest intrusive body (10 km^2) accompanied by radial dyke swarms is located near Cherny Point and is composed of columnarly-jointed granodiorites.

In the centre and in the south of Medny Island there are also a few tonalitic microlaccoliths and dikes.

Granodiorites are massive hypidiomorphic rocks build of plagioclase (50-55%), alkali feldspar (5-7%), hornblend (15-20%), biotite (2-5%) and quartz (15-20%). The accessory minerals (1-2%) are: zircon, magnetite and apatite. The latter often forms rims on magnetite grains. Quartz diorites and tonalites differ from granodiorites in their porphyric fabric, lack of alkali feldspar and higher modal ammounts of quartz (up to 25%) and biotite (up to 10%).

Plagioclase in granodiorites and tonalites corresponds to oligoclase-andesine (An_{17-43}). It is frequently zonal, the outer zones being represented by oligoclase (An_{17-20}) whereas the central - by andesine (An_{36-43}). Or-content also increases from cores to rims of the grains (av. 2.1 Or-Core, 4.10 - rim). In contrast to plagioclase of plagiorhyolites of basalt-plagiorhyolitic association, the one of granodiorites and tonalites has twice as much Sr along with similar distribution of Cr, V, Ba, Zr, Cu. K-Na feldspar in granodiorites corresponds to high orthoclase-cryptopertite. Amphibole is represented by a common hornblend (f = 25.9%). Biotite in tonalites and granodiorites is both primary and secondary, the latter substituting hornblende. Its Fe-index is somewhat lower in tonalites (f = 29-30%), than in granodiorites (f = 32). Epidote and chlorite are present in small quantities.

Volcanic rocks of the described formation, represented mainly by andesitic and andesite-dacitic lavas, are found to be in close spatial association with tonalitic and granodioritic intrusions and can be acacounted as their volcanic analogues (see Table 2). Andesitic lavas are made of plagioclase (An_{55-60}), clino- ($Ca_{40-41}Mg_{43-44}Fe_{14-15}$) and orthopyroxene (bronzite) phenocrysts and green hornblende in hyalopilitic or intersertal groundmass.

According to major and trace element (Co, Ni, Cr, V, Sr) distribution rocks of diorite-granitoid association are near Vinogradov's average values for andesite-diorite group /24/, whereas regarding Ba they are closer to basic rocks. With respect to initial plagiorhyolites of Medny Island (av. 236 ppm) the association is enriched in Sr (370-950 ppm). Quartz diorites, tonalites, aplites, granodiorites and their volcanic comagmats - andesites, andesite-dacites and dacites in relation to chondrites (average REE_{ce}/REE_{y} = 6.6) are identical to calc-alkaline rocks of other Pacific island arcs, particularly of Japan and Solomon Islands /10, 11/.

Radiogenic ages of Komandorskie diorite-granitoid rocks fall within the limit of 8.5-12.5 m.y. (see Table 1), which corresponds to Miocene - Pliocene time boundary and are confirmed by geological observations. Rocks, close in age and composition to the described above are abundant in the Aleutian sector of the island arc on Attu, Agattu, Semichi, Kiska, Adak, Amchitka, Kagalashka, Unalashka where they are represented by andesitic, dacitic, less frequently by basaltic, lavas and granodiorite plutons like Captain's Bay pluton on Unalashka Island with K/Ar ages of 8-15 m.y.

On the petrogenesis of the Aleutian island arc magmatic rocks according to the analysis of Nd and Sr isotopes and REE patterns

Covariation of Nd and Sr isotopes especially after discovering a negative covariation between 143Nd/144Nd and 87Sr/86Sr ratios for magmas of mantle origin recently has become a very effective tool in research done on processes of magma generation and magmatic sources in island arcs /6/.

In the Aleutian island arc the most detailed isotopic and geochemical studies have been carried out for Pliocene-Quarternary andesitic formation /13; 15/. In the Komandorskie segment 143Nd/144Nd and 87Sr/86Sr ratios recently were determined in all the three magmatic association discribed above.

The obtained results are plotted on Nd-Sr diagram (Fig.2). The isotopic characteristics of Nd and Sr in igneous rocks of the successive magmatic formations are lockated mainly in the field encompasing those of ensimatic island arcs and ensimatic parts of active continental margins (Marianas, New Britain, South Sandwich, Ecuador) /9, 13/.

High 87Sr/86Sr values in rocks of the initial magmatic stage causing horizontal shifting of (-Sr on the diagram to the right of the mantle array and not accompanied by likewise shifting in (-Nd can possibly be discribed as a result of isotopic exchange with seawater ((-Sr = +64). Such process is now studied in detail /24/.

Theoretically there are only two possible ways of seawater Sr to contaminate island arc magmatic melts; 1) through involvement during subduction of a water-bearing sediments or hydrothermally altered basalts of the oceanic crust layer 2, i.e. contamination within magmatic sources, 2) by regional hydrothermal-metasomatic alteration of already crystallized magmatic rocks (propilization).

The available data seem to favour the second assumption, for contamination within magmatic sources would hardly provide conciderable variability of (-Sr but rather lead to more of less constant overstating of these values for the same petrographic rock type. Moreover, if overstating of (-Sr is caused by participation in initial island arc magmatism of altered oceanic crust or water-saturated pelagic sediments, characterized by high REE ammounts, then there should be a noteable covariation of (-Sr and (-Nd for rocks of basalt-plagiorhyolitic association which however is not discovered. Among the additional evidence is a direct correspondence between the alteration degree of the tholeiitic basalts and plagio-rhyolites and their 87Sr/86Sr values /2/. Thus unpropilitized tholeiitic basalts on (-Sr - (-Nd diagram would probably plot within the mantle array.

One of the major features of basalt-plagiorhyolitic association is a considerable upward shifting of (-Nd for plagiorhyolites toward higher values (up to +10) in comparison with tholeiitic basalts ((-Nd + + 8.2) and calc-alkaline basalts, andesites, dacites and rhyolites of diorite-granitoid and andesitic associations ((-Nd + 6.5 - + 9). This fact is of extreme importance for constraining the process of generation the first island arc acid melts and definitely indicates to the involvement of cristalline part of the oceanic crust ((-Nd +10 and higher). Indeed, the results of REE mass-balance calculations clearly show that they could be derived by mixing of initial tholeiite-basaltic liquids with 25-30% melts from oceanic crust layer 2 (Tsvetkiv, 1983).

Oceanic crust seems to participate in the generation of subalkaline basalts of basalt-trachydoleritic association. As plagio-rhyolites they have rather high (-Nd values (up to +9.9). At the same time there is a faint positive correlation between (-Sr and (-Nd which, as it was previously shown for the New Britain island arc /6/, is most adequately discribed by contamination of island arc magmas with hydrothermally altered MORB-type crust. REE patterns for the Komandorskie subalkaline rocks do not contradict such a suggestion.

All the isotopic compositions for diorite-granitoid and andesitic associations fall within the mantle array. Again it should be noted that changes in crustal structure of the Aleutian basement from oceanic type (Komandorskie Islands + Central Aleutians) into continental one (Eastern Aleutians - Alaska) do not make any influence on Nd and Sr isotopic patterns. This indicates to a minimal degrees, if any of crustal contamination either of sialic or mafic components during later stages of the Aleutian island arc development, contrary to the initial one when such a process seems to be quite efficient.

As concerned the fate of sediments in genesis of island arc magmas the available data is contradictory /15/. On the one hand high K, Ba, Sr, U, Rb and other LIL-element contents in island arc igneous rocks seem to indicate their considerable involvement into magma generation /12/ but on the other hand currently obtained isotopic values significantly constrains volumetric percentage of sediments participating in melting reactions often up to negligible quantities /25/. Thus, e.g. Rb, Ba and REE distribution patterns for 3/14-78 Komandorskie initial tholeiitic basalt (see Table) are in perfect agreement with a model, proposing its orogin by melting of 99.875% of depleted peridotite (harzburgite) and 0.125% of pelagic sediment /21/. However the measured 87Sr/86Sr values in this sample (0.70363) limit maximal volumes of involved sediments to the first hundreds of a percent.

Presented results of the Komandorskie Islands isotopic studies led us to the following conclusions: 1) all the parental magmatic melts were of mantle origin, 2) magmatic sources throughout the whole Aleutian Island arc remained isotopically homogenous for the whole period of its evolution (for more than 40 m.y.), 3) crustal contamination of magmatic melts played a significant role only during initial stage of the island arc development, during subsequent stages effectiveness of such process was considerably diminished, 4) oceanic sediments were not practically involved in magma generation during the initial evolutional stage. There is strong doubt whether their role increased during the later stages, 5) distinctive variability of 87Sr/86Sr isotopic ratios in rocks of similar petrographic type is most probably due to propilization, 6) variability of 87Sr/86Sr isotopic ratios unless other isotopic systems being taken into consideration do not provide irrefutable evidence of magmatic sources unhomogeneity.

REFERENCES

1. Borsuk A.M., Tsvetkov A.A. Magmatic formations of the western part of the Aleutian Island arc. "Izvestia AN SSSR, Geological series", 1980, N 9, pp. 20-36 (in Russian).

2. Borsuk A.M., Genschaft Yu.S., Tsvetkov A.A. Magmatism in the evolution of the Aleutian Island arc. In: Magmatism and metamorphism as indicators of island arc geodynamic regimes", Moscow, Nauka, 1982, pp. 116-142 (in Russian).

3. Citron G.P., Kay R.W., Sutter J.F. Discovery of an early oligocene magmatic event of Adak Island, Central Aleutian Islands, Alaska, Alaska.-Geology, 1980, vol.8, N 4, pp. 375-379.

4. De Long S.E., Hodges F.N., Arculus R.J. Ultramafic and Mafic inclusions, Kanaga Island, Alaska and the occurrence of alkaline rocks in island arcs.-J.Geol., 1975, vol.83, pp. 151-179.
5. De Long S.E., Fox P.J., McDowell F.W. Subduction of the Kula Ridge at the Aleutian Trench.- Bull.Geol. Soc.Amer., 1978, vol.89, pp.83-95.
6. De Paolo D.J., Johnson R.W. Magma genesis in the New Britain Island arc: constraints from Nd and Sr isotopes and trace element patterns. Contrib. to Miner. and Petrol. 1979, vo.70, pp.367-372.
7. Drewes H., Fraser G.D., Snyder G.L., Barnett H.F. Geology of Unalashaka Island and adjacent insular shelf, Aleutian Islands Alaska. US Geol.Survey Bull., 1961, 1028-S, pp. 583-676.
8. Engel C.G., Engel A.E.J., Havens R.G. Chemical characteristics of oceanic basalts and the upper mantle.- Bull.Geol.Soc.Am., 1964, vol.76, N 7, pp. 325-339.
9. Fifth International Conference on Geochronology, Cosmochronology and Isotope Geochemistry (abstracts). -Geochemical Society of Japan, 1982, 385 p.
10. Jakes P., Gill J.B. Rare earth elements and the island arc thyoleiitic series.- Earth and Planet.Sci. LEtt., 1970, vol.9, N 1, pp.831-862.
11. Kay R.W. Geochemical constraints on the origin of Aleutian magmas. In: "Island Arcs, Deep Sea Trenches and Back Arc Basins/M.Talwani and W.C.Pitman III Eds/ Washington D.C., Am.Geophys.Union, 1977, pp.229-242.
12. Kay R.W. Volcanic arc magmas: implications of a melting-mixing model for element recycling in the crust-upper mantle system. J.Geology, 1980, vol.8, pp.947-552.
13. Kay R.W., Sun S.S., Lee-Hu C.N. Pb and Sr isotopes in volcanic rocks from the Aleutian Islands and Pribilof Islands, Alaska. Geochim. et Cosmochim. Acta, 1978, vol.42, pp.263-273.
14. Kay S.M., Kay R.W., Brueckner H.K., Rubenstone J.L. Tholeiitic Aleutian arc plutonism: The Finger Bay Pluton Adak, Alaska, Contr.to Miner.and Petrol., 1983, vol.82, p.99-116.
15. McCulloch M.T., Pergit M.R. 143Nd/144Nd,

87Sr/86Sr and trace element constraints on the petrogenesis of Aleutian Island arc magmas. Earth and Planet.Sci.Lett., 1981, vol.56, pp. 165-179.
16. Naseskin V.V. Petrogenesis of acid volcanites. Moscow, Nauka, 1975, 197 p. (in Russian).
17. Reed B.L., Lanphere M.A. Alaska-Aleutian batholith. Geochronology, chemistry and relation to circum-Pacific plutonism. Bull.Geol.Soc.Amer; 1973, vol.84, pp.2583-2610.
18. Solovjov S.P. Chemistry of magmatic rocks and some problems of petrochemistry. Leningrad, Nauka, 1970, 311 p. (in Russian).
19. Stern C.R., Huang W.L., Wyllie P.J. Basalt-andesite-rhyolite - H_2O and H_2O undersaturated liquidus surfaces to 35 kilobars with implications for magma genesis. Earth and Planet.Sci.Lett., 1975, vol.28, N 2, pp.396-422.
20. Tsvetkov A.A. Magmatic rocks of the Aleutian Island Arc initial evolutional stage and the associating ore mineralization. In: "Petrology and metallogeny of the natural magmatic associations", Moscow, Nauka, 1982, pp.72-85 (in Russian).
21. Tsvetkov A.A. The Aleutian arc magmatism and the problems of petrogenesis of island arc igneous rocks. "Izvestia AN SSSR, Geol.series," 1983, N 4 pp.3-19 (in Russian).
22. Tsvetkov A.A., Schmidt O.A. Deep-seated inclutions in magmatic rocks of Komandorskie Islands (first findings), Doklady AN SSSR, vol.267, N 4, pp. 743-750 (in Russian).
23. Vinigradov A.P. Average content of chemical elements in the major types of volcanic rocks of the earth crust. Geokhimiya, 1962, N 7, pp. 555-571 (in Russian).
24. Vinogradov V.I., Buyakaite M.I. Isotope composition of strontium in the ocean gloor rocks and in ultrabasites, "Izvestia AN SSSR, Geol.series", 1980, N 3 pp. 63-74 (in Russian).
25. Zhuravlev D.Z., Chernyshov I.V., Tsvetkov A.A., Borsuk A.M., Agapova A.A., Serdiyk N.I. Variations of 143 Nd/144 Nd and 87 Sr/86 Sr isotope ratios in the Aleutian Island arc magmatic rocks with regard

to their petrogenesis. Doklady AN SSSR, 1983, vol. 270, N 6, pp. 1457-1462 (in Russian).

FIGURE 1. Map of Komandorskie Islands magmatic association (compiled from data personally collected by the authers and from O.A.Schmidt's data for sedimentary rocks). A - Bering Island, B - Medny Island, I - recent lacustrine deposits, 2 - Quarternary deluvial-proluvial deposits,
Diorite-Granitoid volcano-plutonic association ($\gamma \delta N_1^3$):
3 - andesitic and andesite-dacitic lavas, 4 - dykes, 5 - intrusions of granodiorites with aplitic veins, 6 - intrusions of tonalites and quartz diorites;
Basalt-Trachydoleritic association ($\varepsilon \nu N_1^{1-2}$):
7 - subalkaline basalts, 8 - trachybasaltic trachydoleritic, teschenitic sills, 9 - trachybasaltic, gabbro-doleritic, trachydoleritic dikes,
Basalt-Plagiorhyolitic association ($\beta \lambda f_3$):
10 - plagio-rhyolitic and plagiorhyodacitic lavas and pyroclastics (a), hypabyssal intrusions (b), 11 - basaltic lavas and pyroclastics (a), dikes (b);
Sedimentary and volcano sedimentary rocks: 12 - polyfacial deposits of Nikolskaya suite: a) conglomeratobreccias, b) tuff-breccias, c) tuff-conglomerates, 13 - Komandorskaya series: a) Preobrazhenskaya suite, b) Gavanskaya suite, c) Gavrilovskaya suite, d) Poludennaya suite; 14 - Faults: a) trased b) assumed, 15 - unconform stratigraphic boundaries.

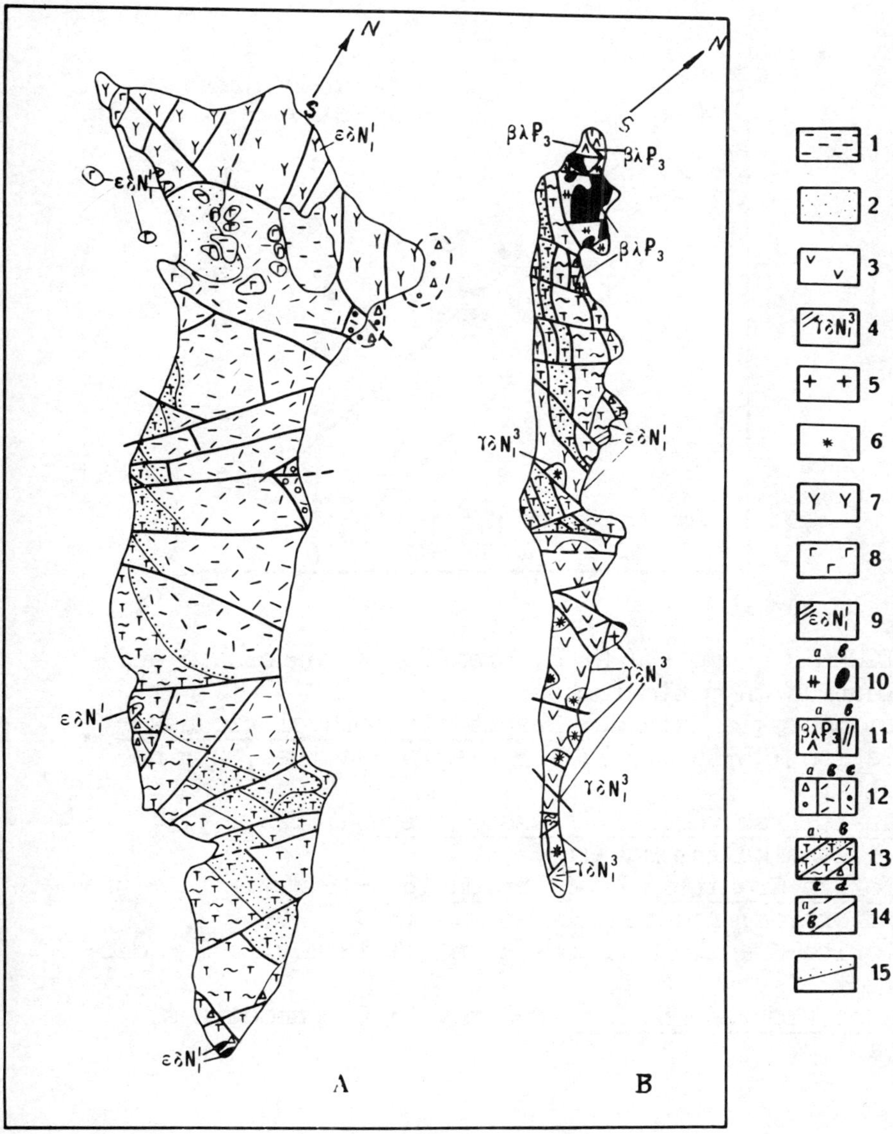

N
S
$\varepsilon\delta N_1^1$
$\beta\lambda P_3$
$\gamma\delta N_1^3$
A
B
1
2
3
4
5
6
7
8
9
10
11
12
13
14
15

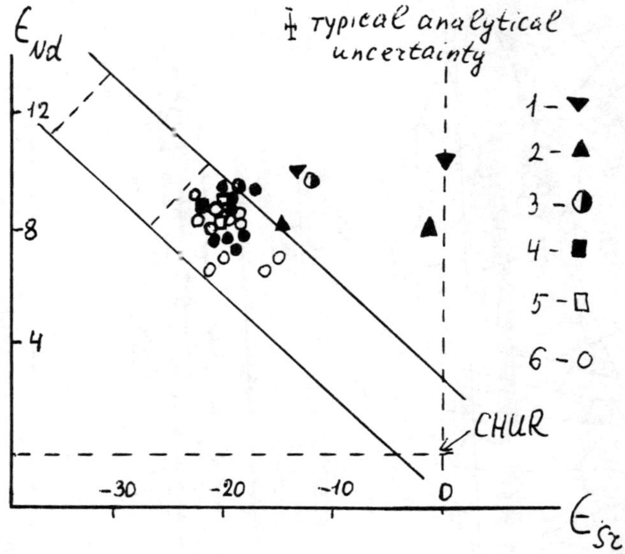

FIGURE 2. (-Nd - (-Sr diagram for Komandorskie and Aleutian magmatic rocks.
Komandorskie Islands: Basalt-Plagiorhyolitic association (28-35 m.y.): 1 - plagiorhyolites, 2 - basalts;
Basalt-Trachydoleritic association (18-25 m.y.): 3 - subalkaline basalts;
Diorite-Granitoid association (8.5-15 m.y.): 4 - granodiorites; Aleutian Islands: /15/:
Diorite-Granitoid association (11-13 m.y.): 5 - diorites;
Andesitic association (0-5 m.y.): 6 - andesites, basalts.

TABLE 1. Radiomatric datings of Kommandorskie magmatic rocks

Magmatic association	Island	Rock	K, %	Ar^{40}, ng/g	Ar^{40} rad. / Ar^{40} tot.	Age, my	location
1	2	3	4	5	6	7	8
Diorite - Granitoid	Medny	Biotite-hornblende tonalite	1.92±0.04	1.15±0.15	3; 5	8.5±2.5	Stock, Kuropatkin Mt., 13/12-77
	Medny	Tonalite andesite				8.6±0.3	Dyke, South Eastern Pt., 10/6-78
	Medny	dacite				10.7±0.8	Lava flow, Glinka Cove, 9/1-78
	Medny	dacite	1.25±0.3	11 ±0.1	5; 15	12.0±2	Same, 9/2-78
	Medny	Granodiorite	1.92±0.03	1.5 ±0.2	13; 21	11.0±3	Stock, Cherny Pt., 11/6-77
	Medny	Aplite	4.32±0.04	3.5 ±0.3	27; 32	11.5±2	Vein, Cherny Pt., 11/7-77
	Medny	Granodiorite	1.53±0.03	1.4 ±0.2	16; 19	12.5±3	Stock, Cherny Pt., 11/5-77
		biotite	6.74±0.05	5.4 ±0.3	11; 21	12 ±2	
		potassic--feldspar	5.27±0.04	4.4 ±0.3	14; 39	12 ±2	
Basalt - Plagiorhyolite Basalt - Trachydoleritic	Bering	Trachydolerite	1.68±0.03	1.8 ±0.2	12; 15	16 ±3	Lava dome, Nakovalnya Mt., 404/77
	Medny	Subalkaline olivine basalt				18.6±0.6	Dyke, Sulkovski Pt., 3/19-78
	Medny	Teshenite	1.54±0.03	2.3 ±0.2	21; 26	21 ±3	Sill, Nkolskoye 406/77
	Bering	Trachybasalt	1.51±0.03	2.6 ±0.3	19; 26	25 ±4	Lava flow, North Western Pt., 408-77
	Bering	Trachydolerite	2.27±0.03	3.3 ±0.3	20; 30	21 ±3	Sill, Gaunta Pt., 415-77
	Bering	Trachydolerite	1.51±0.03	2.2 ±0.2	26; 26	20.5±3	Sill, Svinye Mt., 13/6g-77
	Bering	Teshenite				27 ±0.8	Dyke. Monati Pt., 17/3-78
	Medny	Plagiorhyolite	1.50±0.03	3.0 ±0.3	18; 25	28 ±4	Stock, Rhyolitic Ridge., 5/2-77
	Medny	Plagiorhyolite				31.3±0.9	Lava flow, Preobrazhenskaya Cove, 2/10-78
	Medny	Plagiorhyolite				31.4±1.6	Lava flow, Preobrazhenskaya Cove, 2/11-78
	Medny	Plagiorhyolite	1.67±0.03	4.0 ±0.4	25; 28	34 ±4	Stock, Rhyolitic Ridge, 4/2-77
	Medny	Plagiorhyolite	2.14±0.04	4.9 ±0.4	47; 53	33 ±4	Same, 4/1-77
	Medny	Basalt	1.35±0.03	2.9 ±0.2	32; 33	31 ±3	Dyke, Sulkovski Pt., 3/14-78

dating performed in USGS laboratories, Menlo Park, Ca., under supervision of L. Silberman. The rest of the samples are dated in the Institute of Ore Deposits Geology, Petrology, Mineralogy & Geochemistry of the USSR Academy of Sciences, Moscow. Analist M. M. Arakelyantz.

TABLE 2. Whole rock chemical analyses of Kommandorskie magmatic rocks

	Basalt-Plagiorhyolitic association				Basalt-Trachydoleritic association				Diorite-Granitoid association			
	1	2	3	4*	5	6	7	8**	9***	10	11	12
SiO_2	45.28	47.57	67.95	72.02	44.70	48.60	49.50	53.80	54.70	62.60	64.83	61.44
TiO_2	0.56	0.82	0.05	0.47	0.62	1.81	1.45	0.92	0.42	0.59	0.08	0.50
Al_2O_3	17.54	18.35	13.70	12.33	16.25	15.45	15.43	15.57	17.05	15.84	12.07	16.53
Fe_2O_3	3.02	6.11	1.92	1.23	9.10	5.99	6.22	4.91	2.21	1.66	0.37	4.19
FeO	6.84	1.73	2.43	2.74	3.95	3.32	1.78	1.98	1.05	2.44	0.57	1.52
MnO	0.09	0.15	0.15	0.035	0.17	0.21	0.15	0.12	0.05	0.10	0.02	0.07
MgO	12.25	6.38	1.84	1.91	5.38	5.56	3.54	5.84	2.99	2.54	0.09	2.76
CaO	5.01	8.94	4.19	1.02	11.39	8.91	9.97	4.84	5.52	4.47	0.82	4.83
Na_2O	2.46	3.13	4.31	3.93	2.69	3.58	3.14	5.11	4.03	4.19	4.07	4.19
K_2O	0.45	1.17	2.07	1.24	0.75	1.52	1.99	2.06	1.78	2.26	5.12	2.14
H_2O^-	1.94	3.10	0.14	0.04	0.82	0.79	1.93	1.06	0.32	--	--	0.74
H_2O^+	5.09	1.89	0.94	3.27	1.32	1.93	1.83	2.35	1.13	0.46	0.23	0.78
P_2O_5	0.01	0.18	0.10	0.12	0.27	0.64	1.46	0.33	0.19	0.12	--	0.19
CO_2	0.08	0.61	--	0.11	--	1.03	0.99	--	--	0.76	0.21	--
Total	100.75	100.20	99.75	100.52	99.45	99.34	99.38	99.84	100.54	100.26	100.24	100.00

Basalt-Plagiorhyolitic Association: 1 - fine grained propilitized lithocrystalloclastic tuff, kotenok Pt., Medny Is., 3/7-78; 2 - basaltic lava comprising ultramafic inclusions, Sulkovski Pt., Medney Is., 3/14-78; 3 - plagiorhyolite from a stock to the south of Kotenok Pt., Medney Is., 7/2-77; 4 - plagiorhyolitic lava to the north of Matvei Pt., Medney Is., 6/22-77;
Basalt-Trachydoleritic Association: 5 - subalkaline olivine basalt, lava, northern part of Bering Island, 14/3-78; 6-7 - trachybasalts and trachydolerites (dykes and sills) from the northern part of Bering Is.; 6 - Zabiyaka Pt., 413-77, 7 - Nakovalnya Mt., 404-78; Teshenitic ("beringitic dyke in Zhirovaya Cove, Medney Is., 9/7-77;
Diorite-Granitoid Association: 9 - tonalite, stock in Glinka Cove area, Medney Is., 12/17-77; 10 - granodiorite, stock, Cherny Pt., Medney Is., 11/6-77; 11 - aplite, vein in granodiorite, cherny Pt., Medney Is., 11/7-77; 12 - andesitic lava, Glinka Cove, Medney Is., 10/6-78

Additionally determined: *Cl 0.03%, F 0.028%. **Cl 0.05%, F 0.013%. ***Cl 0.04%, F 0.21%

PETROLOGICAL AND GEOCHEMICAL CRITERIA FOR THE IDENTIFICATION OF (POTENTIAL) ORE BEARING NIGERIAN GRANITOIDS

P BOWDEN and Judith A KINNAIRD

Department of Geology,
University of St Andrews,
Fife, Scotland.

ABSTRACT

The Nigerian anorogenic granites which developed at high levels in the Pan African continental crust, mainly as roots of volcanoes, are potential hosts for metasomatic reactions by residual fluids. Those granites emplaced as ring dykes in polygonal or circular fractures with steep sided contacts display limited but often locally important subsolidus assemblages. These new mineralogical associations define the escape routes of rising fluids, zones of weakness which aid pressure release, brecciation and ore dumping, when periods of retention and interaction of fluid cannot be maintained. The associated mineralization with ring dyke marginal reactions consist dominantly of sulphide assemblages with chalcopyrite, sphalerite, galena, pyrite, pyrrhotite, arsenopyrite and molybdenite. As quartz dominated over fluorite in the sulphide assemblages, pervasive greisenization associated with cassiterite developed with disseminated sphalerite, chalcopyrite and galena. These sulphide associations are found abundantly in roof zones adjacent to the volcanic cover, or Precambrian basement. It is in granite cupolas that fluids can be retained and where metasomatic reactions are more intense,

Proceedings of the 27th International
Geological Congress, Volume 9, pp. 85–119
PETROLOGY (IGNEOUS AND METAMORPHIC ROCKS)

modifying host granite, related volcanic rocks and country rock compositions.

According to published experimental work, both metaluminous and peralkaline granites exsolve a peralkaline fluid phase from a residual silicate melt at depth. The early alkaline stage in the Nigerian Younger Granite province is represented by sodic metasomatism of perthite, growth of ordered albite, and other critical mineralogical and chemical changes. The dominant ore associated with the widespread albitization can be divided in two types of dispersed mineralization: the columbite- type found in biotite apogranites; and the pyrochlore-type found in peralkaline apogranites. The alkaline stage is followed by loss of Na, growth of intermediate microcline, and can be termed microclinization or generally potash metasomatism. The microclinization stage is more restricted in its development but does have important ore associations. Finally with changing conditions in the fluid (Burt 1981) greisenization becomes more widely pervasive, or is confined to pockets or fissures and fracture in the roof zones of biotite granite. Greisenization can be found superimposed on various mineral assemblages. The mineralogical expression and the extent of greisen development depends on the intensity of albitization and microclinization and the timing of fracturing and fissuring in the roof zone. Important oxide and sulphide ore associations formed at this period.

Fluid reactions can be monitored and interpreted by the nature of the new subsolidus assemblages. The new silicate, oxide and sulphide minerals developed in response to rock-fluid interaction provide the criteria for assessing the ore-bearing potential of the Nigerian alkaline granitoids.

INTRODUCTION

In general all granitic structures developed at high levels in the continental crust, as roots of volcanoes, are potential hosts for metasomatic reactions by residual fluids. The reactions are recorded as new subsolidus mineral assemblages with trapped fluid inclusions providing valuable petrological criteria for assessing the ore-bearing potential.

In any one Nigerian anorogenic ring complex irrespective of whether it is multicentred (Jos-Bukuru) or single centred (Ririwai, Tibchi) there are two contrasting but related modes of granite emplacement into their volcanic cover and Precambrian country rocks. These contrasting styles can be recognised at the present erosional level. As steep-sided units intruded along polygonal or circular ring fractures, or as shallow outward-dipping cupolas. The first mode of emplacement has been described in previous literature (Jacobson _et al_., 1958) as 'ring-dyke' form, and according to accepted hypotheses represents the most common structural expression of high level emplacement of granitic rocks.

Examples of the visible petrological changes in hand specimen related to mineralization of margin zones of ring dykes are briefly discussed in this paper, with localities taken from Kinnaird (1981), displayed in Fig 1.

The second mode of emplacement has been referred to as 'cupola' form (Bowden and Jones, 1978) and represents the terminal phase of magmatic activity related to the evolution of any one centre. It is only in these granite intrusions with shallow outward-dipping

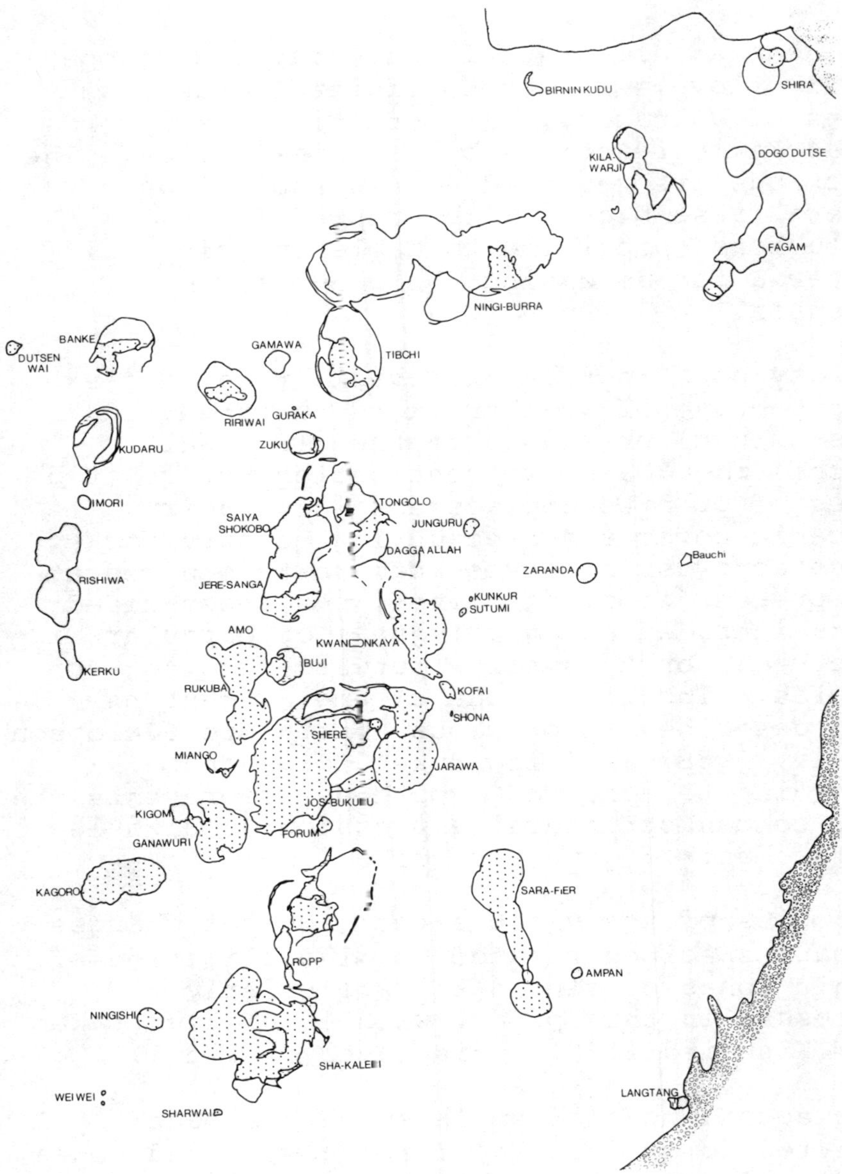

Fig 1 Location map of the Nigerian anorogenic ring complexes (After Kinnaird et al., 1983).

contacts with the centres own volcanic cover, or the Precambrian basement, that the fluids can be retained in the roof zone, and are more intense. The endocontact zone of the granite cupola and the exogenic region within the volcanic cover or Precambrian country rocks become potential sites for substantial rock-fluid interaction. Initially the reactions are widespread and pervasive but as the solidifying granite cools and contracts, a series of microfractures, and later Q F and P joints provide more local access and channelways within the intrusive body for circulating fluids.

According to the experimental work of Currie (1968) and Mustart (1972) discussed in Taylor et al. (1979) both metaluminous and peralkaline granites exsolve a peralkaline fluid phase from a residual silicate melt at shallow depth. It is the behaviour of this fluid phase in a low pressure environment that dictates the formation of the new subsolidus assemblages and the ore deposits in Nigeria associated with the younger granites. From the mineralization viewpoint there are many parallels with alkaline undersaturated rocks and carbonatites, particularly with the abundance of sphalerite, rare-earth minerals, zircon and complex titanium silicates, uranium, thorium, columbite and pyrochlore. Such a distinctive alkaline mineralization suite is superimposed on a more normal, group of ore minerals such as cassiterite, wolframite, chalcopyrite and galena.

It is worth emphasizing that there is considerable mineralogical variation from centre to centre in the subsolidus assemblages, and that sequential metasomatic changes of albitization, microclinization and greisenization are not recorded in every locality in the Nigerian anorogenic province.

A model is developed in this paper which combines all the various subsolidus mineral assemblages found in biotite granite cupolas in Nigeria. Recognition of these important features in outcrops and under the microscope is critical to any assessment of the ore-bearing potential.

STRUCTURAL CONTROL OF POSTMAGMATIC METASOMATISM

Ring-dyke form

Granites emplaced as ring dykes in polygonal or circular fractures with steep-sided contacts are zones which aid pressure release and provide ready routes of escape for rising fluids which may be responsible for ore dumping, but where long periods of retention and interaction of fluid cannot be maintained. Such examples are known at Zarara Quarry in the Banke complex and other areas including Bongweli, near Pingell (Daggah Allah complex), Rafin Jaki near Tildan Fulani (Jos Bukuru complex), Gindi Akwati (Ropp complex), south of Richa (Sha Kaleri complex). Often petrological changes are limited to narrow zones although the degree of reaction may be locally intense. In most instances the granitic rocks which crystallized in ring-dyke fractures are porphyritic with mainly feldspar crystals showing arrested growth forms set in a fine grained matrix. It is this matrix which has substantially reacted with fluids while the phenocrysts have preserved part of their magmatic crystallization history. An example of greisenized matrix with feldspar phenocrysts partly transformed to microcline but retaining their original crystalline form is demonstrated in plate 1. With further sudden pressure release, local marginal brecciation and ore dumping can occur (plate 2). The latter example was collected from Zarara quarry at the eastern end of the Banke complex. Here miarolitic cavities containing elongate prismatic quartz crystals with microclinite linings, are developed within and between brecciated blocks of ring-dyke porphyry recemented by late stage vein quartz. The associated mineralization consists dominantly of sulphide assemblages. Other examples are known in the Nigerian younger granite province where locally cassiterite and pervasive greisenization is associated with sulphide mineralization. Two distinctive sulphide associations have been recognised (Kinnaird, 1981) depending on the dominance of fluorite or quartz, in other examples of marginal ring dyke mineralization. Associated with fluorite, Kinnaird (loc. cit.) has noted the occurrence of massive sphalerite, chalcopyrite, galena, with pyrite, pyrrhotite, stannite, arsenopyrite and molybdenite. In contrast quartz sulphide associations contain disseminated fine grained sphalerite with chalcopyrite and galena. These sulphide associations may develop more substantially in marginal zones of the 'cupola'-form.

'Cupola' form

From a structural viewpoint, it is only those granitic intrusions which have shallow outward dipping contacts beneath the centres' own volcanic cover, or within a Precambrian basement, that the fluids can be retained and where metasomatic reactions may be intense. The whole granite cupola endocontact zone and the exogenic region within the volcanic cover

Plate 1 Greisenization of the fine grained matrix with feldspar phenocrysts as arrested growth forms relatively intact. Saiya Shokobo complex.

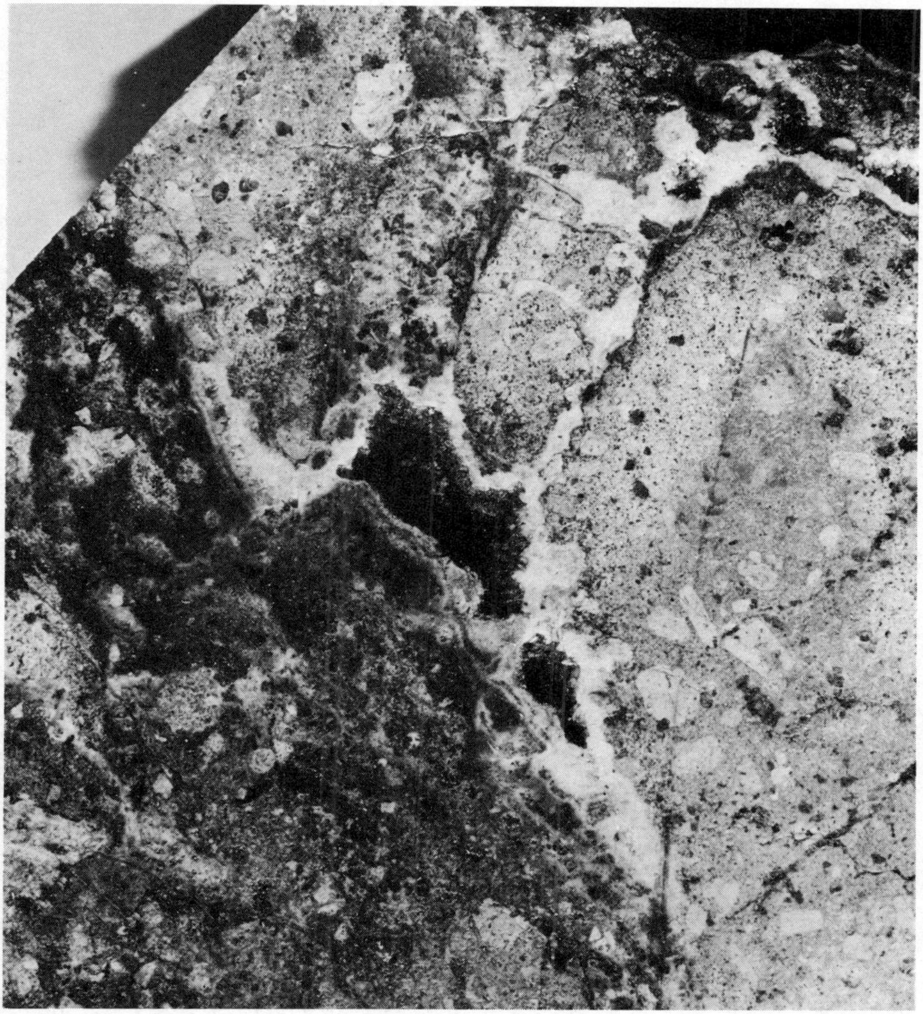

Plate 2 Brecciated and mineralized prophyry ring dyke, Zarara quarry, Banke complex, Nigeria. Brecciated fragments cemented by quartz with miarolitic cavities lined with microclinite and partly filled by siderite and elongate prismatic quartz.

or the basement became potential sites for substantial rock fluid interaction, metasomatic changes, and mineralization.

PETROLOGICAL CRITERIA FOR POSTMAGMATIC METASOMATISM IN GRANITE CUPOLAS

Three petrologically contrasting assemblages can be broadly equated with postmagmatic periods of sodic metasomatism, potassic metasomatism and acid metasomatism. These periods of chemical activity and mineralogical reconstruction are responsible for introducing and redistributing various ore minerals of economic importance.

Sodic metasomatism

Minerals characteristic of sodic metasomatism include albite, aegirine, alkali amphiboles (riebeckite, lithian arfvedsonite), micas (compositions ranging from annite through protolithionite and zinnwaldite to lepidolite) and a range of accessory minerals including chevkinite, astrophyllite, narsarsukite, and cryolite. The actual mineral assemblages which form in response to sodic metasomatism depend on the intensity of rock-fluid interaction, with the strongly peralkaline compositions representing the maximum effect. Such reactions can be variously described as Na-silicate alteration, albitization, and Na for K exchange and are displayed chemically in Fig 2 by the series of closed and open triangles trending towards the albite mineral pole. The closed triangles are arfvedsonite-albite granites with pyrochlore: the upright open triangles and the closed stars are zinnwaldite-albite granites and lepidolite-albite granites, with columbite (see text-figure captions).

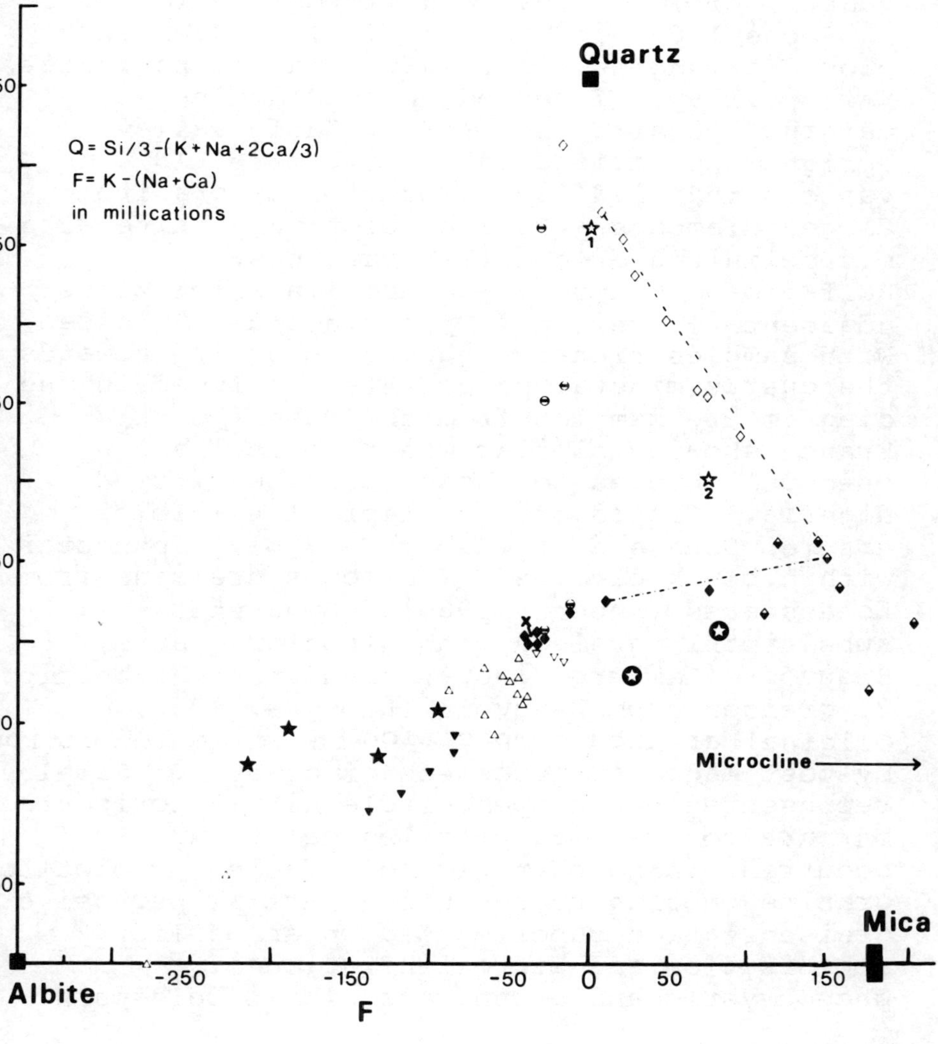

Fig 2 see overleaf for a full description.

Fig 2 QF diagram for samples from the Ririwai complex, Analytical data published in Kinnaird _et al_. 1984, expressed in millications and plotted in relation to the mineral poles quartz, albite, mica, and microcline (after De La Roche 1963, and Charoy 1979). Inverted closed triangles - strongly peralkaline facies, Kaffo valley. Inverted open triangles - marginal peralkaline facies, Kaffo valley. Upright open triangles - drill core L13, in various modified facies of biotite granite. Closed diamonds - Ririwai biotite granite with microclinized or argillic variants. Half-closed diamonds - microcline-rich zones adjacent to greisen. Open diamonds - greisen, with samples richer in quartz plotting towards the quartz mineral pole. Also included in the diagram are samples from the Massif Central France showing similar mineralogical and chemical changes to those described from Nigeria. Closed stars - lepidolite-albite granie, Echassières (Aubert, 1966). Open star within black circles - miceceous greisens from Echassières (Aubert, 1966). Open star - subscript 1, greisen with globular quartz, Beauvoir (Lameyre, 1966); open star - subscript 2, greisen from Beauvoir (Lameyre, 1966). Original granite composition before modification by post magmatic processes given by X. Samples represented by an open circle with a horizontal bar are contrasting petrological trends occurring in porphyritic roof facies of biotite granite cupolas representing partial pervasive greisenization superimposed on an earlier albitization and microclinization of phenocrystic and groundmass alkali feldspars.

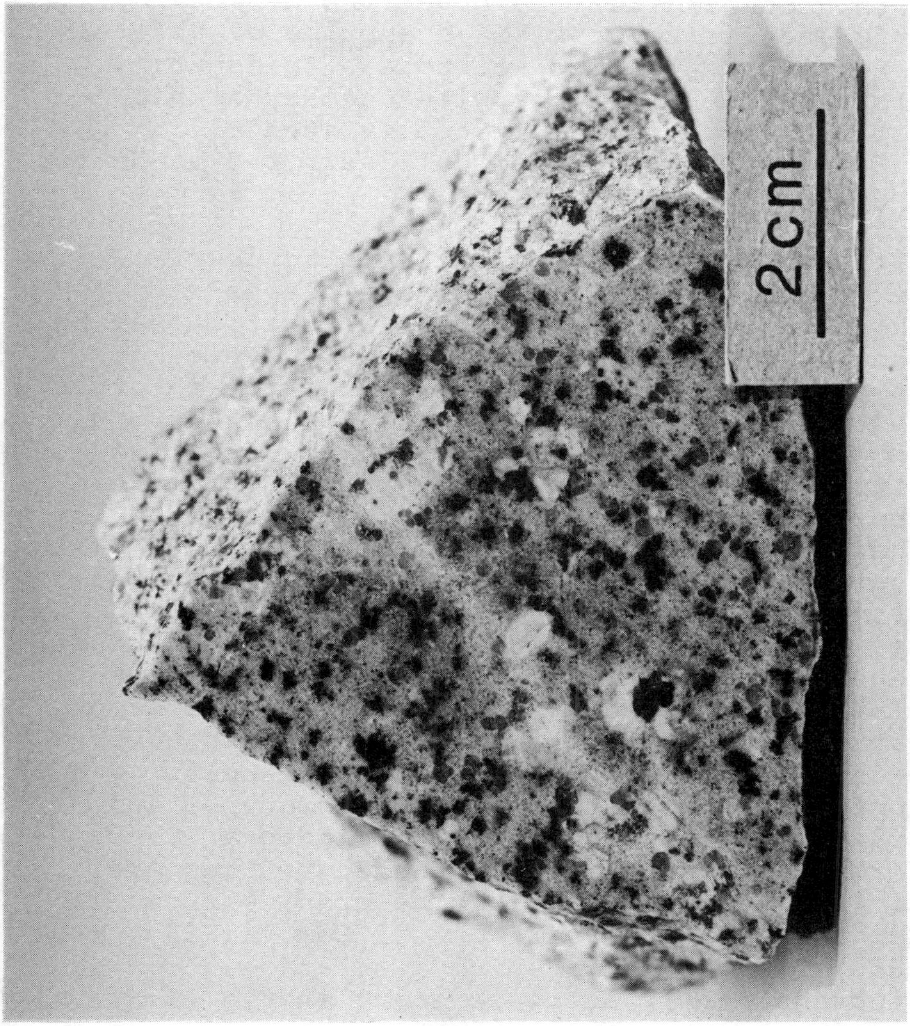

Plate 3 Vein of albite associated with albitization of perthite in specimen from Rayfield Gona albite biotite granite, Jos-Bukuru complex.

Sodium silicate alteration can be seen under the microscope by the green rims to sodic ferrohedenbergite, the blue rims of arfvedsonite to less sodic amphibole, and the pervasive replacement of potash feldspar by albite. The reactions witnessed by alkali feldspar are frequently abbreviated to 'albitization' and chemical described as Na for K exchange.

If Na for K exchange continues to lower temperatures then mineral assemblages characteristic of propylitic alteration may be generated (Rose and Burt, 1979). Typical associations include albite, epidote, chlorite and siderite. Evidence for the albitization process in biotite granite is well displayed at Ririwai in drill core L13, and illustrated in the QF mineralogical-chemical diagram (Fig 2). The cluster of closed diamonds represents typical compositions of the Ririwai biotite granite. The upright open triangles are sections of the drill core taken at different levels with mineralogical variants towards albitite recorded at >400 m depth. The same petrological and mineralogical changes have been recorded elsewhere in sheet-like roof-intrusions at Jos-Bukuru (plate 3) and Afu. Locally zinnwaldite- albite granite has formed. Compositionally these granites are similar to lepidolite-albite granite from Echassières, Beauvoir, Massif Central (Aubert, 1966, Lameyre 1966) and illustrated diagramatically as close stars in Fig 2. In most exposures of biotite granites from the Nigerian anorogenic province, the petrogical evidence for albitization is not always well preserved, because of continued lower temperature fluid reactions. However albitization in cupola roof zones is an early stage process, and is recorded by important petrological variations depicted in Figs 3A and 3B.

The early fluid must have been rich in albite components as well as other constituents (Na, Fe etc) and some important and unusual trace element assemblages (U, Th, Zr, Nb, HREE). From experimental studies (Luth, 1967, Kovalenko 1976, Manning, 1982) there may have been a continuum from residual magma to hydrothermal fluid for alkaline provinces without involving immiscibility or phase separation. However should immiscibility occur, or degassing, or boiling, then the resultant fluid compositions would be drastically different. In the Nigerian province the granites, particularly albitized varieties, contain high U values (Bowden *et al*., 1981). The sequence of mineralization U-Nb, allows one to predict the early loss of CO_2 concomitant with the formation of albitites (Fig 3A and B). CO_2 degassing of the high temperature residual fluid could transport U and Nb into the roof zones of granite cupolas. Important U concentrations are also found in the volcanic cover to the Niger-Nigeria Younger granite complexes (Bowden *et*. *al*. 1981). Futhermore subsequent lower temperature postmagmatic reactions could mobilize the uranium and carry it to higher levels. Geochemically high U values are only recorded in granites retaining their record of albitization (e.g. albitites Fig 3B). Brecciation of the albite rich facies may then occur locally, (see plate 4) In our opinion Mo is transported in a similar way (CO_2 degassing) but is more inert to lower temperature rock-fluid reactions. Other geochemically similar elements like the HREE, Th, Zr could be transported and concentrated by similar processes.

Plate 4 Brecciated and greisenized albite rich facies of Rayfield Gona biotite granite, Jos Bukuru complex containing abundant sulphide mineralization

Plate 5 Microclinite developed adjacent to the Ririwai lode, Ririwai complex.

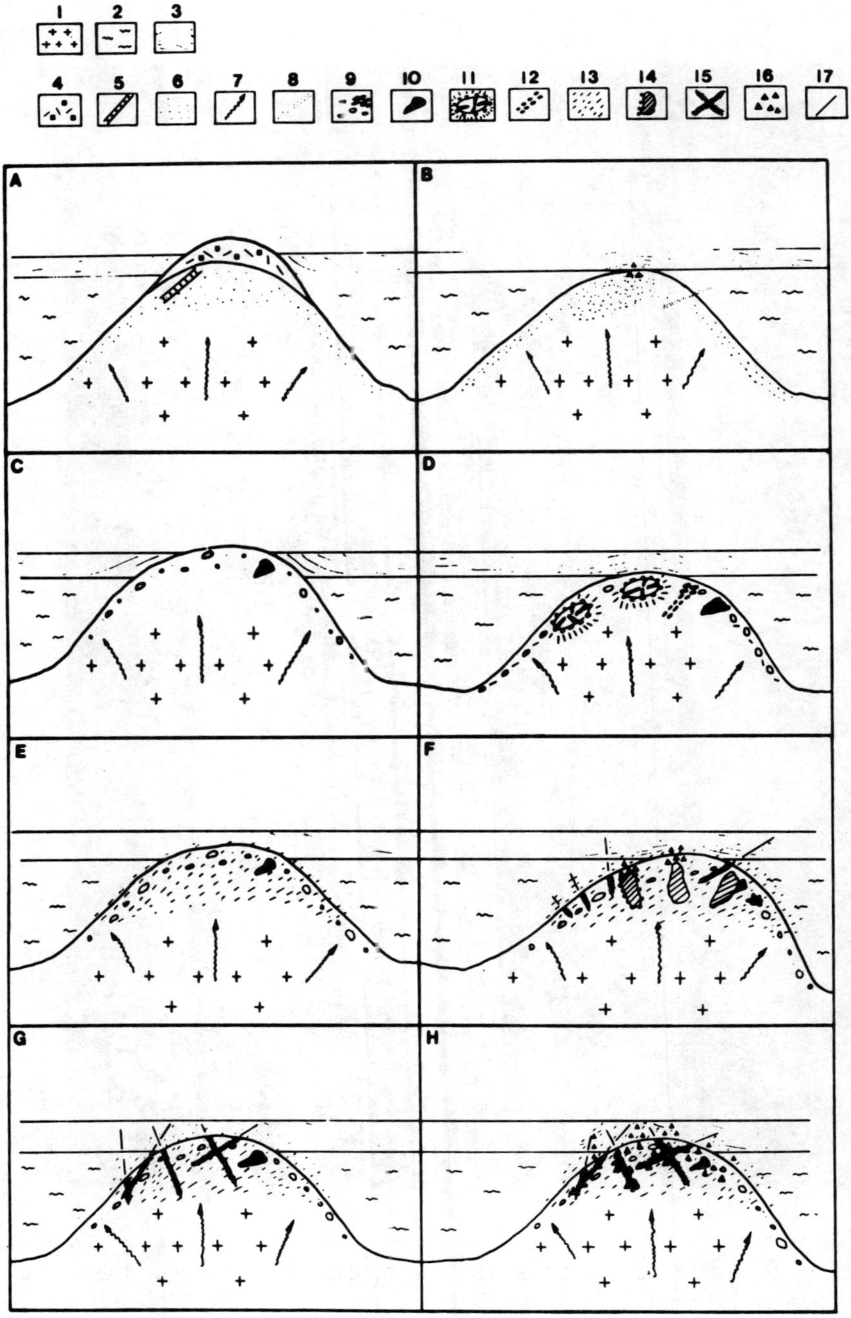
1 2 3
4 5 6 7 8 9 10 11 12 13 14 15 16 17
A B C D E F G H

Fig 3 Schematic explanation of petrological variations in the roof zones of Nigerian biotite granite cupolas. The sequence A-C- E-G represents a series of roof facies variations generated in response to fluid reactions with changing μNaF, μKF, and μHF (Burt, 1981) conditions during falling temperature. Contrasting petrological sequences, generated by successive periods of boiling in a multicomponent fluid as a result of decreasing confining pressure, are indicated by path variants A-B, C-D, E-F, and G-H. Substantial ore formation of economic importance is associated with model D, F and H. Key 1 - biotite granite, biotite syenogranite cupola 2 - Pan African basement 3 - volcanic cover of same anorogenic centre. Petrological variants caused by fluid interaction 4 - albite quartz genthelvite pegmatite 5 - quartz albite aplite 6 - albitization (sodic metasomatism) 7 - movement of residual fluids 8 - albitite veins, (note albitite concentration in 3B) 9 - microclinization (potassic metasomatism) 10 - gem pockets of beryl 11 - miarolitic cavities lined with microcline and filled with alpha quartz 12 - microclinite veins 13 - pervasive greisenization 14 - greisen pocket associated with local concentrations of microclinite 15 - quartz-vein greisenization filling local fissures with quartz, and limited zones of wall rock reaction. 16 - breccia zones generated by vapour pressure release on boiling of a multicomponent fluid. 17 - quartz vein in Pan African basement and volcanic cover contains wolframite.

Potassic metasomatism

Minerals characteristic of potassic metasomatism include mica compositions ranging from originally annite to siderophyllite, and the generation of intermediate to ordered microcline. Some amphibole compositions towards ferroactinolite may also be valuable petrological criteria. Such mineral reactions can be described as K-silicate alteration (Rose and Burt, 1979), microclinization and K for Na exchange. More than one period of potassic metasomatism may be recognised depending on the mineralogical variants observed in the subsolidus assemblages.

K-silicate alteration of original granitic micas can be seen under the microscope by zoning towards pleochroic schemes dominated by deep green to brown. Mica overgrowths at the expense of other minerals such as perthite can be demonstrated in a number of complexes in Nigeria. It is interesting to note that any zircons trapped in biotite during near solidus crystallization are dissolved in fluids depositing the rim of ferrous siderophyllite. If these pervasive reactions continue to lower temperature then mica compositions change to lithian siderophyllite accompanied by the breakdown of alkali feldspar to give topaz and sericite. Such pervasive changes in trioctahedral mica compositions follow the siderophyllite-polylithionite join (Rieder, 1970).

There are two distinct generations of microcline which can be recognized under the microscope. One, the conversion by recrystallization, during reconstructive transformation of orthoclase to microcline. This is witnessed in thin section by the typical cross hatched or tartan twinning. The other microcline resulting from pervasive potassic metasomatism of albite and perthite is not readily identified under the microscope without experience. It appears as a cloudy low birefringent mineral difficult to distinguish from orthoclase unless confirmed by x-ray diffraction. Structurally the feldspar is intermediate microcline which may develop substantially to form pockets (see plate 5). When microcline is generated this process is often referred to as microclinization. The effect seems to be restricted to the margins or fissures and fractures in the roof zones of biotite granites and related to an early stage of vapour separation. It can be recognised by the presence of microclinite linings to miarolitic cavities (plate 6). The process of microclinization of the original perthite can be monitored by XRD studies. Chemical analyses of quartz-microcline assemblages shows that almost all Na_2O has been lost. The mechanism has been interpreted as K for Na exchange (Martin and Bowden, 1981). At the same time as the K for Na exchange, iron is released (from the feldspar lattice) and forms a haematite coating to individual crystals thus giving the microclinite assemblages a distinctive red appearance. In previous literature we have referred to this reaction zone as 'reddened wall rock' (Bowden and Jones 1978). The process of microclinization (K for Na exchange) has been arrested at different stages in the Nigerian younger granite province. At Ririwai the host rock has not been extensively desilicated at the present erosional level, so that only localised zones of microclinite occur adjacent to fissure- filling greisens (plate 7). This trend of microclinization is displayed on the QF diagram (Fig 2) as a series of half-filled triangles. From an economic standpoint the process of microclinization is an essential prerequisite for the ore

Plate 6 Miarolitic cavity developed in roof facies of biotite granite at Ginshi hill, Ririwai complex. Cavity partially filled by long-prismatic quartz, coating of siderite, and fringed by local microclinite (white) and pervasive greisen (dark).

deposition of abundant sphalerite and cassiterite during greisenization.

Acid metasomatism

Acid metasomatism reactions contrast with the previous discussion in that it concerns instability of granitic minerals in the presence of uHF fluids and the growth of new subsolidus assemblages. Feldspar reactions include the breakdown of perthitic alkali feldspar to sericite and topaz; the destabilization of albite to form fluorite, cryolite, topaz, with some montmorillonite; and the destruction of microcline to form micaceous aggregates, zones of chlorite, and more rarely kaolinite. Mica compositions generated by acid metasomatism of perthitic granite show compositional variations from siderophyllite, a characteristic green mica, appearing initially as overgrowths to more annitic compositions, and ultimately as independent crystals of more uniform optical properties similar to siderophyllite first described from Pikes Peak Colorado c.f. Barker et al., 1975). Other mica compositions include protolithionite, zinnwaldite and lepidolite associated with acid metasomatism of albite roof rocks, and lithian siderophyllite generated from microcline-rich facies. Some of these reactions released silica and lead to the formation of siliceous variants. All the petrological changes discussed above have been referred to as greisenization in the literature and are regarded as a series of complex pneumatolytic-hydrothermal processes (Shcherba, 1970). Greisenization can be found superimposed on various earlier mineral assemblages. The petrological characteristics and the extent of greisen development depends upon the intensity and retention of earlier stages of albitization

Plate 7 Localised zones of microcline with quartz forming border zones to vein greisens. Ririwai lode, Ririwai complex. Long axis specimen 25 cm.

Plate 8 Multiple greisen zones as arrested stages of pervasive greisenization, Banke complex.

and/or microclinization, and the timing of fracturing and fissuring in the roof zone. Greisenization in Nigeria has affected three distinctive mineralogical assemblages; perthitic granite, albitized variants, and microclinized variants. The chemical composition of these mineralogical variations are displayed on the QF diagram (Fig 2) for the Ririwai complex (Nigeria), together with additional data from a petrologically similar style of postmagmatic metasomatism at Echassieres (France). Greisenization can be generally pervasive (Fig 3E), form locally pervasive pockets associated with microclinite (Fig 3F), or be marginally confined to quartz-filled fissures and fractures in the roof zone (Fig 3G). In all cases chlorite is a by-product of these reactions.

Pervasive greisenization

Greisenization that is pervasive uses incipient microfractures in the host rock to penetrate and transform the original mineral assemblage. Occasionally the acid fluids may also greisenize the surrounding country rock and/or the volcanic cover (Fig 3). One particular feature of pervasive greisenization is that the greisen retains the texture of the original host rock. This is particularly relevant when either foliated Pan African country rock or perthitic granite is greisenized. All stages can be observed from zones of arrested pervasion (sheeted or multiple greisens, plate 8) to partial or complete coalescence (plate 9). More commonly pervasive greisenization has occurred in Nigeria in the roof zones of granite cupolas superimposed on albitic variants forming albite-greisens (Shcherba, 1970, Table 39). Typical minerals include columbite, fluorite, albite, zinnwaldite, lepidolite, cassiterite, wolframite, zircon, thorite, Th-rich monazite, and topaz. Greisenized albitic variants plot in the QF diagram (Fig 2) towards a vector connecting the albitized samples and the quartz mineral pole.

Other instances of pervasive greisenization has occurred following microclinization of the albitized roof zone. Topaz is considerably less abundant in this assemblage but substantial cassiterite has been recorded in representative samples from Uwar Gida on Ginshi Hill (Kinnaird et al., 1984). The chemical variations are displayed in the QF diagram (Fig 2) as open circles with horizontal bars, and define an important trend. They represent pervasive greisenization of a porphyritic biotite microgranite in which the groundmass has been greisenized but remnants of albite and microcline are retained within the feldspar phenocrysts.

Pervasive greisenization of microcline-rich variants in which the albite component is missing (such as microclinite) leads to the formation of micaceous greisen. The plotting positions of greisenized microclinites are illustrated by two samples from Echassieres shown as white stars in black circles on the QF diagram (Fig 2). Local brecciation, depicted in Fig 3F yields co-existing mineral associations of microclinite and greisen pockets containing ore. According to Burt (1981) these assemblages formed from boiling of an aqueous fluid (under release of pressure) giving an acid vapour (enriched in μHF) and residuum (enriched in μKF) leading to the bimodal formation of greisen and microcline. This example is well displayed at Saiya Shokobo, and dramatically interpreted in Fig 3F.

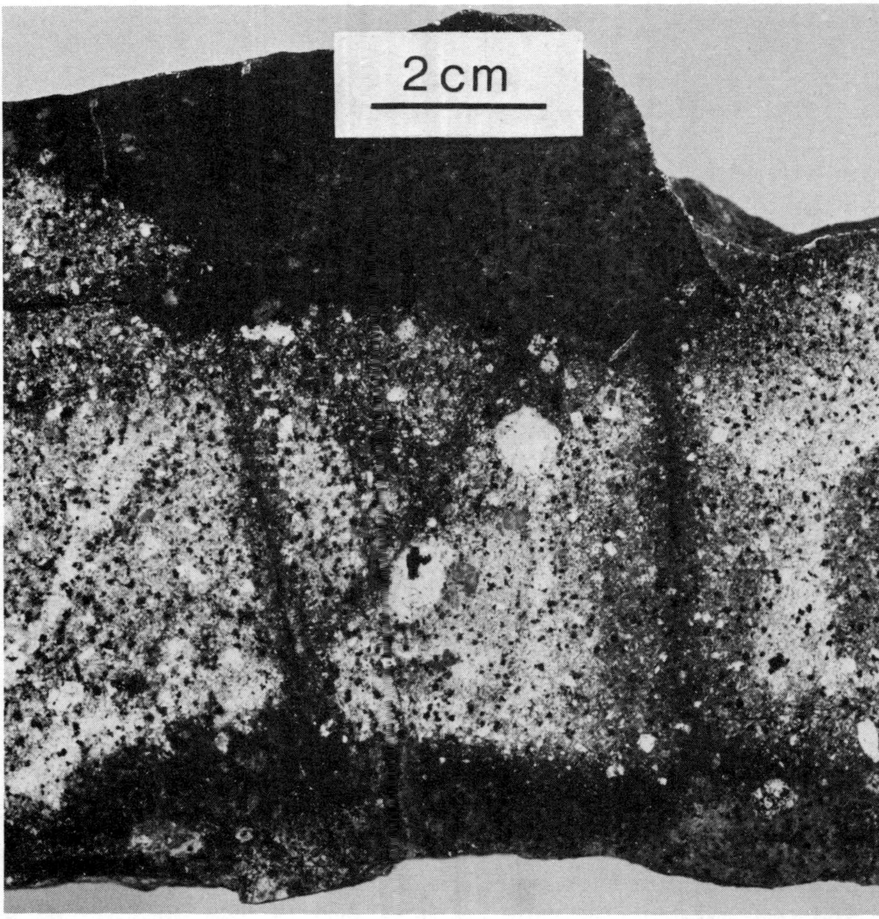

Plate 9 Partial pervasive greisenization in roof zone of porphyritic biotite granite superimposed on albitization, and microclinization of phenocrystic and groundmass alkali feldspars with limited fracturing and brecciation Saiya Shokobo complex.

Plate 1Ø Comb textured quartz veins carrying crystals of wolframite in outer roof zone of Tibchi biotite granite.

Evidence for brecciation following pervasive greisenization has been noted, particularly at Banke. Blocks of greisen breccia cemented by quartz containing wolframite are common in Banke both in the biotite granite roof zone and in the pervasively greisenized basement at Ginya Hill. Fluid inclusion populations in the vein quartz (B93) give lower homogenization temperatures (230 °C), are very dilute (1-2% NaCl) and have dominant vapour filled inclusions in secondary healed fissures. Associated with the development of a vapour phase is widespread ore dumping of cassiterite and wolframite. Late stage boiling following pervasive greisenization has probably occurred in many subvolcanic centres. Brecciated albitic roof facies of the Rayfield Gona biotite granite at Jos Bukuru (see plate 4) indicates that this process can occur at any stage during fluid evolution when the fluid pressure exceeds the lode pressure. Again massive ore dumping of sulphides (sphalerite, chalcopyrite, molybdenite) with cassiterite occurs at this stage. For the Nigerian complexes it is clear that major cassiterite/sphalerite mineralization only forms from the vapour phase when boiling and/or brecciation is recorded. This observation has important implications for the transport and deposition of Sn and Zn.

Vein greisenization

Greisenization that is termed vein type is restricted to the border zones of fissures or joints which allowed limited penetration of fluids into the host rocks (Fig 3G). Vein greisens are zoned with increasing degree of wall rock alteration and with greisen borders to comb-textured quartz veins. A typical sample collected from the Ririwai lode is shown in plate 7. The mineralogical and chemical parameters are displayed in Fig 2 as open diamonds, and the trend to quartz-rich compositions indicated by a dashed line connecting microcline rich border zones. Topaz is only a minor accessory and is rarely observed in thin section. Fluorite is more common. Mica compositions approach zinnwaldite. The greisens are quartz-mica-fluorite often with abundant cassiterite and/or sphalerite. However, it is in the central comb-textured quartz lining the fissure that contains large blocks of sulphide ores particularly sphalerite in addition to oxide ores such as cassiterite. Fluid inclusion populations show clearly that a fluid had boiled, with some inclusions containing mainly vapour and other dilute (1-2% NaCl) 2 phase (liquid and vapour) fluids. It appears that boiling provoked massive ore dumping in the fissure-filling veins. It is clear from underground sampling by drill cores and exploratory mining that the vein system at Ririwai follows a dominant E-W joint pattern and that the mineralizing fluids were derived from another granite body below (Kinnaird _et al_., 1984). Further evidence for the rapid release of a vapour phase at a late stage of mineralization is based on the occurrence of brecciated zones at Banke, Saiya Shokobo and Afu, and represented diagrammatically in Fig 3H. Wolframite veins may develop in the roof zones and the surrounding country rocks (plate 10).

The schematic diagram Fig 3 depicting the principal postmagmatic petrology in biotite granite cupolas provides a useful summary compilation emphasizing the importance of successive periods of boiling in a multicomponent fluid as a result of decreasing pressure to generate ore deposits. The major petrological variants are, in our opinion, dependent on the various stages of metasomatism, fluid interaction etc, and provide valuable criteria for assessing the ore- potential of the Nigerian younger granite province.

Further it seems clear that without the separation of a vapour phase, ore deposits cannot be generated (Roedder, 1977). Thus the schematic model displayed in Fig 3 is dependent on physical and chemical changes in fluid composition to generate the ore deposits. Should drastic phase changes not occur the fluid evolution and petrological variations follow the falling temperature path A-C-E-G. It is only the sudden pressure variations that paths A-B, C-D, E-F, G-H could appear. From the economic standpoint it is the constraining pressure release that generates the abundant ores, and the characteristic mineral assemblages.

Acknowledgements The work reported in this paper was Financed by grant R2679 from the Overseas Development Ministry, London. The grant provided support for a number of research personnel. In particular we would like to thank S I Abaa, C.A. Abernethy, J N Bennett, E C Ike and U M Turaki for providing some of the samples and to Richard Batchelor, and the University of Edinburgh for the analyses used to complete Fig 2.

Professor E K Walton is thanked for his support and encouragement, Jim Allan for the excellent photographs and Catherine Finlay for for secretarial assistance.

We would like to acknowledge the generous help and assistance in Nigeria provided by various

Mining Companies, the Universities, and the Geological Survey. In particular we wish to thank D Dent-Young, I Cox of ATMN and the engineering staff at Barakin Ladi: L Gibb of Gold and Base Limited, and the managers, engineers and geologists at Ririwai mines: our colleages at Ahmadu Bello University, Zaria, and the University of Ibadan: C N Okezie and his staff at Kaduna. We would also like to thank the Nigerian Mining Corporation for their interest in our work, especially R Lukman and S O Ford. We are grateful to the Ministry of Mines and Power and I Gombe for permission to work in Nigeria and for their encouragement.

REFERENCES

Aubert G. (1966). Les coupoles granitiques de Montebras et d'Echassières (massif central Francais) et la genèse de leurs minéralizations en étain, lithium, tungstène et beryllium. Thèse docteur sciences naturelles, Université de Clermont, 345 pp.

Barker, F, Wones D.R., Sharpe, W.N., and Desborough, G.A. (1975). The Pikes Peak Batholith, Colorado Front Range, and a model for the origin of the gabbro-anorthosite-syenite-potassic granite suite. Precambrian Research. 2, 97-160.

Bowden P. and Jones J.A. (1978). Mineralization in the younger granite province of northern Nigeria. Symposium Metallization Associated with Acid Magmatism, vol. 3, 179-190.

Bowden P., Bennett J.N., Kinnaird Judith A., Whitley J.E., Abaa S.I., and Hadzigeorgiou-Stavrakis P.K. (1981). Uranium in the Niger-Nigeria younger granite province. Mineral. Mag. 44, 379-389.

Burt D.M. (1981). Acidity-salinity diagrams application to greisen and prophyry deposits. Econ. Geol. 76, 332-843.

Charoy B. (1979). Définition et importance des phenomenes deutériques et des fluides associés dans les granites. Conséquences métallogeniques. Sci de la Terre, Fr., Memoire 37, 364 pp.

Currie K.L. (1968). On the solubility of albite in supercritical water in the range 400 to 600 °C and 750 to 3500 bars. Amer. J. Sci. 266, 321-341.

De La Roche H. (1963). Sur l'expression graphique des relations entre la composition chimique et la composition minéralogique quantitative des roches crystallines. Présentation d'un diagramme destiné à l'étude chimico-minéralogique des massifs granitiques ou granodioritiques. Application aus Vosges cristalline. Sci de Terre, Fr. 9, 293-337.

Jacobson R.R.E., Macleod W.N. and Black R. (1958). Ring complexes in the younger granite province of northern Nigeria. Mem. geol. Soc. Lond. 1, 117 p.

Kinnaird J.A. (1981). Geology of the Nigerian anorogenic ring complexes. 1:500,000 Geol. map. John Bartholemew and Sons. Edinburgh. (copies available on request to the author).

Kinnaird J.A., Bowden P. and Ixer R.A. (1984). The Ririwai lode - a zinc- tin rich greisen-wallrock assemblage from norther Nigeria. Tr. I. M. M. (section B) in press.

Kovalenko N.I. (1976). The reactions between granite and aqueous hydrofluoric acid in relation to the origin of fluoric bearing granites. Geokhimiya no 4, 1976, 503-515.

Lameyre J. (1966). Leucogranites et muscovitisation dans le massif central Francais. Annales de la Faculté des Sciences de l'Université de Clermont no. 29, 264 p.

Luth W.C. (1976). 'Granitic rocks'. In D.K. Bailey and R. Macdonald (eds.). The Evolution of the Crystalline Rocks. Academic Press, London pp 335-417.

Manning D.C. (1982). An experimental study of the effect of fluorine on the crystallization of granitic melts. In A.M. Evans (ed.) Metallization Associated with Acid Magmatism. John Wiley & Sons pp 191-203.

Martin R.F., and Bowden P. (1981). Peraluminous granites produced by rock-fluid interaction in the Ririwai nonorogenic ring complex, Nigeria: mineralogical evidence. Canad. Min. 19, 65-82.

Rieder M. (1970). Chemical composition and physical properties of lithium-iron micas from the Krusne Hory mountains (Erzgebirge) Part A Chemical Composition. Contr. Mineral. Petr. 27, 131-158.

Roedder E. (1977). Fluid inclusions as tools in mineral exploration. Econ. Geol. 72, 503-525.

Shcherba G.N. (1970). Greisens. Internat. Geol. Review 12, pp 114-150; 239-255.

Taylor R.P. Ø., Strong D.F., and Kean B.F. (1980). The Topsails igneous complex: Silurian-Devonian peralkaline magmatism in western Newfoundland. Canad. J. Earth Sciences 17, 425-439.

MAGMATIC ASSOCIATIONS OF CARPATHIAN-BALKAN AREA AND THEIR METALLOGENIC SPECIALIZATION

E. DIMITROVA[1], Y. YANEV[1], D. BAHNEVA[2], B. KAMENOV[3], B. MAVRUDCHIEV[3]
1 Geological Institute, Bulgarian Academy of Sciences;
2 Higher Institute of Mining and Geology;
3 University of Sofia "Kliment Ohridski"; Bulgaria.

The analysis of magmatic associations in the Carpathian-Balkan (CB) Area is based on the Map of magmatic associations in CB Area in scal 1:1 000 000 (The Carpathians-Balkans regions, 1983) compiled by the authors for the purpose of metallogenic zonation. In the course of the work the authors used the most important achieve ments of soviet petrologists in the study of magmatic associations (Kuznetzov, 1964; Shteiberg et al., 1976; USSR magmatic formations, 1979; Dobretsov, 1981). In accord with these authors the term magmatic association is understood as a comagmatic association of rocks which occurs repeatedly in time and space and retains some existing common features. In this way a magmatic association is product of single-type petrogenetic processes, determined by a definite endogenic regime and geodynamic crustal development. The term magmatic complex is applied to occurrences of a given association in limited structural zones, within a relatively short time interval (epoch or two adjacent epochs). The subdivision of magmatic associations in CB Area is based on the most important features of the rock components: petrographic, petrochemical (acidity and alkalinity) and facies composition, degree and type of differentiation. Two groups - volcanic and intrusive, are separated, according to depth of formation. The name of the magmatic association reflects the composition of the dominating and most typical rocks in a given association, described in the order of increasing acidity (the names of the rocks are according to the Classification and nomenclature of magmatic rocks, 1981. The metallogenic specialization indicates the spatial and temporal relations of a given association to endogenic mineralizations. The controversial problems concerning

Proceedings of the 27th International
Geological Congress, Volume 9, pp. 121–136
PETROLOGY (IGNEOUS AND METAMORPHIC ROCKS)

the sources of ore components, the form and conditions of their mobilization are not discussed. In the systematization of the divided associations the authors have used the scheme proposed in USSR magmatic formations (1979): the rocks are grouped into 4 families according to mafic-salic content (Table 1,2,3).

The separated magmatic associations characterize CB Area as a whole as characteristically mafic-salic and salic (Figs 2,3,4) as far as mafic and ultramafic formations are of limited occurrence (Fig 1). Normal and subalkaline[x] associations dominate. The alkaline associations are represented by only one - "as", which indicates the lack of end members in the alkaline magmatic evolution. For this reason association groups according to alkali content are not divided in the systematization of the magmatic associations in CB Area. A distinct lateral (north-south) variation of the chemical composition (Table 1,2,3) is typical feature of the associations in this area: in the Carpathides the magmatism is calcalkaline, rarely Na-subalkaline, and in the Balkanides - calcalkaline and K-subalkaline. In respect of the degree and type of differentiation the mafic-salic associations are complex or discontinuously differentiated while the ultramafic and salic ones (with the exception of "gd") are undifferentiated.

The mafic-salic associations are bearers of the most important polymetallic and Cu-Mo mineralizations, accompanied by a number of additional but very important elements. Amongst the salic formations only "gd" is of definite metallogenic significance. The differentiated character of these associations determine the sulphuro- and in parts oxiphile type (Marakushev, 1979) of the endogenic mineralizations while the silicatophile type is practically lacking. It is empirically established (Isokh, 1971) that the most interesting endogenic mineralizations, products of diverse genetic types and

[x] Subalkaline rocks are those which have alkali content between alkaline and calcalkaline rocks (Classification and nomenclature of magmatic rocks, 1981)

containing of wide range of ore metals (Table 2), are usually related to continuously differentiated associations which in CB Area are the "ba, gg, gm, a, dg, al, m" associations (from gabbroto granites) Fe mineralizations appear in the mafic members. The weak development of final acid members (aplites, alaskites) in CB Area determines the low, practically only mineralogic significance of W-Sn mineralizations (for instance in the "gg" association of the South Carpathians and Babat). In case of incomplete differentiation (up to monzonites and syenites - for instance in the "gm" association) Cu and Mo are main metals. The lack of contrasting intrusive associations determines the leading role of Cu, not of Mo. A definite relation between the type of alkalinity of the associations and gold content is marked (1971). Thus for the calcalkaline "a" association (Na > K), Au and in parts Ag specialization is very typical (in East and West Carpathians, and paricularly in Apuseni Mts). In the K-subalkaline "al" association (K>Na) from the Rhiopes, developed in similar environment, the significance of Au is limited.

Despite of the controversy on the tectonic position of the magmatic associations a necessary requirment to modern association analysis is to find their place in definite tectono-magmatic cycles and their relation to endogenic regimes. In CB Area the tectonic typization of the divided association families (Figs. 1,2, 3,4) is based on the endogenic regimes of Belousov 1975 - Table 4. A possible interpretation of the magmatic activity in light of the geosyncline concept and plate tectonics is also proposed (Janev and Bachneva, 1980). Similar parallels are made for many regions by other authors (Bogatirov et al., 1983).

The magmatic activity of the early geosynlinical regime of the Phanerozoic cycles is mainly of mafic character (Precambrian cycles are not considered, i.e. the so-called ancient magmatic complexes, exposed in the crystalline basement are affected by intensive polyphase and polyfacies metamorphism). The "sk" associations (Table 1, Fig. 1) is the most typical association of this stage, the amount of its products decreasing from the Caledonian-Hercynian towards the

Alpine cycle. Amongst the intrusive associations the "u" formation has some metallogenic significance but the age and the geologic position of its complexes is not always clear. The weak intensity and a-typical occurrence of initial magmatism, the lack of typical ophiolites, as well as the considerable volume of granite magmatism in the following, inversion stadium, confirm the opinion of Pave (1969) that the area under consideration was developed upon a continental crust. This development probably took place under the conditions of the so-called Europian type of endogenic regime (Dobretsov, 1981). The low intensity of the initial magmatism determines its small metallogenic significance. Fe and Mn volcanogenic-hydrothermal mineralizations (Table 1, Fig 1) are relatively better presented and are related to the Pc and Pz complexes.

The inversion stage of the geosynclinal regime is characterized by a rapid increase in the intensity and degree of differentiation of the magmatism from earlier to later tectono-magmatic cycles. If we assume that the largest granite massifs in CB Area, refferred to the "g" associations, developed and were related to the Precambrian consolidation of the crystalline basement, than the earliest magmatic occurrences of the inversion stage were of salic intrusive character. Moreover, insignificant accumulations of some silicatophile elements (Be,B) are related to them only. In the Caledonian-Hercynian cycle the salic and intrusive character of the magmatism from the inversion stage is retained but the part of differentiation ("gd" association) increases. Related to them are Sn, W, Mo mineralizations in the acid members of the associations (Table 3, Fig. 2). In the Alpine cycle the magmatic activity of this stage is already of mafic-salic character and shows wide development of both volcanic and intrusive associations. They form volcano-plutonic series ("ar"-"gg", "ba+tl"-"gm") and show a wide petrographic diversity. The intensity and the wide range of differentiation determine the metallogenic importance of the magmatism, particularly for Cu in the volcanogenic and Cu, Mo - in the intrusive associations.

The orogenic regime in CB Area is typified by the defelopment mostly of volcanic mafic-salic and salic associations (Fig. 4). Its intensity rapidly increases from earlier to later cycles. The Caledonian-Hercynian cycle terminates with the occurrence of a single volcanic and intrusive associations are related to volcanoplutonic series ("a" - "dg" and "al" - "m") - - Table 2. These series have a metallogenic specialization, comparable in importance with the associations of the inversion stage of the Alpine cycle, but with main elements Pb, Zn, Cu, Au, Ag, Hg (Table 4). In this stage, however, the variety of genetic types of mineralizations decreases - the magmatic types disdappear, the part of skarn and volcanogenic-hydrothermal mineralizations decreases. It is significant also, that the orogenic magmatic activity rejuvenates from nort to south: in the Rhodopes the magmatic activity took place in the Upper Eocene - Oligocene, in the Carpathides - in the Neogene. Most probably this is related to later nappes in the Carpathides, as compared to the Balkanides which leads to a thickening of the crust, assumed to be a reason for magma generation (Stille, 1954; Dobretsov, 1981).

The analisis of magmatic associations in CB Area indicates also some reglarities in its lateral migration (The Carpathians-Balkans region magmatism and metallogeny, 1983). Thus, in the early endogenic regimes of a given tectonomagmatic cycle or in the given endogenic regime, but in different cycles, the magmatic activity shows a trend to migrate from the fold belt to the median massifs. The metallogenic zones follow this lateral trend. This is well expressed in East, South Carpathians and in the Balcanides.

A typical final volcanism of basic composition marks the end of the Alpine cycle ("b" association) and terminates the magmatic activity in the CB Area.

REFERENCES

Belousov, V.V. 1975. Geotectonics principles. Moscow, Nedra. 260 p.

Bogatikov, O.A., A.M. Borsuk, Ju.I.Dmitrijev, V.I. Kovalenko, I.D. Ryabchikov. 1983. Magmatic formations in the evolution of Lithosphera. - USSR Academy of Sci. Izvestija., geol. series, 1,3-16.

Dobretsov, N.L.. 1981. Great petrological processes. Moscow, Nedra. 236 p.

Isokh, E.P. 1971. Classification scheme of granitoid formations with various ore-contents and the example of its implementation in West Uzbekistan.-B: Petrology and geochemistry problems of granitoids. Sverdlovsk. UFC Acad. Sci., p. 34-43.

Classification and nomenclature of magmatic rocks. 1981. Moscow Nedra. 160 p.

Kuznetzov Ju.A. 1964 Main types of magmatic formations. Moscow. Nedra. 388 p.

USSR magmatic formations. 1979, vol. 1,2. Leningrad, Nedra. 320 p. and 280 p.

The Carpathians-Balkans regions magmatism and metallogeny. 1983. Sofia, BAS. 300 o.

Marakushev A.A., 1979. Petrogenesis and ore formation. Moscow. Nedra 264 p.

Pave, A.V. 1969. Oceanic crust of the geological past. -Geotectonics, 6,3 - 23.

Shteiberg, D.S., O.A. Bogatikov, M.B. Borodajevskaja, L.S. Borodin, A.A. marakushev. 1976. Magmatic formations problems.- B: Petrology problems. Moscow, Nauka, p. 75-93.

Janev, j., D. Bachneva. 1980. Alpine magmatizm of the Carpathians- Balkans region at the model of a tectonic plate.- B: Geodynamics on the Balkans. Sofia, Techniques, p. 63-75.

Stille, H., 1939. Zur Frage der Herkunft der Magmen.- Abh.Preuss. Akad. Wiss. Math.-nat. kl., 19, 3-32.

Stille, H., 1954. Unterschiebungs Palingenese.- C.R. 19 session congr.inter. 1952. sec 15, fasc. 17, Alger, 127-137.

Table 1. Ultramafic and Mafic Associations in the Carpathian-Balkan Area

Associations with index; age of the complexes	Petrographic and facies features	Metallogenic+ specialization and genetic types of mineralizations
Dunite-peridotitic (u); Pc(?), Pz(?), Mz(?) - Fig. 1	Undifferentiated: dunites, peridotites (harzburgites predominate), less commonly pyroxenites and gabbro; calcalkaline with high Mg index; intensive syn- and allometamorphic serpentinization; fissure massives along regional faults or sill-like bodies, related to certain stratigraphic horizons of the crystalline basement	Cr (Co,Ni) - magmatic; asbestos, talc, Mg - metamorphogenic
Spilite-keratophyric (sk); Pc, Pz, Mz - Fig. 1	Contrasted: porphyric diabases (spilites) dominate, less commonly porphyrites, quartz-porphyries; gabbro and serpentinite lenses; calcalkaline with superimposed Na-metasomatism, submarine volcanism with dominating pyroclastites in association with shale and flysch sediments; mostly greenschist metamorphysm rarely ranging to amphibolite facies	Fe-Ti - magmatic; Fe-Mn; Cu-pyrites with Sb (Pb,Zn) - volcanogenic-hydrothermal
Ophiolitic group (of); T_2, J-Cr_1 - Fig. 1	Undivided intrusive and volcanic associations, mostly of maphic composition: aphyric and porphyric basalts (diabases, spilites) and their pyroclastics, ultrabasites, gabbro, diorites; less commonly rhyolites, essexites, teshenites; tholeiitic and calc-alkaline with deviations to Na- or K-alkaline composition; deep-water eruptions in association with limestones, radiolarites, red shales, jaspers or flysch-like sediments; sills and cross-cutting bodies; greenschist metamorphism	Fe-Ti(V) - magmatic; Mn, Fe (Cu,Ba) - volcanogenic-hydrothermal
Trachybasalt-latitic (tl)++; Cr_2 - Fig. 2	Differentiated: from picrites to trachytes with dominating trachybasalts and latites, rarely alkaline basaltoids (analcite, leucite), trachyandesitobasalts, trachytes; K-subalkaline to K-(rarely Na-) alkaline; hydrothermal alterations: zeolitization, K-feldspatization, argillization, formation of secondary quartzites; small volcanoes with high explosive index or fissure eruptions; subvolcanic bodies and dikes; submarine to shallow-water volcanism, synchronous to tephroidal flysch	Pb, Zn, Au, Cu (Bi, Sb, Ag) - hydrothermal veins
Basalt-basanitic (b); N_2-Q - Fig. 4	Undifferentiated: sub- and alkaline picrites, trachy- and alkali basalts, basalts and andesitobasalts (aphyric, rarely doleritic), Na-subalkaline to alkaline, rarely K-subalkaline and calcalkaline with increase of SiO_2 and alkalies from the foreland of the system towards the median massifs; subvolcanic bodies and dykes, monogenic, rarely small stratovolcanoes in subaeral or lacustrine-lagoonal environment	(Sb,As - hydrothermal)
Gabbro-leucogranitic (gl); Pc, Pz - Fig. 1	Contrasted: gabbro with ultrabasic cummulates, trondhjemites, leucogranites with contaminated varieties of dioritic composition; calcalkaline to K-subalkaline; synkinematic bodies, co-magmatic with the rocks of "sk" association; greenschist metamorphism	Cr (Co,Ni) - magmatic

+ The second-order endogenic mineralizations in Tables 1,2,3,4 are shown in brackets

++ Only in the Balkanides (see Table 2)

Table 2. Mafic-salic Associations in the Carpathian-Balkan Area

Carpathides[+]			Balkanides[++]		
Associations with index; age of the complexes	Petrographic and faces features	Metallogenic specialization and genetic types of mineralizations	Associations with index; age of the complexes	Petrographic and faces features	Metallogenic specialization and genetic types of mineralizations
Andesite-rhyolitic (ar); Cr_2 - Fig. 2	Differentiated: andesites, dacites, rhyolites with dominant development of acid rocks; calcalkaline; submarine and subaeral eruptions with high explosive index and wide occurrence of ignimbrites, linear volcanic centres and calderas; lava sheets, irregular bodies; flows, extrusions, dikes; ash deposits bentonitized	not clear; bentonites	Basalt-andesitic (ba); Cr_2 - Fig. 2	Differentiated: from basalts to rhyolites with dominating andesites and porphyritic basalts; calcalkaline to K-subalkaline; hydrothermal alterations: propilitization, formation of secondary quartzites; large polygenic, polyphase stratovolcanoes with high explosive index; submarine volcanism synchronous to tephroidal flysch	Cu porphyry and vein ores (Ba,Pb,Zn,Au, Ag,Se,Te,Bi, Hg); Cu-pyrites; Mn-Fe - volcanogenic-hydrothermal
Andesitic (a); Pg_2, N_{1-2} - Fig. 4	Differentiated: from aphyric basalts to rhyolites with dominating andesites and andesitobasalts; calcalkaline; hydrothermal alterations - argillization, adularization, propilitization, formation of secondary quartzites; stratovolcanoes in subaeral, fresh-water and rarely marine environment; subvolcanic bodies and dikes; the sedimentary association is refered to the upper molasse; products of initial phases alternate with pyroclastites of "r" formation	Cu-Mo porphyry ores; Au, Ag,Pb,Zn,Cu, Hg (Mo,Fe,Bi, Ba,Sb,Te,S) - hydrothermal veins and metasomatic bodies; Fe,S-volcanogenic-hydrothermal alunites, bentonites - hydrothermal-metasomatic	Andesite-latite-trachyrhyolitic (al); Pg_2-N_1(?) - Fig. 4	Differentiated: from absarokites to trachyrhyolites and from andesitobasalts (aphyric) to rhyolites with predominance of latites and trachyrhyolites; K-subalkaline, rarely calcalkaline; alterations: zeolitization, sericitization, adularization, propilitization, secondary quartzites; stratovolcanoes, volcanoes with endogenic domes, calderas with ignimbrites; subvolcanic bodies and dike swarms; alternation of acid and intermediate products in marine, locally subaeral environment; sedimentary association referred to lower molasse	Fe - skarns; Cu-Mo porphyry ores; Pb,Zn,Cu (Ag,Au,F,Mn,Mo W,Bi,Ba,Sb) - hydrothermal veins and metasomatic bodies; (Mn,P, Be - volcanogenic-hydrothermal); zeolitites, bentonites, alunites - hydrothermal-metasomatic
Diorite-granodioritic (dg); [illegible]	Differentiated: from gabbro to granodiorites with [illegible] intermediate	Same as "a" association + Cu,Pb,Zn (Mo, [illegible]	Monzonite-syenitic (m); Pg_2-N_1(?) -	Differentiated: from gabbro to syenites with predominance of monzoni-	Same as "al" association

	volcanic bodies intruded into large stratovolcanoes of the comagmatic "a" association			hypabyssal bodies, dikes and sills, intruded in stratovolcanoes of the comagmatic "al" association and in the crystalline basement; insignificant hornfelses and skarns	
Basalt-rhyolitic (br); C-T, Fig. 4	Contrasted: aphyric basalts ("melaphyres") and rhyodacites (quartz porphyries), rare andesites, plagiorhyolites, trachybasalts, latites; calcalkaline and Na-subalkaline, with superimposed Na-metasomatism, less commonly argillization; small extrusions and polyphase central volcanoes, rarely subvolcanic bodies and dikes; subaeral volcanism or occurrences in small isolated basins with molasse deposits				U,Ba,Mg,Cu,Mo (Pb,Zn,Sb) - hydrothermal
Gabbro-monzonitic (gm); Cr_2 Fig. 2	Long but incompletely differentiated: from gabbro and monzogabbro to leucosyenites with dominating monzonites, syenites, alkaline quartzsyenites; monzonites and syenites predominate, less common essexites and subalkaline pyroxenites; K-subalkaline to K-alkaline; hypabyssal to subvolcanic isometric bodies of small size intruding the volcanoes of "ba" and "tl" comagmatic associations; dike rocks of similar composition; thermal contact metamorphism with hornfelses				(Fe,Cu,Pb,Zn-skarns);Cu,Mo (Au,Pb,Zn,Bi, Co,Fe,Hg,Ba)-hydrothermal veins
Gabbro-diorite-granitic (gg); Cr_2-Pg_1 - Fig. 2	Long but incompletely differentiated: from pyroxenites and gabbro to granites; monzodiorites, quartzmonzodiorites less common; in the Carpathides the range of variations is smaller, quartzdiorites and granodiorites dominate, calcalkaline with rare transitions to K-subalkaline; in the Balkanides - tholeiitic in the early phases and calcalkaline in the final ones with transitions to K-subalkaline; linear hypabyssal bodies of small and intermediate size; dike rocks of similar composition; thermal contact metamorphism - hornfelses, magnesium and calcium skarns; K-feldspatization, biotitization, argillization, sericitization silicification; comagmatic with "ar" and "ba" associations				Fe-Ti(V)-magmatic; Fe,Cu, Mo(Pb,Zn,W,B) -skarns; Cu-Mo porphyry ores with Pb, Zn(Au,Ag,Se, Re); Cu,Ba(Mo Au,W,Pb,Zn,F) hydrothermal veins
Alkali gabbro-syenitic (as); **Pz** - Fig. 4	Differentiated: two petrochemical series: 1) in the Carpathides - calc to Na-alkaline - diorites, granites, syenites, nepheline syenites, monchikites; 2) in the **Balkanides** - K-subalkaline to K-alkaline - quartzsyenites, granosyenites, shonkinites, monzonites, quartz-bostonites and tinguaites, alkaline syenites; small hypabyssal stocks, veins; narrow thermal contact aureoles				Rocks rich in accessory minerals with rare-earth elements

[+]Carpathides - West, East and South Carpathians, Apuseni, North Dobrodgea, Pannonian Massif

[++]Balkanides - Stara Planina, Timok-Srednogorie zone, Serbo-Macedonian and Rhodope Massifs

Table 3. Salic Associations in the Carpathian-Balkan Area

Associations with index; age of the complexes	Petrographic and facies features	Metallogenic specialization and genetic types of mineralizations
Rhyolitic (r); Pz, Pg_3-N_1 - Fig. 4	Undifferentiated: from rhyolites to dacites (commonly ignimbrites); in Carpathides - plagiorhyolites, in Rhodopes - sanidine rhyolites; calcalkaline to K-subalkaline; hydrothermal alterations: argillization, zeolitization and formation of secondary quartzites; areal volcanism with fissural, less commonly central volcanoes with calderas, volcanic grabens and horsts; subaeral or shallow-water volcanic activity in association with upper molasse.	Zeolitites, kaoline, alunite, bentonites - hydrothermal-metasomatic
Granite-migmatitic (g); Pc(?), Pz(?), Mz(?) - Fig. 3	Undifferentiated: biotite granites with transitions in the endozones to hornblende-biotite or porphyraceous varieties (with large K-feldspar endoblasts); calcalkaline, rarely K-subalkaline; thick aplite-pegmatite veins and bodies; deep polyphase migmatite plutons, formed as a result of granitization of the crystalline basement	Muscovite, feldspars (Be) - pegmatitic; (Mo, W - hydrothermal)
Granodiorite-granitic (gd); Pz - Fig. 3	Differentiated: granodiorites predominate; calcalkaline syn-, late- or postkinematic, stocks or batholitic polyphase intrusions: 1-st phase - granites-granodiorites, 2-nd - aplitoid, red granites (K-subalkaline), 3-rd - aschistic and diaschistic vein rocks; thermal contact metamorphism with development of hornfelses	Fe - skarns; Mo, W, Cu, Pb, Zn, Ag, Au (Sn, Ba, Sb, Bi, Hg, Fe) - hydrothermal

Table 4. Comparative Scheme of Phanerozoic Tectonomagmatic Development of the Carpathian-Balkan Area

Geosynclinal concept			Plate tectonics concept		Magmatic Associations in the CB area		Metallogenic specialization and genetic types of mineralizations
Magmatism after Stille (1939)	Magmatism of endogenic regims after Belooussov (1975)		Geodynamic environment	Geotectonic position	Caledonian-Hercynian cycle	Alpine cycle	
Initial	Geosynclinal regim	Ophiolitic stage (eugeosynclinal regim); compensated subsidence stage (miogeosynclinal regim)	Rift environment (oceanic) - accreation and spreading of the oceanic basin	Continental rifts Mid-oceanic ridges Oceanic islands Back-are rifts	sk Pz_1 gl Pz_1 u Pz(?)	of $T-Cr_1$	Fe-Ti(V); Cr(Ni, Co) - magmatic; Fe;Mn;Ba;Cu - pyrites with Sb (Pb,Zn) - volcanogenic-hydrothermal
Synorogenic		Inversion stage (eu-and miogeosynclinal regim)	Island-arc environment (West Pacific type) and active continental margins (East Pasific or Andian type)	On top of subduction zones in island arcs or in active continential margins	gd Pz_2	ar Cr_2 gg Cr_2-Pg_1 ba Cr_2 tl Cr_2 gm Cr_2	Fe-Ti(V) - magmatic; Fe,Cu,Mo(Pb,Zn, W,B) - skarns; Cu,Mo,Pb,Zn,Au, Ba(W,Co,Bi,Ag, Hg,Re,Se,Te,Sb, F) - hydrothermal; Cu - pyrites; Fe-Mn - volcanogenic-hydrothermal
Subsequent	Epigeosynclinal orogenic regim		Collision: closing of oceanic basins	In continential margins (in front of island-arc magmatism)	as Pz_2 br C-T	a Pg_2; N_{1-2} dg N_{1-2} al Pg_2-N_1(?) m Pg_3-N_1(?) r Pg_3-N_1	Fe,Cu,Pb,Zn(Mo)-skarns; Pb,Zn,Cu,Mo,Au, Ag,Hg(U,Mg,Ba, W,Fe,Bi,Sb,Se, Te,F) - hydrothermal; Fe(S,Mn,F,Be) - volcanogenic-hydrotermal
Final			Postcollisional(?)			b N_2-Q	(Sb,As - hydrothermal)

FIGURE 1. Magmatic associations of the early stage of the geosynclinal regime in Carpathian-Balkan Area and their metallogenic specialization. 1 - complexes of the dunite-pyroxenitic association; spilite-keratophyric association: 2 - Late Riphean - Lower Paleozoic (?) complexes, 3 - Lower Paleozoic complexes; gabbro-leucogranitic association: 4 - Late Riphean complexes, 5 - Lower Paleozoic complexes; 6 - complexes of the ophiolitic group associations; 7 - metallogenic specialization of the magmatic complexes. Remarks: in all figures the main tectonic units of CB Area are denoted as follows - foreland: RP - - Russian platform, MP - Moesian platform, DR - - North Dobrudja; Carpathian orogene: F - foredeep, KE - external (flysch) megazone, KI - internal (Mesozoic) megazone, A - Apuseni, Balkanides: FB - Fore-Balkan and transitional zone, SP - Stara Planina zone, S - Timok-Srednogorie zone; D - Dinarides; median massifs: PM - Pannonian Massif, M-D - Mecsek-Debrecen zone, M - Muresh zone, T - Transilvanian Depression, SM - Serbo-Macedonian Massif, RM - - Rhodope Massif.

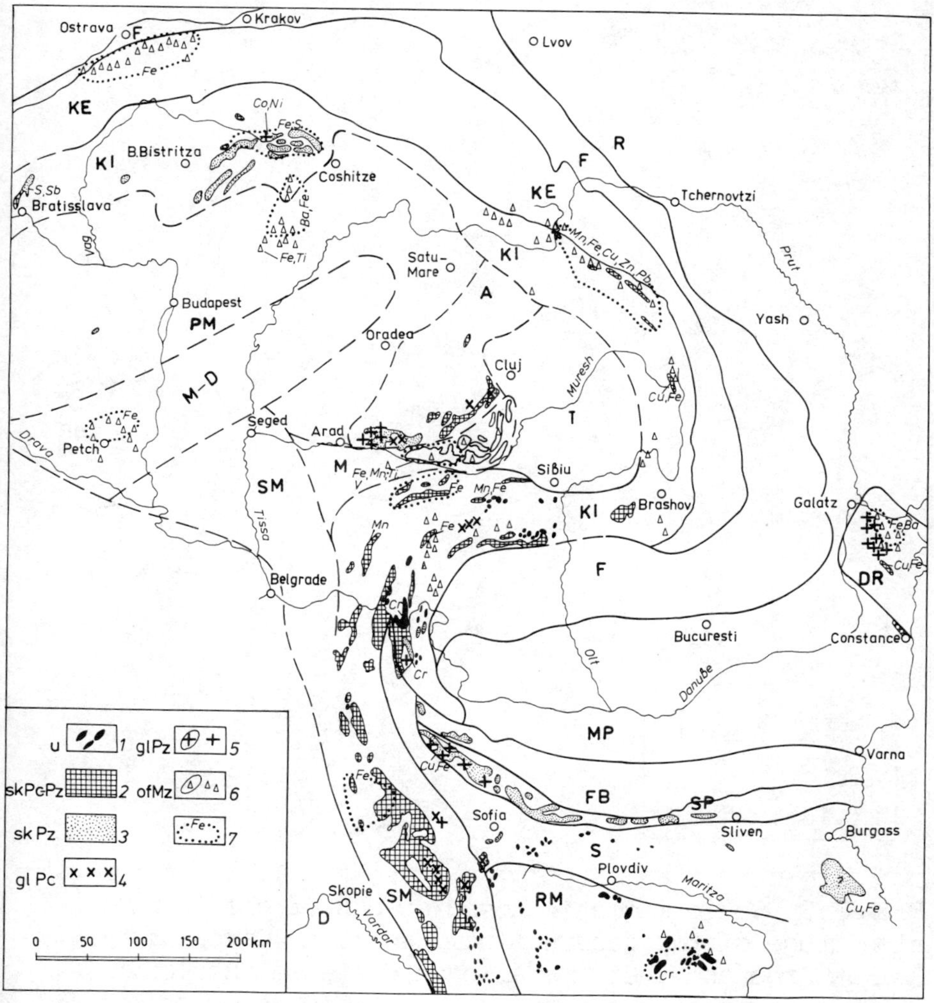

Ostrava
Krakov
Lvov
KE
KI
B.Bistritza
Coshitze
Bratisslava
R
F
KE
Tchernovtzi
Prut
Satu-Mare
KI
A
Budapest
PM
M-D
Oradea
Cluj
Muresh
Yash
T
Seged
Arad
Petch
Drava
M
SM
Tissa
Sibiu
KI
Brashov
Galatz
DR
F
Belgrade
Bucuresti
Constance
Olt
Danube
MP
Varna
FB
SP
Sofia
Sliven
Burgass
S
Plovdiv
Maritza
RM
Skopie
SM
D
Vardar
u 1
glPz 5
skPcPz 2
ofMz 6
sk Pz 3
7
gl Pc 4
0 50 100 150 200 km

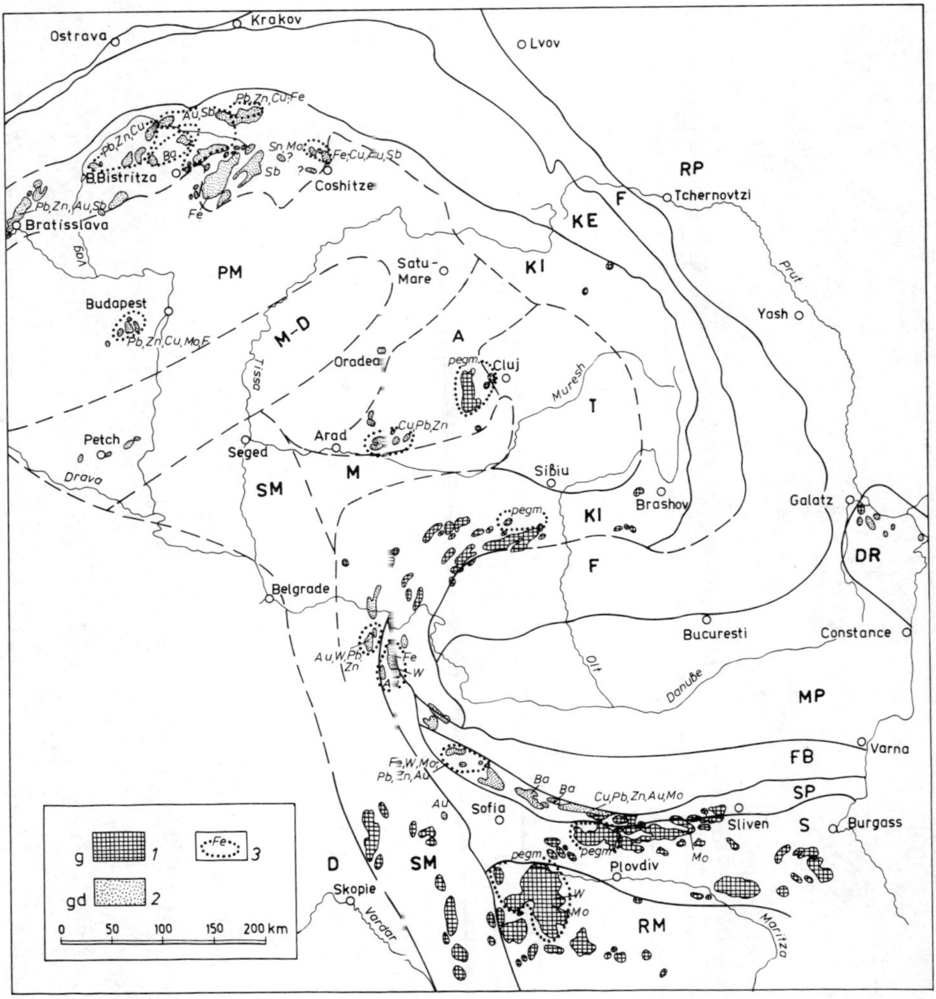

FIGURE 2. Salic magmatic associations of the inversion stage of the geosynclinal regime in Carpathian-Balkan Area and their metallogenic specialization. 1 - complexes of the granite-migmatitic association; 2 - complexes of the granodiorite-granitic association; 3 - metallogenic specialization of the magmatic complexes.

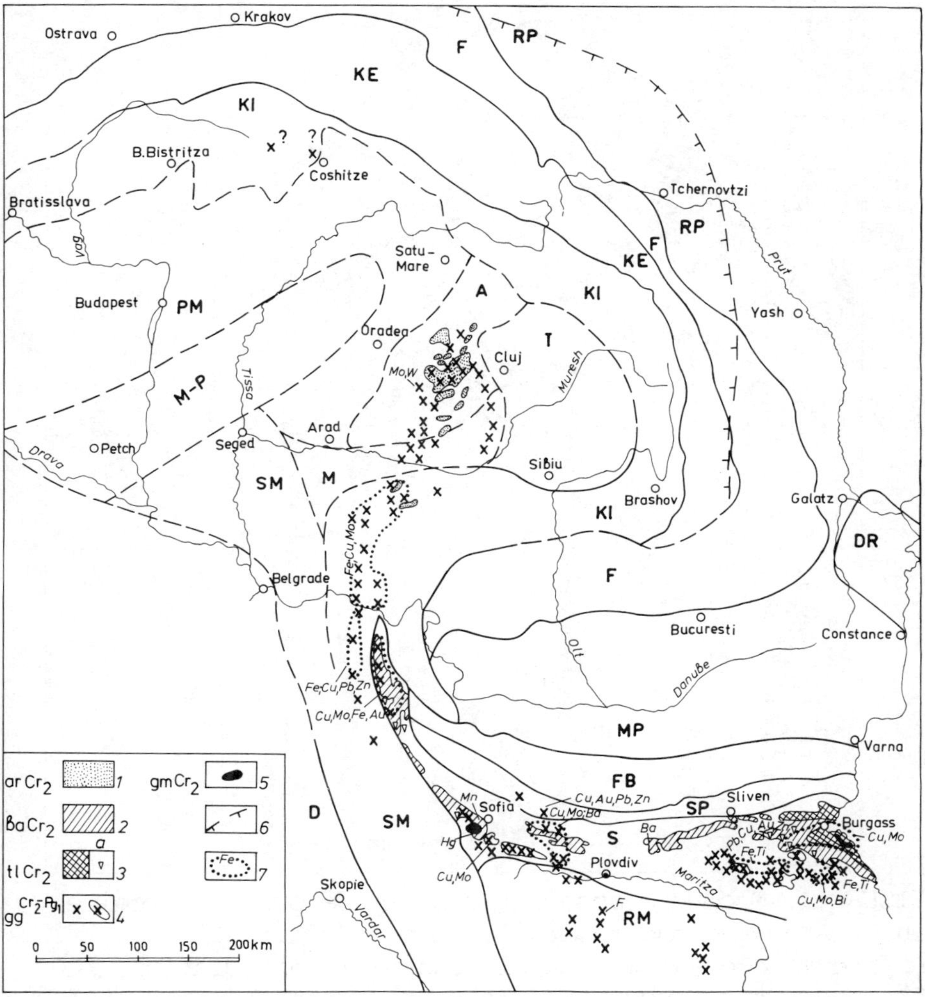

FIGURE 3. Salic-mafic magmatic associations of the inversion stage of the geosynclinal regime in CB Area and their metallogenic specialization, Complexes of the: 1 - andesite-rhyolitic association, 2 - basalt-andesitic association, 3 - trachybasalt-latitic association (a - scale exaggerated), 4 - - gabbro-diorite-granitic association, 5 - gabbro-monzonitic association; 6 - boundary of distribution of "ar" association tuffs; 7 - metallogenic specialization of the magmatic complexes.

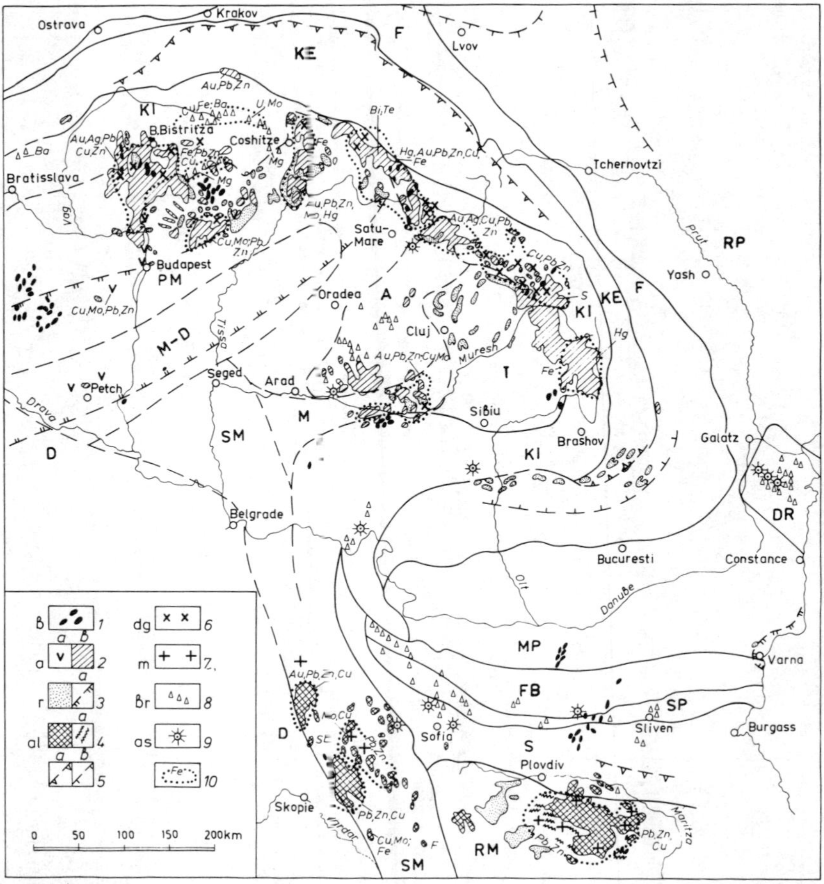

FIGURE 4. Magmatic associations of the orogenic regime in CB Area and their metallogenic specialization. Alpine complexes: 1 - basalt-basanitic association, 2 - andesitic association (a - Paleogene, b - Neogene), 3 - rhyolitic association (a - boundaries of pyroclastites), 4 - andesite-latite-trachyrhyolitic association (a - dikes); 5 - boundaries of the pyroclastites of "a" and 'al" association: a) Paleogene, b) Neogene; 6 - minor intrusions of the diorite-granodioritic association; 7 - minor intrusions of the monzonite-syenite association. Hercynian complexes: 8 - basalt-rhyolitic association, 9 - alkaline gabbro-syenitic association; 10 - metallogenic specialization of the magmatic complexes.

BASIC PRINCIPLES AND CRITERIA FOR COMPARISON OF OCEANIC IGNEOUS ROCKS AND THEIR ANALOGIES ON CONTINENTS

Yu.I. DMITRIEV, E.V. SHARKOV
Institute of Geology of Ore Deposits, Petrography, Mineralogy and Geochemistry of the USSR Academy of Sciences, Moscow.

The method of comparative alalysis is one of the most effective and often used means of the scientific study, which reveals features of similarity and difference between objects or phenomenons under investigation, as well as their relations and variations in space and time. To provide correct, reliable results, comparative studies should carry out on the basis of methodologically sound principles, that can be demonstrated using as an example comparison of magmatic rock associations of the Earth's continental and oceanic regions. Continents and oceans are the largest morphostructures of our planet and one of the most important tasks of the global petrology is to disclose peculiarities of magmatism in their confines.

According to general principles of the comparative analysis it is rational to compare only that objects, which have at least a few features of resemblance. The output of the new information will be the largest, when similarity or difference between objects compared is not extreme. The results of the comparative study are also the best, when objects are equally investigated that is they are described by facts of the same quantity and quality, which are uniformly sistematized and analyzed. Theoretical conclusions based on these facts should be confirmed, if possible, by experiments.

The best choice for comparative study of continental and oceanic magmatic rocks are associations of basaltic rocks, which are widely spread on continents, practically make up near completely the oceanic crust and are in good accord with principles of similarity and equal study. Manifestations of ultrabasic, andesitic and acid magmatism, especially of high alkalinity are rare in morphostructures of the oceanic floor, contrary to continental regions, and there is possibility that in

Proceedings of the 27th International
Geological Congress, Volume 9, pp. 137–149
PETROLOGY (IGNEOUS AND METAMORPHIC ROCKS)

some cases acid, andesitic and alkalic rocks, which occur mainly on oceanic islands, arc genetically connected with remnants of the sialic crust in the basement of these islands (Kergelen, Yan-Mayen, Gulf of Guinea).

The similarity of continental and oceanic basaltic associations stems primerly from their belonging to derivates of picritic melts, generated in both cases during partial melting of the upper mantle (11, 22).

Basalts have been studed more thoroughly than other oceanic magmatic rocks and nevertheless our knowledge about regularities of basaltic magmatism in the oceanic lithosphere is essentially less comprehensive in comparison with available information on genesis of continentla basaltic associations.

Oceanic basalts were studed less due to their occurence under thic layer of sea-water and sediments, so that only expensive deep-sea drilling, as well as dradging in trenches and tectonic fractures, representing insignificant part of the oceanic floor, provide possibility for the direct study of solid oceanic crust. As to notions about tectonic patterns and particularly about deep-seated structure of the oceanic crust, they are based, as a rule, mostly on geophysical data. Difficulties in the study of oceanic floor basalts can be compared with those, that hinder investigations of continental platform basements under the thick sedimentary cover.

The continental baslats have been better investigated not only because of their easier accessibility, but also due to their formation during billions years of the geologically documented history of the Earth, under various geodynamic regimes, including the regime of compression and folding. In the oceanic regions, on the other hand, there are basalts not older then 200 m.y., which were formed, with small exceptions, under the geodynamic regime of extension.

Unequal wealth of facts about continetal and oceanic basalts can be partly compensated by use of such instrument of study as a method of analogy. Reasoning by analogy means, that if two objects under comparison have a numb4r of similar characteristics, ther is fair chance for coincidence of other parameters, which are known only for one of objects. The method of analogy was successfully used for evaluation of ore-bearing potential of the oceanic floor basaltic associations. On the basis of general petrological and geochemical similarity of continental and oceanic basaltic rocks, it was predicted, that ore-deposits of chromites, Pt-minerals, sulfides, titanomagnetite, which were known in relation with basic igneous rocks of continental platforms, might as well associate with basaltic rocks in oceanic lithosphere (12, 32). Results of study recently discovered modern mettalliferous hydrothermal systmes in mid-oceanic ridges (31) prove, that such srognosises made by analogy can be sound and fruitful.

Composition and structure of basaltic association under comparison are imprinted by a number of factors and significance of each factor become less obvious with increasing of their quantity. Among thes factors are: a) stage of evolution and regional peculiarities of the mantle, which produced parental magmas of basalts; b) depth of magma generation and crystallization; c) degree of differentiation of melts and their interaction with country-rocks, d) geodynamic regime, speed and time of magma ascent in the upper parts of the Earth's crust.

To reveal the most important differences between oceanic and continental basalts it is necessary to compare associations of these rocks, which were formed under conditions of the closest recemlance.

First of all it is necessary to select continental and oceanic basaltic associations, which were formed in same geodynamic environment. Manifestations of magmatism in the oceanic lithosphere are known along constructive boundaries of lithospheric plates (mid-oceanic ridges, rift zones) and transform faults, which are formed and evolved under clearly prevailing regime of extension of oceanic crust above ascending masses of the heated mantle material. This model agrees with such features of magma-active oceanic morphostructures as lower density of the upper mantle, reduced thickness of the crust, high values of the heat-flow and large elevations of the oceanic floor in their confines.

The areal basaltic magmatism on continental platforms (traps, flood-basalts) manifests itself under similar conditions (4, 17) and it is not surprisingly, that mid-oceanic ridges and continental rift-zones sometimes are regarded as morphostructures similar in their formation and geodynamic regime (13). According to geophysical parameters the nearest equivalents of mid-oceanic ridges are such areas of recent basaltic magmatism on continents as province of Deccan traps (105-100 to 45-42 m.y.) on Indian platform and basalts of Colombia and snake river-basins in the western part of North America, where the last eruptions of basaltic lavas had occured only 2100 years ago. In these regions, like in magma-active zones of oceanic lithosphere, the thickness of the crust is lower, upper mantle is less dense, heat flow is anomalously high and there are large fallouts in the consolidated basement under basalts (1, 21, 28, 33).

Even more perfect analogy, from the tectonic point of view, exists between continental flatforms and oceanic plates beyond mid-oceanic ridges and continental environs. Due to this similarity these segments of oceanic lithosphere are discribed as oceanic platforms or talassocrations.

Thus, it is possible to conclude, (15) that geostructural and geodynamic condition of basaltic magmatism in morphostructures of the oceanic floor and on continental flatforms were positively similar.

The reemblance between basaltic associations of continental platforms and oceanic floor is also result of their formation from the same types of basaltic magmas: theleiitic (normal alkalinity), subalkalic and alkalic. Tholeiitic basalts occur as very extensive trap-fields on continental platforms and they near entirelly constitute oceanic floor under the sedimentary layer. Subalkalic and alkalic basalts on continental platforms usually gravitate to deep-seated faults on the periphery of trap-provinces and to rift-zones. On the oceanic floor, subalkalic and alkalic basalts occur in transform fault zones and on oceanic islands, which were controlled in their distriburion mainly by fracture tectonics of the oceanic crust.

To avoid masking of oceanic or continental environment influence on coposition of basalts, by differences in chemistry of parental magmas (tholeiitic, subalkalic, alkalic) it is necessary to compare only derivates of identical basaltic melts.

Study of the evolution of magmatism in the Earth's history reveals, that this process has definite trend and is irreverable (20). For example, komatiite-basaltic magmatism declined, wheras calc-alkaline and alkalic rock - associations became more and more wide-spread in the course of geological time with decreasing energy potential of the Earth, increasing deficit of incompatible, volatile and riogenic elements in the upper mantle and growing thickness and heterogeneity of the Earth's crust. Comparative analysis of continental magmatic associations of the same type, but of different age, also reveals definite trends in variations of their petrochemistry during successive epochs of magmatism. It was discovered (4), that evolution of trap-magmatism on northern ancient platforms (Siberian, East European), whereas on southern platforms (African, Indian, South-American) average compositions of traps from proterozoic formations became more and more ferruginous (homodrome trend).

The differences in composition of magmatic rocks of the same type but of different age were also revealed during petrological investigations in continental rifts. This differences, as it seems don't reflect regional conditions of magmatism. For example, in the thoroughly studed permian graben Oslo (7) there are tholeiitic, subalkalic and alkalic effusive rocks in association with monzonites syenites and alkalic granites. Petrochemical parameters of these rocks are very similar to corresponding rocks of the East-African rift-zone. In much more older (2000± 100 m.y.), early proterozoic rift system of the Baltic shield (16) extrusives are represented by rocks of picrite-basaltic and basalt-trachytic series, whereas differentiation of intrusions of tholeiitic baslatic magma resulted in formation of lange stratified peridotite-pyroxenite-anorthosite bodies (Monchetundra, Kivakka, Kemisuchalko) and differentiated peridotite-pyroxenite-gabbro-nepheline syenite intrusives

(Gremyacha, Virmes, Elatosero), were formed from basaltic magma with alkaline affinity. By their petrochemical characteristics these rocks are similar to magmatic rock associations of Ethiopian and Afar rifts.

In contrast to paleozoic graben Oslo, there are no independant alkalic rock series in precambrian rift-system of the Baltic Shield (East-European platform) and nepheline syenites here are differentiates of Subalkalic basaltic magma. The same situation is common for other precambrian rift-zones. In phanerozoic rifts derivates of subalkalic basaltic magma are relatively less abundant among igneous rock associations.

Alkalic rocks are totally unknown and subalkalic rocks are rare in the oldest continental rift-zones - archean greenstone belts.

Thus, the characteristic feature of magmatic evolution in continental rift systems is the progressively increasing role of subalkalic and alkalic rocks in igneous rock associations formed in these morphostructures from precambrian to junger ones.

All these facts cause the conclusion, that it is necessary to compare associations of oceanic and continental magmatic rocks, which were formed close in time. This requirement restricts possibilities to correlate oceanic and continental igneous rocks because there are no oceanic baslats older than Jurassic. Orphyolites, which are usually regarded as relics of the old oceanic curst, often are strongly altered by secondary processes and, possibly, were formed not only in typical oceanic environment, but also in zones of interaction between oceanic and continental lithospheric plated, on so called active oceanic margins. Ophiolitic complexes formed in such conditions have more or less obvious "island arc" features (19).

To comply with requirment to compare oceanic and continental magmatic association of the similar are means to use for comparison oceanic tholeiitic basalts and postpaleozoic traps of continental platforms on the one hand as well as oceanic subalkalic and alkalic basalts along with analogues, by type and age, basalts of continental rifts on the other hand.

When selecting associations of continental and oceanic

magmatic rocks for comparative study, it is important, in addition to criteria of similarity already mentioned, to choose associations belonging to the same petrochemical province, which are known on continents (10) and in oceanic regions (3).

The comparative study of oceanic and continental magmatic associations gains the best results when these associations are adjacent in space or were formed inside one magma-active zone, as it is assumed for mezo-cenozoic traps on the western periphery of the Indian platform adn adjoining parts of the Indian ocean (30).

The observance of similarity requirements directed by comparative study methodology means also the necessity to compare rock associations belonging to the same facial fype. Extrusive rocks should be compared with extrusive rocks and intrusibe rocks with intrusive ones. It is essential to respect this recommendation because even average compositions of extrusive and intrusive rocks belonging to the same basaltic formation sometimes differ noticeably.

Taking into account all discussed principles of the comparative study of oceanic adn continental magmatic rocks, let us compare traps of continental platforms with tholeiitic basalts of the oceanic floor and magmatic rocks of continental rifts with analogues rocks of oceanic islands. This comparison is aimed, as was mentioned, to reveal specific features of magmatic rock associations, reflecting their formation in the continental or oceanic lithosphere.

The comparative petrological study of oceanic basalts and basalts belonging to trappean formations of continental platforms shows, that oceanic basalts are less crystallized rocks, with lower content of olivine, but with higher concentration of magnesium in this mineral. The plagioclase in oceanic basalts contains more anorthite component and these rocks have lower content of Fe-Ti oxide ore minerals.

All these features are evidences of higher Mg-content in the initial melts of oceanic basalts, as well as lower concentration of volatiles in them.

Comparison of oceanic and continental basalts on the basis of geochemical and petrochemical data supports this conclusion, showing higher content of MgO, CaOm, TiO_2 and lower concentrations of K, P, Cs, Rb, Y, Sr, Ba, U, Th in tholeiitic, subalkalic and alkalic oceanic basalts (9, 26). According to some petrochemical parameters, oceanic basalts have more uniform composition, than their continental analogous. For example, variations of Fe_2O_3/Fe_2O_3+FeO and K_2O/K_2O+Na_2O ratios in oceanic basalts are essentially smaller, than in continental basalts. Oceanic and continental basalts have also different isotopic characteristics. Average $^{87}Sr/^{86}Sr$ ratio in basalts of oceanic floor is ~0.7026, whereas in continental basalts this ratio is ~0.705 (5).

There are differences in variations of some petrochemical parameters during differnetiation of parental magmas of oceanic and continental tholeiitic basalt high-alumina variations (Al_2O_3/Fe_2O_3+FeO+Mg $\geqslant$1) belong in the same time to the rocks with higher Mg-content. During differentiation, with relative enrichment of melts by iron (FeO'/FeO+MgO), content of Al_2O_3 in the sequence of differentiates decreases (24). On the contrary in differntiated series of tholeiitic basaltic rocks of continental platforms, especially in intrusive rock series, the content of Al_2O_3 increases simultaneously with rise of Fe-concentration and the late differentiates, such as leucobasalts, belong to the most high-alumina members of the series. (6).

Derivates of subalkalic basaltic magma (ankaramites, olivine basalts, hawaiites, mugearites, trachyandesites, trachytes) and alkalic basaltic magma (ankaratrites, nephelenites, olivine tephrites, phonolites, carbonatites) occur on oceanic islands and seamounts. These rocks associate with tholeiitic basalts of transitional type, which differ from typical oceanic tholeiitic basalts by somewhat higher content of alkalies and titanium. Differentiation of the corresponding parental magma results in formation of such series as picrites - olivine basalts - ferro-basalts - rhyodacites and rhyolites.

Associations of subalkalic and alkalic rocks of oceanic islands and seamounts in general are very close to associations of the same type in continental rifts and existing differences are less pronounced than in the case of oceanic and continental tholeiitic basalts. The most obvious differences in manifestations of subalkalic and alkalic magmatism in continental and oceanic regions are represented by different ratios of certain rock-types and their complexes in magmatic associations.

It is well known, that subalkalic and alkalic rocks play clearly subordinate role among volcanics of oceanic islands and seamounts, whereas in continental rifts (Baikalean, Central European, East African) rocks of the high alkalinity are domineering. Ultra-alkalic rocks ($Na_2O+K_2O>10-12\%$) are considerably less common in oceanic regions than on continents and they occur, as a rule, on islands with submerged continental crust in the basement (islands of Gulf of Guinea, Trindade, Fernando de Noronha). High potassium alkalic ricks (Katungites, mafurites, leucitites) are practically unknown on oceanic islands and, on the contrary, are quite common in the areas of alkaline magmatism on continental platforms.

The special interest fo, comparative study, as already was noted, present cases, when manifestations of magmatism of the same type in oceanic and continental morphostructures are directly related in space.

Such relations exist between subalkalic and alkalic associations of magmatic rocks along, so called, "Cameroun Line", which was traced by chain of alkalic volcanoes on islands of Gulf of Guinea and father to the East on African continent. The major characteristics of magmatic association remain unchanged along this line, but in the continental segment of the zone pantellerites appear in association with subalkalic basalts (2, 29).

By petrochemical and geochemical parameters, including data on Sr and Nd isotopes, as well as on rare earth abundances, rocks of tholeiitic, subalkalic and alkalic series of oceanic islands and continental rifts are closely resemble each other (18) and only chemically less variable composition of oceanic rocks distinguish them form their continental counterparts.

The comparative study of magmatic associations of oceanic floor and continental platforms shows, that specific features of the magmetism in sialic curst are represented by: a) more prominent role of alkalic associations, especially high-potassium, b) more extensive occurence of andesitic and acid rocks, c) lower basicity of basalts and their enrichment by large-ionic, lithophile elements.

All these peculiarities can be explained and are explained by three main reasons: a) different composition of upper mantle in continental and oceanic segments of the Earth's lithosphere (5), b) different depths and, consequently, different thermodynamic conditions of magma generation under continents and oceans (8, 27), c) interaction between mantle melts and continental sialic crust (14, 25).

There are little doubts, that compositions of magmas and, as result, of igneous rocks are controlled by all these factors acting together, but it is important to find out, which one is key-factor.

Essentially greater thickness of the continental crust and less heated state of the lithosphere, especially in the frames of ancient shields, have to cause the melting of the upper mantle under higher pressures and lower temperatures, with generation of magmas enriched by alkalies and other incompatible elements. In magma-active oceanic zones, the upper mantle melts at considerably more shallow levels, that even under similar thermal conditions leads to the higher degree of the upper mantle partial melting and, correspondingly, to generation of more magnesium melts.

It is necessary to stress the importance of contamination of mantle magmas by material of the sialic crust in genesis of continental igneous rocks. The reality and signigicance of this process were proved by petrographical and petrochemical data (4)m as well as by special isotopic studies (23, 30).

There are no doubt at present, that the comparative study of continental and oceanic magmatic rock associations is and will be one of the most effective scientific means to reveal general regularities of manifestations and evolution of magatism on our planet.

REFERENCES

1. Witte, L.V., Fotiady, E.E., 1976. Genetic classification of the continental Earth's crust. Geol. and Geophys.,9: 15-22 (In Russian).
2. Borodin, L.S. (ed), 1974. Main provinces and formations of alkalic rocks: Moscow (Nauka), pp.1-376. (In Russian).
3. Dmitriev, L.V., Sobolev, A.V., Sushevskaya, N.A., 1979. Conditions of generation of the initial melt of oceanic tholeiites and variations of its composition. Geochimiya, 2: 163-178. (In Russian).
4. Dmitriev, Yu.I., 1983. Traps and mobilism. In problems of petrology, mineralogy and ore genesis: Moscow (Nauka), pp. 100-106. (In Russian).
5. Yoder, H., 1979. Genesis of basaltic magma: Moscow (Mir), pp. 1-238. (In Russian).
6. Classification and nomenclature of magmatic rocks, 1981: Moscow, (Nedra), pp. 1-159. (In Russian).
7. Continental rifts, 1981: Moscow (Mir), pp. 1-483. (In Russian).
8. Lutz, B.G., 1980. Geochemistry of Oceanic and continental magmatism: Moscow (Nedra), pp. 1-247. (In Russian).
9. Marakushev, A.A., Frolova, T.I., Yakovleva, E.B., Emeljanenko, P.F., 1977. Magmatism and the Earth's crust. Vestn. MGU, Ser. geol. 1:3-29. (In Russian).
10. Masaitis, V.L., 1964. Magmatic trappean provinces on Siberian platform. Tr. Vses. Nauchn.-issled.geol. instit., nov.ser., 97: 137-158. (In Russian).
11. Ryabchikov, I.D., Bogatikov, O.A., Piloyan, G.O., Babansky, A.D., 1983. Process of generation of oceanic tholeiitic magmas. In magmatic and metamorphic rocks of the oceanic floor and their genesis : Moscow (Nauka), pp. 17-24 (In Russian).
12. Smirnov, V.I., 1975. On metallogeny of the ocean. Geol.rudn. mestorozhd., v.17, 1: 3-13 (In Russian).
13. Beloysov, V.V., (ed), 1978 Tectonosphere of the Earth: Moscow (Nauka), pp. 1-531. (In Russian).
14. Frolova, T.I., Perchuk, L.L., 1982. Volcanicity of oceanic islands and island arcs. Vest. MGU, Ser. Geol., 6: 27-39. (In Russian).
15. Khain, V.E. 1973. General geotectonics. Moscow (Nedra) pp. 1-514. (In Russian).

16. Sharkov, E.V., 1984. Early proterozoic continental rift magmatism of the Karelo-Cole region. Geotectonica, 3 (In press). (In Russian).
17. Allegri C.I., Dupre P.R., Rousseau D., Brooks C. Subcontinental versus suboceanic mantle, II. Nd-Sr-Pb isotopic comarisons of continental tholeiitic with mid-ocean ridge tholeiites and the structure of the continental lithosphere.- Earth Planet.Sci.Lett., 1982,57,25-34.
18. Basaltic volcanism of the Terrestrial Planets (Basaltic Volcanism Study Project 1976-1979) Pergamon Press, 1981.
19. Beccaluva A. ophiolites and actualism: petrological constrains. -Ofioliti, 1982,v.7, N 213, p.109-112.
20. Bogatikov O., borsuk A., Dmitriev Yu., Kovalenko V., Ryabchikov I. The evolution of Magmatism in the Earth's History.- Geolisky Zbornik, Geologia Carpatica, 34, 1, Bratislava, 1983, p.3-13.
21. Bose M.K. Deccan basalts.- Lithos, 1972, v.5, N2, p.131-145.
22. Cox K.G. A model for flood basalt volcanism.- J. Petrol., 1980, 21, 629-650.
23. De Paolo D.S. Trace element and isotopic effects of combined wall rock assimilation and gractional crystallization.- Earth Planet. Sci.Lett., 1981, 53, 189-202.
24. Dmitriev Yu.I. Basalts of the Middle America trench (Guatamala trensect) and their correlation with oceanic and island Arc tholeiites.- In Anbouin I. von Huene et al. Initial Reports of the Deep Sea Drilling Project, 1982, N 67, Washington, p.577-584.
25. Green D.H. The origin of basaltic and nephelinitic magmas in the Earth's mantle.- Tectonophysics, 1969, v.7, N 5-6.
26. Hamilton W. columbia River Basalt in the Rigyind Quardrangle Western Idaho.- Geol.Surv.Bull., 1963, 1141-L.
27. Kushiro I., Kuno H. Origin of Pimary Basalt Magmas and Classification of Basaltic Rocks.- J. of Petrology, 1963, v.4, N 1.
28. La Fehr T.R., Pakiser L.C. Gravity, volcanism and crustal deformation in the Eastern Snake River Plain, Idaho. U.S. Geol.Surv. Profess.Paper, 45-D, 1962, p.76-78.
29. Liotard I.M., Dipny C., Dostal I., Cornen G. Geochemistry of the volcanic island of Annobon, Gulf of

Guinea.- Chem.Geol. 1982, v.35, N 1/2,p.115-128.
30. Mahoney I., Macdougall I.D., Lugmair C.W., Murali V.V., Sankar Dal.M., Gopalen K. Origin of the Deccan Trap flows at Mahabaleshwar inferred from Nd and Sr isotopic and chemical evidence.- Earth and Planet. Sci. Lett., 1982, 60, 47-60.
31. Mottl M.I. Metabasalts, acial hot springs, and the structure of hydrothermal systems at mid-oceanic ridges.- Geol.Soc. Am.Bull., 1983, v.94, p.161-180.
32. Schrader Ed.L., Furbish W.I. Mineralogy and geochemistry of ore minerals from basalts related to spreading centers of the Eastern pacific with speculations on ore-forming processes.- In Rosendahl B.R., Hekinian R. et al. Initial Reports of the Deep Sea Drilling Project, 1980, v.54, Washington, p.789-806.
33. Thompson R.N. Primary basalts and magma genesis II. Snake River Plain, Idaho, USA. Contrib. Mineral. Petrol., 1975, N 3, v.52, p.213-232.

EXPERIMENTAL INVESTIGATION OF THE PHYSICO-CHEMICAL CONDITIONS OF THE METASOMATIC PROCESSES

ZHARIKOV V.A., IVANOV I.P., ZARAISKY G.P.
Institute of Experimental Mineralogy, USSR Academy of Sciences. Chernogolovka, Moscow Region, USSR.

Statement of the problem and Experimental approaches.

Metasomatic processes produce such alterations in wall rocks by the agency of hydrothermal solutions which involve changes in the chemistry of the rocks. In the general case, interaction between solutions and rocks will result in a metasomatic column, i.e. the total of simultaneously formed zones, each having its own set of thermodynamic parameters of state. the number and amount of inert components, the intensive parameters of temperature, pressure, chemical potentials (activity) of perfectly mobile components (16,17,2,4,8). It has been shown that zoning in rocks is due to a consequence of the component differential mobility which shows up in the successive transition of the components from the inert to a perfectly mobile state, as metasomatic alterations proceed. This transition (a change in the thermocynamic regime) initiates mineral exchange reactions by which phase boundaries develop. The general trend is a decrease in the number of minerals with increasing intensity of the process to the point where, under certain circumstances, metasomatism may result in the development of monomineralic zones. The occurence of several mineral reactions at the alteration front will bring about a metasomatic column featuring additional zones composed of the "reaction' minerals (this type of zoning can prevail in some of the metasomatic type).

The total metasomatites formed in a metasomatic column under constant parameters or factors of state häs a common name a metasomatic facies, while those developed under different parameters of sate, i.e. the total of metasomatic facies make up a metasomatic formation (1,8,9). As many

Proceedings of the 27th International
Geological Congress, Volume 9, pp. 151–180
PETROLOGY (IGNEOUS AND METAMORPHIC ROCKS)

as about forty metasomatic formations have been recognized (9)-distinctly defend natural groups of metasomatic facies which are characterized by a definite common mineral assamblage and a definite common geological setting.

The experimental research proceeded along two major lines based on the above characteristics of metasomatizm:

a) The mineral equilibrium studies on systems with perfectly mobile components. The run conditions provide the definite meaning og the state parameters: the masses of inert components and intensive parameters - the temperature, pressure, activity of perfectly mobile components. The stable mineral assemblages in these experiments represent one zone in a metasomatic column. By changing the intensive parameters, facial varieties of parageneses were obtained for some definite zone (containing the same inert components) in different metasomatic columns occurring in various metasomatic facies or formations. Also, by changing the number of inert components, while holding the intensive parameters constent, it is possible to obtain the whole metasomatic column, although the mineral equilibrium method does not seem particularly helpfull in this respect. The alternative experimental approuch has proved more effective.

b) Direct modeling of metasotic zoning at definite intensive parameters (the temperature, pressure, concentration and the activity of components in the reacting solution). By varying the compositiong of the sample rocks and the intensive parameters for the system it is possible to determine the boundary conditions for different metasomatic facies and formations.

Both approaches have been endeavoured at the Institute of Experimental Mineralogy USSR Acad.Sci. in a reasonable combination, because each of the methods has its advantages and limitations. Laborious experimental studies were paralleled by the development of mathematical metasomatic modeling for both equilibrium heterogeneous systems and metasomatic zoning, to obtain preliminary and additional information, including that which would not be available from experiment, on the physico-chemical conditions of metasomatism. The combination of experimental and mathematical modeling has proved effective in solving many of the complex metasomatic problems.

Experimental study of mineral equilibria in systems with perfectly mobile components.

The experimental studies of mineral equilibria in systems with perfectly mobile components are most close in method to the mineral equilibrium investigations. The major difference is that in the present experiments the activity (concentration) of perfectly mobile components was set independent of the mineral reactions in the system. Hence, the activity of perfectly mobile components became an independent intensive parameter of the system and defined their thermodynamic regime. Changes in the activities of perfectly mobile components (from run to run) may result in the complete replacement of one mineral equilibrium by another, as do variations in other intensive parameters (e.g. temperature, pressure).

The regime of perfectly mobile components can be achieved in several ways:

a) by introducing a solid buffer, whose heterogeneous reaction keeps the activity of a perfectly mobile component constant (for given T and P). The first solid buffers were suggested (20) by Eugster ("oxygen buffers") and later he and other scientists developed solid buffers to define the activities of HCl, HF and other components;

b) by introducing a buffer solution which contains perfectly mobile components in amounts for greater than the mineral system would be able to take, and thereby maintain (at given T and P) the activities of perfectly mobile components constant throughout the runs. The methods that have been applied are: the "big Capsule" and the "reactor -cold buffer". In the first method, the buffer solution at a given concentration of perfectly mobile components (normally, the mass ratio of perfectly mobile components in the solution and run sample of the mineral system should be greater than 10:1, to 50:1) is placed into the capsule at the T, P of the run so that it either touches the sample or is separated from it by a membrane permeable to the solution. A major difficulty in this method, although technically simple and widely used, is that the solubility of the mineral system becomes impermissibly high for large volumes of the buffer solution. To promote mineral reactions, which are slow at temperatures below 400°C, an electromagnetic initiator has been desighned.

In the second more complicated mothod, the capsule containing the sample and a small volume of the buffer solution is contacting a larger volume of the solution loaded into the cold zone. The temperature gradient on the boundary between the capsule and cold zone is sort of a membrane which is only permeable to perfectly mobile components. This method requires data on the thermodiffusional partitioning of the perfectly mobile components between the "hot" capsule and "cold" buffer and hence the knowledge of the specific patterns of this partitioning and their accurate determination which does presents certain difficulties.

It is no simple experimental task - to study mineral equilibria as a function of perfectly mobile components which in addition to H_2O, CO_2, O_2 can include potassium, sodium, calcium etc. That is probably why only one system, Al_2O_3-SiO_2-H_2O-HCl-KCl-NaCl, has been studied in detail. Some of the reactions within this system were studied by Hemley(22,23,24,25). We have carried out careful study of the potassium portion of the system (10.13) over the temperature range 200-600°C, the KCl/HCl activity from lga_{KCl}/a_{HCl} 0 to 4 and P=100MPa (1kb). The major boundary reactions between the potassium and sodium portions of the system were also studied of the temperature and activity of the relevant virtual perfectly mobile components: albite + HCl + KCl = muscovite = quartz + NaCl; albite + + KCl = potassium feldspar + NaCl; albite + potassium feldspar + HCl = quartz + muscovite + KCl + NaCl.
And finally, the equilibrium between quartz and the alumosilicates from the rear zone metasomatites-andalusite, pyrophylite and kaolinite were investigated in the coordinates temperature - $AlCl_3$/HCl activities.

The position of the equilibrium was determined by the shift of a monovariant reaction approached from both sides-i.e. by the conventional method, and also by the special PH-indication method which is to measure accurately changes of the acid concentration in the quench solution relative to the starting concentration (also from both sides).

To obtain preliminary estimates of the experimental conditions, and particularly the compositions of equilibrium solutions and quench solution characteristics, a mathematical model of the experiment was calculated on a computer using programs compiled by Karpw (15) and Shvaro (19). As in the above experiments, the model system was formed and the calculated amounts of minerals from the specified assemblage were introduced into starting aqueous salt solution for the given T and P. The equilibrium in the monovariant hydrolysis reaction was also calculated from the both sides. The preliminary information on the equilibrium parameters was used to define the experimental and theoretical models. The computer has made it possible to extrapolate accurately experimental data into the low+temperature region (to 25°C), where direct experiments on hydrolysis reactions can not be accomplished because of the extremely low reaction rates.

The results for the system Al_2O_3-SiO_2-H_2O-HCl-KCl-NaCl are presented in figs 1, 2, 3. The complete experimental diagram for the system Al_2O_3-H_2O-HCl-KCl was published earlier (7). In fig. 1 the new results for the position of the major hydrolysis reactions are plotted on a temperature versus KCl/HCl activity (P=100MPa) diagram:

(6) 1.5 kaolinite + KCl = muscovite + HCl + 1.5 H_2O
(7) 1.5 pyrophyllite + KCl = muscovite + 3 quartz + HCl
(8) 1.5 andalusite + 1.5 quartz + KCl + 1.5 H_2O = muscovite
(9) muscovite + 3 quartz + KCl = 1.5 potassium feldspar + + HCl.

The equilibrium was determined by the monovariant reaction and pH-metry methods. The monovariant lines differ insignificantly from those published earlier, the difference being notable only in the low-temperature part of the line kaolinite - muscovite and is 0.5 0.6 (lg kcl) (HCL) in the high-value region.

Fig. 2a presents the experimental results for reaction (17) : albite + potassium feldspar + HCL = 0,5 muscovite + 3 quarts + NaCl. This reaction with potassium buffer (potassium is an inert component) attains equillibrium depending on the temperature and NaCl/HCL activities (P = 100MPa).

In fig.2b the equilibrium for the reaction 1.5 albite + HCL + 0.5 KCL = 0.5 muscovite + 3 quartz + 1.5 NaCl is plotted on th4 T-lgKp diagram where Kp is the concentrational equilibrium constant of the reaction. The parameters of the reaction define the boundary between the metasomatites from the potassium and sodium series in alumosilicate rocks. Also presented in this plot is the reaction equilibrium extrapolated toward low tempa-ratures by a computer from the thermodynamic data for the minerals and solution components which were checked against the experimentally-determined position of the reaction equilibrium.

Fig.3 is the experimentally-determined quartz stability field, bounded by the reactions:

(1) 0.5 kaolinite + 3 HCL quartz + $AlCl_3$ + 2.5 H_2O

(2) 0.5 pyrophylite + 3 HCL = 2 quartz + $AlCl_3$ + 2 H_2O

(3) 0.5 andalusite + 3 HCL = 0.5 quartz + $AlCl_3$ + 1.5 H_2O

It is seen readily that the quartz stability field deli-niated by the incongruent dissolution reactions of kaoli-nite, prophyllite and andalusite falls within the range of acid and ultra-acid solutions with increasing tem-perature, the boundaries of this field displace towards higher HCL concentrations. The slope of the curves suggests that the solution is enriched in the hydrated forms of aluminium ($Al(OH)_3$ $Al(OH)_2^+$). This inference has been confirmed in thermodynamic calculations: $C_{Al(OH)3/3} = n.10^{-7}, n.10^{-5}, n.10^{-4}$ m/l for 200, 350 and 450°C, respectively; and $C_{Al(OH)_2^+} = n.10^{-11}, n.10^{-8}$, $n.10^{-7}$ m/l for the same temperatures.

Using the present experimental results, data available in the literature and thermodynamic data for minerals and solution components, the complete mineral equilibrium diagrams have been calculated for the system Al_2O_3-SiO_2-H_2O-HCL-KCL-NaCl for the temperature and pressure range of 100-600°C and P = 25-100MPa as a function of the potassium, sodium and hydrogen ion activities.

All the boundary hydrolysis reactions that were studies in this system are presented in Table 1. The complete topological analysis of the system (which was performed prior to a systematic experimental investigation) shows that all the important equilibria have been covered by the experiment. The table gives the hydrolysis reaction constants in dependence of temperature (saturated vapour pressure) for the temperature range 25-360°C (for which reliable thermodynamic data are available).

Fig.4 presents equilibrium phase diagrams for the system. Fig.4a plots three-mineral parageneses (Si is an excess component, Al, K-inert components) versus temperature and activity ratios of sodium and hydrogen ions. Fig.4b is the same ternary diagram but for the case of inert sodium (excess Si, Al is an inert component as in Fig.4a). The boundary reactions in the diagrams separate the temperature facies (hydration reactions) from acidity facies (hydrolysis reactions) or alkali-metal facies (at constant pH) of metasomatites. The phase equilibrium diagrams (to the richt of every diagram) present mineral parageneses for each of the diagram fields which are stable at certain T, lg $(Na^{+})/(H^{+})$ and lg $(K^{+})/(H^{+})$.

Fig. 5 plots the mineral equilibria in the system Al-Si--watersaline solution (with excess quartz) as function of the activity ratios $(Na^{+})/(H^{+})$ and $(K^{+})/(H^{+})$. The diagrams 5a and 5b are for 300°C and 200°C, respectively. Solid, dashes and dotted lines represent different alkali chloride concentration in the solution (see figure captions). These diagrams are also based on the above experimental results and thermodynamic data (wholly reliable to 300°C). The experimental results combined with thermodynamic data make it possible a) to derive all possible types of metasomatic zoning, as a function of temperature, acicity, relative thermodynamic mobility and activity of potassium and sodium in the solutions; and b) to quantify these types and also the solution in equilibrium with each of the metasomatic column zones. Table 2 presents to the major types of metasomatic columns and their boundary conditions (the temperature range, type of "components differential mobility", relationships for boundary reaction constants). As an example, Fig.6 presents the composition of a solution in equilibrium with column III-A, at T = 300°C, $C_{KCl} = 10^{-1}$m, $C_{HCl} = 10^{-4}$M. Without going into much detail, it will be noted that the solutions are enriched in silica (10^{-2}M), which occurs, like alumina, as an nondissociated molecule, and the general tendency for all the components is to occur in increasing concentration of nondissociated molecules.

Experimental modeling of metasomatic zoning

Consider now the results from the study on metasomatism of silicate rocks by experimental modeling of metasomatic zonation. This new line of research has been activly forced since 1966 at the Institute of Experimental mineralogy, USSR Acad. Sci.; as for other laboratories, no more than some beginnings have so far been made in this direction. It seems relevant to mention some of the studies on metasomatism of alumosilicate rocks: experiments by Kazitsin et all 14 on the reprocudtion of orgillisation, a study on alkali metasomatism in schists and granites, a series of experimental studies on bimetasomatic interaction of carbonate, iron-magnesian-silicate and alumosilicate rocks (see review in 5,26).

In the present report, we will only consider the results obtained at this Institute on the modeling of processes occuring in wall rock metasomatism of igneous rocks, closely resembling the above mineral equilibria in metasomatites from the system K_2O-Na_2O-Al_2O_3-SiO_2 - aqueous solution.

In most runs, diffusion technique was used (fig. 7): finelyground rocks (particle size <0.1-0.2mm) were packed into platinum or gold capsules (d=5 mm, l=40 and 50 mm), which were placed vertically into hermetically sealed containers of 150 cm^3. The containers were filled with the reaction solution, mineral buffers and other required components (solid carbon dioxide and oth.). The experiments were run in autoclaves under pressures of 50, 100, 200 to 400 MPa, accurate to better than 5 % and at temperatures to 600^oC, accurate to $\pm$ 5^oC.

A relatively constant composition of the solution (and the activity of perfecthly mobile components)was achieved through a large buffer volume (the ratio solution: solid $\approx$100:1).

The solution reacted with the starting rock through the opening in the capsule and penetrated into it by diffusion. The outcoming diffusion flux carried rock components. As the result, the homogeneous starting rock in the capsule has transformed into a column of metasomatic zones which recorded the changes in the starting rock due to interaction with the solution in the buffer volume acting sort of environment to exert influence on the starting rock. Fig.7 presents photomicrographs for some of more distinct experimental metasomatic columns. By varying the factors of state - composition of starting rocks, composition and activity of species in the solution, temperature and pressure, - all the major types (formations) of wall rock metasomatites have been obtained: quartz-feldspar, quartz-siricite, quartz-tourmaline metasomatites, greizens, gumbeites, beresites, propylites, argillisites, alkali metasomatites etc.

It was also possible to extablish the major factors which define the replacement of one formation by another. The boundary conditions for some of the formations have been detailed.

Fig.8 presents a comprehensive diagram for the major types of metasomatic columns developed at T=500°C and P=100MPa from the above equilibria in the system Al_2O_3-SiO_2-H_2O-HCl-KCl. In other words, the composition of the reacting solution correspondend to the equilibria in the rear bimeneralic zones (see diagrams in Fig.4 and 5). The richt-hand diagram in Fig.8 shows the columns produced with excess silica (the buffer solution contained excess quartz), the left-hand diagram presents the columns which were formed from solutions undersaturated in silica. For a better presentation, both the left and richt diagrams are given in mirror projection, and the coordinates of the origin X (distance along a column) are somewhat displaced so that the ordinate which represents lg(HCl)~ lg(KCl)/(HCl) is located in the middle and represents the solution.

It is clearly seen in the diagram tha the structure of metasomatic columns changes with the acicity and potassium activity in the fluid. Highly acicic environments (solutions of 1.0M 0.1M HCl and 0.01M KCl/0.1M HCl) give rise to metasomatic columns which correspond to secondary quartzites. At increasing KCl:HCl, "greyzen" columns develop to be later changed by columns of the quartz-potassium feldspar metasomite type. The boundary between the lather is at lg((KCl)/(HCl))$\approx$1.0 (at T = 500°C and P = 100 MPa). When the solution is undersaturated with silica-saturated solutions the rear zone is composed of alumosilicates, whereas with silica-saturated solutions the rear zone is invariably quartz-either mono- or bineralic, which is a characteristic feature of metasomatites formed at the acidic stage. And finally, at high (KCl)/(HCl), potassium feldspar metasomatites are formed, not necessarily in the presence of alkaline solutions:

at high KCl concentration the solution may be acidic. A very important feature has been brought out in experiments by vaying the potassium acidity and activity in the solution it is possible to obtain the major varieties of apogranitoid metasomatites.

The structure of experimental columns reflects the typical zoning in secondary quartzites, greyzens and potassium feldspar metasomatites. The more rare types such as topaz greyzens and tourmaline metasomatites were also obtained experimentally with more complex solutions.

The topaz greyzen was obtained from fluoride solutions ($T=500^{o}C$, $P=200MPa$, the solution of 0.1MHF + excess SiO_2 (solid), Al_2O_3(sold) as a metasomatic column of a typical structure:

Q + Tz|Q + Tz + Ms|Q + Ms + Ksp + Pl granite

The introduction of boron alone was found insufficient for tourmaline metasomatites to form: increased activity of iron and magnesium is also essential. Depending on the ratio of the alkali activity to solution acidity, the tourmaline appeared in parageneses with albite, potassium feldspar or muscovite, i.e. in metasomatites of different types. At $T=500^{o}C$, $P=100MPa$ and under the action of the solution 1.0M H_3BO_3 + 0,1M NaCl + 0.1M $FeCl_2$ + quartz (excess) a metasomatic column of the following composition was produced:

Q + Tu|Q + Tu + Ab|Q + Tu + Pl + Chl granodiorite.

Experimental model of metasomatic zoning were zoning were used to determine in great detail the boundary conditions for the development of bereside and other related medium- and low-temperature processes of wall-rock metasomatism (11,12).

The starting rock was quartz diorite (Q +Pl+Am+Bi+Chl), the solution species were KHl, HCl, SiO_2 and carbon dioxide and sulphur. The runs were conducted at T^{o}= 200-400°C and P = 50-400MPa. In the composition and structure, experimental zoned columns corresponded to beresides, argillisites, gumbeites, quartz-sericite and quartz-potashfeldspar metasomatites.

To define the stability conditions for the mineral parageneses in such a complex system by the mineral equilibrium method would have required extensive experimenting. The direct modeling of metasomatic zoning has proved by far more effective in determining the boundary values for the major factors in metasomatism: T, X_{CO_2} the acidity and activity of potassium in the Solution, which xharacterize the formation conditions for all the five metasomatic formations. The experimental results are plotted in fig.9 as diagrams of the stability fields of argillistiter (A), beresides (B), gumbeites (G), quartz-sericite (QS), and quartz-potash feldspar metasomats (QF versus the established values for the major metasomatism factors. It will be noted that the experimental points in the diagram (shown in fig.9a only) refer to metasomatic columns of some definite type (as those, in fig.7b) rather than to seperate mineral parageneses. The lines in the diagrames deliniate the stability fields in the defenite columns which characterize the metasomatic type of fomation. These lines in fact represent the main boundary reaction by which the replacement of metasomatites occurs. For example, the line between fields A and B represent the reaction 1.5 Kl + KCL = Ser + HCL + 1.5 H_2O, the line seperating QS and B is for the reaction of carbonate appearance : Chl + CO_2 = AnK + Ser + H_2O; the line between B and G is for the reaction Ms + 3Q + KCl = 1.5 Ksp + HCL, the line between G and QF corresponds to the ankerite decarbonation equilibrium etc. The diagrams in fig.9 illustrate the dependence of the stability fields of the metasomatic types in question on the major factors: temperature, activity (concentration) of carbon dioxide, potassium and solution acidity, and thet do not seem to call for a more detailed consideration. It will only be noted that a comparison with the diagrams of stat in fig.1-4 reveals in the more complex system the boundary stability reactions for kaolinite, sericite and potash feldspar to have shifted into the higher-temperatur region as the acidity decreases. This is clearly the effect of carbon dioxide: the difference disappears at lower CO_2 mole fractions. The inference from the above diagram is theat the reason for the chance of metasomatic formations can be either temperature variations at constant (KCL)/(HCL) in the solution or the variations in this ratio under isothermal conditions. Examples of outwardy similar metasomatic columns obtained by varying different factors are given in fig.10 in the coordinates T versus X (distance along the column) at lg([KCL]/[HCL]) = 3 (fig.10a) and lg([KCL]/[HCL]) versus X,T = 300°C (fig.10b). However on close examination of the columns and seemingly insignificant differences in the mineral and zone compositions, the effects of particular factors are identified reliably. For example, at relatively low temperatures (250°C), the zones of experimental columns contain the Mg-Fe-breunerite-type carbonate which is replaced at higher temperature by ankerite and dolomite. However, an increase in the potassium activity or a decrease in the solution acidity under isothermal conditions while not changing the carbonate composition will promote the primary appearance of sericite and potash feldspar rather thean plagioclase in the middle and forward zones of the columns.

Quartz-sericite metasomatites have been found to be highly sensitive to increasing CO_2 pressure (fig.9). This suggests that the wide natural occurence of ankerite-free quartz-sericite metasomatites is teh result of either a low CO_2 partial pressure (below 1-2MPa) or the preence of highly acid (pH 3) or low-K ([LK^{+}]<0.01m) solutions. Beresides and gumbeites occur through a wide range of compositions but are limited in temperature to 350-370°C, the most likely values of X_{CO_2} being not higher than 0.1-0.2.

In conclusion, consider some general characteristics of metasomatic zoning established in the present experimental study. The direct modeling of metasomatic processes seems to be somewhat less accurate for determing the physico-chemical parameters of mineral formation than the traditional methods used in such studies. At the same time, the special adventage of the method is that it brings the experimental model closer to natural phenomena and produces highly informative results. The metasomatic column provides information not only on the equilibrium characteristics of mineral formation but on kinetic and dynamic factors of metasomatism, where there is a close analogy between an experimental model and natural object, the information obtained for the model can be applied with confidence to the object. A very impotant constraint in this respect is attainment of local equilibrium in natural and experimental columns.

The experiments on modeling metasomatic zoning have confirmed the main concepts of Korzhinskii's theory (17). The most adequate explanation for structural features of experimental columns is provided by the component differential mobility and the attainment of local equilibrium within the zones and at their boundaries while on the whole the metasomatic alteration is non equilibrium process. Experimental xolumns are always distictly zoned, the zones appear simultaneously and grow with time. Due to chemical replacement reactions between zones, the solid phase takes up some species from the solution and releases some others with the result that growth or loss of the solid phase occurs at the zone boundaries with the related changes in the zone chemistry. The most detailed information on the process was provided by the microprobe analysis of zoned columns. The zone boundaries are most distict in high-temperature columns (fig.7). With decreasing temperature, zoning becomes less pronounced due to the presence of relict nonequilibrium minerals. Nevertheless, the tendency towards the formation of sharp replacement fronts persists and shows up as an abrupt production or loss of minerals at the zone boundaries, such as Cal, Ank, Ser, Ksp, Py. Minerals of variable composition (Chl, Ank, Brn and others) undergo gradual changes within the zones and sharp changes at the zone boundaries.

The feature common to all experimental columns is a decrease in the number of minerals from the forward to rear zones and a decrease in zone thicknesses in the same direction. The rear zones in the more developed columns are normally composed of one mineral. The number of phases in the zones is always less than the number of chemical components in the starting rock and solution- an indication to the perfectly mobile behaviour of some of the components, according to the phase rule. The uccession of the replacement front along the column corresponds to the differential mobility of components for given factors of state defined by experimental conditions. Although they somewhat differ among the discussed columns of acidic metasomatosis depending on particular experimental conditions the generalized sequence of components mobility is as follows (quartz, being an excessive species, is not included):

qartz-feldspar metasomatites: Mg, Fe > K, Na > Al
quartz-sericite metasomatites: Ca>Na, K>Mg > Fe > Al
gumbeites : Na, Ca > K > Mg > Fe > Al
beresites : K > Na, Ca > Mg > Fe > Al
argillizetes : Na > Ca > Mg > Fe > Al

These sequences are similar to those found by analysing natural matasomatic zoning (4).

It will be noted that experiments on rocks and solutions of different composition help to spot out the factors responsible for the components' differential mobility. It depend on the component contents in the rock and solution, but the greatest effect is produced by pH variations in the intersecting solution.

At high alkalinity (pH = 9.5 - 12.0) this dequence of mobility may be reserved : magnesium and iron (and sometimes calcium) become more inert than silicon and aluminium.

In experimental and natural columns the main zoning produced by changes in component mobility is paralleled by an additional zoning which while not changing the number of phases changes their mineral composition. This type of zoning is shown in the schemes of metasomatic columns.

Earlier it was noted (4,5,26) that experimental studies have revealed convergence of metasomatites, i.e. the formation of compositionally similar metasomatic columns under different conditions, as for example in the replacement of argillisites by beresites and gumbeites (fig.10) with increasing temperature or changing solution composition. Convergence presents a problem which should be accounted for when analyzing natural phenomena. However, a careful study of subtle features of zoning may help reconstruct the actual evolution path of the process, as is shown above.

An important qualitative characteristic of metasomatosis to be only determined experimentally is the growth rate of zoned columns, which slows down as the diffusion mass transfer proceeds. The experimental relationship between the movement of replacement fronts and time is described satisfactorily by equations of the type $X = w\sqrt{t}$, where X is the distance; t, the time; w, the rate of the zone boundary movement. As is seen in the schematic representation of experimental zoning, the thickness of internal the mote heavily altered portions of acidic metasomatic columns, which are normally called "greyzens", "beresides" etc., did not exceed 10-15 mm in run durations of 14 days. Time extrapolation gave the values : 16 cm in 100 yrs , 50 cm in 1000 yrs, 1.6 m in 10 th. yrs. These values should be reduced by a factor of 10-20 to correct for porosity (which was 20-40% in experiments as compared with 1-2% porosity in granoid rocks). It is believed that common aureoles of altered rocks take tens of thousand years to form

Abbreviations for Minerals

1. And - andalusite; 2. Ab - albite; 3. Kl - kaolinite; 4. Q - quartz; 5. Mu - muscovite; 6. Prh - pyropylillite; 7. Kfs - potassium feldspar; 8. Bi - biotite; 9. Prg - paragonite; 10. Pyr - pyrite; 11. Ser - sericite; 12 - Cc - calcite; 13. Chl - chloride; 14 Am - amphibole; 15. Ank - ankerite; 16. Pl - plagioclase; 17. Br - brinorite; 18. Can - kankrinite; 19. Aeg - aegirine.

REFERENCES

1. Zharikov V.A. - Geology and metasomatic phenomena in the skarm-polymetallic deposits, West Karamazar. - Trudy IGEM USSR Academ Sci. 1959, N 14, 371 p.
2. Zharikov V.A. - Some patterns of metasomatic processes. "Metasomatic wall-rock alterations and their bearing on ore fomation" Nedra, 1965, p. 21-34.
3. Zharikov V.A. - Some topical problems in experimental mineralogy. - Zaponsky Vses. Mineral. Ob., 1976, 105, N 5, 543-561.
4. Zharikov V.A. - Physico-chemical investigation in wall-rock metasomatism. Geochimiya, 1982, N 12, 1254-1287.
5. Zharikov V.A., Zaraisky G.P. - Experimental study of metasomatism: state-of-the-art, prospects. G.R.M. 1973, 15, N 4, 3-18.
6. Zharikov V.A., Ivanov I.P. - Experimental and theoretical study of equilibria in system with perfectly

mobile components. Geochimiya, 1972, N1, 35-50; Translations in Geochemistry international, 1972.
7. Zharikov V.A., Ivancv I.P., Fonarev V.I. - Mineral equilibria in the system $K_2O-Al_2O_3-SiO_2-H_2O$. Nauka, 1973, 160p.
8. Zharikov V.A., Omelyanenko B.I. - Some problems involved in studies on wall-rock alteration within the framework of metalogenic investigations. In: Study of the patterns of mineralization occurence related to metalogenic investigations". Nedra, 1965.
9. Zharikov V.A., Omelyanenko B.I. - Classification of metasomatites. In: Metasomatism and ore formation. Nauka, 1978, 9-28.
10. Zharikov V.A., Fedkin V.V. - Studies in physicochmical mineralogy. In: "Contributions to Physicochemical petrology" , v. X, Nauka, 1982, 5-35.
11. Zaraisky G.P. Shapovalov Yu.B. - Experimental Study of phusico-chemical conditions of beretisation. - Doklady Acad. Nauk USSR, 1978, 238, N 1, p. 207-210.
12. Zaraisky G.P., Shapovalov Yu.B., Belyavskaya O.N.- Experimental study of acidic metasomatosis. M. Nauka, 1981, 220 p.
13. Ivanov I.P., Borisov M.V., Redkin A.F. - Thermodynamic and experimental modeling of local equilibrium in metasomatic columns of acidic leaching. In : "Problems of physico-chemical petrology". M., Nauka, 1979, v. 2, 145-176.
14. Kazytsin Yu. V., Chernoruk S.G., Nechiporenko G.O., Dubak O. Yu. - Diffusion metasomatic argillisation column. Doklady Acad. Nauk USSR, 1967, 173, N 1, 126-128.
15. Karpov I.K. - Computer programs for physico-chemical modeling in geochemistry. Nauka, Novosibirsk, 1981.
16. Korzhinskii D.S. - The ouline of metasomatic processes. In : Basic problems in the study of magmatogene ore deposits, Moscow, Acad. Nauk, 1953, p 332-452.
17. Korzhinskii D.S. - The theory of metasomatic zoning. Nauka, 1969, 110 p.
18. Redkin A.F., Ivanov I.P. An experimental study of the hydrolysis reaction in the system $Al_2O_3-SiO_2-H_2O-$ HCL. Doklady Acad. Nauk, v 250, N4, p. 963-965.
19. Shvarov Yu.V. Calculation of th4 equilibrium composition for a multicomponent heterogeneous system.

Doklady Acad. Nauk, 1976, 229, N 5.
20. Eugster H.P. Geterogeneous reactions involving oxidation and reduction at high pressure and temperature. Journ. Chem. Phys. 1957, v. 26, p. 1760.
21. Winkler H., Johannes W. Experimentalle Metasomatose an einem Granotcontact.-Naturwissenschaften, Hf. 24, 1963, S. 730-731.
22. Hemley J.J. Hydrolysis of K-feldspar and mica at elevated temperatures and pressures. J. Phys. Chem. 1959, v.63, N2,
23. Hemley J.J., Meyer C., Rickter O.H. Some alteration reaction in the system Na_2O-Al_2O_3-SiO_2-H_2O. U.S. Geoch. Surv.Prot. Paper 424-D 1961, 338-340.
24. Hemley J.J., Jones W.R. Chemical aspects of hedrothermal alteration with emphasis on hydrogen metasomatism. Econ. Geol. 1964, v. 59, N 4.
25. Meyer C., Hemley J.J., Wall rock alteration in book "Geochemistry of Hydrothermal or3 deposits" Ed Barnes H.L. N.Y. 1967, p 670.
26. Zharikov V.A. Zaraisky G.P. Experimentalle Untersuchungen der Metasomatose: Zustand. Perspektiven. Zeitschr. für amgewadr. Geologic 1975, Bd 21, hf 9.

Table 1. Boundary hydration-dehydratior and hydrolysis reactions of minerals in diagrams. Fig.4a,b and 5a,b.

Reaction No.	Reaction equations	lgKp	Temperature dependence (°C), LgK
	Hydration-dehydration reactions		
4	$2Q+Kl=Pph+H_2O$	lgf_{H_2O}	
5	$Pph=3Q+And+H_2O$	-"-	
18	$Ab+0.5\ Kl=Prg+2Q+H_2O$	-"-	
21	$Q+Ms=And+Fsp+H_2O$	-"-	
22	$Q+Prg=And+Ab+H_2O$	-"-	
	Hydrolysis reactions		
10	$Ab+H^++0.5\ H_2O=2Q+0.5\ Kl+Na^+$	$lg(Na^+/\ /H^+/)$	$LgKp=1.21\cdot10^{-5}T^2-1.656\cdot10^{-2}.T+7.771$
11	$Prg+H^++1.5\ H_2O=1.5\ Kl+Na^+$	-"-	$lgKp=1.70\cdot10^{-5}.T^2-2.169\cdot10^{-2}T+8.495$
12	$3Q+Prg+H^+=1.5\ Pph+Na^+$	-"-	$lgKp=1.30\ 10\ .T^2-1.269\cdot\ 10^{-2}T+6.491$
13	$Prg+H^+=1.5Q+1.5\ And+Na^+ + 1.5\ H_2O$	$lg(/Na^+/\ /\ /H^+/)$	-
14	$Ad+H^++5Q+1.5\ And+Na^++ 0.5\ H_2O$	-"-	-
16	$1.5Ab+H^+=3Q+1.5\ Prg+Na^+$	-"-	$lgKp=6.80\cdot T^{-5}.T^2-3.726\cdot10^{-2}T+9.161$
17	$Ab+0.5Fsp+H^+=3Q+0.5Ms+Na^+$	-"-	$lgKp=1.10\cdot10^{-5}T^2-1.480\cdot10^{-2}T+7.866$
6	$Ms+H^++1.5H_2O=1.5Kl+K^+$	-"-	$lgKp=-0.5\cdot10^{-5}\cdot T^2-0.696\cdot T+4.326$
7	$3Q+Ms+H^+=1.5\ Pph+K^+$	-"-	$lgKp=-1.0\cdot10^{-5}\ T^2+0.204\ 10^{-2}T+2.322$
8	$Ms+H^+=1.5Q+1.5And+K^++1.5\ H_2O$	-"-	-
9	$1.5Fsp+H^+=3Q+0.5Ms+K^+$	-"-	$lgKp=-0,9\cdot10^{-5}T^2-0.173\cdot10^{-2}T+4.604$
19	$Fsp+H^+=5Q+1.5And+K^++0.5\ H_2O$	-"-	-
20	$1,5\ Ab+Ms+H\ =3Q+1.5\ Prg+K^+$	-"-	$lgKp=16.2\cdot10^{-5}\cdot T^2-6.904\cdot10^{-2}\cdot T+8.480$
23	$Ms+Na^+=Prg+K^+$	$lg(Na^+/\ /\ /K^+/)$	$lgKp=2.20\cdot10^{-5}T^2-1.486\cdot10^{-2}T+4.179$
24	$Ab+H^++0.5K^+=3Q+0.5Ser+1.5Na^+$	$lg(/Na^+/^{1.5}//H^+/\cdot$ $\cdot/K+/^{0.5}$	-
25	$Fsp+Na^+=Ab+K^+$	$lg(/Na^+/\ /\ /K^+/)$	$lgKp=2.0\cdot10^{-5}\cdot T^2-1.305\cdot10^{-2}T+3.261$

Table 2. Major types of metasomatic zoning in alumosilicate rocks

Temperature range °C	Columns	Inert components	Equilibrium constants for boundary reactions
1	2	3	4
	Ultra-acidic and acidic metasomatites (K-series)		
	I-A		
	Q+Kl+Ser+Ab	Na,K,Al	$lg([Na^+]\,[H_2O]^{0,5})$
25-160	Q+Kl+Ser	K,Al	$lg([K^+]\ /\ [H^+]$
	Q+Kl	Al	$lg([Al(OH)_2^+]\,[H_2O]^{0.5}/[H^+]$
	Q		
	I-B		
	Q+Kl+Ser+Prg	Na,K,Al	$lg([Na^+]\ /\ [H^+].[H_2O]^{1,5})$
160-270	Q+Kl+Ser	K,Al	$lg([K^+]\ /\ [H^+])$
	Q+Ser	Al	$lg([Al(OH)_2^+].[H_2O]^{0,5}/[H^+]$
	Q		
	I-C		
	Q+Pph+Ms+Prg	Na,K,Al	$lg([Na^+]\ /\ [H^+])$
270-390	Q+Pph+Ms	K,AL	$lg([K^+]\ /\ [H^+])$
	Q+Pph	Al	$lg([Al(OH)_2^+]\ /\ [H^+])$
	Q		
	I-D		
	Q+And+Ms+Prg	Na,K,Al	$lg([Na^+]\,[H_2O]^{1,5}/\ [H^+])$
390-470	Q+And+Ms	K,Al	$lg([K^+]\,[H_2O]^{1,5}/\ [H^+])$
	Q+And	Al	$lg([Al(OH)_2^+]\ /\ [H^+].[H_2O]^{0,5})$
	Q		
	I-E		
	Q+And+Ms+Ab	Na,K,Al	$lg([Na^+]\,[H_2O]^{0,5}/[H^+])$
470-520	Q+And+Ms	K,Al	$lg([K^+]\,[H_2O]^{1,5}\ /\ [H^+])$
	Q+And	Al	$lg([Al(OH)_2^+]\ /\ [H^+].[H_2O]^{0,5})$
	Q		
	I-F		
	Q+And+Ksp+Ab	Na,K,Al	$lg([Na^+]\,[H_2O]^{0.5}\ /\ [H^+])$
Above 520	Q+And+Ksp	K,Al	$lg([K^+]\,[H_2O]^{0.5}\ /\ [H^+])$
	Q+And	Al	$lg([AL(OH)_2^+]\ /\ [H^+].[H_2O]^{0,5})$
	Q		

1	2	3	4
	Ultra-acidic and acidic metasomatites (Na series)		
	II-A		
	Q+Kl+Ab+Ser	K,Na,Al	$\lg([K^+] / [H^+] [H_2O]^{1,5})$
25-160	Q+Kl+Ab	Na,Al	$\lg([Na^+] / [H^+] [H_2O]^{0,5})$
	Q+Kl	Al	$\lg([Al(OH)_2^+] [H_2O]^{0,5} / [H^+])$
	Q		
	II-B		
	Q+Kl+Prg+Ser	K,Na,Al	$\lg([K^+] / [H^+] [H_2O]^{1,5})$
160-270	Q+Kl+Prg	Na,Al	$Lg([Na^+] [H_2O]^{0,5} / [H^+])$
	Q+Kl	Al	$\lg([Al(OH)_2^+] [H_2O]^{0.5} / [H^+])$
	Q		
	III-C		
	Q+Pph+Prg+Ms	K,Na,Al	$\lg([K^+] / [H^+])$
270-390	Q+Pph+Prg	Na, Al	$\lg([Na^+] / [H^+])$
	Q+Pph	Al	$\lg([Al(OH)_2^+] / [H^+]$
	Q		
	II-D		
390-470	Q+And+Prg+Ms	K,Na,Al	$\lg([K^+] [H_2O]^{1.5} / [H^+])$
	Q+And+Prg	Na,Al	$\lg([Na^+] [H_2O]^{1,5} / [H^+])$
	Q+And	Al	$\lg([Al(OH)_2^+] / [H^+].[H_2O]^{0,5}$
	Q		
	II-E		
	Q+And+Ab+Ms	K,Na,Al	$\lg([K^+] [H_2O]^{1,5} / [H^+])$
470-520	Q+And+Ab	Na,Al	$\lg([Na^+] [H_2O]^{0,5} / [H^+])$
	Q+And	Al	$\lg([Al(OH)_2^+]\ [H^+].[H_2O]^{0.5}$
	Q		
	II-F		
	Q+And+Ab+Ksp	K,Na,Al	$\lg([K^+].[H_2O]^{0.5} / [H^+])$
Above 520	Q+And+Ab	Na,Al	$\lg([Na^+].[H_2O]^{0,5} / [H^+])$
	Q+And	Al	$\lg([Al(OH)_2^+] / [H^+].[H_2O]^{0.5})$
	Q		
	Acidic metasomatites (K - series)		
	III -A		
	Q+Ms+Ksp+Ab	Na,K,Al	$\lg([Na^+] / [H^+])$
25-520	Q+Ms+Ksp	K,Al	$\lg([K^+] / [H^+])$

1	2	3	4
	Q+Ms	Al	$\lg([Al(OH)_2^+] [K^+] / [H^+]^4)$
	Q		
	Acidic metasomatites (Na - series)		
	IY-A		
	Q+Prg+Ab+Ms	K,Na,Al	$\lg([K^+] / [H^+])$
160-470	Q+Prg+Ab	Na,Al	$\lg([Na^+] / [H^+])$
	Q+Prg	Al	$\lg([Al(OH)_2^+]^3 [Na^+] / [H^+]^4)$
	Q		
	Weakly acidic metasomaties		
	Y - A		
	Q+Ksp+Ab	K,Al	$\lg([K^+] / [Na^+])$
Above 25	Q+Ksp	Al	$\lg([Al(OH)_2^+] [K^+] / [H^+]^2)$
	Q		
	Y- B		
	Q+Ab+Ksp	Na,Al	$\lg([Na^+] / [K^+])$
Above 25	Q+Ab	Al	$\lg([Al(OH)_2^+] [Na^+] / [H^+]^2)$
	Q		

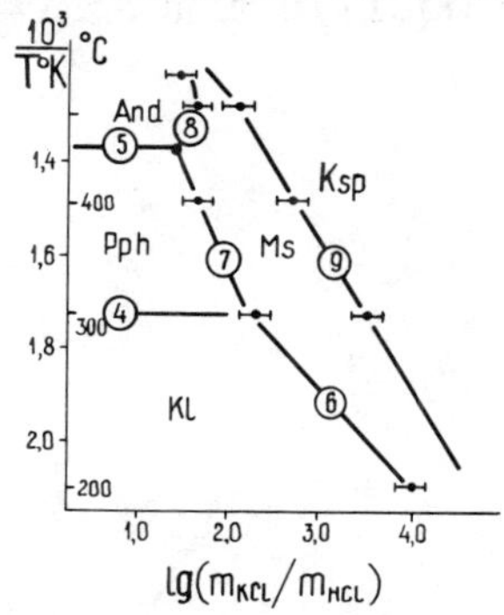

FIGURE 1. A refined T-lg (m_{KCL}/m_{HCL}) experimental diagram for hydrolysis equilobria in the system K_2O-Al_2O_3-SiO_2-H_2O-HCL with excess quartz, P= 100 MPa. Numbered reaction are:

4. 2Q + Kl = Pph + H_2O
5. Pph = 3Q + And + H_2O
6. 1.5 Kl + KCL = Ms + HCL + 1.5 H_2O
7. 1.5 Pph + KCL = 3Q + Ms + HCL
81.5 Q + 1.5 And + KCL + 1.5 H_2O = Ms
9. 3Q + Ms + KCL = 1.5 Ksp + HCL

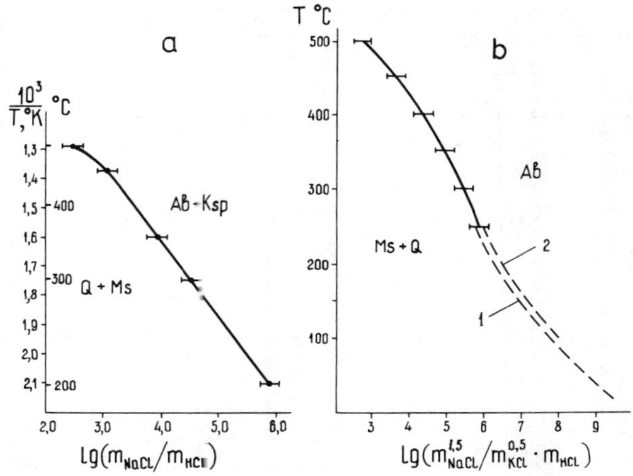

FIGURE 2. a) An experimental T-lg(m_{NaCl}/m_{HCL}) diagram for the reaction 3Q + Ms + NaCl = Ab + o.5 Fsp + HCL, p= 100Mpa.

b) An experimental T-lg($m^{1.5}_{NaCl}/m^{0.5}_{KCL}$ m_{HCL}) diagram for the equilibrium reaction 3Q + Ms + 1.5 NaCl = 1.5 Al + 0.5 KCL + HCL, p =100 Mpa. The calculated curves in the low-temperature region: 1 - equilibrium at saturated water vapour pressure. 2 - equilibrium at P=100 MPa.

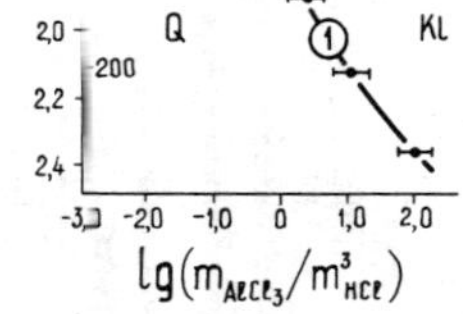

FIGURE 3. An experimental T-lg($m_{AlCl}{}^{3}/m^{3}_{Hcl}$ diagram for hydrolysis equilibria in the system Al_2O_3-SiO_2-H_2O-HCL; eactions: 1. 0.5KL+3HCL+Q+$AlCl_3$+2.5H_2O;

2. o.5Pph+3HCL=2Q+$AlCl_3$+2H_2O;

3. 0.5And+3HCL=0.5Q+$AlCl_3$+1.5H_2O

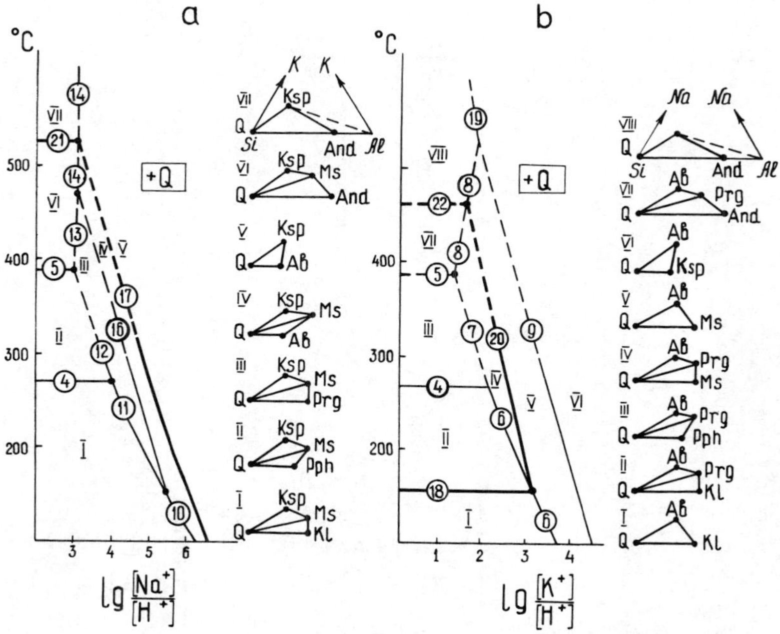

FIGURE 4. S diagram of stat for the system Al_2O_3-SiO_2-H_2O-HCL-Kcl-NaCl.
a) the dependence of phase equilibria on T and Lg $\frac{(Na^+)}{(H^+)}$ for the system K-Ab-Si - aqueous saline solution with excess quartz. Arabic numbers indicate monovariant reactions whose equations are given in Table 1. Roman numbers denote stability fields of three-mineral paragenesis, whose composition-paragenesis are shown on the right. See text for details
b) The dependence of phase equilibria on T and lg $\frac{(K^+)}{(H^+)}$ for the system Na-Al-Si - aquous saline solution with exess quartz. Notations are the same.

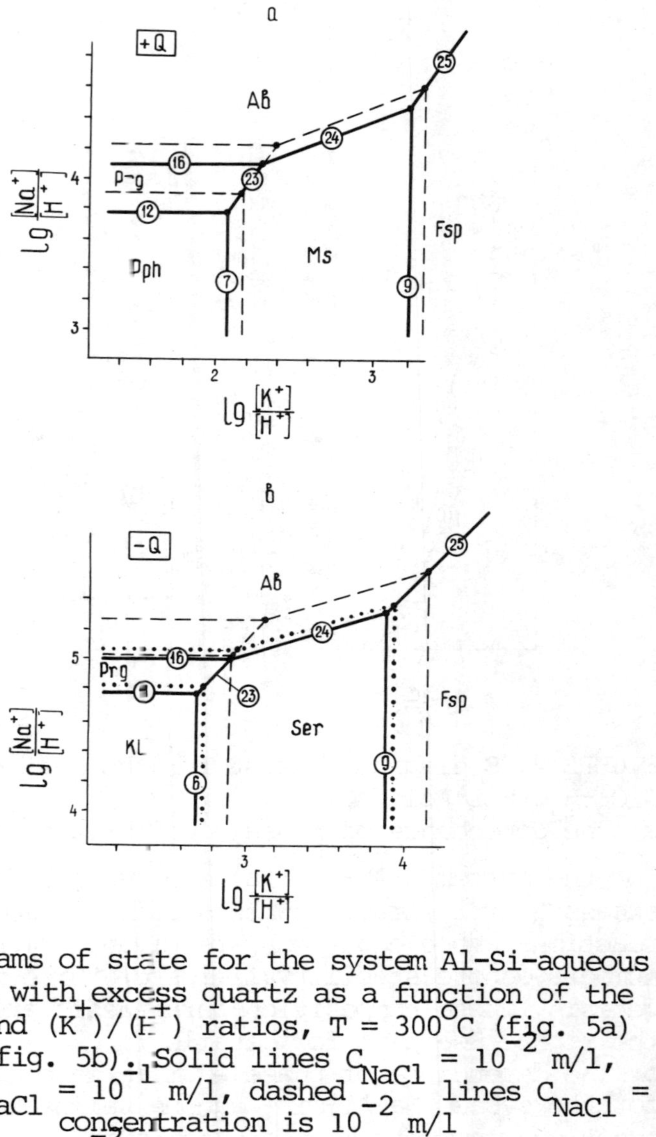

FIGURE 5. Diagrams of state for the system Al-Si-aqueous saline solution with excess quartz as a function of the lg $(Na^+)/(H^+)$ and $(K^+)/(H^+)$ ratios, T = 300°C (fig. 5a) and T = 200°C (fig. 5b). Solid lines $C_{NaCl} = 10^{-2}$ m/l, dotted lines $C_{NaCl} = 10^{-1}$ m/l, dashed lines C_{NaCl} = 1.0 m/l. KCl concentration is 10^{-2} m/l throughout; $C_{HCl} = 10^{-2}$ m/l in starting solution. The reaction equations are given in Table 1.

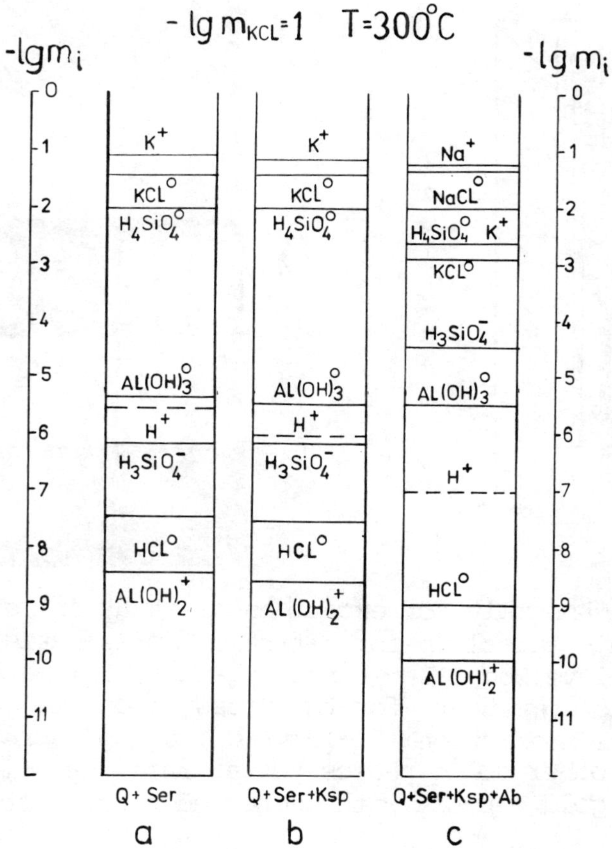

FIGURE 6. Component concentration in the solution which is in equilibrium with the paragenesis Q+Ser (a), Q+Ser+Ksp (b), Q+Ser+Kps+Ab (c) at T=300°C and starting solution concentration 10^{-1} m KCL.

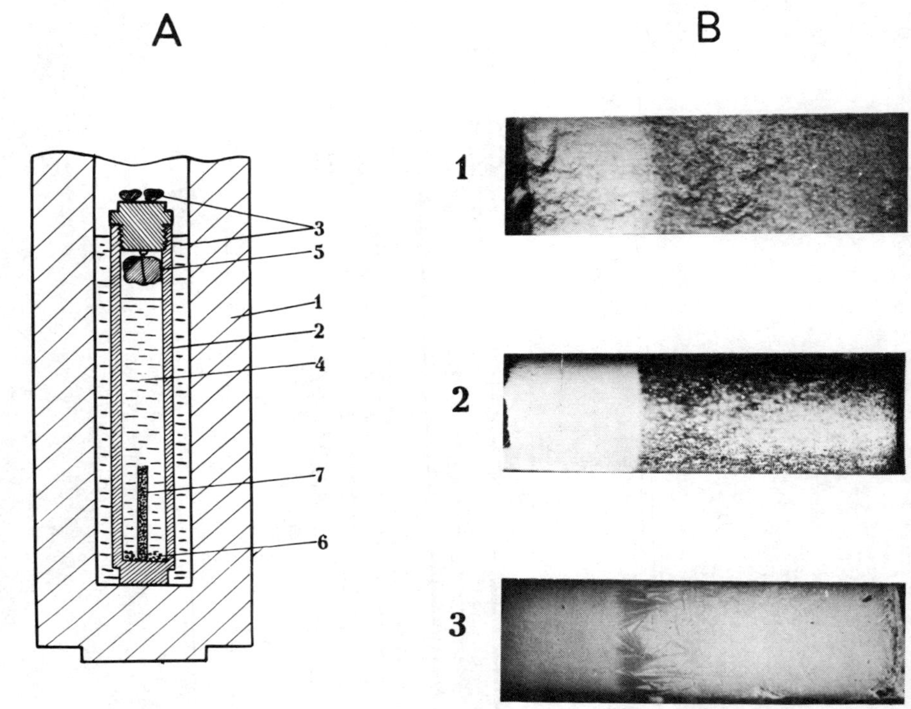

FIGURE 7. Assembly for experimental study of metasomatic zoning. a - autoclave for diffusion metasomatisis runs: 1 - autoclave body; 2 - titanum gasket; 3 - water and dry CO_2 to compensate for the pressure on the gasket; 4 - reaction solution; 5 - dry CO_2; 6 - powdered quartz, Al_2O_3 or other solid phases to saturate the solution, 7 - open gold capsule with compacted powered starting rock.

b - Photomicrographs of experimental metasomatic columns

1. Column of greysen- A distinct contact is seen between "quartz-muscovite greysen" zone and "greyzenized" granite Binocular photo, magnofocation x5^{10} .
2. Contact between Q+Ms and Q+Ab+Ms+Bi zones in apogranitic column of andalusite greysen. (T=500°C, p=100Mna, solution:0,05 MKCL+0,1MHCL)
3. Contact between Can+Am+Aeg and Ab+Ksp+Am+Aeg zones in sodium metasomatite column (T=500°C, p=100Mna, solution: 0.37 M NaOH+0.14 Na_2CO_3, starting material Q+ Bi shists)

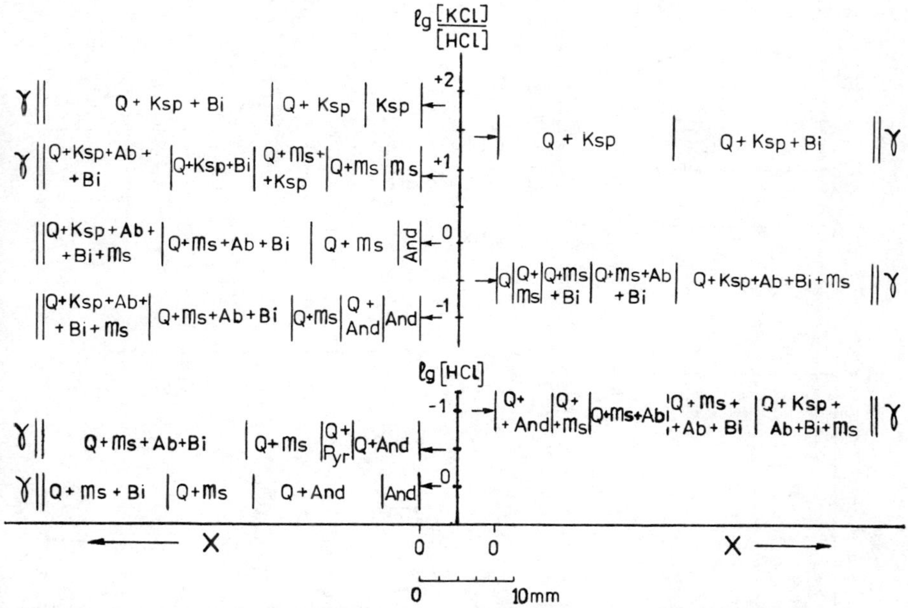

FIGURE 8. A comprehensive diagram for the relationship between the structure of experimental metasomatic columns and composition of reaction solution, T=500°C, P=100 Mpa. The ordinate is logarithms of HCL concentration and concentration ratios of KCL and HCL in the reaction solution. The abcissa x is the distance (mm) from the beginning of the columns (x=0).

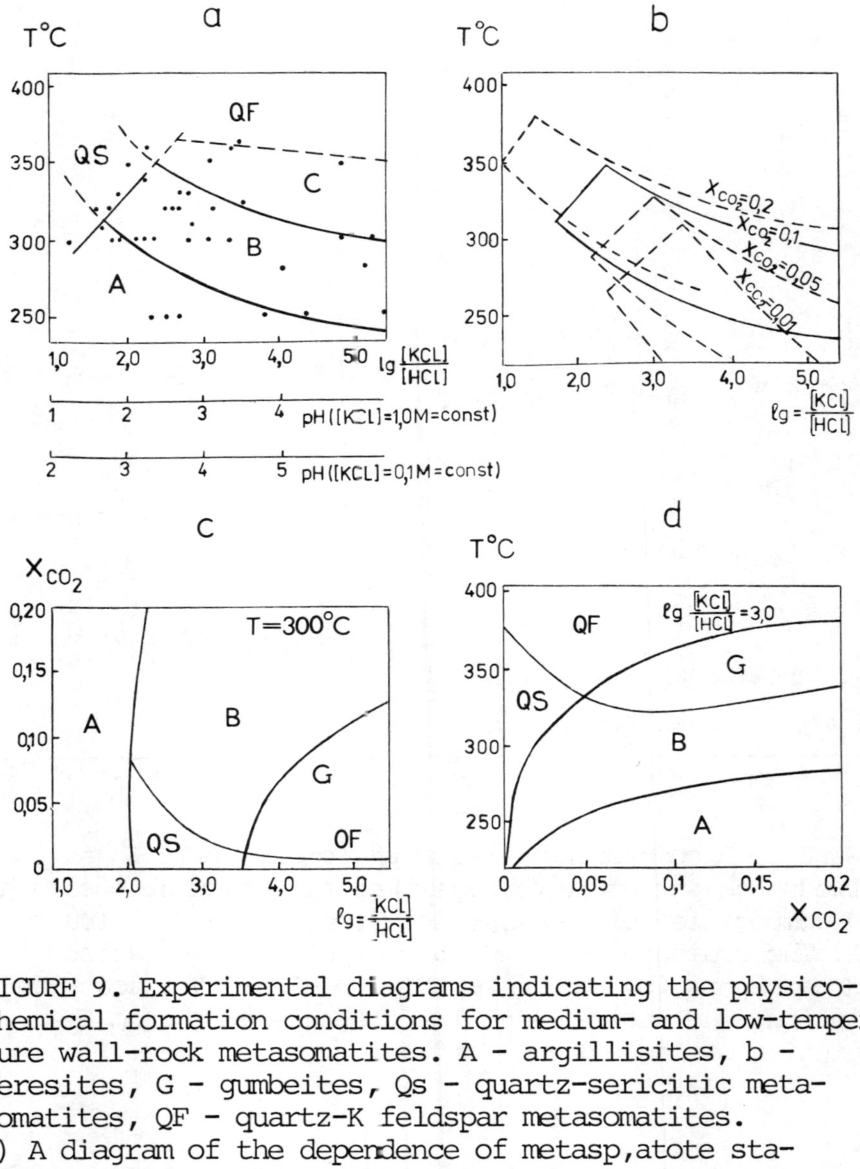

FIGURE 9. Experimental diagrams indicating the physico-chemical formation conditions for medium- and low-temperature wall-rock metasomatites. A - argillisites, b - beresites, G - gumbeites, Qs - quartz-sericitic metasomatites, QF - quartz-K feldspar metasomatites.
a) A diagram of the dependence of metasp,atote stability fields on temperature and concentration ratio (KCL)/(HCL), x_{CO_2}=0.1, P=P_{H_2O}+P_{CO_2}=100 MPa (for all diagrams).
b) A T versus lg (KCL)/(HCL) diagram for different carbon dioxide contents in the fluid: X_{CO_2}=0.01; 0.05; 0.1

0.2. $P_{H_2O}+P_{CO_2}$ =100 MPa.
c) A X_{CO_2} versus lg(KCL)/(HCL) diagram for T=300°C and P= 100MPa.
d) A T versus X_{CO_2} diagram for lg (KCL)/(HCL) =3.0 and P =100 MPa.

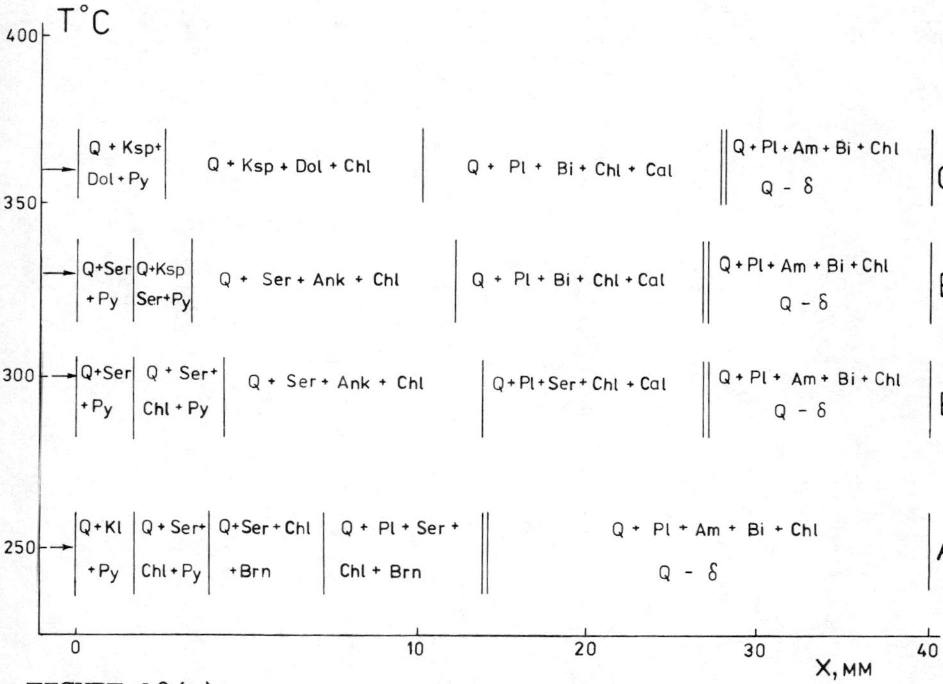

FIGURE 10(a).

FIGURE 10. Structure of experimental metasomatic columns a) as a function of temperature (at lg(KCL)/(HCL)=3; b) as a fuction of lg (KCL)/(HCL) (at T=300°C). In all the runs X_{CO_2} = 0.1, $P=P_{H_2O}+P_{CO_2}$ = 100 MPa; X - the distance along the column (mm). For the clarity of presentation, the scale of the rare part of the column is doubled.

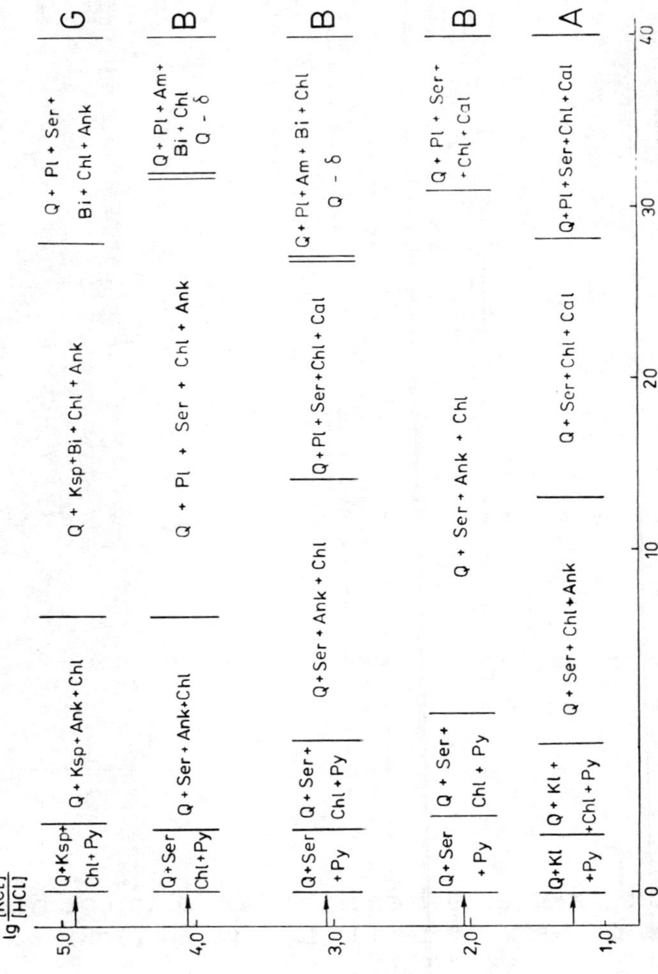
lg [KCl]/[HCl]
5,0
4,0
3,0
2,0
1,0
0
10
20
30
40
X,мм
Q+Ksp+ Chl+Py
Q + Ksp + Ank + Chl
Q + Ksp+Bi + Chl + Ank
Q + Pl + Ser + Bi + Chl + Ank
G
Q+Ser Chl+Py
Q + Ser + Ank+Chl
Q + Pl + Ser + Chl + Ank
Q + Pl + Am+ Bi + Chl Q - δ
B
Q+Ser +Py
Q + Ser+ Chl+Py
Q+Ser + Ank + Chl
Q+Pl + Ser+Chl + Cal
Q + Pl+Am + Bi + Chl Q - δ
B
Q+Ser +Py
Q + Ser+ Chl + Py
Q + Ser + Ank + Chl
Q + Pl + Ser+ +Chl + Cal
B
Q+Kl +Py
Q + Kl + +Chl + Py
Q + Ser+ Chl+Ank
Q + Ser+Chl + Cal
Q+Pl+Ser+Chl+Cal
A

FIGURE 10(b).

ULTRABASITES OF THE BALKAN PENINSULA AND THEIR ORE POTENTIAL

ZHELYASKOVA-PANAYOTOVA M., KOLCHEVA k., KARAMATA S., MAJER V., SCARPELIS N.
Belgrade University, Yugoslavia; Zagreb University, Yugoslavia; Athens University, Greece; Sofia University, Bulgaria

In the Balkan Peninsula ultrabasites occur on the territories of Yugoslavia, Albania, Greece, Bulgaria and European Turkey. A review on their character, occurrences and associated mineralizations was published by Hiessleitner (1951-52). Recent publications provide much concrete information as well as several surveys covering individual countries and regions (Woodtli, ed., 1964; Scarpelis and Economou, 1979; Karamata et al., 1980; Panayiotou, ed., 1980 Augustithis, ed., 1981; Dunning et al., ed., 1982; Zhelyaskova-Panayotova et al., 1982). A systematic study of this new information ill give a more correct idea of the regularities in the distribution of ultrabasites in space and time, of their role in the geological history of the Peninsula and of their economic importance as carriers of a number of productive mineralizations.

The rock complexes enclosing ultrabasites are of ages ranging from the Precambrian to the Upper Cretaceous. Dating of ultrabasites is a debatable question. The division accepted by the present authors is based on the relations between ultrabasites and regional metamorphism (are they affected or unaffected by it), on their association with dated complexes, and on the effects of tectonic events or magmatic actions of identified age. The groups distinguished in this way are as follows: Rhodope premetamorphic and postmetamorphic ultrabasites occuring in the crystalline-schist complex, Paleozoic and Mesozoic ultrabasites belonging to the ophiolitic associations.

The premetamorphic ultrabasites are best studied in the Central Rhodope Mountains. They also occur in the area of the town Kardjali, near the town of Elhovo (Zhelyazkovski massif, Krainevski ultrabasic bodies

Proceedings of the 27th International
Geological Congress, Volume 9, pp. 181–198
PETROLOGY (IGNEOUS AND METAMORPHIC ROCKS)

near the village of Hliabovo, etc.) as well as on the northern slope of the Belesica Mountain (Fig. 1). They are small bodies of varying size and rounded, lenticular shape occuring near each other in boudinage structures and representing probably boudins of former larger bodies. The individual boudins range in size from tiny bodies several meters in diameter to masses 1000 x 500 m large (Zabardo). They are confined to the lower stratigraphic levels of the Precambrian crystalline characterized by irregular distinct alternations of gneisses, amphibolites and marbles. Their association with amphibolites is particularly close. Usually, the ultrabasite bodies are partially or completely surrounded by orthoamphibolites, which are altered basites. Similar amphibolites are found also seperately as lenticular interbeds in gneisses and marbles thinning rapidly out. Although infrequently, small bodies of amphibolized eclogites are also to be found in similar geological environment.

The ultrabasite rocks often show a specific spheroid structure. Their main rock-forming minerals are olivine (Fo_{91}) and orthopyroxene (Fs_8), less frequently clinopyroxene (diopside) which in rare cases only prevails in quantity. The accessory chromispinelids are usually metamorphosed. Tremolite-actinolite occurs extensively as larger or smaller prismatic, sometimes needle like grains, forming in some cases monomineral zones at the peripheral parts of the ultrabasite bodies. Serpentinization is poorly manifested, but rapidly increases in intensity in the tectonically weakened zones. The premetamorphic ultrabasites contain about 3.5% of water, which increases up to 16% in the serpentinized zones. According to their chemical and normal compositions the ultrabasites correspond to harzburgites (most frequently found) and dunites. Lherzolites are considerably less frequent, and olivine websterites and clinopyroxenites are very rare. On the diagramme showing the relationship between SiO_2 and $100FeO^x/ Fe^x+MgO$ all ultrabasites fall within a common field limited by a minimum SiO_2 value of about 35% and a maximum one of about 50%, and by ratios of $100Fe^x/FeO^+ +MgO$ of 13 and 22, respectively. These values make them similar to the so called metamorpic periotites (Coleman, 1979). They are characterized also by a wide range of variations of NiO, CoO and Cr_2O_3, their contents being considerably lower or higher than the clark values.

The occurence of this group of ultrabasites in boundinage structures amid the rocks of the high-crystalline complex, the low water content and espacially the typical mineral parageneses present in them are evidence that they suffered the same metamorphic transformations which affected the enclosing rocks. The paragenesis tremolite-enstatite-forsterite together with the presence of the kynatite-sillimanite-garnet paragenesis in the two-mica gneisses and the forsterite-diopside one in the marbles indicate that all rocks underwent the same stage of metamorphism corresponding to the high-grade amphibolite facies. These facts compare favourably with the experimental studies and the caculated stable mineral associations as products of the possible reactions in the CaO-MgO-SiO_2-H_2O system (Evens, Trammsdorff, 1970). It is also evident that under metamorphic conditions the accessory components, such as nickel, kobalt and chromium, are perfectly mobile.

Poor chromite mineralization is associated with the Central Rhodope premetaorphic ultrabasites. It consists of small accumulations of iron-rich chromspinelids and of chromium and nickel-bearing magnetite ore manifestations referred to the metamorphogenious formations. Of no productive value are also the magnetite mineralizations in chloride schists formed at the expense of ultrabasites, as well as the anthophyllite-asbestos and vermiculite mineralizations.

The Rhodope postmetamorphic ultrabasites occur mostly in the framework of the Northern and Southern Rhodope at the place where the complex of crystalline schists outcrops on the surface (Fig. 1). They usually form groups of 2 or 3 up to 10 or 15 bodies varying, generally small size. The largest are exposed on the surface over an area of about 10 to 13 km^2. They are plate-like bodies, frequently of monoclinal dip, showing elongated cross-sections on the surface. Their contacts with the enclosing rocks are sharp, with sign of movements along them which are obscured by allometamorphic transformations. The rocks inside the bodies are usually less changed.

The enclosing rocks consist of gneisses, marbles, amphibole and chloride schists. The almost invariable association with amphibolites, which, being members of the crystalline-schist complex, are very closely related to ultrabasites, is also very typical. Rodingites and rodingite-like rocks are frequently found in the pripheral and central parts of some larger massifs (Golyamo Kamenyane, Dobromirci, Zhilti-chal, etc.). In Many cases ultrabasites are crosscur by pegmatites. The rocks around them are talcized, vermiculitized and chlorized. As is known, there are no dated Paleozoic and Mesozoic deposits in the Rhodopes. This creates difficulties in determining the age of ultrabasites. The only comment which can be made is that they were brought in this environment probably after the main stages of metamorphism and later suffered the action of some granitoids. The age of the granitoids themselves is a problem complicated enough in itself, not yet solved either.

The Sinikli, Soufflion, Dobromiski, Jakovicki and other massifs in the Western Rhodopes, Parvenckii, Markovski, etc. in the Northern Rhodopes are the largest massifs in this group.

The rocks suffered transformations everywhere. Nevertheless they frequently contain relics of primary rock-forming minerals - olivine (Fo_{91}), enstatite (Fs_8) and diopside-diallage. The accessory chromsponelids bear traces of alteration. The mineral composition, relict structures and petrochemical analysis have made it possible to carify the petrographic composition and the rock relationship inside the massifs. The great similarity of the petrographic varieties is very characteristic. Harzburgites (predominating), dunites, rare lherzolites and wehrlites are distinguished. They gradually blend into each other. Pyroxenites are of three types: pegmatoid vein pyrexenites, fine-grained replacement veins of porphyroid structure and amoebashaped bodies. Their contacts with the other ultrabasite rocks are sharp and distict. According to their composition, diosidites, bronzitites and enstatitites are distinguished. In some cases a regular rock sequence from the base upwards can be observed: dunites - banded dunite-harzburgites - harzburgites and lherzolites. The rocks are magnesium-rich, M/F~9. The TiO_2 content is invariably low (about 0.10%), the contents Cr_2O_3~0.30, NiO~0.25% and CoO~0.15% are relatively constant. From chemical alalysis of three of the vest studied massifs in the Eastern Rhodopes (Jakovica, Dobromirci and Bruševci) the relationship between SiO_2 and 100 FeO^x/FeO^x+ MgO varies only slightly and is near the point characterizing "the mean bulk composition of metamorphic dunites" (Coleman, 1977).

The postmetamorphic ultrabasites are chromite-bearing. A distinct functional relationship is established between the composition of the oreforming chromespinelid and that of the enclosing ultrabasite rock varieties. Magmochromite ores occur in the dunites, chrompicotite ones in the banded dunite-harzburgite complex, aluminarich chrompicitite to alumochromite ores in the harzburgites. The largest chromite bodies are also most frequently found in the dunites.

With the exception of several exotic pyroxenite viens, practically all ultrabasic rocks of this group suffered complicated alterations of various grades. Besides serpentinization, which is the most characteristic, amphibolization, chloritization, carbonatization, talcization etc. are frequently observed. These alterations are manifestly associated in space with tectonic dislocations and with the action of pegmatites and granitoids. They determined the formation of anthophyllite-asbestos, talc and vermiculite deposits. Platinum and gold mineralizations have been reported. The ultrabasites themselves are good decorative material.

<u>Paleozoic ultrabasites</u> are known on the territoties of Yugoslavia and Bulgaria (Fig. 1). Yugoslavian geologists refer them to two groups: Riphean-Cambrian and Paleozoic. On the vasis of the great similarity of their petrographic compositions, character of metamorphic transformations and geological occurrence it seems for this paper justified to combine them in a single group of Lower Paleozoic ultrabasites.

In the dated Riphean-Cambrian complex from the Serbian-

Macedonian massif and in the Northeast Serbia there are small bedded bodies of ultrabasic rocks reaching up to 2-3 km x 300-500 m in area. They are accompanied as a rule by gabbro and basaltoid rocks, as for instance in Crna Trava and Machkatica (Serbia), and other places. The ultrabasic bodies consist mostly of lherzolites, less frequently of wehrlites, harzburgites and dunites with relics of typical magmatic textures (poikilitic, hypidiomorphic). The rocks usually are intensely serpentinized. The small bodies squeezed between various Paleozic rocks suffered the greatest changes - brecciation, foliation. The ultrabasites contain sometimes lean chromite mineralizations. Small talc depoids are formed as the result of later hydrothermal activity. They are traced further into Halkidiki (Serbian-Macedonian massif), where they consist of foliated very altered (antigoritized, talcized, chloritized) ultrabasites. A peculiarity of the chromite-bearing massifs is that the chromite mineralization, referred to the podiform type, is alumina rich)Gomati massif).

On Bulgarian territory the so called Vlasinsko-Osogovo complex forms the extension of the Riphean-Cambrian complex of Northeast Serbia and Serbian-Macedonian massif. The ultrabasite bodies there also occur as isolated lenses or elongated bed-shaped bodies amid phyllites frequently associated with amphybolized gabbro. They are small, reaching up to 1 km in length and up to several dozen meters in width. They are found in the Kjustendil District (near the vilaages of Frolosh, Topolnica and Diakovo), near the village of Dren, Radomir District and other places. The rocks forming these bodies in most cases correspond to lherzolites according to their chemical and modal composition. Relics of olivine (forming frquent inclusions in clinopyroxene), ortho-Pyroxene and some amphobole are observed in rare cases. The rocks are usually strongly serpentinized as well as talcized and carbonatized. Besides the talc mineralization there are also tremolite-asbestos and copper-nickel ore mineralizations of non-productive value (near the village of Frolosh).

The ultrabasites characterized as Riphean-Cambrian show no essential differences as compared to the ultra basite manifestations found in the area of the town Berkovica, in the Iskar Gorge, Bilo Planina Mountain and other places forming part of the Lower Paleozoic complex in the Stara Planina Mountain known as the diabase-phyllitoid complex. The ultrabasic bodies are small (up to 1 km x 50-100 m) and rare, and are usually found as isolated, elongated, bedded lenses amid phyllitized aleurolitic and aleuro-pelitic Lower Paleozoic rocks. They are frequently associated with lenses and blocks of gabbro and spilites. In seperate cases (the Berkovica area) it has been found that the ultrabasic rocks form bodies of varying size inside gabbro. In the ultrabasites themselves there are amoe-shaped lenses of anorthosite-like rocks, partially rodingitized. At the same places gabbro is crosscut by numerous dykes of basic composition resembling the complex of parallel dykes. The ultrabasites, as well as gabbro and spolites are foliated and strongly altered. Together with the sedimentary rocks embedding them they are affected by greenschist metamorphism. Ther are very rare small relics of clinopuroxene forming poikilitic envelopes around idiomorphic olivine almost entirely replaced by serpentine. The relict paragenesis contains also a brown red amphibole (pargasite). The newly formed minerals consist of tremolite-typee amphiboles comprising in some places up to 70% of the bulk of the rock, as well as serpentine, chlorite, less frequently talc and carbonates. Relics of chromspinelids, ilmenite and long-prismatic apatite are also to be found. The primary texture is hypidiomorphic. According to their chemical and normal compositions the ultrabasites correspond to lherzolites and sometimes websterites with very low M/F values (~4), increased TiO_2 content (~ 0.50% at an average). Harzburgites have been found in seperate cases only. On the basis of the mineral and chemical composition of these ultrabasites, their textures and especially their relationship with the rocks of the gabbro series it seems that they can be considered as part of the cumulative complex of the ophioliticassociation. Their occurrence as chaotically scattered lenses in a single aleuro-peliti sedimentary complex leads to the idea that these are fragments of older ophiolites trapped in the form of olistolites and olistoplaques in the Lower Paleozoic sediments (unpublished data of Ivanov, Kolcheva and Moskovski).

The large massifs of Deli Jovan and Zaglavka in East Serbia occur in analogues geological environment. According to Tersic (1960) ultrabasites occur in the northern part of the Deli Jovan massig as beds and layered bodies up to 5-6 km long and up 1 km wide amid rocks where gabbro predominates. They are the end members, the most basic differentiates - dunites, plagioclase-bearing dunites and wehrlites, blending into olivine gabbro. The olivine content ranges from 5 to 90%. The amount of plagioclase (An_{95} to An_{75}) varies widely. The rocks show various degrees of serpentinization, amphibolization and carbonatization. Lean mineralizations of alumina-rich chromite ores (Cr_2O_3: Al_2O_3: Fe_2O_3 = 8:9:4) traced as beds or schliers of no productive importance are associated with the Deli Jovan ultrabasites.

The problem of the age of the completely serpentinized harzburgites and of the less frquent dunites exposed along the western margin og the Moesian platform (near the villages of Planinica, Stubik, Brestovac and other places) is not yet solved. Those are small outcrops of tectonically deformed bodies emplaced as diapits along faults. Judging from certain peculiarities (local transitions toward plagioclase peridotites), they are similar to the Deli Jovan ultrabasites and also carry small chromite mineralizations.

The Mesozoic ultrabasites are the most extensively found on the Balkan Peninsula (Fig. 1). They represent members of the ophiolitic associations occupying the central sections of the large Tethys ophiolitic belt, which, passing through the Apennines and Alps, is marked on the Balkan Peninsula in the Dinarides and Hellenides spreading eastwards into Anatolia. The Mesozoic ultrabasites in the Dinarides, Albanides and Hellenides form part of two well-expressed ophiolitic belts of NW orientation. Extending round the eastern and western flanks of the Pelagonian massif they join together, then further north they branch off again forming already three distinct belts of the same northwestern orientation.

The ultrabasites are always associated with gabbro and

diabases (spilites), i.e. they belong to the typical ophiolites. The period of their formation is probably limited within the interval between the Middle-Upper Triassic and the Upper Jurassic. Their emplacement into the rocks surrounding the association, however, took place probably at the end of the Middle and at the beginning of the Upper Jurassic. They consist both of tectonic periodитites and of coumulites. The size of the tectonic ultrabasic massifs as seen on the present-day erosional surface varies greatly, ranging from several to hundreds of m^2, reaching in the largest bodies hundreds to thousands of km^2 (Zlatibor, Krivaja-Konjuh, the Merita Mountain, etc.). Usually these are plates of small thickness overlying the associated sedimentary rocks (containing inclusions of spilites, diabases and gabbro). The largest ultrabasic plate of Zlatibor is only about several hundred meters thick. The Krivaja-Konjuh massif is somewhat thicker (~1 km) and represents a concave plate marking a synclinal structure.

The grade of serpentinization varies. The central parts of the large massifs are moderately serpentinized (20-50%), while the peripheral zones and the strips along tectonic faults are strongly, almost completely serpentinized. The small ultrabasic bodies completely altered.

The ultrabasic of the Western Dinarides are mostly of lherzolite composition, and the massifs of the Eastern and Southern Dinarides, as well as of the Hellenides are chiefly dunite-harzburgites representing a depleted Upper Mantle material. The amount of the accompanying basic (igneous and volcanic) rocks, considerably larger in the eastern parts, indicate a higher degree of depletion.

The lherzolites prevailing in the western part (starting from the Banija massif as far as Krivaja-Konjuh) consists of olivine (Fo_{91-89}), enstatite (Fs_{9-12}) with clinopyroxene exsolution lamellae (diallage of high MgO:FeO ratio and somewhat increased Al_2O_3 content). They contain also subordinate amounts of pargasite amphibole, accessory picotite and various secondary minerals. The normal mineral composition of lherzolites varies within narrow limits: olivine 60-75%, enstatite 15-25%, clinopyroxene 15-25%. They have parallel structure determined by banding and more frequently by foliation ("gneissoid" structure). The texture is porphyroblastic with strongly deformed grains as well as kataclastic. A characteristic feature of the massifs in the Northwestern Dinarides is the absence of any productive chromite mineralization in spite of their size, which in some cases is extremely large. Further to the south, on Albanian territory, occur two zones of ultrabasites, a western and an eastern zone. The latter comprises ultrabasite massifs with productive chromite mineralization (Bulqiza, Kukës, Tropoja, etc.). The massifs consist of serpentinous harzburgites with local occurrences of dunites and very rare basites. They are crosscut by infrequent pyroxenite and gabbro-pegmatite dykes.

The ultrabasic bodies from the southeastern part of the Mesozoic ophiolites consist of harzburgites and dunites in varyin proportions. Besides the gradual transitions between then there are also (e.g. in Brezovica and Ljuboten) irregular amoeba-like dunite bodies in the harzburgites and boudin-like lenses of harzburgites in the dunites. North of the Goles massif dunites are very rare, and south of it the harzburgite: dunite ratio is about 3:1. Harzburgites consists of olivine (Fo_{90-92}) and enstatite (Fs_{8-12}) with clinopyroxene exsolution lamellae and accessory chromspinelid. Dunites contain olivine of the same composition, up to 2-3% of enstatite and chromspinelid. Dunites are always more serpentinized than harzburgites. The rocks are of massive structure, in some cases (in the massifs more to the north) harzburgites show indistinct parallel structure (e.g. in the Troglav massif).

The cumulative perodotites of the Mesozoic ophiolites occur in gabbro and rest over tectonic ultrabasites. They consist of a great variety of rocks: dunites, plagioclase dunites, lherzolites and wehrlites, frequently of poikilitic texture. The magmatic sequence of mineral formation, olivine - pyroxenes - plagioclases, is well expressed. The basic rocks are re-

represented by gabbro and spilitized diabases and basalts. The Dren massif in Macedonia (2 x 60 km) is very interesting. Its lower parts consist of gabbro and melanocratic gabbro blending into cumulative peridotites of great variety of structures. These are crosscut by quartz-dioritic dykes. Higher up (the middle zone) the massif consists mostly of diabase subparallel dykes accompanied by some keratophyric dykes. In the uppermost zone there are pillow-lavas.

The Vourinos massif in North Greece presents another very interesting example of cumulative ultrabasites. According to Jackson et al. (1975), the Vourinos massif is characterized by a well expressed banded complex about 1500 m thick. The lower part of the sequence consists primarily of olivine cumulates with thin olivin-clinophyro-xenite zones. Higher up there is a thick series ($\leqslant$ 500 m) of olivine cumulates containing thin chromite interlayers. Further up the sequence, already a typical directed rhythmicity is observed. The separate rhythms start with olivine cumulates blending upwards into olivine-clinopyroxene and, further up, into clinopyroxene cumulates with some admixture of postcumulative plagioclase. The separate thythms are about 33 m thick. The amount of olivine decreases regularly from the bottom to the top of the sequence. In the uppermost sections of the Vourinos massif the rhythms start with two-pyroxene cumulates blending upwards into thick cumulates with plagioclases and two pyroxenes. Rhythmic structure (frequent alternation "chromitites - dunites - harzburgites-dunites - chromitites", etc.) is observed in the dunite-harzburgite massifs of southwestern Chalkidiki. Their textures are medium-grained, kataclastic, recrystallized with deformed olivine and pyroxene. The cumulative ultrabasites are of little value as a source of mineral raw materials.

The economic mineral deposits associated with the ultrabasites of the Balkan Peninsula are quite varied. They belong to the liquid magmatic, postmagmatic, weathering crust, sedimentary and metamorphic mineralizations (Fig.2).

The largest are the podiform chromite deposits, rich in chromium characteristic of the peridotitic tectonites of the Mesozoic group of ultrabasites - Radusa, Deva, Bulqiza, Kozani (Vourinos) etc. A number of massifs in the Hellenides carry alumina-rich chromite mineralizations, a raw material for refractory products. Of the same type is the small Paleozoic deposits of Gomati as well as the deposits of Dobromirci, Goljamo Kamenjane, Jakovica, Soufflion, Sinikl etc. associated with the Rhodope postmetamorphic group of ultrabasites. Chromite mineralization is practically absend from the premetamorphic ultrabasites. Only small mineralizations, accumulations of iron-rich chromspinelids, have been found. The origin of the specific chromium- (up to 14% Cr_2O_3) and nickel-rich (up to 3.5% NiO) metamorphic magnetite ores in the area of Ardino is attributed to the mobilization of matter during the regional metamorphism.

Liquation copper-nickel sulphide mineralizations are associated primarily with the basite members of the ophiolithic formations. However, lean mineralizations have been also found in the peridotite zones of the Paleozoic (Frolosh) and Mesozoic (Petkovic, Smolikas) ophiolithic complexes.

Asbestos mineralizations are confined to the contacts and to the zones of intensive tectonic and postmagmatic reworking of all groups of ultrabasites. However, specific differences are also to be found. Anthophyllite-asbestos occurs in the Rhodope ultrabasites (Goljamo Kamenjane, Jakovica, Avren, etc.). Unique deposits of picrolite asbestos (Brusevci and Djebel) are also known. The Paleozoic ultrabasites contain only lean tremolite-asbestos mineralizations (Kopilovci, Frolosh). Productive chrysotile-asbestos deposits are associated with the Mesozoic ultrabasites in Stragari, Bosansko Petrovo Selo, Korlache, Rujishte, Puka, Kozhani, etc.

Talc deposits formed in all cases when the ultrabasites suffered metamorphic alterations and were enriched in silica (and in carbon dioxide as well). Processes of intense talcization have been traced in many areas in the Rhodope Mountains (Djebel, Ovcharovo, Pletene, Krainevo) and in the Paleozoic ultrabasites

(Crna Trava, Mackatica, etc.). The largest talc deposits in Mesozoic ultrabasites are frequently associated with chrysotille-asbestos deposits (Zarkovac, and Muzici). The talc deposits of Izvor, Kesandra and Zemblak, and a number of deposits in the areas of Larisa and Thessaloniki, as well as on the Tenos Island are under explotation.

Vermiculite deposits are very rare. They have been reported in the contact zones of pegmatites crosscutting the Rhodope ultrabasites.

Large magnesite deposits, directly or indirectly associated with ultrabasites (primarily Mesozoic), are know in the Balkan Peninsula. They belong to the following genetic types: infiltration deposits (in the bottom parts of ultrabasitic weathering crusts) - Zlatibor Maljen etc., of little economic value; hydrothermal deposits (vein and metasomatic) occurring in ultrabasites as products of the action of later carbonated fluids - Vardos and Gerakini (Western Chalkidiki), Mantoudi (the Euboea Island), Golesh, etc., as well as the sedimentery deposites formed in young freshwater basin ty the influence of magmatogene fluids which mobilized magnesium passing through the ultrabasites - Bela Stena, Strezovce.

All ultrabasites in the Balkan Peninsula, which were exposed on the surface by the geological processes, suffered mechanical and chemical weathering. The latter affected most intensively the Mesozoic massifs during the Lower Jurassic - Upper Cretaceous interval and in Tertiary. These processes resulted in complete sections of weathering crusts of the areal, linear and karst types, among which those formed over tectonic peridotites are of greatest economic value. Fossilized weathering crusts, which are workable silicate-nickel cobalt-bearing deposits, are known in the Vardar zone, in the Dinarides and Hellenides. The largest among them are those of Golesh, Vrnjacka Banja, in the area of the town of Kukës, in the valley of the Shkumba River etc. Later on in many cases the changing geological environment started destructive processes affecting the weathering crusts themselves. In some areas the transgressing sea washed out and redeposited in situ forming oolitic iron-nickel oxide ores (Shumadija). In other cases the washed out material after some transport was redeposited under marine conditions. The redeposited materials are nickel-iron ores with chromium and cobalt of great economic interest. There are large workable deposits of this type in Rzanovo, Cervenaka, Bushtrica, Pagonda, near the town of Larymna and in other places.

As a conclusion of the review on the ore potential of ultrabasites on the Balkan Peninsula it should be noted that the rocks themselves more and more often draw the attention of Balkan geologists as workable raw materials. Even ancient Greeks highly valued the decorative properties of ultrabasite rocks. Now they are used also in the procudtion of forsterite refractories, sitals and fertilizers as inert components, etc. Apparently, however, the most promisint use of ultrabasic rocks in the near future will be the wasteless production of magnesium, nickel, cobalt and iron oxides and glass.

REFERENCES

AUGUSTITHIS, S.S. (edit.) 1981. Unesco and International Symposium on metallogeny of mafic and ultramafic complexes: The eastern Mediterranean - Western Asia area, and its comparison with similar metalogenic environments in the World. - vol. 1-3, Athens.

DUNNING, F.W., MYKURA, W., SLATER, D. (edit.) 1982. Mineral deposits of Europe, vol. 2: Southeast Europe. 304. London.

EVANS, W. and TRAMMSDORFF. 1970. Regional metamorphism of ultramafic rocks in Central Alps: Parageneses in system CaO-MgO-SiO_2-H_2O - SMPM, 50, 3, 481-492.

COLEMAN, R.G. 1977. Ophiolites. Springer-Verlag, Berlin, Heidelberg, New York, 229.

HIESSLEITNER, G. 1951-52. Serpentin und Chromerz - Geologie der Balkanhalbinsel und eines Teiles von Kleinasien. Jb. geol.Bundesanst., Sonderband 1, pts 1 and 2, 683, Wien.

JACKSON, E.D., GREEN, H.W., MOORES, E.M. 1975. The Vourinos ophiolite, Greece: cyclic units of lineated

cumulates overlying harzburgite tectonite. Geol. Soc. Am. Bull., 86, 390-398.
KARAMATA, S., MAJER, V. and PAMIC, J. 1980. Ophiolites of Yugoslavia. - In: G. Rocci/edit./"Tethyan Ophiolites", spec. issue Ofioliti, 1, 105-125, Bologna.
PANAYIOTOU, A. (edit.) 1980. Ophiolites, 781, Nicosia.
SCARPELIS, N., ECONOMOU, M. 1979. Genesis and metasomatism of Chromite ore from Gomati Area, Chalkidiki, Greece. - Annal. Geolog. d. Pars Helléniques, 716-728, Athenes.
TERZIC-PETKOVIC, M. 1960. Petrolosko proucavanje gabrova masiva Del Jovana. - Geol. anali Balk. Poluostrva, XXVII, 327-416, Beograd.
WOODTLI, R. (edit.) 1964. Methods of prospection for Chromite, 244, Paris.
ZHELUASKOVA-PANAYOTOVA, M., IVANOV, Z., KOLCHEVA, K., CHAIDUTOV, I. 1982. Ophiolites and ultrabasites on the territory of Bulgaria. Bulg. Acad. Sci., 98, Sofia.

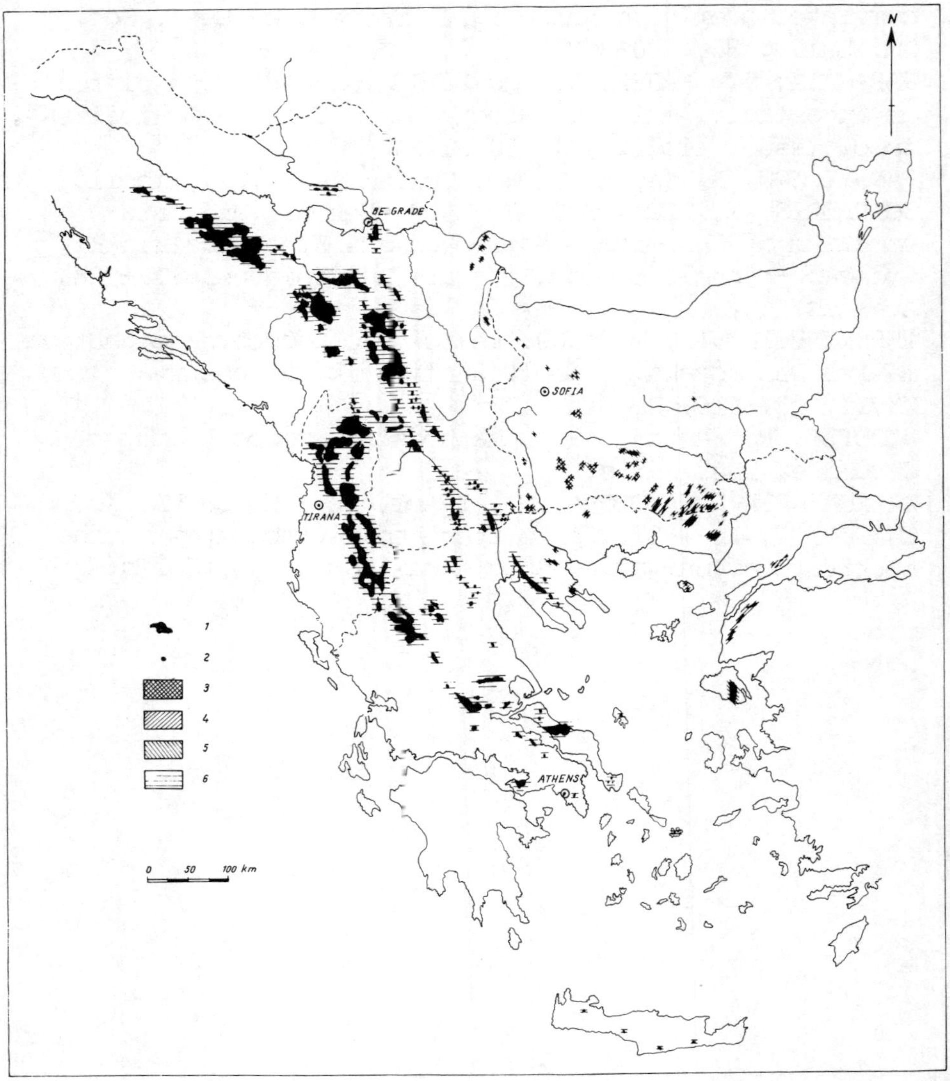

FIGURE 1. Map of the distribution of ultrabasic rocks on the Balkan Peninsula.
1 - ultrabasic rocks; 2 - small bodies of ultrabasic rocks; Ultrabasic groupes; 3 - Rhodope premetamorphic ultrabasites; 4 - Rhodope postmetamorphic ultrabasites; 5 - Paleozoic ultrabasites; 6 - Mesozoic ultrabasites.

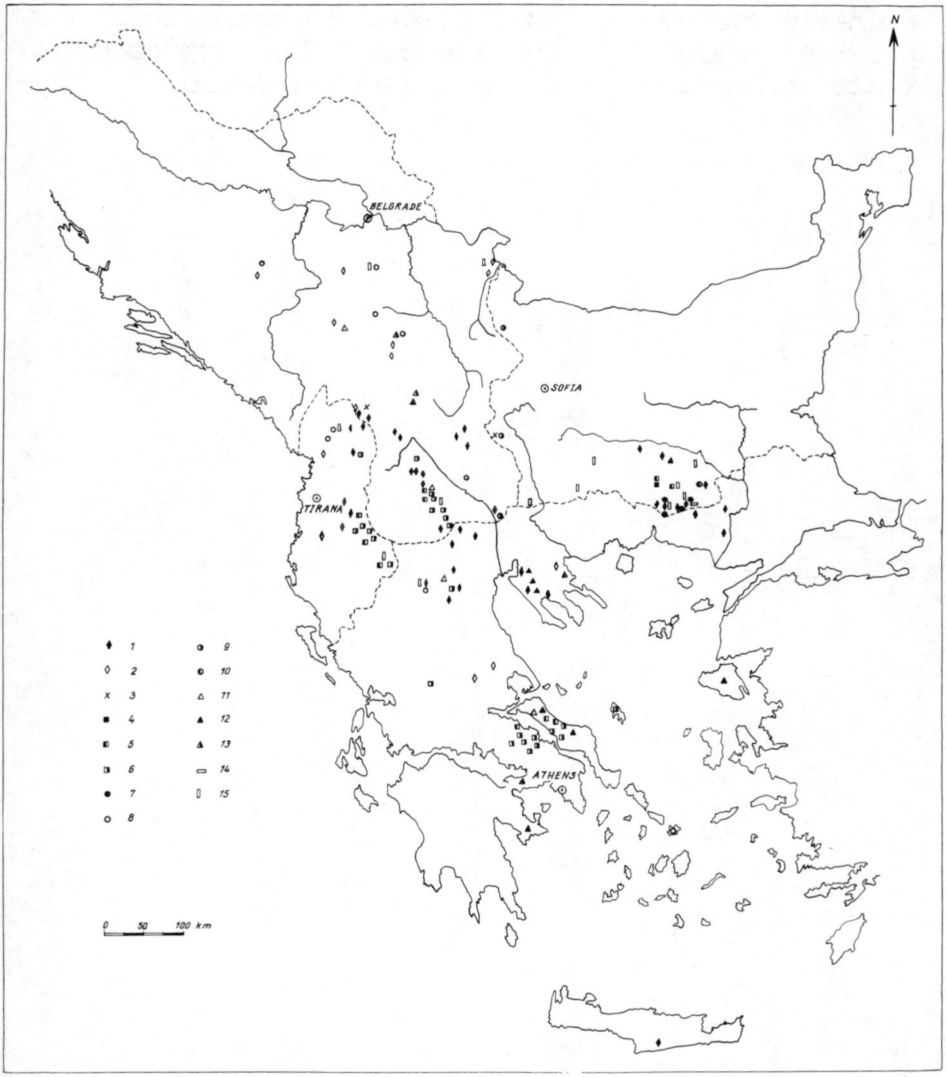

FIGURE 2. The principal mineral deposites related with the ultrabasic rocks of Balkan Peninsula.
1 - Cr-rich chromite ores; 2 - Al-rich $\pm$ Fe-rich chromite ores; 3 - Cu-Ni sulphide ores; 4 - Fe-Cr-Ni metamorphic ores; 5 - Ni-bearing silicate ores; 6 - Redeposited Fe-Ni silicate oxide ores; 7 - Anthophyllite-asbestos; 8 - Chrysotile-asbestos; 9 - Tremolite-asbestos; 10 - Picrolite-asbestos; 11 - Magnesite from

weathering crusts; 12 - Hydrothermal magnesite in ultrabasic rocks; 13 - Hydrothermal sedimentary magnesite in freshwater sediments; 14 - Vermiculite; 15 - Talc.

POTENTIAL ORE-BEARING CAPACITY OF ACID MAGMATIC ROCKS

KOVALENKO V.I., NAUMOV V.B., BOGATIKOV O.A.
Institute of Geology of Ore Deposits, Petrography, Mineralogy and Geochemistry and Analytical Chemistry, Academy of Sciences of the USSR, Moscow.

This paper deals with criteria for the identification of potential ore mineralization and conditions of formation of acid magmatic rocks, with which a large number of endogneic ore deposits is associated (primarly tin, tungsten, molibdenum, rare elements and rare earths, sometimes gold, copper, polymetals, iron and some others).

In the analysis of conditions of formation specific attention was given to the generalization of the available (primarily original) data on microinclusions of mineralizers (melts, fluids) as well as to the comparison of such data with natural and experimental petrological facts. Criteria for the identification of potential ore-bearing acid magmatic rocks are different for their different geochemical types /4, 10, 23/: lithium fluorine (with ongonites /4/ and topaz-bearing rhyolites /25/ as volcanic analogues), standard and alkaline-granitic /26/. A geochemical type cover- a group of rocks similar in chemical and mineral composition, distribution and behaviour of petrogenic and rare elements, capable of forming independent intrusive bodies or volcanic fields with a specific combination of late- and post-magmatic formations, including ore-bearing ones. It should be noted that the lithium-fluorine and the standard geochemical types of acid magmatic rocks belong mainly to the groups of normal (often praluminous and subalkaline leucogranites and rhyolits /12/, whereas lithium-fluorine granitoids and thier volcanic analogues typically have higher contents of lithium, rubidium, fluorine, tin tantalum and other rare lithophile elements, relative to their average contents in acid rocks. The geochemical type of the alkaline granites includes mainly agpaitic acid rocks. Mineral occurences of rare elements (tnatalum, niobium, lithium, rubidium, beryllium /8, 25/, tungsten, tine molibdenum, sometimes copper, are associated with

Proceedings of the 27th International
Geological Congress, Volume 9, pp. 199–219
PETROLOGY (IGNEOUS AND METAMORPHIC ROCKS)

rocks of the lithium-fluorine type; occurrences of the rare earths, niobium, zircon and some other rare elements /8,4/ are associated with alkaline granites. Granites of the lithium-fluorine type often formed from residual magmas during an extreme differentiation of magmas of the standard geochemical type granites /4/.

Criteria for the identification of mineralized acid magmatic rocks

The following criteria for the identification of potential ore mineralization in magmatic rocks have been suggested basing on a great amount of factual data related to relationships between magmatism and mineralization (primarly from the Mongolo-Okhotski belt territory). For rare-metal tin and tungsten occurences in association with the lithium-fluorine leucogranites, ongonites and often with rapakivi granites /8,4/ those are:

1) abundance of hypabyssal facies leucogranites representing the late products of the differentiation of granitoid complexes pertaining to the final stages of folded belt evolution

2) presence of the microcline-albitic granitoid types as well as of thier varieties, containing lepidolites, topaz, amazonite, all featuring heterogeneous structures and textures, characteristic of the rare-metal granites

3) ilmenite-monazitic or monazite-clumbitic with fluorite and topaz character of accesory minerali zation

4) higher, relative to their average contents in granites, concentrations and notably dispersions of the content of the typomorphic Li, Rb, F, Sn, Nb, Ta, Hf, sometimes Be, W, lower concentrations of Sr, Ba, REE, lower Nb/Ta and Zr/Hf ratios; eupoium minimum in the rare earth spectrum; the La/Yb ratio close to unity

5) typomorphic rare elements enrichment of the late stage rocks and of the apical parts of dome-like massifs

6) presence of specific metasomatites in the massifs: zwitters quartz-lepidolitic greisens, feldspathides and albitites

7) traditional prospecting criteria: original occurrences of ore mineralization, concentrate aureoles of

of exocontact rocks enriched in Li, Rb, Sn; dome-like forms of massifs; small degrees of erosian of intrusions.

Let us also list the criteria for the identification of potential tungsten and tin ore mineralization massifs of the standard geochemical type leucogranites /8, 4, 10/, i.e. those which do not have higher concentrations of the lithophile rare elements:

1) leucocratic character of granites
2) two-mica and miscovitic mineral assemblages in rocks
3) presence of quartz-muscovite greisens and quartz veins
4) higher contents of tungsten in greisens and other metasomatic formation
5) higher F/Li ratios in siderophyllite
6) dome-like forms of massifs
7) genetic association with granitoids of the lithium-fluorine geochemical type
8) formation of massifs under the conditions of the hypabyssal and the near-surface depth facies.

The following are the major criteria for the identification of potential rare-metal mineralization /8/ in alkaline granitoids:

1) abundance among granitoids of the varieties with aegirine
2) presence of the microcline-albitic granitoid varieties
3) abundance of elpidite and other zircon-, titan- and niobosilicates of alkalies, as well as of tantalo-niobates, rare-earth silicates, cryolite, etc.
4) high values of the agpaitic coefficient in rocks, substantially higher than unity
5) higher, relative to the average contents in granites, concentrations of Li, Rb, Be, Zn, Sn, REE, Nb, Ta, Zr, sometimes F, lower concentrations of Ba, Sr, and higher Nb/Ta, Zr/Hf ratios, europium minimum in the rare earths spectrum and a high La/Yb ratio
6) accumulation of the above-listed typomorphic rare elements in the late-stage rocks
7) occurence of alkaline rock bodies with heterogenous structures (banded, maculose, etc.) and the presence of schlieren alkaline- granitic pegmatites
8) presence of radioactive anomalies over massifs of

rare-metal alkaline rocks detected by air-borne methods; concentrate aureoles of tantalo-niobates with zircon; dome-like and circular forms of massifs with small degree of erosion late formation character, relative to other magmatic rocks.

Physico-chemical parameters of the formation of ore-bearing acid magmatic rocks

Owing to the significant progress made during the recent years in studing melt and fluid microinclusions in minerals, it is now possible to characterize the physico-chemical parameters (such as temperature, pressure, fluid composition, water content in melts, etc.) of the "magmatic melt/hydrothermal sulution" ore-bearing system for the above-described geochemical types of acid magmatic rocks responsible primarily for the formation of tin, tungsten, molibdenum, rare element and rare earths deposits, more rarely - of the cooper, gold, iron and other deposits.

All the currently available data on the temperatures (homogenization) and pressures of the fluid in the acid melt /solution system are presented in Fig.1, which is a supplemented version of the originally published one /14/. The dashed line represents the solidus lines in the "granite-H_2O" and "ongonite-H_2O-Hf" systems /9/. Lines were also drawn which characterize the two-phase equilibrium "vapour-liquid-halite" /a/ /29/. The projections of critical points in the H_2O-CO_2 and H_2O-CH_4 systems /29/ are shown for containing volatiles solutions. Besides, Fig.1 shows the lines corresponding to the temperatures (at a gradient of 50^oC/km) and the pressures (at a geostatic load of 270 bar/km) at the depths of 3 and 7 km. Such depths were chosen because many inverstigators estimate the depth for the formation of rare-metal deposits not be in excess of 2-3 km, with reference to a number of geological criteria. According to the data from many gologists, the depth of the formation of granitoid massifs associated with rare-metal deposits does not exceed 3-5 km.

The analysis of this diagram reveals the following major properties of the system "acid ore-bearing melt/

solution":

1) The temperature of acid magmas vary from those in exces of 1000°C towards 550°C. With the increase in the fluid pressure, which can range from 100 bar to 6 kbar, temperature of magmas decrases along the line of the water-enriched granitic solidus. The lowest-temperature magmas plotted in Fig.1 below the solidus of the aqueous granite are the ongonitic, alkaline-granitic and lithium-fluorine granitic magmas, for which a low-melting character has also been established experimentally /4,9/. It is not incidentally that they fall into the solidus continuation in the system "ongonite-H_2O-HF". A substantially wide temperature range for the crystallization of ore-bearing acid magmas has also been noted in specific magmatic bodies (e.g. ongonites /19/). Undoubtedly, such a broad range of crystallization represents a favourable factor for teh extreme differentiation of the magmas giving rise to the lithium-fluorine granites and ongonites.

2) Between the fields of the magmas and the fluids of the rare-metal deposits of the ore productive stages there exists a field of high-temperature pre-ore fluids coexisting either with crystallized granites or with magmas that froquently implies the boiling of the latter. It can possibly be suggested that quite often before the precipotation of ores the fuilds must have undergone a durable thermal, dynamic and geochemical evolution, beginning with their seperation from the magma or from the rock which crystallized it.

3) Judging from the geothermal gradient the temperatures of the magmas and the related ore fluids are 150°-300°C higher than possible the temperatures at the depth of 7 km for magmatic rocks and of 3 km for deposits, which demands a substantial heat supply and movement of large magma volumes from deeper horizons of the crust.

4) Fig.1 clearly shows that in many cases crystallization of acid melts would occur at high fluid pressures, substantially exceeding the geostatic pressure. Aqueous fluids generated at the magmatic stage also are under high pressures relative to the pressure caused by the overlying rocks. In this connection,

such fluids have high density and can dissolve larg amounts of many elements including the ore elements. It is clear that resulting excessive pressure gradients may account for the movement of the hydrothermal solutions responsible for the formation of ore deposits. Theoretically one could anticipate the excessive fluid pressure phenomenon in the crystallizing magma, since volume per units of mass of the acid magma with a given water content is smaller than the volume of the same mass of crystals and of the fluid /24/.

5) Generally, regressive evolution of the system "acid magma/fluid" is directed towards the decrease in fluid pressure, which is perfectly consistent with the location of ore deposits above magmatic bodies and accounts for the uprising of ore-bearing magmatogenic fluids. At the same time, the decrease in the fluid pressure during the transition from the magmatic conditions to the conditions of the postmagmatic ore-formation process, as well as the related drcrease in density and dissolving potential of the fluid, are favourable for the precipitation of ore material. For example, for the Industrailnoye tin deposit (USSR), the magma temperature is 820^0--740^0C, the fluid pressure - 2,5 kbar; the same parameters for the ore-formation process: 430^0-340^0C and 1,7-1,2 kbar. For the Baga-Gazryn (Mongolia) tin-tungsten deposit the parameters are: 830^0C and 2,4 kbar for melts, 310^0-340^0C and 150-250 kbar for fluids; for the Kyzyl-Tau (Mongolia) tungsten deposit these are: 680^0C and 5,6 kbar for melts, 350--300^0C and to 2 kbar for fluids.

Complex investigations of the fluid inclusions in minerals of the deposits under consideration and in leucogranites have shown that the ore fluids can be subdivided into three major types (Fig.2): low-concentration, aqueous /3/, aqueous-carbonate /2/ and high-concentration, chloride /1/. The latter are identified by the presence of captured minerals in the inclusions (halite, sylvite, potassium chlorides, manganese and iron), whereas the second group is identified by the presence of liquid carbon dioxide or of gas hydrate ($CO_2 \cdot 5,75\ H_2O$). It should be noted that the presence of halite indicates the NaCl concentration in excess of 350 g/kg H_2O, while the presence of the liquid carbon dioxide phase testifies to the CO_2 concentration not less than 60 g/kg H_2O.

The generalized diagram "temperature - solution concentration"presented in Fig.2 is based on 300 determinations from 55 ore deposits and occurences published in international literature (63% of determinations on quartz, 14% on fluorite, 9% on cassiterite, 14% on topaz, scheelite, apatite, beryl, hyhnerite and calcite). Field No.1 is characterized by very high concentrations of solutions: over 25--30 wt% NaCl, therefore, when such solutions cool to the room temperature, the captured minerals, whose number is normally 1 to 5, sometimes 7 to 10, will always crystallize. The abundant among the solid phases are chlorides, that is why the solutions of Field 1 are substantially chloride. Field 3 includes rather numerous analyses (made by the cryometric method without splitting yp the inclusions) with relatively low concentration solution, in most cases not exceeding 8-10 wt% NaCl. Limited data so far available on eutectics temperatures of these solutions, apparently indicate their dominantly fluoride compositions. Field 2, which is intermediate, is characterized by substantially carbonate solutions with CO_2 concentration of 5 to 25 wt%. The concentrations of electrolytic salts in such solutions usually are low (not exceeding 3-7 wt% NaCl). It is clear that the solutions which constitute Field 1 are on the whole more high-temperature ones. The temperature difference can reach 100^0--150^0C. Typical ore elements in corresponding deposits are shown for each field of fluid composition.

The currently available data (unfortunately incomplete) on the compositions and concentrations of the pre-ore stage solutions of rare-metal deposits, suggests that the major physico-chemical features of fluids are formed at the magmatic and early post-magmatic stages. This is evident from the succession of the ore-bearing fluid composition from its magmagene predecessors. For example, at the Industrialnoye (USSR) tin deposit, the ore stage solutions fall into the Field 2 in Fig.2 and are characterized by the temperatures of 465^0--275^0C, the pressures of up to 1 kbar, electrolytic salts concentrations of 62-29 wt%NaCl and the densities of 1,73-0,70 g/cm^3. The pre-ore fluids, including the magmagene ones, exhibit the temperatures of 920^0-760^0C, the pressure of up to kbar, the concentrations of up to 75 wt% NaCl and the density of up to 1,9 g/cm^3 /18/. Another example is the Spokoinoye tungsten deposit (Eastern Zabaikalje, USSR). The study of fluid inclusions in quartz, wolframite and fluorite of greisens and ore veins indicated participation in its formation of solutions with high CO_2 concentrations (Field 2, Fig.2). A detailed investigation /21/ of the liquid and fluid inclusions in quartz of the Spokoinoye granites made it possible to detect essentially carbonate fluids at as early as at the magmatic stage of the massif formation.

Generally, there is likely to be an association of the Field 3 fluids primarily the lithium-fluorine granites. Aqueous fluids, sometimes containing substantial amount of carbon dioxide, are mainly known in association with alkaline granites.

Despite the above-listed fluid composition variations, the basic volatile component of natural magmas and ore-bearing fluids is water. Fig. 3 presents a summary of all the results available at the moment (115 determinations) on the concentration and the pressure of water, obtained on liquid inclusions in minerals of acid magmatic rocks. Lines have also been drawn in Fig.3 to characterize the dependence of the water solubility on pressure at the temperatures of 700^0 and 1000^0C, established experimentally for the Eldzhurtinskyi granite (Tyrnyauzskyi region, Northern Caucasus) /11/.

The results obtained indicate substantial variations in the concentration of water dissolved in acid magmatic melts: from 0.1 to 13.9 wt%. The majority of analyses are characterized by high concentrations of about 2-8 wt%. It can be seen from Fig.3 that many compositions are poorer in water, relative to the water saturation isotherms for the Eldzhurtinskyi granite melt. It means that during the isolation of the inclusions, the melts were often undersaturated in water, though sometimes enriched in it. Beyond the saturation lines are only the melts of higher alkalinity and the melts with high fluorine contents. Experimental data on the melts of such composition, suggest a possibility of higher solubility of water in them /9/. However, the water concained in magmas is quite sufficient for saturating them during the late stages of melts and fluids often co-existed. There is a tendency in natural rhyolitic and ongonitic magmas for a decrease of water content with the increase in SiO_2 content /20/.

Considerable variations are also observed for the water pressure. It can vary from 0,1 to 6,1 kbar with a distinct tendency for direct correlation between the water pressure and its concentration in the melt (Fig.3). For a number of locations, e.g. the Eldzhurtinskyi granite intrusive, the Ary-Bulakskyi ongonitic intrusive (Zabaikalje, USSR) and some others, there is a tendency for the increase in water concentrations and pressures with the decrease in magma temperature /14,17,2/. Note also a significant fluid pressure excess at the magmatic stage, relative to the pressure of the geostatic load of overlying rocks.

On the basis of the data presented in Fig.1 and in Fig.3 and using equation /61/ from 31, as well as the composition of biotites from ore-bearing granites (x=Fe/Fe+Mg=0,5-0,95), according to the data of /13/, one can estimate oxygen fugacity in magmagene fluids. At 700^0-900^0C, P=6 kbar, x=0,95, it varies around 10^{-15}, reaching 10^{-14} at x=0,5, T=900^0C and P=1 kbar. Under these redox conditions, the paragenesis of the potassic feldspar, magnetite and biotite characteristic of many ore-bearing granites is stable.

Possible origins for ore-bearing acid magmas

Analysis of the data presented here and of the abundant published data on the compositions of the ore-bearing acid magmas, shows that the rare-metal mineralization can be associated with two major groups of acid magmas:

1) Extremely differentiated magmas of the ongonitic type (lithium-fluorine type) and of the alkaline-granitic type, effectively concentrating ore components.

2) Less differentiated magmas with ore element concentrations close to the average (tin and tungsten-bearing leucogranites of the standard geochemical type).

Associated with the magmas of the first group are the magmatic deposits of rare elements themselves (rare granites, pegmatites, etc.) and the post-magmatic hydrothermal occurrences of tin, tungsten and molibdenum. Associated with the magmas of the 2nd group are hydrothermal tungsten deposits, more rarely - tin and molibdenum deposits. Consequently, the processes of magmatic differentiation (crystallization and emanation) and distillation are of major significance for the concentration of the ore components of the first group, and for the second one the distillation processes must be important.

Magmatic differentiation

Experimental studies in the systems: "granite-H_2O-HF", "ongonite-H_2O-HF" /9/, "albite-orthoclase--quartz-H_2O-F" /27/, which have a critical role for the understanding of the evolution of the magmas of the lithium-fluorine granites, have shown that the presence of fluorine in granitic systems promotes the following effects:

1) The solidus temperature decreases to 550-575^{0}C at 1 kbar bulk pressure with the liquidus temperature decreasing substantially at the same time;

2) The quartz crystallization field broadens up to the formation of normative albite-rich residual

melts ($Quartz_{15}$, $Albite_{58}$, $Orthoclase_{27}$);

3) The co-existing albite, potassic feldspar, lepidolites, topaz, fluorite crystallize from the melt;

4) The melt crystallization temperature range broadens to 200^0-500^0C, relative to normal granites;

5) The melt viscosity decreases;

6) Due to a correlation between the fluorine content and the bulk composition of acid magma, it is possible that emanative differentiation of the latter may take place enriching more fluorine-rich magmas in many rare elements /4,26/.

Some similar effects in granitic magmas bring about a high agpaitic coefficient, equivalent to the addition to magma of a standard metasilicate, for alkaline granitic magmas (accompanied by crystallization of alkaline amphiboles, aegirine replacing lepidolites and topaz) /4/ and an increased boron content.

An important parameter of crystallization differentiation of magmas are the partition coefficients. For ongonites, the equilibrium values of bulk partition coefficients (K_b) between crystals and melts are considerably lower than unity /8,7/: lithium 0,14-0,28, cesium 0,30-0,50, beryllium 0,08-0,40, fluorine 0,02-0,31, niobium 0,05-0,09, tantalum 0,02-0,04, tungsten 0,01-0,20. In this connection, the above-listed rare elements accumulate in residual magmas of the lithium-fluorine granites. In the residual magmas of the alkaline granites, these elements are supplemented by rare earths, zircon, hafnium and by some others. Due to the established decrease of the K_b effective values and with the increase in concentrations of some elements in magmas a more intensive concentration of rare elements takes place, relative to the processes that occur at K_b = const. We have called this process "progressing concentration" /5/. It can play a significant role in the formation of rare-metal deposits. The essence of this process can be understood from Fig.4, presenting the partition coefficient of rubidium between potassic feldspar and melt: fractionation of the acid magma at $K < 1$ leads to enrichment of rubidium in the residual melt, that causes a decrease of K, which in its turn, is likely to bring about further enrichment of the element in the residual magma and so forth, with the continuous decrease of K and progressive increase of the concentration of the elements in the magma.

Similar relationships have been established for lithium, tin, tungsten, beryllium, fluorine and some other elements including ore elements, not only for individual, but also for bulk partition coefficients. They determine the high probability of the progressive concentration of rare elements in the course of fractionation of the acid magmatic melts of the first group, possibly with participation of fluids. This is an essential pre-requisite for the formation of rare element deposits in pegmatites and in pegmatites and in rare-metal granites, and it leads, in some cases, to the occurrence (in ongonites) of magmatic phenocrysts of cassiterite and tungsten.

This process is responsible for the above-listed empirical criteria for the identification of ore-producing potential in granitoids: the increased concentrations of ore elements in rocks, minerals-concentrators; the enrichment of elements in the late phases of massifs; europium anomaly in the rare earths' spectrum; the positive role of fluorine and of the excess of alkalies - aluminium (in alkaline granites); the presence of albites, lepidolites, topaz, of irregular textures, structures in granites, etc.

Naturally, the major physico-chemical parameters of the first group magmas (the lowest temperatures, high fluid pressures and high water concentrations in melts - see Fig.1 and 3) also reflect the extreme differentiation processes under consideration.

The termination of the crystallization in such magmas undoubtedly resulted in substantial growth of chemical potential of fluorine, water, and the increase of fluid pressure, its density and solubility. All this contributed to the intensive rare-element, tin and tungsten transition to late magmatic (distillation) and post-magmatic ore-bearing fluids. It is not accidentally that the ore-bearing fluids of Field 3 are associated with the lithium-fluorine granites, as shown in Fig.2. Due to the partition of fluorine to the melt, relative to co-exisiting fluid /9/, the latter often display low concentrations (Field 3, Fig.2).

The degree of mineralization of such fluids does not however totally depend on the fluorine content in magma, but it is also determined by the presence of chlorine, CO_2, and of other volatiles. If chlorine (or CO_2) also relatively accumulated in the residual melt, then due to the preferential partition of chlorine to the fluid, more concentrated fluids falling into Fields 1 and 2 in Fig.2 may separate from the magmas of the first group. Indeed such cases occur (e.g. fluids of the lithium-fluorine granites of the Baga-Gazrynskyi intrusive in Mongolia fall into Field 1).

Probably, the extremely low values of K_b of ore components between crystals and melts also contributed to the transition of these components form magmas to post-magmatic fluids. Comparison of concentrations of elements in melts with those of the rocks that formed them /1/ suggests that during the disappearance of the melt, such elements as beryllium, tingsten, tin, lithium, tantalum and others, substantially fractionate into the fluid phase. It is not accidental that these elements are characteristic of not only the magmatic rare-metal granites and ongonites, but of the hydrothermal deposits as well.

<u>Magmatic distillation</u>

The role of this process has been in part considered for the extremely differentiated ore-bearing magmas. Specific for the magmas of the group is a less extreme differentiation of the latter which suggests that ore-bearing fluids might have separated from some higher-temperature melts. Such fluids mostly plot within Field 1 of Fig.2, suggesting essentially chloride composition of such fluids. It is relevant that Field 1 in Fig.2, together with the fluids of the rare-metal deposits, exhibits a wide array of fluids related to deposits of iron, copper, molibdenum, sometimes polymetals, which have not been discussed in detail in this paper. This is appropriate to be mentioned since the major elements of these deposits relate to the chalcophile family, characterized by the proximity of the initial magmas to the saturation by ore phases, as well as by the partition coefficients between crystals and melts being close to 1 and by high affinity to chlorine /7,22,30/. The very presence of fluids related to some tin and tungsten deposits within Field 1 (Fig.2) indicates the existence in natural of the conditions under which geochemical behaviour of tin and tungsten appears to be similar to that of the chalcophile elements, at least as far as their affinity to chlorine is concerned. The conditions mentioned exist mainly during the crystallization of primarily magmas of leucogranites of the second, less differentiated group of rocks ("standard" granites), characterized by ore element concentrations close to the average. Formation of ores from potential ore-bearing granitoid magmas of the second group probably was favoured by an increased water content, which contributed to magma differentiation, weak as it might be, relative to that of the first group magmas. An increased water content is characteristic of magmas of many intrusives of the "standard" geochemical type leuzogranites (Fig.3), whereas units with H_2O-rich melts are likely to occur in other bodies, showing low water content in magmas. It can be suggested that such units of magmatic chambers are primarily responsible for the separation of the ore-bearing fluids. We would like to emphasize once again that magmatic differentiation of the standard leucogranites had not been so extreme as that for the first group melts and did not yield magmagene rare element concentrations. However, it probably facilitated the accumulation of chlorine in the melt, which separates from the higher-temperature melts, relative to fluorine.

In both groups of the ore-bearing magmas were observed excess pressures of fluids, which were, in the first place, characterized by a higher solubility of the ore-matter from magmas and rocks and, secondly, moved into the low-pressure zones (primarily to the apical and the over-intrusive zones of massifs) and precipitated ore components there.

Conclusions

The evolution of acid magmas tendin- towards the accumulation of fluorine, alkalies, boron and water, is favourable for the formation of the rare-metal and tin-tungsten deposits, both at the late magmatic and at post-magmatic stages. The ore elements concentration is facilitated by an extended melt differentiation with a progressive accumulation of ore elements in residual magmas and fluids, as well as by the increased fluid pressures in such magmas. During the evolution of the leucogranitic magmas with a substantial chlorine participation (probably CO_2 as well), tin and the tungsten deposits were probably formed as a result of the extraction by chloride fluids of tin and tungsten from magmas (or rocks), as it normally happens with the chalcophile element deposits.

This paper deals only with some general aspects of the potential ore-deposit production by leucogranitic magmas. Equal importance is attached to the evolution of ore-bearing fluids in the process of their migration, as well as to the conditions of precipitation of components from them. This, however, is beyond the scope of the present paper.

It is natural that all the above-described processes take place most frequently not only in different massifs, but also in different (often continental) tectonic-magmatic zones and geodynamic environments /3,2/. However, it is likely that in some cases the magmas of the lithium-fluorine granites and ongonites residues of the differentiation of the standard type granitic melts. In these cases the process of an extende differentiation of magma is likely to involve a successional separation of ore-bearing fluids from the melts representing various stages of differentiation up to the residual ore-bearing magmas, and their spatial juxtaposition. It is relevant that in the Rudnye Mts., in Zabaikalje, in Mongolia, the tungsten-bearing standard-type granites represent earlier stages of the same tectono-magmatic cycle /4/ than the tin-bearing and rare-metal lithium-fluorine granites.

REFERENCES

1. Antipin V.S., Kovalenko V.I., Kuznetsova A.I., Persikova L.A. The behaviour of tin and tungsten in ore-bearing acid magmatic rocks on the basis of the partition coefficients analysis. Geokhimiya, 1981, No.2

2. Bogatikov O.A., Borsuk A.M., Dmitriev Yu.I., Kovalenko V.I., Ryabchikov I.D. Magmatic formations in the evolution of the lithosphere. "Izvestiya AN SSSR, Geological series", 1983, No.1 pp.3-17

3. Zonenshain L.P., Kuzjmin M.I., Moralev V.M. Global tectonics, magmatism and metallogeny. Moscow, "Nedra", 1976, 232 p.

4. Kovalenko V.I. Petrology and geochemistry of rare-metal granites. Novosibirsk, Nauka Sibirskogo Otdeleniya, 1977, 207 p.

5. Kovalenko V.I., Antipin V.S. Element partition coefficients in ongonites. "Doklady AN SSSR", 1980, v.225, No.1, p.182-186

6. Kovalenko V.I., Zaitsev N.S., Yanshin A.L. Genetic types of the rare-metal mineralization in Mongolia. "Geologiya rudnykh mestorozhdenyi", 1979, v.21, No.5, pp.27-36

7. Kovalenko V.I., Ryabchikov I.D., Bogatikov O.A. The problem of associationg mineralization with acid magmatism. V-the MAGRM Congress, Theses of Reports, Tbilissi, 1982, pp.77-78

8. Kovalenko V.I., Koval P.V., Antipin V.S., Yakimov V.M. The problems of the potential ore presence in magmatic rocks of the tectonic-magmatic activation zones. Book: "Petrology and Petrochemistry of the Ore-bearing Magmatic Formations", Moscow, "Nauka", 1981, pp.72-136

9. Kovalenko N.I. Experimental investigations of the

rare-metal lithium-fluorine granitic formations, Moscow, "Nauka, 1979, 152 p.
10. Koval P.V., Yakimov V.M., Naigebauer V.A., Goreglyad A.N. Regional petrochemistry of the Mezozoic intrusions in Mongolia. Moscow, "Nauka", 1982, 207 p.
11. Kadik A.A., Lebedev E.B., Khitarov N.I. Water in magmatic melts. Moscow, "Nauka", 1971, 207 p.
12. Classification and nomenclature of magmatic rocks. Moscow "Nedra", 1981, 160 p.
13. Lapide I.L., Kovalenko V.I., Koval P.V. Micas of the rare-metal granitoids. Novosibirsk: "Nauka, Sibirskoye otdeleniye", 1977, 104 p.
14. Naumov V.B. Determination of the concentration and the pressure of volatiles in magmatic melts from inclusions in minerals. Geokhimiya, 1979, No.7, p.997
15. Naumov V.B., Ivanova G.F., Motorina E.M. Conditions of the formation of tungsten, tin-tungsten and molibdenum-tungsten deposits. Collection: Basic parameters of the environmental processes of the endogenic ore formation. Novosibirsk, "Nauka", 1979, v.2, p.53
16. Naumov V.B., Ivanova G.F. On the association of the raremetal mineralization with the acid magmatism, using the data on inclusions in minerals. "Geologiya rudnykh mestorozhdenyi", 1980, No.3, p.95
17. Naumov V.B., Kovalenko V.I. The concentration and the pressure of water in acid magmas, according to the data on inclusions in minerals. "Doklady AN SSSR", v.261, No.6, p.1417
18. Naumov V.B., Sokolov A.L. Genetic relationships between granites and tin ore-veins of the Industrialnoye deposit, based on the data on inclusions in minerals. "Geologiya rudnykh mestorozhdenyi", 1981, No.4, p.74
19. Naumov V.B., Kovalenko V.I., Kosukhin O.N. Crystallization parameters of the ongonitic magmas, based on the data on liquid inclusions. "Doklady AN SSSR", 1982, v.267, No2, p.435-437
20. Naumov V.B., Kovalenko V.I., Clocchiatti R., Solovova I.P. Crystallization parameters and phase compositions of liquid inclusions in quartz of ongonites. Geokhimiya, 1983.

21. Reif F.G., Bazheev E.D. Magmatic process and tungsten mineralization. Novosibirsk, "Nauka", 1982, 168 p.
22. Ryabchikov I.D. Thermodynamics of the fluid phase of granitoid magmas. Moscow, "Nauka", 1975, 232 p.
23. Tauson L.V. The geological types and potential ore content of granitoids. Moscow, "Nauka", 1977, 280 p.
24. Burnham W.C. Magmas and hydrothermal fluids. Geochemistry of hydrothermal ore deposits. Wiley, New York, 1979, pp.71-136
25. Burt D.M., Sheridan M.F., Bikun J.V., Christiansen E.N. Topaz rhyolites - distribution, origin and significance for exploration. Econ. Geology, 1982, v.77, p.1818-1836
26. Kovalenko V.I. The genesis of rare-metal granitoids and related ore deposits. In: Metallization, associated with acid magmatism, 1978, v.3, p.235-247
27. Manning D.A.C. The effect of fluorine on liquidus phase relationships in the system Qz-Ab-Or with excess water at 1 kbar Coutrib. Mineral.Petrol., 1981, v.76, p.206-215
28. Pichavant M. An experimental study of the effect of boron on a water saturated haplogranite at 1 kbar vapour pressure. Coutrib. Mineral. Petrol., 1981, v.76, p.430-439
29. Pichavant M., Ramboz Cl., Weisbred A. Fluid immiscibility in natural processes use and misuse of fluid inclusion date Chemical Geology, 1982, v.37, p.1
30. Ryabchikov I.D. Mobilization of ore metals by supercritical fluids from crystallizing magmas. Chemistry and Geochemistry of solution at high temperatures and pressures. Editors D. Rickard and F. Wickman, Pergamon Press, 1981, p.529-536
31. Wones D.R., Eugster P. Stability of biolite; experiment, theory, and application. Amer. Mineralogist, 1965, v.50, p.1228-1272

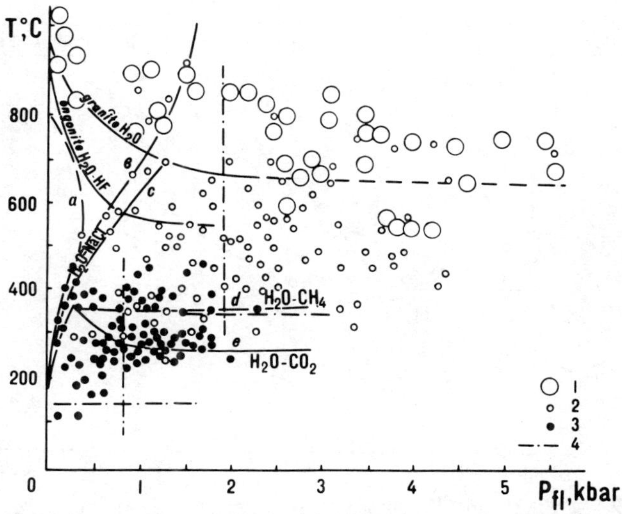

FIGURE 1. Temperature and pressure values obtained by examination of melt and fluid inclusions in minerals.
1-magmatic melts
2-magmatic and post-magmatic fluids in granitoids and metasomatic facies thereof
3-hydrothermal solutions of the ore- and post-ore stages of the rare-metal deposits
4-temperatures (at a gradient of 50^{0}C/km) and pressures (at a geostatic load of 270 bar/km) at the depths of 3 and 7 km

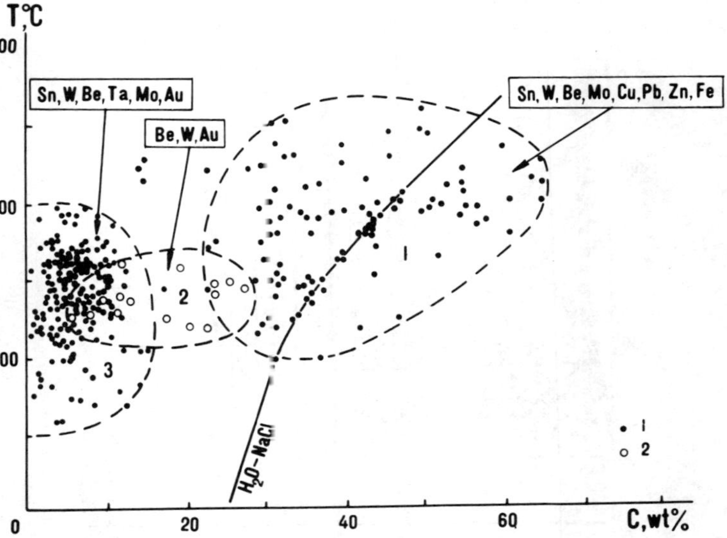

FIGURE 2. Diagram "Temperature/solution concentration" for the ore and the post-ore stages of the rare-metal deposits
1-electrolytic salts solutions
2-carbon dioxide solutions
The elements indices show the ore elements characteristic of the deposits related to either one of the three types of the ore-bearing fluids

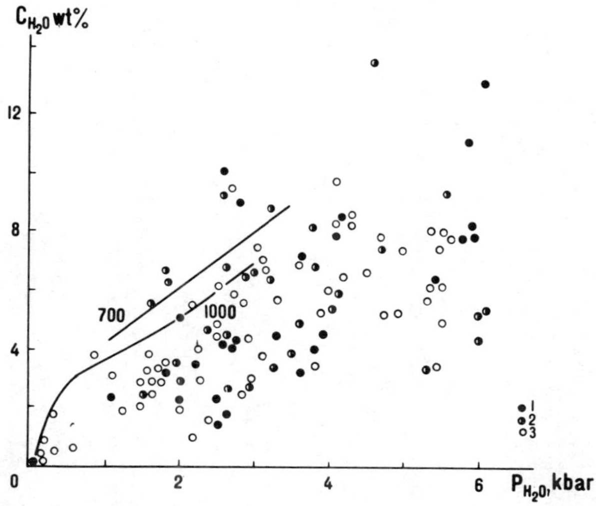

FIGURE 3. Concentrations and water pressures, obtained from melt inclusions in minerals of acid magmatic rocks
Geochemical rock types: 1-lithium-fluorine
2-alkaline
3-standard

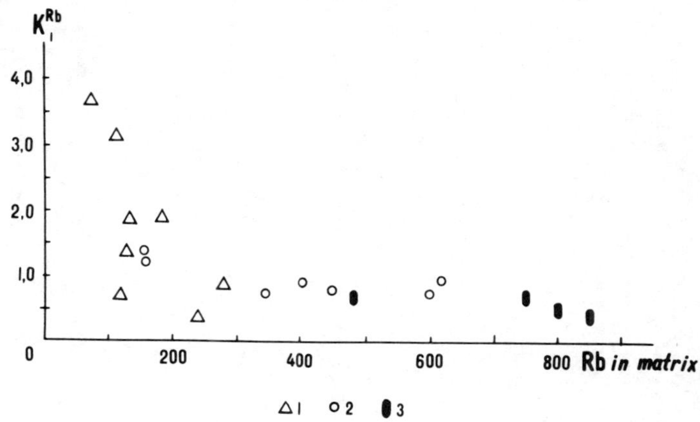

FIGURE 4. The dependence of rubidium partition coefficient between potassic feldspar and the acid magmatic melt upon the rubidium content in the melt
1-comendites
2-rhyolites
3-ingorhyolites

THE TIHAMA ASIR IGNEOUS COMPLEX, A PASSIVE MARGIN OPHIOLITE

R.G. COLEMAN
Geology Department, Stanford University

ABSTRACT

New igneous crust has formed in a narrow hinge zone on the eastern side of the Red Sea near Jizan, Saudi Arabia. Extension of continental rocks on the Red Sea passive margin in late Oligocene and early Miocene gave rise to bi-modal volcanism and the formation of intracontinental lacustrine basins. The mafic igneous rocks represent partial melting in the mantle and the leucocratic igneous rocks form by partial melting of the extended continental crust. Continued extension by faulting permitted mantle magmas to invade the thinned continental crust. The Tihama Asir igneous complex within the exposed hinge zone consists of multiple intrusions of dikes and layered gabbros whose parent magma was tholeiitic and is similar to magmas developed at mid-ocean ridges (MORB). Fractionation of the magma occurs when the magma is trapped within the extended continental crust and forms hypabyssal magma chambers similar to those found in East Greenland. Partial melting of the continental crust by the invading MORB magmas gives rise to the silicic peralkaline magmas that are characteristic of extended continental crust. The similarity between the Tihama Asir complex and East Greenland igneous complexes provides new insights on the importance of mantle igneous processes in the transition of a passive continental margin to oceanic crust.

INTRODUCTION

Formation of new igneous crust within extended passive margins has been the subject of much interest since the advent of plate tectonics (Dewey and Bird, 1970). The presence of volcanic and sub-volcanic provinces along

Proceedings of the 27th International
Geological Congress, Volume 9, pp. 221–239
PETROLOGY (IGNEOUS AND METAMORPHIC ROCKS)

passive continental margins has long been known and carefully described in such areas as East Greenland (Brooks and Nielsen, 1982), the British Tertiary intrusive centers of the Hebridean Islands (Wager and Brown, 1967 and Walker, 1975) and in general by Gibson (1966). The actual mechanism of continental breakup and the evolution of new igneous crust along passive margins is now receiving attention from structural geologists and petrologists and it is the purpose of this paper to

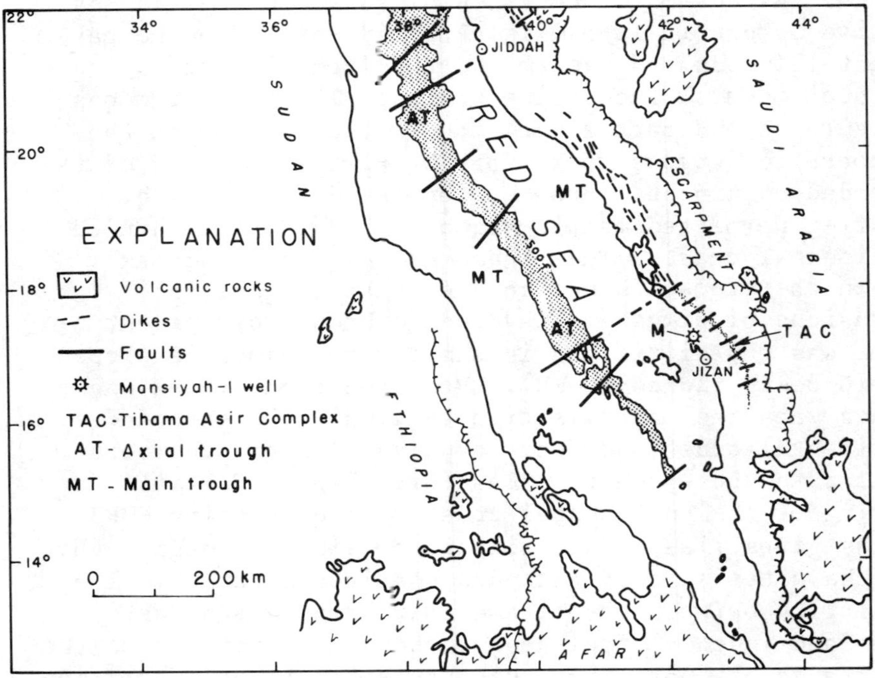

FIGURE 1. Index map showing location of the major features of the Red Sea area.

describe the evolution and development of new igneous crust along the eastern side of the Red Sea coastline 40 Km east of Jizan in Saudi Arabia (Fig. 1). Previous studies in the area have provided much of the information on which this discussion is based (Gillman, 1968; Blank, 1977; Gettings, 1977; Coleman et al., 1977;

Coleman et al., 1979; Taylor, 1980; Ghent et al., 1980, Fairer, 1981; and Schmidt et al., 1982).

GEOLOGY AND PETROLOGY OF THE TIHAMA ASIR COMPLEX

The coastal plain in the vicinity of the Jizan consists of a series of coalescing alluvial fans derived from streams actively eroding an escarpment some 65 Km inland (Brown, 1970). This coastal plain is essentially a featureless alluviated surface that appears to have been undergoing active submergence since Miocene time (Gillman, 1968). A narrow hinge zone (4-6 Km) striking NW parallel to the Red Sea contains the Tihama Asir igneous complex (Fig. 2). This narrow zone consists of older Precambrian metamorphic rocks overlain by Phanerozoic sediments and invaded by dikes, layered gabbros, and extrusive volcanics that in certain discontinuous zones are made up entirely of these igneous rocks.

Precambrian metamorphics of the Sabya formation are the oldest exposed rocks and form the eastern boundary of the Tihama Asir igneous complex (Fairer, 1981). These Precambrian rocks have undergone greenschist to amphibolite facies metamorphism and are deformed into a north-south trending folded sequence. Unconformably overlying these Precambrian metamorphic rocks is Wajid sandstone of Ordovician-Cambrian age that forms a series of antithetically block faulted ridges that dip towards the Red Sea. Just south of the Tihama Asir complex near the Yemen border Jurassic marine sedimentary rocks unconformably overlie the Wajid sandstone and have also been deformed by the antithetic block faulting. The igneous complex is dominated by a sequence of sub-parallel dikes that have a northwesterly strike parallel to the axial trough of the Red Sea and dip easterly towards the escarpment (Fig. 2). The dikes are infrequent within the Precambrian metamorphic rocks exposed northeast of the hinge zone and increase in abundance southwestward towards the axis of the hinge zone. In the hinge zone dikes have screens of Wajid sandstone, mafic and silicic volcanics as well as the lacustrine beds of the Baid formation of Miocene age.

In restricted areas, the dike swarms have no screens of pre-existing country rock and develop chilled margins one against the other such as seen in some ophiolite complexes (Coleman, 1979). Recent work by Schmidt et al. (1982) has revealed that the screens of Baid formation lacustrine sediments are interlayered with bi-modal volcanics and contain Miocene vertebrate fossils. These sediments and volcanics which pre-date the dikes are now collectively referred to as the Jizan group (Schmidt et al., 1982).

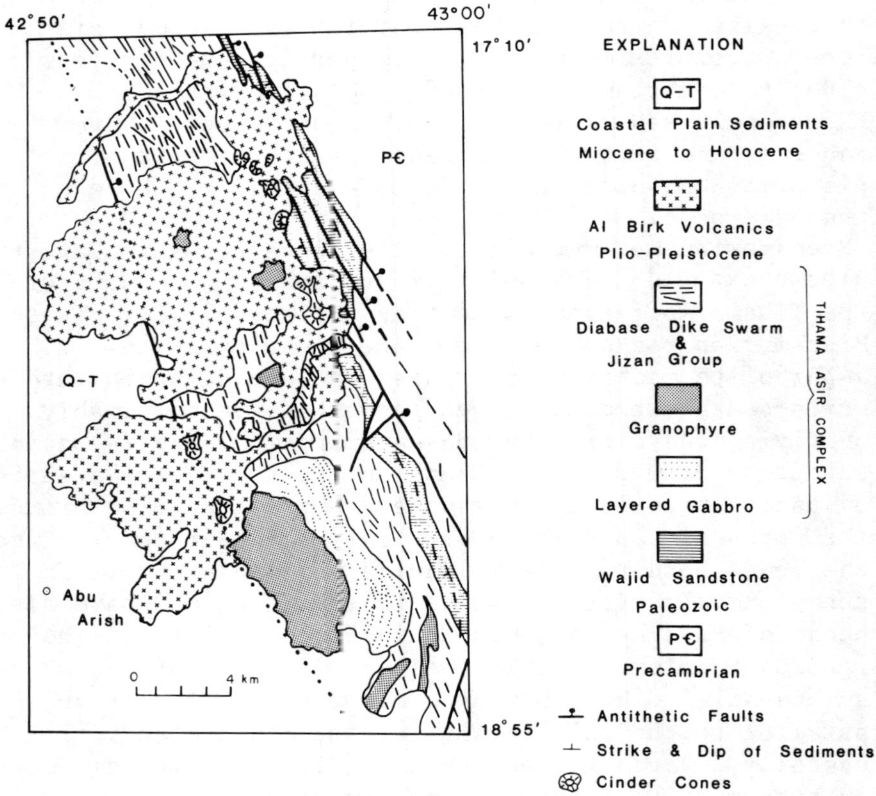

FIGURE 2. Geologic map of the Tihama Asir igneous complex. Compiled from Coleman et al. 1977, 1979; Fairer, 1981; and Gillman, 1968.

Dike widths range from ½ to 15 m with the late cross-cutting dikes usually less than 1 m wide. In general, the older dikes dip 50-60° NE whereas the younger dikes are nearly vertical. Bedding in the Jizan group volcanics and sediments form screens in the dike swarm and dip 50-60° SW towards the Red Sea axial trough. Rotation of the Jizan group and the dikes towards the Red Sea results from the same antithetic block faults that have affected the older underlying rocks (Fig. 3). The dike rocks are predominantly diabase and consist of calcic plagioclase, clinopyroxene, and iron-titanium oxides and are contemporaneous with less abundant leurocratic dikes having rhyolitic composition and granophyric to trachytic textures.

A funnel shaped layered gabbro intrusive developed within the dike swarm at Jabal al Tirf also dips steeply (40 to 60° SW) towards the Red Sea. The layering produces a semicircular pattern at the surface and when projected downward describes a funnel shaped body. Earlier reports by Coleman et al. (1977, 1979) interpreted the layering as having been horizontal initially; however, rotation of the fault blocks to the horizontal still requires that layering in the gabbro had primary inclined dips. Recent papers on the layered gabbro bodies from east Greenland describe nearly identical tilted funnel shaped bodies whose initial magmatic layering dipped as much as 30° (Brooks and Nielsen, 1982; Meyers, 1980). Fifty kilometers north of the Jabal al Tirf another funnel shaped layered gabbro (Masliyah) is present within the hinge zone. The igneous layers of this body also dip steeply towards the Red Sea and are terminated against a NW trending antithetic block fault. These sporadic gabbro bodies in the hinge zone apparently develop on the down thrown side of the antithetic faults where magma ascends along the fault trace (Fig. 3). Weak zones in the hanging wall are invaded and enlarged into funnel shaped magma

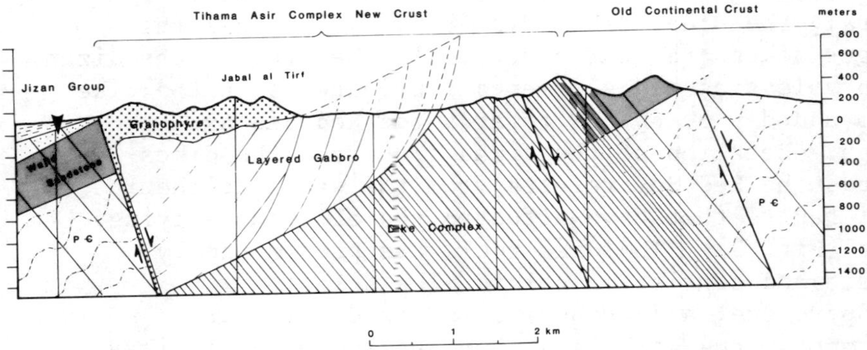

FIGURE 3. Generalized cross section through the Tihama Asir complex. The relationships between various units are idealized and can not be verified from surface geology. The dike complex most likely has screens of Jizan group volcanics and Wajid sandstone. Much more granophyre is present than shown on the section as some of the dikes are granophyric in composition. The overall structural relationships are considered to be quite similar to East Greenland and the sub-surface relationships are based on Myers' (1980) interpretation.

chambers as movement continues on the fault and magmas continue to enlarge upward and outward. Layering in the Jabal al Tirf gabbro shows strong variations in dip and strike and in certain instances unconformities (Fig. 3) are developed in the sequence of layers similar to the gabbro intrusions of East Greenland (Meyers, 1980). These unconformities can be related to continued rotation resulting from the antithetic block faulting within the hinge zone. Near the top of the layered sequence massive, coarse grained gabbros develop pegmatitic segregations very similar to the upper border group of the Skaregaard gabbro intrusion of East Greenland (Wager and Brown, 1967). Extensive hornfelsing of the intruding dikes extends up to 200 m away from the basal contact of the gabbro.

Rhythmic layering results from accumulation of plagioclase, clinopyroxene, and olivine grains in individual units up to 1 m thick. Most layers are clinopyroxene olivine gabbro or clinopyroxene gabbro but both norite and troctolite may be present. Ultramafic cumulates or evidence of progressive upward fractionation are absent and it is considered that these rocks formed in an open system magma chamber (Coleman et al., 1977, 1979). Chemical variation found in the layered gabbros show that separate batches of partially fractionated tholeiitic melts were introduced into the Jabal al Tirf magma chamber as successive magmatic pulses rather than closed system fractionation of a single magmatic pulse (Coleman et al., 1979).

The presence of large sills and dikes of peralkaline granophyre and rhyolite within the Tihama Asir complex provide an igneous link to partial melting of the continental crust. The Tihama Asir leucocratic rocks are present in all stages of its igneous development and can not be the products of differentiation of the mafic magmas as they make up at least 30 percent of the complex. The large mass of granophyre capping the Jabal al Tirf gabbro is a fairly uniform igneous mass approximately 200 m thick (Fig. 3). The granophyric magma apparently rose along the same fault zone that fed the underlying gabbro. It is a medium-grained leucocratic rock consisting mainly of graphic intergrowths of quartz and potassium feldspar. Zoned plagioclase (An_{30-40}) where present is rimmed by granophyric intergrowths. Acmitic pyroxene and riebeckite are sparingly present but attest to the peralkaline nature of the magma. These same rocks plot in the experimental minimum of the ternary system Ab-Qz-Or providing evidence that the melts represent low temperature compositions of partial melting of the continental crust.

Alkali olivine basalt lavas cover unconformably the tilted Tihama Asir complex and represent a short lived Holocene eruption coeval with spreading in the Red Sea axial trough. Small individual cores marking vent areas for these lavas occupy a northwest trend within the igneous complex and have produced approximately 57 Km^3 of lava. These lavas are distinct from the rift zone dikes and gabbros and represent a much deeper magma source as shown by their mantle xenoliths (Ghent et al., 1980).

The mafic rocks of the Tihama Asir complex can be derived from a tholeiitic parent magma similar to the axial trough basalts of the Red Sea. The layered gabbro was emplaced at shallow depth and the extensional system of antithetic block faulting allowed intimate contact between the rising basaltic magma and Precambrian crust. Therefore silicic melts may have formed by melting this older crust and were emplaced as rhyolite dikes and granophyric sheets.

TECTONIC EVOLUTION OF THE TIHAMA ASIR COMPLEX

Previous discussions of the Tihama Asir complex (Coleman et al., 1977,1979) have shown that it occupies an important structural boundary between the Precambrian continental crust and attenuated crust formed by extension and volcanism as the Arabian plate moved away from the African plate. There is general disagreement as to how much of the eastern Red Sea continental margin is made up of new igneous crust (oceanic crust and/or ophiolite) and how much is older attenuated crust (Lowell and Genik, 1972; Girdler and Styles, 1974; Blank, 1977; Gass, 1977; Coleman et al., 1979; Schmidt et al., 1982; and Cochran, 1983). The geological and geophysical evidence will be reviewed in light of recent studies and an attempt will be made to reconcile the differences and develop a model that contains elements of both rift and drift.

The newly proposed stratigraphic sequence for the Tertiary rocks along the southern Red Sea coastal plain of Saudi Arabia provides a consistent picture of sedimentation that can be related to the formation of the Red Sea (Schmidt et al., 1982,1983). The presence of a shallow marine sea of Paleocene-Eocene age extending southward from Jordan to Jeddah, Saudi Arabia as a narrow gulf parallel to but east of the present Red Sea is the only early Tertiary event that may relate to the development of the Proto-Red Sea (Madden et al., 1979; Schmidt et al., 1982). This narrow downwarp could have resulted from early crustal extension across a broad region from eastern Egypt to Jordan. In the Tihama Asir area there is a long time gap between the Jurassic sedimentary deposits and the early Miocene tectonic and volcanic events that gave rise to the Jizan group. Alkali olivine basalts of late Oligocene age now form plateaus that rest on lateritic surfaces of early Tertiary age some 70 km inland from the Tihama Asir along the escarpment edge almost 2 km above the coastal plain. The flat lying laterite surface and the overlying horizontal plateau basalts represent remnants of a low lying surface that apparently extended across the coastal plain according to Schmidt et al. (1983). The recent discovery that the Baid formation consists of lacustrine deposits containing vertebrate fossils of early Miocene age (21-25 Ma) and that these sediments are interlayered with contemporaneous bi-modal volcanic flows and pyroclastic deposits (Jizan group) requires an early Miocene period of extension and volcanism (Schmidt et al., 1983; Madden et al., 1983). This important discovery negates the model of Gass (1970) which required early uplift of this stage to produce the eruptions of peralkaline olivine basalts of the plateau areas. Invasion of dike swarms and development of cone-shaped gabbros with huge sheets of granophyre within the Jizan group lacustrine sediments and bi-modal volcanics during the early Miocene demonstrates the continuation of crustal extension and development of new igneous crust.

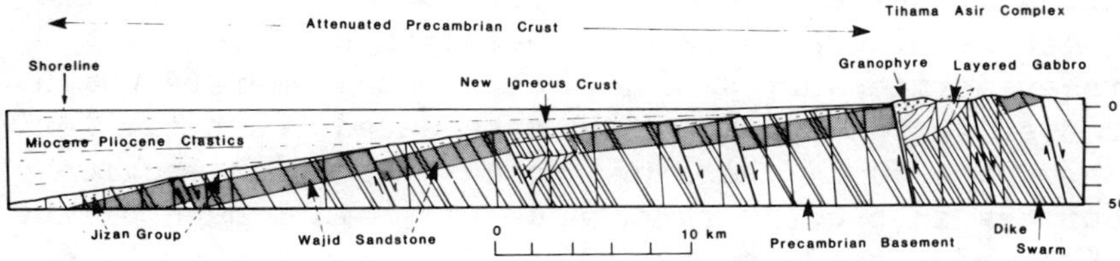

FIGURE 4. Extended cross section across the Red Sea coastal plain from the Tihama Asir complex to the Red Sea shoreline. Sub-surface control taken from Gillman (1968) and includes seismic reflection and the

exploratory well at Mansiyah. The style of faulting is considered to be lystric and antithetic resulting from crustal extension as described by Black et al. (1974) in the Afar depression. Evaporites are present near the base of the clastics but are not shown in this diagram. The presence of new igneous crust in the middle of the diagram is hypothetical, but is considered to be typical of the extended crust. The Miocene-Pliocene clastics are made up of continental red shales accompanied by sandstones and conglomerates.

The formation of antithetic block faulting apparently was contemporaneous with these early Miocene volcanic events. Rotation of the Tihama Asir complex and its associated older rocks towards the Red Sea on these antithetic faults during and after its formation is evidence that crustal extension was contemporaneous with the igneous intrusions (Fig. 4). Reconstruction of the stratigraphy in this area after the formation of the Tihama Asir is obscured by the coastal plain cover. Seismic studies on the coastal plain and a deep exploratory oil well (Mansiyah, 22 Km north of Jizan) provide stratigraphic control on the depositional sequence following the tilting of the Tihama Asir complex towards the Red Sea (Gillman, 1968). These seismic profiles across the coastal plain show that it is underlain by a sedimentary basin that rapidly deepens to 4000 m at the Red Sea shoreline over a distance of 45 Km. The stratigraphy of the basin can be reconstructed, in part, from the Mansiyah well log. A clastic continental series approximately 2200 m thick overlies 1200 meters of evaporites and below this is an infra-evaporite series of shale and conglomerate which must overlie the subsided and attenuated crust similar to that exposed in the hinge zone. The thick section of evaporites is present throughout the main trough of the Red Sea and extends out to the axial trough (Ross and Schlee, 1973). Sediments interlayered with the evaporites and directly above them are considered to be late Miocene in age. This unconformity between the salt and overlying sediments is easily found by seismic reflection studies and is commonly referred to as the

S-reflector (Ross and Schlee, 1973). If we can project this same age to the salt unconformity in the Mansiyah well it is obvious that at least 2,200 meters of clastics have been deposited on the extended crust and evaporites since late Miocene. Geophysical interpretation shows that continental crust with screens of new igneous crust form a westward dipping homocline that extends across the coastal plain (Gillman, 1968). The homocline is interupted by a series of block faults which are here interpreted as continuation of the antithetic block faults seen in the hinge zone (Fig. 4). A recent gravity survey across the Tihama Asir complex and coastal plain extending offshore to the Farasan Islands shows a strong positive gradient over the hinge zone declining over the sedimentary basin and rising again near the axial trough (Gettings, 1977). Gettings has interpreted the gravity profile as marking the beginning of new oceanic crust at the hinge zone and extending across the coastal plain and main trough of the Red Sea. Cochran (1983) has questioned the gravity modeling of Gettings and suggests that the magnetics and gravity can be interpreted to represent local intrusions within a crust being thinned and extended by rotational faulting. A deep seismic refraction profile extending from the Farasan Islands to Riyadh adds further geophysical constraints on the nature of the crust and mantle along the Red Sea margin (Healy et al., 1982). Intpretation of these results shows an extremely steep gradient across the coastal plain from crustal thicknesses of 40 km to 17.5 km. From the coastal plain to the western edge of the main trough the crust thins to less than 8 km. Velocities in the mantle average about 8 km/sec. whereas the averages in the thinned crust vary from 6.0 to 6.4 km/sec. The overlying sedimentary basin of the coastal plain reaches a maximum thickness of 5 km in the vicinity of the Mansiyah well with velocities of 4.2 km/sec. Taking this new seismic data all together it appears that the crust thins abruptly along the coastal plain to 17.5 km and at the axial trough has only a thickness of 8 km. No known oceanic crust reaches thicknesses of 17.5 km nor has average velocities of 6 to 6.4 km/sec. Interpretations of the intermediate seismic velocities of the rocks underlying the main Red Sea trough have been equivocal and have led to controversy (Drake and Girdler, 1964; Tramontini and Davies, 1961; Coleman, 1974; Healy et al., 1982; Cochran, 1983). However, it would appear most reasonable to interpret this segment of crust underlying the coastal plain and the main trough of the southern Red Sea as a mixture of old continental crust and newly formed igneous crust (Cochran, 1983).

The structural evolution of the Tihama Asir complex is intimately related to events leading up to the movement of the Arabian plate away from Africa. As pointed out earlier, rotation by antithetic block faulting appears to post-date the Jizan group rocks; however, at least some faulting and extension are required to form this significant association of bi-modal volcanics and vertebrate-bearing lacustrine sediments. The widespread occurrence of northwest trending hypabyssal dikes of the same general age inboard from the hinge line indicate that crustal extension was acting over a broad zone along the western margin of the Arabian plate (Blank, 1977; Kellog and Blank, 1982). These inboard dikes extend northward to the Gulf of Aquabah where they are offset by the left lateral motion of the Dead Sea Rift and then continue well into the western side of the the Sinai Peninsula (Bartov et al., 1980). Blank (1977) has suggested that these dikes represent traces of early Miocene block faulting and perhaps form from magmas developed in the asthenosphere as it rose along the Red Sea rift. None of these dikes are rotated by anithetic block faulting so their position may mark the easternmost boundary of continental extension.

The block faults within the hinge zone are interpreted as antithetic listric faults extending below the zone of brittle fracture similar to those described by Black et al. (1974) along the flanks of the Afar depression. Movement along these faults apparently was sporadic but continuing from the time when the Jizan group volcanics were erupted until the final rotation of the Tihama Asir complex some time in the Miocene. Field evidence shows that both the gabbros and granophyres invaded the rift zone along these antithetic faults (Figs. 3 & 4). Unconformities within the layered gabbro indicate that at least some rotation accompanied the development of these funnel-shaped gabbros. Dike swarms intruded by the layered gabbro dip away from the Red Sea axial trough and show similar rotation but these are cut by vertical dikes of the same magmatic affinities attesting to continued magmatic and tectonic activity (Fig. 3). East Greenland igneous rocks occupy a similar structural position on the passive margin Atlantic margin. In Greenland, Cretaceous tilting of the dike swarms and gabbroic intrusions have been related to a marginal hinge zone that may also be the result of antithetic listric faults formed by crustal extension during the initial opening of the Atlantic (Brooks and Nielsen, 1982A, 1982B; Myers, 1980).

Rapid downwarp and thick post-Miocene clastic deposits of the coastal plain have covered the seaward extension of the attenuated continental crust and it is impossible to confirm a predicted flattening of the antithetic lystric faults and increase in dip of the bedding towards the Red Sea (Fig. 4). Cochran (1983) has suggested that the initial Red Sea rift was 100 km wide and that crustal extension amounted to nearly 75% in the southern Red Sea by a combination of listric faults and interspersed dike injection sometimes enlarging to major, en echelon igneous complexes. Kellogg and Blank (1982) have suggested that the antithetic faulting is a response to lateral movement of plastic basal crust and upper mantle being dragged by athsenosphere as it moves away from the spreading center.

CONCLUSIONS

The proto-Red Sea formation as dictated by the geologic facts was definitely initiated by crustal extension. This extension gave rise to bi-modal volcanism and the formation of intracontinental lacustrine basins. These early mafic volcanics are peralkaline and derived from deep melting in the mantle and the leucocratic volcanics and intrusions result from partial melting at the base of the extending continental crust. Continued extension of Arabia away from Africa was accommodated, in part, by discontinuous formation of new Red Sea crust as characterized by the Tihama Asir igneous complex. The new igneous crust resembles oceanic crust, in part, as it formed from tholeiitic magmas developed by shallow melting in the mantle. Continued extension by lystric faulting permitted zones of new igneous crust to form. The geophysical seismic velocities on rocks underlying the Red Sea main trough show that they have velocities intermediate between ocean crust and the Precambrian crust (Cochran, 1983; Healy et al., 1982). The magnetic patterns within the main trough of the Red Sea show irregular non-linear patterns that appear to be more characteristic of intermittent igneous complexes similar to the Tihama Asir complex than those generally recognized for oceanic crust (Cochran, 1983).

Petrologic relations within the Tihama Asir complex indicate that the parent magma was tholeiitic and similar to magmas developed at mid-ocean ridges. Considerable fractionation of the magma is related to its being trapped in continental crustal blocks to form small hypabyssal magma chambers or melting continental crust that forms the leucocratic cogenetic bi-modal suites. It would appear to be reasonable to assume, as Cochran (1983) has done that the Miocene stage of evaporite deposition in the Red Sea represents continued continental extension and formation of new igneous crust in zones of extension where magmas of MORB parentage develop restricted, en echelon igneous complexes within increasingly thinned continental crust towards the Red Sea axial trough. Royden et al. (1980) have proposed a similar general model where dike intrusion increases seaward from the continental margin. At the end of Miocene, spreading commences in the axial trough of the southern Red Sea and slowly propogates northward towards the Gulf of Aquabah where left lateral motion and continuing continental extension accommodates the movement of Arabia away from Africa (Bartov et al., 1980).

Previous comparisons of the Tihama Asir complex with oceanic crust and/or allochthonous fragments of oceanic crust are still valid (Coleman et al., 1977, 1979). However, the presistent association of leucocratic volcanic and intrusive rocks are evidence of partial melting of subjacent continental crust cogenetic with the emplacement of the MORB tholeiitic magmas. It is suggested that during the period of Miocene extension at the Red Sea continental margin that two separate processes were active: 1) widespread lystric faulting related to brittle extension at the surface and ductile flow in the lower parts of the crust, 2) intrusion of tholeiitic magmas of MORB parentage along these lystric faults and sporadically developing restricted zones of new igneous crust. Concomitant melting of continental crust along its extended base gave rise to the bi-modal igneous suite.

The striking similarity between the Tihama Asir complex and the East Greenland igneous complexes further enhances this analogy and provides new insights regarding the initial continental break-up. Therefore it is suggested that igneous associations such as we find in the Tihama Asir complex and East Greenland are typical for passive margin transformation (oceanization) of continental crust into oceanic crust.

REFERENCES

Bartov, Y., Steinitz, G., Eyal, M., and Eyal, Y. (1980). Sinistral movement along the Gulf of Aquabah - its age and the relation to the opening of the Red Sea: Nature. 285, 220-222.

Black, R., Morton, W.H., and Hailn, T. (1974). Early structures around the Afar triple junction, Nature. 248, 496-497.

Blank, H.R. (1977). Aeromagnetic and geologic study of Tertiary dikes and related structures on the Arabian margin of the Red Sea. In: Red Sea Research 1970-1975, L.S. Hilbert, (ed). Saudi Arabian directorate of Mineral Resources Bull. 22, pp. 61-618.

Brooks, C.K. and Nielsen, T.F.D. (1982A). The Phanerozoic development of Kangerdlugssuag area, East Greenland, Meddelelser om Gronland, Geoscience. 9, 3-30.

Brooks, C.K. and Nielsen, T.F.D. (1982B). The E. Greenland continental margin: a transition between oceanic and continental magmatism. J. Geol. Soc. London. 139, 265-275.

Brown, G.F. (1970). Eastern margins of the Red Sea and the coastal structures in Saudi Arabia. Phil. Trans. R. Soc. London. A267, 75-87.

Cochran, J.R. (1983). A model for development of Red Sea, Amer. Assoc. Pet. Geol. Bull. 67, 41-69.

Coleman, R.G. (1979). Geological background of the Red Sea: Initial Repts. Deep Sea Drilling Project. 23, 813-820.

Coleman, R.G., Gregory, E.T., Brown, G.F. (1983). Cenozoic volcanic rocks of Saudi Arabia. Saudi Arabian Deputy Ministry for Mineral Resources Open File Report USGS-OF-03-93

Coleman, R.G., Hadley, D.G., Fleck, R.G., Hedge, C.T., and Donato, M.M. (1979). The Miocene Tihama Asir ophiolite and its bearing on the opening of the Red Sea, In: Evolution and Mineralization of the Arabian-Nubian Shield, Al-Shanti, A.M. Convenor, Oxford Pergamon Press Ltd. 1, pp.173-186.

Coleman, R.G., Fleck, R.S., Hedge, C.E., and Ghent, E.D. (1977). The volcanic rocks of southwest Saudi Arabia and the opening of the Red Sea. In: Red Sea Research, 1970-1975, Hilbert, L.S. (ed). Saudi Arabian Directorate General of Mineral Resources, Bull. 22, D1-D30.

Dewey, J.F., and Bird, J.M. (1970). Mountain belts and the new global tectonics, Jour. Geophys. Res. 75, 2625-2647.

Drake, C.L., and Girdler, R.W. (1964). A geophysical study of the Red Sea: Royal Astron. Soc. Geophys. Jour. 8, 473-495.

Fairer, G.M. (1981). Reconnaissance Geology of the Wadi Sabya Quadrangle 17/42D, Kingdom of Saudi Arabia, Saudi Arabian Deputy Ministry for Mineral Resources Open-File Report USGS-OF, 1 Sheet, Scale 1:100,000.

Gass, I.G. (1977) The age and extent of the Red Sea oceanic crust, Nature. 265, 722-724.

Gass, I.G. (1970). The evolution of volcanism in the junction area of the Red Sea, Gulf of Aden and Ethopian rifts: Royal Soc. London Philos. Trans. A267, 369-381.

Gettings, M.E. (1977). Delineation of the continental margin in the southern Red Sea region from new gravity evidence. In: Red Sea Research 1970-1975. Min. Res. Bull. 22. Dir. Gen. Min. Res., Saudi Arabia. K-1 K-11.

Ghent, E.D., Coleman, R.G., Hadley, D.G., 1980, Ultramafic inclusions and host alkali olivine basalts of the southern coastal plain of the Red Sea, Saudi Arabia, Amer. Jour. Sci. 280-A, 499-527.

Gibson, I.L. (1966). Crustal flexures and flood basalts, Tectonphysics. 3, 447-456.

Gillman, M. (1968). Prelminary results of a geological and geophysical reconnaissance of the Jizan coastal plain in Saudi Arabia, Second Regional Tech. Symposium Society of Pet. Engrs. AIME, Dhahran, Proceedings. 189-208.

Girdler, R.W. and Styles, P. (1974). Two-stage Red Sea floor spreading, Nature. 247, 7-11.

Healy, J.H., Mooney, W.D., Blank, H.R., Gettings, M.E., Kohler, W.M., Lamson, R.J., and Leone, L.E. (1982). Saudi Arabian seismic deep refraction profile: Final Project Report. Saudi Arabian Deputy Ministry for Mineral Resources Open File Rrport USGS-OF-02-37, 141.

Kellog, K.S., and Blank, H.R. (1982). Paleomagnetic evidence bearing on Tertiary tectonics of the Tihama Asir coastal plain, southwestern Saudi Arabia. Saudi Arabian Deputy Ministry for Mineral Resources Open File Report USGS OF-02-06, 37.

Lowell, J.D. and Genik, G.J. (1972). Sea floor spreading and structural evolution of southern Red Sea, Amer. Assoc. Pet. Geol. Bull. 56, 247-259.

Madden, C.T., Schmidt, D.L.,. Whitmore, F.C., Jr. (1983). Masritherium (Artiodactyla, Anthracotheriidae) from Wadi Sabya, Southwestern Saudi Arabia: An earliest Miocene age for continental rift-valley volcanic deposits of Red Sea margin. Saudi Arabian Deputy Ministry for Mineral Resources Open File Report USGS-OF-03-61, 24.

Madden, C.T., Nagvi, I.M., Whitmore, F.C., Jr., Schmidt, D.L., Langston, W., Jr., and Wood, R.C. (1979). Paleocene vertebrates from coastal deposits in the Harrat Hadan area, at Taif region, Kingdom of Saudi Arabia, U.S. Geological Survey, Saudi Arabian Mission, Proj. Report 269, 29.

Myers, John S. (1980). Structure of the coastal dyke swarm and associated plutonic intrusions of East Greenland. Earth and Planetary Science Letters. 46, 407-418.

Ross, D.A. and Schlee, J. (1973). Shallow structure and geologic development of the southern Red Sea, Geol. Soc. Amer. Bull. 84, 3827-3848.

Royden, L. and Keen, C.E. (1980). Rifting processes and thermal evolution of the continental margin of eastern Canada determined from subsidence curves. Earth Planetary Science Letters. 51, 343-361.

ALKALINE MAGMATISM AS AN INDICATOR OF PERIODIC ACTIVATION IN THE EARTH'S MANTLE

KONONOVA V.A., ANDREEVA E.D., SVESHNIKOVA E.V., YASHINA R.M.
Institute of the Geology of Ore Deposits, Petrography, Mineralogy and Geochemistry of the USSR Academy of Sciences, Moscow, USSR

Alkaline rocks do not occur as widely as do many other types of magmatic rocks. Yet they constitute an important link in the Earth's geologic history. Reconstruction of the history of alkaline magmatism in particular massifs and regions, using experimental information on the physical and chemical parameters relevant to the formation of alkaline rocks can be utilized to draw a picture of the state of deeper interiors during specified segments of the Earth's geologic history. The present paper is concerned with three points: (1) composition of the principal groups of alkaline magmatic series, (2) periodicities in their manifestations in the platforms and folding areas, (3) correlation between the principal stages in the evolution of alkaline magmatism and epochs of greatest tectonic activity.

Alkaline rocks are here understood /after O.A. Vorobyeva (7), the authors of "The alkaline rocks..." (5), "Classification and nomenclature..." (3)/ to be magmatic silicate compounds that contain feldspathoids and/or alkaline dark-coloured silicates - pyroxenes and/or amphiboles. Alkaline rocks usually occur in nature in magmatic series represented by a number of rocks that exhibit a temporal pattern of alternation. Each series has a specific composition controlled by geologic-tectonic and physical-chemical conditions for the formation of different deep alkaline magmas. The principal alkaline magmatic series are listed in Table 1, the variability in the composition of their constituent rocks is demonstrated in the diagram of Figure 1. The range of alkaline rock compositions (Figure 1, field III) considerably exceeds the variability in rock composition for the normal (Fi-

Proceedings of the 27th International
Geological Congress, Volume 9, pp. 241–257
PETROLOGY (IGNEOUS AND METAMORPHIC ROCKS)

gure 1, field I) and subalkaline (Figure 1, field II) series, which is reflected in their mineral composition. The entire diversity of alkaline magmatic series can be comprised within four groups: alkaline ultrabasic, alkaline basic, alkaline intermediate, and alkaline acid rocks[x]. One can distinguish potassium-sodium (the more common) and potassium series[xx].

Effusive, hypabyssal, mesoabyssal equivalents can be definitely established as belonging to one of the troups. Below we underline the most important features found in the composition and in tectonic concitions of evolution for the commonest groups of alkaline magmatic series. Potassium-sodium alkaline ultrabasic series (Table 1, series 1 through 3) are formed in rifts and deep fault zones in crations, within marginal blocks of cratons. Typical features of the alkaline ultrabasic rocks composition are: high concentration of Mg, Ca, Fe, Ni, Ti, P, Zr, Nb, Ta, TR, alkalies, volatile components (CO_2 in the first place); high magnesia content of the minerals (diopside, salite, phlogopite, olivine etc.), a wide occurrence of calcium-containing minerals (melilite, cancrinite, calcite etc.); the feldspathoids are represented by nepheline, characteristic petrogeochemical rock types are diopside ijolite, salite ijolites, särnaites.

Potassium-sodium alkaline basic series (Table 1, series 6 through 8) mostly occur within consolidated folding areas, can be encountered in rift zones within a platform, and in oceanic islands. They are mainly represented by alkaline basic rocks (tephrite, theralite, essexite, etc.). Typical petrogeochemical rock types are fassaite ijolite, fassaite urtite, subalkaline olivine gabbro; typomorphic rock-forming minerals are represented by fassaite, olivine, amphibole, (kaersutite, barkevikite), plagioclase having

[x]We do not consider the alkaline acid rock series.

[xx]The relation between Na_2O and K_2O is >4 for sodium series rocks, 1 to 4 for potassium-sodium series, and <1 for potassium.

the composition labradorite, bytowhite, andesite. The composition of feldspathoids is diverse: nepheline, analcite, sodalite, hauyne. Typically these alkaline series are highly differentiated, the later series members increasing in SiO_2, Al_2O_3, alkalies, as well as Rb, Sr, Ba, Zr, TR, P. The common volatiles are H_2O, F.

Potassium-sodium alkaline intermediate (syenitoidal) series (Table 1, series 11 through 18) are mainly represented by intrusive facies, while their volcanic analogues including phonolites and trahytes are limited in occurrence. The composition of intrusive series and their degree of differentiation strongly vary according to the tectonic setting (older shields, near-platform rises, median massifs, post-orogenic faulted rises). However, certain common specific features to these series should be emphasized: a common line in the variation of chemical properties as we go from early intrusive phases to later ones (the rocks getting more basis and alkaline); a leucocratic composition of the rocks with strong variations in the concentration of SiO_2 (45-65%), alkalies (9--19.5%), Al_2O_3 (12-26%), with calcium and magnesium kept in a subordinate position, a variety of volatiles (F, Cl, CO_2, etc.). Typical petrogeochemical rock types are lūjavrite, foyaite, mariupolite, miaskite, khibinite, rischorrite. The typomorphic minerals are feldspathoids (nepheline, sodalite, etc.), alkaline feldspars and pyroxenes (aegirine, aegirine-hedenbergite, etc.), amphiboles (arfvedsonite, kataphorite), biotite, lepidomelane.

Potassium series of alkaline ultrabasic (Table 1, series 4, 5), alkaline basic (Table 1, series 9, 10), and alkaline intermediate rocks (Table 1, series 19, 20) are not as wide-spread as the potassium-sodium series. They are represented by volcanogenic and shallowlevel manifestations, while meso-abyssal ones are rare. Potassium alkaline series occur near rifts and other structures of tectonic-magmatic activation on older platforms and in consolidated folding areas. Typical for their composition are minerals like leucite, kalsilite, magnesian pyroxenes, olivines, micas, as well as the elements like potassium, rare alkalies, strontium, barium, phosphorus, zirkonium, volatiles (F, Cl); high concentrations of magnesium are typical for the potassium alkaline ultrabasic series.

Keeping in mind this classification of alkaline magmatic series, we now discuss the evolution of alkali magmatism within two major continental structures: an old platform (East European one) and a folding area with a long geological history (central part of the Asia), located south of the Siberian platform.

The East European platform is unique in that its alkaline rocks of different ages are exposed in the Baltic and Ukrainnian shields, enabling one to reconstruct the evolution of alkali magmatism since the pre-platform epoch during the formation period of the crust extending over 2,500 m.Y. The principal stages of alkaline magmatism in the East European platform are as follows:

The end of the Lower Proterozoic Stage: appearance of the earliest (in this region) massif Siilinjarvi (Finland) made up of glimmerites together with syenites, carbonatites (about 2,500 m.y. old);

The Middle Proterozoic Stage: several massifs with subalkaline gabbro, nepheline and alkaline syenites, sometimes together with ijolites. Their age is 1900 to 1700 m.y. (Gremyakha-Vyrmes, Elet'ozersky, Malotersyansky, Priazovsky massifs in the USSR; Almunge, Norra-Cherr massifs in Sweden). During this time, occasional miaskite (Sakharioksky in the USSR) and alkaline granitic (Keivy in the USSR) bodies appear.

The Late Proterozoic Stage: the Berdyaush pluton was formed up in the eastern margin of the East European platform, consisting of rapakivi granites, alkaline gabbroids (xenoliths) and nepheline syenites (dykes). The age of later one are 1,330 m.y. old.

Late Riphean-Paleozoic stage: massifs of alkaline ultrabasic rocks with carbonatites formed in the western margin of the Baltic shield under platform conditions - Fen (Norway), Alno (Sweden). Their age is about 560 m.y. Similar massifs located within the eastern part of the Baltic shield are younger (500 to 400 m.y. old), though data on ijolite-pegmatite veins in teh central olivine core of one of these (Kovdor massif) are close to 600 and 700 m.y.

The Middle Paleozoic Stage is characterized by considerable rebuilding of the East European platform: graben-like depressions are formed (Pripyatsko--Donetsk, Priazovsk, Kontozepck ones) filled with volcanogenic sequences of various compositions (phonolites, trachytes, melilitic basaltoids). Simultaneously (about 400 m.y. ago) massifs of alkaline ultrabasic rocks with carbonatites terminated their formation in the Kola Peninsula, and huge plutons of agpaitic nepheline syenites were formed (Lovozero and Khibiny ones);

Upper Paleozoic Stage: volcanic and intrusive activity in the Oslo rift (essexites, monzonites, alkaline and nepheline syenites, etc.), the Särna massif of cancrinite - nepheline syenites in Sweden, similar in composition volcanites within Conbass (shonkinites, trachydolerites). Their ages range between 270 to 290 m.y.;

Jurassic stage: the youngest alkaline magmatism in the East European platform represented by dyke fields (about 160 m.y. old) in Norway and the Donbass area (USSR).

To sum up, the history of alkali magmatism in the East European platform falls into two epochs:
a) the pre-platform epoch (early and middle Proterozoic), in which occasional massifs of miaskites and potassium-sodium alkaline basic rocks were formed about 2,500 and 1,900 m.y. ago on the consolidated rises in time of orogenic tectonic activity;
b) the platform epoch, characterized by sharply increased volume and number of centres of magmatism, in particular, at about 560, 400, and 260 m.y. ago. The compositions vary widely (potassium-sodium alkaline ultrabasic, basic, and intermediate), but the typical ones are numerous manifestations of alkaline ultrabasic rocks with carbonatites, as well as massifs of agpaitic nepheline syenites.

Periodic (multiple) manifestation of alkaline magmatism is characteristic for the folding areas with a long continental formation stage. An example is the south Siberia region, where the mountain-folding units arose during Upper Riphean (Baikalides) and Cambrian (early Caledonides) time. Later a long continental stage of development of this area has continued during entire Phanerozoic time (over 700 m.y.). The following periods of alkaline magmatism are typical for this stage (Fig. 2):

Late Riphean period (700 to 600 m.y. ago): predominantly alkaline ultrabasic magmatism. It is connected with activation of the through fault systems on the boundary of the platform and adjacent rises of Baikalides. It was in the time when the Paleoasian oceanic basin began to form in the Eurasian continent. Occasional massifs of aegirine foyaites with ijolites were formed on consolidated rises of Baikalides during the same time;

Silurian to Devonian period (430 to 400 m.y. ago): sub-alkaline olivine-basaltic, alkaline basaltoidal, and alkaline gabbroidal magmatism took place simultaneously with the formation of continental volcanogenic structures (depressions, basins, grabens) that arose during fragmentation of consolidated folding areas. It was in the time of the opening of the Paleothetys. In the same time the alkaline ultrabasic massifs with carbonatites were formed on the platform;

Carboniferous period (330 to 300 m.y. ago¾: alkaline sialic (syenite or nepheline-syenite) magmatism occurring during faulting of postorogenic (epi-Devonian) block-dome rises and activation of deep faults in the time of the Hercynian orogenesis in the adjacent Ural-Tien-Shan-Mongolian folding belt;

Late Paleozoic (290 to 240 m.y. ago) and early Mesozoic (220 to 170 m.y. ago) periods: the intracontinental volcanic activity gradually increases in area. It was formed upon a Baikalian-Caledonian folded basement and represented by trachybasalt-trachyandesite-trachyte-trachyrhyolitic volcanism and the related syenite-alkaligranitic magmatism;

Late Mesozoic period (150 to 110 m.y. ago): this period is different from the preceding ones in that subalkaline trachybasaltic-trachyandesitic volcanism was dominant. It was connected with the formation of a large (up to 500 km^2) volcanic area in the eastern Zabaikalye and East Mongolia along the western periphery of the Pacific mobile belt. Volcanic-intrusive complexes of potassium basaltoids were formed on the platform during this time interval.

Cenozoic period (25 to 0.1 m.y. ago): the typical volcanism is subalkaline trachybasaltic and potassium basaltoids, whose enormous area is associated with the formation of the Central Asian (including Baikal-Khubsugulian) rift system.

One will easily notice the synchronousness of the periods of alkaline magmatism in folding areas and adjacent Siberian platform (see Fig. 2). A major tectonic rebulding of the Eurasian continent is typical for such periods. However, the composition of alkaline rocks varies in particular units, reflecting variability in the regimes of deep magma formation.

Taking account of this information on the periodicity of alkaline magmatism in major continental structures (old platforms, folding areas), we have for the first time attempted to correlate the history of alkaline magmatism to epochs of terrestrial tectonomagmatic activity. Two remarks must be made before we proceed to discuss the results. (A) in evaluating the formation time of alkaline massifs it has been taken into account that many of them, especially poly-formational ones, have a long, multistage history of formation. In addition, alkaline massifs in folding areas (the Urals, Norwegian Caledonides, Grenville region in Canada and Greenland) have been strongly metamorphosed, resulting in rejuvenation of their ages. Because of this, we have made a thorough revision of age determinations, leving only best documented dates of their times of origin. One example can be using for illustration - the Ilmen-Vishnevogorsk miaskite massif (the Urals). Studies of the isotope composition of strontium in miaskites and of lead in zirkons from miaskites and carbonatites by the isochronic method in collaboration with I.V. Chernyshev (IGEM AND SSSR), W. grauert and V. Kramm (Mineralogical Institute, University of Munster, West Germany) have revealed the existence of two stages in the formation of this massif. According to strontium isochrones from rocks corroborated by lead isotopes from zircons, the massif must have arisen in bourder of the Ordovician - Selurian time about 440 m.y. ago. This figure has been adopted as the age of the massif. Changes in the isotopic equilibrium of lead and strontium about 240 m.y. ago, as well as the K-Ar age of minerals (nepheline, mica) and rocks coincide with the intensive metamorphism of Hercynian orogeny that terminated the formation of the Ural fold system. This figure refers to rejuvenation of the Ilmen-Vishnevogorsk massif. (B) Analysing the evolution of alkaline magmatism we used A.S. Monin and O.G. Sorokhtin's (4) hypothesis about the periodicity of the Earth's tectonic evolution and changing of convective structures (single-cell and double-cell ones) in the mantle. According to this hypothesis, the Earth's history is divided into five megacycles whose boundaries have been estimated as 2.6×10^9, 1.8×10^9, $0,95 \times 10^9$, 0.45×10^9 years, and which correspond to the periods of single-cell convective structures with maximum rates of convective mass transport in the mantle, formation of connected super-continents, peaks of tectonomagmatic activity. Each megacycle includes several cycles characterized by double-cell convective structure and weaker magmatic and tectonic activity. We proceed to give a brief discussion of the principal stages of terrestrial alkali magmatism, using the plot of Figure 3 which summarizes all available information on alkaline magmatism of all the continents, only keeping in mind that the regions are not uniformly studied.

Archean megacycle (4,650-2,600 m.y. ago). The first oldest massifs of continental crust are being formed. Alkaline rocks appear at the boundary to the next early Proterozoic megacycle; these are pseudoleucite syenites that make up the Kaminac and Puba massifs in the Canadian shield.

The early Proterozoic megacycle (2,600-1,870 m.y. ago) begins with a powerful world-wide diastrophism of the Kenoran epoch (single-cell convective structure). Continental crustal blocks are formed which are included in the shields of nearly all older platforms. The date 2,500 m.y. marks the formation of occasional miaskite massifs (Siilinjarvi in the Baltic shield), pseudoleucite syenites, phonolites, trachytes etc. (Lake Kirkland in the Canadian shield). An age of 2,200 m.y. can as yet be ascribed only to the Palabora massif (Africa) consisting of alkaline ultrabasic rocks with carbonatites.

The middle Proterozoic megacycle (1,870-1,060 m.y. ago) begins during Karelian time with a powerful diastrophism that reached its peak of activity about 1.8×10^9 years ago. All the older continents seem to have coalesced during this time, into a single megacontinent, Stille's Megagea. Massifs of potassium-sodium alkaline basic rocks (Gremyakha-Vyrmes and others on the East European platform), carbonatites (Karghil on the Canadian shield), nepheline syenites and phonolites (Pilansberg, Africa) formed in stable blocks adjacent to Karelian orogenic features. Outbursts of tectonomagmatic activity occurred during the third terrestrial megacycle at dates 1,580 an 1,330 m.y. ago. It is precisely these time intervals where occasional, though sometimes fairly large, massifs were formed, consisting of nepheline syenites (Ilimaussak in Greenland), alkaline ultrabasic rocks with carbonatites (Spitzkop and others in Africa) etc.

The Late Proterozoic megacycle (1,060-530 m.y. ago) is characterized by the formation of predominantly potassiumlsodium alkaline ultrabasic, alkaline basic, and alkaline intermediate sienitoidal rock series in nearly all old platforms except those in Australia. The worldwide Grenwille diastrophism which occurred at the beginning of the megacycle caused profound metamorphism of enormous areas, including Early Proterozoic alkaline massifs within the northern continents - in the eastern Canadian shield, in Greenland, in the western Baltic shield. Massifs of alkaline ultrabasic rocks with carbonatites were formed everywhere towards the end of the fourth megacycle, at the Riphean-Paleozoic boundary. The North Atlantic alkaline province was formed with rock ages between 650 and 580 m.y. (2), marking the riftogenetic epoch prior to opening of an oceanic basin (Paleoatlantic).

The current (fifth) megacycle began with the Caledonian diastrophism (530 - 0 m.y. ago) which attained its peak of activity in the latest Early Paleozoic (450 m.y. ago), the time of formation of the integrated Pangea continent. Its subsequent breakup within the older European continent created new oceanic basins - the Paleothetys, the Ural basin. Prior to the formation of the Ural ocean, as demonstrated by paleotectonic reconstructions, there was the early Gercinian riftogenesis when the volume of alkaline magmatism increased sharply, as did the number of provinces and massifs.

There is some evidence for a certain difference in the periodicity of the fifth megacycle alkaline magmatism within northern and southern continents: the northern ones show alkaline magmatism during each of the tectonic epochs (Caledonian, Hercynian, Cimmerian), while no noticeable amount of Hercynian alkaline rocks has yet been discovered in the southern continents.

To sum up, the following facts are noteworthy in the history of terrestrial alkaline magmatism:

- very few alkaline massifs appeared within the most stable nuclei (shields) of older platforms 2,800 to 2,500 m.y. ago. Therefore one can accept sufficient differentiation in the mantle during Early Proterozoic time;
- alkaline rocks and their series were formed both in platforms and in consolidated folding areas. They were not infrequently formed at one and the same time, but compositionally the arising alkaline rock series were always different. This is obvious in folding systems of the Phanerozoic megacycle, but occasional massifs of alkaline gabbroids, feldspathoid syenites

etc. are also found in Proterozoic folding systems, which are usually regarded as components of older platforms;

- sharp indrease in scale of occurrence and diversity of composition is observed for alkaline rocks in the Phanerozoic tectonic megacycle, a noticeable increase in the importance of potassium alkaline rocks in the Mesozoic-Cenozoic volcanism. Experimental data permitting modeling of conditions prevailing in deep magma formation have proved that alkaline magmas can be formed by partial melting of upper mantle (6). Isotopic studies have confirmed mantle origin for substances that composes most alkaline rocks. Even for Ural miaskites whose origin has long been debated, data on $^{87}Sr/^{86}Sr$, 0.7030 to 0.7033, leave no doubt as to their mantle origin. However, the composition of alkaline melts, as shown in a majority of experimental petrogenetic models, is fairly diverse, depending on the depth of magma formation and the pressure, among other physical and chemical factors. Thus, according to the well-known Green-Ringwood model, if pressures of order 20 kbars (depth 100-70 km) are required to produce melts like olivine nephelinites, then in the others similar conditions melts like olivine basanite must form at about 15 kbars (depth 70-35 km). We recall that alkaline ultrabasic rock series with carbonatites on old platforms can be regarded as derivates of olivine nephelinites, and weries of alkaline basic rocks of consolidated folding areas as derivates of alkaline basaltoids. Apparently, periodic activation of the mantle on the different depths under the picks of tectonomagmatic activity of the Earth and was responsible for the mutial arised the main alkaline series in the Earth history.

REFERENCES

1. Andreeva E.D., Kononova V.A., Sveshnikova E.V., Yashina R.M. Alkaline magmatic series, their origin, and causes of their diversity. In: Problems of Petrology, Mineralogy, and Ore Genesis. Moscow, "Nauka",

1983 (in Russian).
2. Doing R. An alkaline rock province linking Europe and North America. Can. J. Earth Sci., 1970, 7, No. 22.
3. Classification and nomenclature of magmatic rocks. Moscow, "Nedra", 1981 (in Russian).
4. Monin A.S., Sorokhtin O.G. Geophysics of the Ocean. Moscow, "Nauka", 1979 (in Russian).
5. The alkaline rocks. J. Wiley and Sons, 1974.
6. The evolution of the igneous rocks. Princeton Univ. Press. Princeton, New Jersey, 1979.
7. Vorobyeva O.A. Principal features in the location and formation of alkaline rocks. In: Problems in the Geology of Mineral Deposits, Petrology, and Mineralogy. Moscow, "Nauka", 1969, v. II (in Russian).

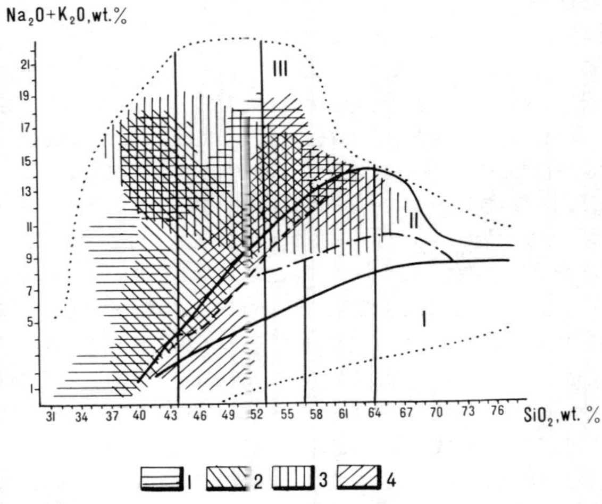

FIGURE 1. Composition of alkaline magmatic series in the SiO_2 -(Na_2O + K_2O) diagram. Field boundaries are given after "Classification and nomenclature..." (1981). Normal (I), subalkaline (II), and alkaline (III) rock series. Alkaline magmatic series: potassium-sodium alkaline ultrabasic rocks (1), potassium-sodium alkaline basic rocks (2), potassium-sodium alkaline intermediate rocks (3), potassium alkaline rocks (4).

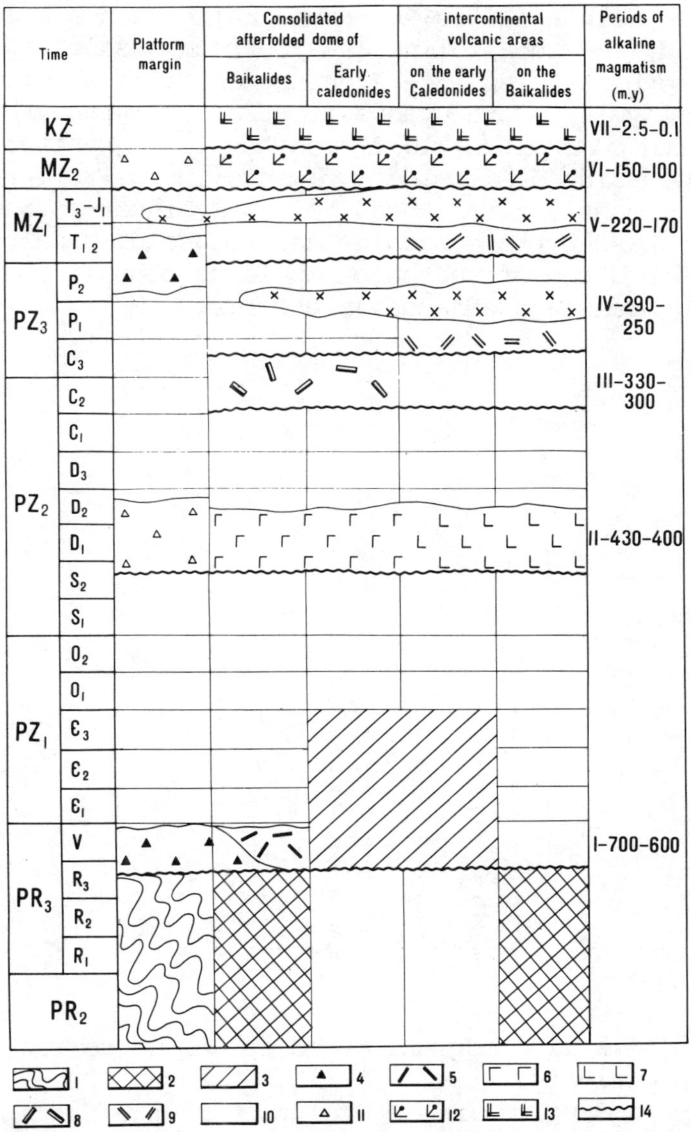

FIGURE 2. Principal geologic periods of the development for alkaline and subalkaline magmatism in folding areas to the south from the Siberian platform.

1-3 - granitic-metamorphic basement: 1- Archaean - -Proterozoic; 2 - pre-upper Riphean; 3 - Lower Paleozoic; 4-14 - alkaline magmatic series of different composition: 4 - alkaline ultrabasic; 5 - aegirine foyaites with ijolites; 6 - subalkaline and alkaline gabbroid; 7 - subalkaline basalt-trachybasalte and alkaline basaltoids; 8 - syenite-nepheline syenite; 9 - trachybasalt-trachyandesite-trachyte-trachyrhyolite with comendites; 10 - syenite-alkaligranite; 11 - potassium basaltoids (volcanic-intrusive); 12 - dominantly trachy-basalte; 13 - hihger-potassium-content basalt-trachybasalte and potassium basaltoids; 14 - main dates of tectonic rebulding consolidated folding areas and adjacent platforms.

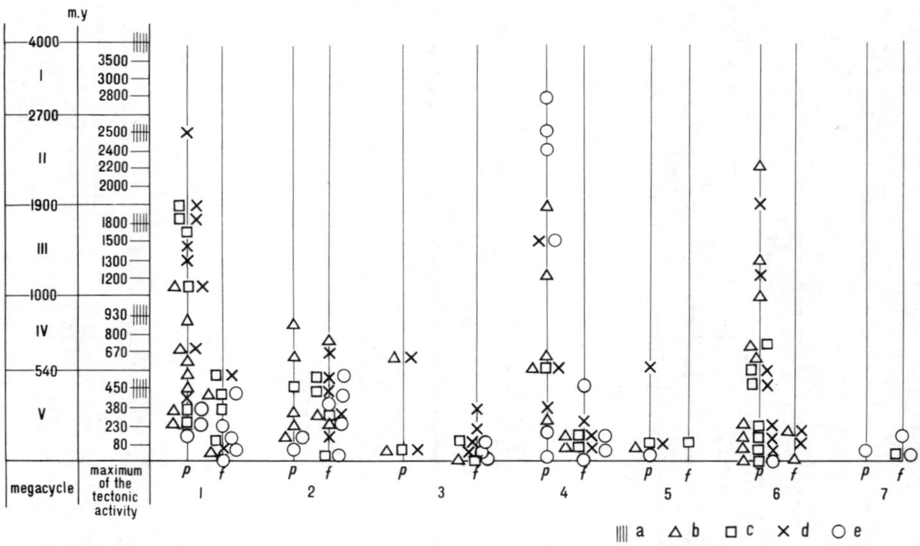

FIGURE 3. Alkaline magmatism and epochs of terrestrial tectonic activity.

I - V - megacycles in the evolution of the Earth; a - epochs of single-cell mantle convective structures.

Alkaline magmatism: b - Na-K alkaline ultrabasic rock series; c - NaK alkaline basic rock series; d - Na-K alkaline intermediate rock series; e - group of K alkaline rock series.

The symbols in the plot denote occurrence of an alkaline rock series without indicating the relevant number of massifs and provinces. Location of alkaline magmatism in relation to continental features (p - ancient platform, f - folding area around the ancient platform); continents: 1 - European; 2-4 - Asian (s - Siberian platform, 3 - Indian platform, 4 - Alp-Himalayan folding belt); 5 - North American; 6 - South American; 7 - African; 8 - Australian.

Table 1. Principal Alkaline Magmatic Series

Alkaline content	No.	Magmatic series	Hypothetical initial magma	Possible depth of formation
Magmatic series of alkaline ultrabasic rocks				
K - Na	1.	meymechite, melilitite, phonolite, carbonatite	olivine melanephelinite or nepheline picrite	surface sheets and volcanoes
	2.	olivinite, uncompahgrite, turjaite, melteigite, ijolite, feldspathoidal and alkaline syenite, carbonatite		down to 5 km
	3.	pyroxenite, melteigite, ijolite, malignite, feldspathoidal and alkaline syenite, carbonatite		down to 9-11 km
K	4.	ugandite, mafurite, katungite, carbonatite	ugandite	surface effusions
	5.	volgidite, fitzroite, saedrisite		near surface

K - Na and Na	6.	subalkaline olivine basalt, tephrite, phonolite	subalkaline olivine basalt or olivine tephrite	surface sheets and volcanoes
	7.	analcite dolerite, teshenite, analcite syenite		near surface
	8.	subalkaline gabbro, theralite, urtite, nepheline syenite		9-11 km
K	9.	potassium limburgite and picrite, potassium basalt, leucitite, leucite phonolite, trachyte	potassium olivine basalt or leucite tephrite	surface effusions
	10.	potassium pyroxenite, shonkinite, pseudoleucite and nepheline syenite		near surface
Magmatic series of alkaline intermediate rock				
K - Na and Na	11.	phonolite, trachyte	phonolite oversaturated with alkali (Na $\gg$ K)	surface sheets
	12.	phonolite, tinguaite, aegirine ijolite-porphyry		near surface
	13.	umptekite, foyaite, rischorrite, ijolite-urtite	rich in Cl, F, S, P	4-5 km
	14.	urtite, foyaite, lujavrite, naujaite		2-3 km

K - Na and Na	15.	ijolite, juvite, foyaite	phonolite saturated with ($Na \approx K$) rich in CO_2, P, F, Cl	5-7 km
	16.	pulaskite, amphibole foyaite, aegirine foyaite, mariupolite		5-7 km
Na = K	17.	syenite, pseudoleucite and nepheline syenite, tinguaite	phonolite oversaturated with alumina ($K \geqslant Na$) CO_2, F	near surface
	18.	syenite, miaskite, tinguaite		9-11 km
K	19.	leucite phonolite, trachyte	phonolite enriched with potassium ($K > Na$), F	surface sheets
	20.	pulaskite, nepheline, pseudolencite and kalsilite syenite		shallow level down to 2 km

Note: The most typical examples of the above series are the following:

1) East African rift; Maimecha-Kotuy province, USSR; 2) Karelian, Maimecha-Kotuy provinces, USSR; 3) East-Sayan, Sette-Daban provinces, USSR; 4) Rift of East Africa; 5) Australia; 6,7) Minusin depression,USSR; 8)Kija-Shaltyr and other massifs in the Kuznetsky Alatau, USSR; 9) Daubaba in Talass Alatau, USSR; 10) Irisu in Talass Alatau; Inagli in Aldan, USSR; 11) Kenyan rift; 12) South Africa; 13) Khibiny in Kola Peninsula, USSR; 14) Lovozero in Kola Peninsula, USSR; 15) Kharlinsky, Bayankolsky, Tuva province, USSR; 16) Korgeredabinsky, Tuva province, USSR; 17) Lughigolsky, Mongolian Peoples' Republic; 18) Vishnevogorsky, Ilmenogorsky in the Ural Mountains, USSR; 19) Rome province; 20) Synnyrsky in Zabaikalye, USSR.

GENESIS OF ARC MAGMAS : A CASE IN JAPANESE ARCS

I. KUSHIRO

Geological Institute, University of Tokyo, Tokyo, Japan

ABSTRACT

The lateral variations in alkalies/silica, alkalies/alumina, FeO/MgO, alkalies, REE, H_2O, F, Ba, U, Th, and Sr isotope ratio observed for the Quaternary volcanic rocks across the Japanese arcs, especially in the northeast Japan are discussed in the light of our recent experimental studies. It is concluded that most of these lateral variations can be explained by (1) increase in the depth of generation or segregation of primary magmas (40 - 70 km in depth) with increasing distance from the volcanic front toward the Japan Sea, (2) decrease in the degree of partial melting in the diapirs which may have ascended from the lower part of a wedge of the mantle above the subducted slab, and (3) density filtering for heavier primary tholeiitic magmas near the volcanic front. Decrease in the volume of the volcanic rocks as well as in the number of volcanoes away from the volcanic front may be explained by the decrease in the number of diapirs. It is also suggested that the harzburgite/lherzolite ratio in the upper mantle beneath the island arcs may be larger than that in the oceanic regions.

INTRODUCTION

Quaternary volcanic rocks in the Japanese Islands have been studied in detail by a number of investigators during the last half a century, and their petrological and geochemical characteristics are well understood among volcanic rocks in island arcs and continental margins in the world. The genesis of the volcanic rocks in the Japanese Islands as well as in other island arcs is, however, still controversial. We have been trying to

Proceedings of the 27th International
Geological Congress, Volume 9, pp. 259–281
PETROLOGY (IGNEOUS AND METAMORPHIC ROCKS)

solve some basic problems on the genesis of arc magmas, based on both the petrological observations on volcanic rocks in the Japanese Islands and the experimental studies on representative volcanic rocks. The problems which have been studied during the last several years are(1) lateral variations in volume and chemical composition of volcanic rocks across Japanese arcs, (2) depth and mechanism of generation of arc magmas, (3) mechanism of ascent of arc magmas, (4) existence of a volcanic front which is a sharp boundary between the volcanic and non-volcanic zones, (5) origin of calc-alkalic andesite, and (6) origin of a large amount of felsic magmas near the volcanic front. Although these problems are listed separately, they are more or less related one another.

In the present article for the symposium, our recent experimental studies are briefly reviewed among other studies. The results of these experimental studies give useful information on the genesis of arc magmas. Based on these studies discussions are made mainly on the problems (1) - (3) listed above. In the following sections, the facts on the across-arc lateral variations in the Japanese arcs are summarized first, followed by discussions with our recent experimental results.

LATERAL VARIATIONS

In the Japanese arcs, chemical and mineralogical compositions of the Quaternary volcanic rocks vary systematically across the arcs. A number of studies have been carried out on these lateral variations. Gill (1981) summarized such lateral variations in island arcs and active continental margins in the world including the Japanese Islands. Followings are some principal variations obtained for the Quaternary volcanic rocks mainly in the northeast Japan (northern Honshu and Hokkaido), which must be taken into account for the discussions on the genesis of arc magmas.

Volume of volcanic rocks

Sugimura et al. (1963) estimated the volume of volcanic rocks in the northeast Japan. Their results are shown in

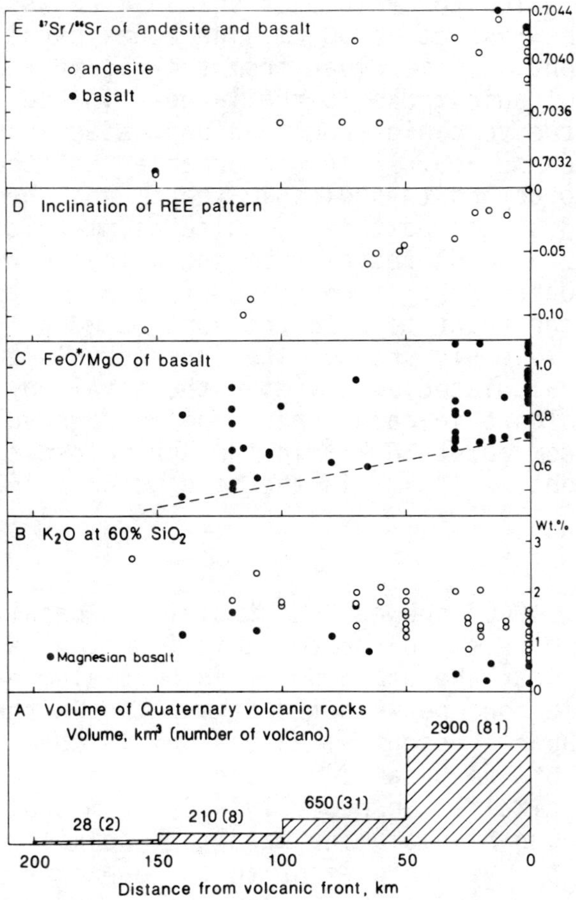

FIGURE 1. Some of the lateral variations observed for Quaternary volcanic rocks across the Japanese arcs (mainly in the northeast Japan). A, Volume of volcanic rocks and number of volcanoes (Sugimura et al., 1963); B, K_2O in rocks with 60 weight per cent SiO_2 (Aramaki and Ui, 1983) and in magnesian basalts (Kushiro, 1983); C, FeO*/MgO (total iron as FeO*) in magnesian basalts in northeast and west Japan (Kushiro, 1983); D, Inclination of chondrite-normalized REE pattern (Fujitani and Masuda, 1981); and E, $^{87}Sr/^{86}Sr$ ratios of andesite and basalt (Notsu, 1983). All the data except C are for northeast Japan.

Fig. 1A, in which the volume of volcanic rocks is summed up within the respective 50 km wide zones parallel to the volcanic front. It is clear from the figure that the volume of volcanic rocks is the largest in the zone closest to the volcanic front and decreases sharply toward the Japan Sea. It is remarkable that the volume becomes zero beyond the volcanic front, and therefore, the volume-distance pattern is quite asymmetric. Number of volcanoes also decreases from the volcanic front toward the Japan Sea; it decreases from 81 in the zone nearest to the front to 2 in the zone between 150 and 200 km from the volcanic front. The average volume per one volcano is calculated by dividing the total volume by number of volcano in each zone. The average volume decreases from about 36 km^3 in the zone closest to the volcanic front to 14 km^3 in the zone between 150 - 200 km.

Basalt type

Kuno (1959, 1960) showed that tholeiitic basalts occur in the zone closest to the volcanic front, alkali basalts in the zone near the Japan Sea, and high-alumina basalts in the middle zone between the former two in the northeast Japan. Essential variations in chemical composition of the basalts are decrease in silica and increase in alkalies (especially K_2O) from tholeiitic basalts through high-alumina basalts to alkali basalts (i.e., from the volcanic front to the Japan Sea). Later, Kuno (1968a) mentioned that calc-alkalic andesites also show the variations similar to the associated basalts.

Alkalies

The K_2O content in volcanic rocks for constant SiO_2 (without distinguishing rock series) increases with increasing distance from the volcanic front or with increasing depth of the Wadati-Benioff zone in the northeast Japan among others (e.g., Kawano et al., 1961; Dickinson, 1968; and Hatherton and Dickinson, 1968 and 1969). Recently Aramaki and Ui (1983) have mapped K_2O and Na_2O of volcanic rocks in the Japanese islands more extensively than before and showed that the increase of K_2O for constant SiO_2 (60 wt %) from the volcanic front toward the Japan Sea is more systematic than that of Na_2O. Fig. 1B shows their results on K_2O for the northeast Japan. The K_2O contens in magnesian basalts are also shown for comparison. The Rb content in volcanic rocks with similar SiO_2 contents also increases with increasing distance from the volcanic front (Notsu, 1983). Other incompatible elements such as Ba, Th and U are also lower in volcanic rocks near the volcanic front than in those near the Japan Sea similar to K_2O (Masuda et al., 1975; Katsui et al., 1978; Masuda, 1979).

Alkali/alumina ratio

Sugimura (1960) proposed Θ index ($\Theta = SiO_2 - 47(Na_2O + K_2O)/Al_2O_3$) and showed that this index decreases from the volcanic front toward the Japan Sea. Miyashiro (1974) pointed out that this variation is the same as the increase of the $(Na_2O + K_2O)/Al_2O_3$ ratio with increasing distance from the volcanic front.

H_2O

Sakuyama (1979) estimated the H_2O contents in magmas in the northeast Japan based on the assemblage of phenocrysts and the order of their crystallization. The volcanic rocks (mainly andesite and dacite) near the volcanic front do not contain any hydrous minerals and often contain quartz, whereas those near the Japan Sea often contain hornblende and biotite but do not contain quartz. From these observations and the experimental results on the crystallization temperature of hornblende, biotite and quartz for andesite - dacite compositions under hydrous conditions (Robertson and Wyllie, 1971; Wyllie, 1977), it is shown that the H_2O content in the magma increases from the volcanic front toward the Japan Sea. The absolute amounts of H_2O could not be determined but the maximum H_2O content was estimated to be about 3 wt % for basaltic magmas erupted near the Japan Sea.

Fluorine

Ishikawa et al. (1978) measured the fluorine contents in a number of volcanic rocks in the northeast and west Japan. Their results show that the F content in basalts increases from the volcanic front toward the Japan Sea or from tholeiites to alkali basalts. As fractional crystallizaion advances, the F content increases in tholeiitic series rocks, whereas it stays nearly constant in high-alumina basalt series and alkali series rocks.

Rare earth elements (REE)

Masuda et al. (1975) determined REE in basalts of three different types of Kuno (1960) and associated calc-alkalic andesite and dacite in Hokkaido, and showed that the La/Yb ratio in these rocks increases from the volcanic front toward the continent. Fujimaki and Kurasawa (1980) measured REE in representative basalts in the Japanese Islands and also showed that while the heavy REE do not vary, the light REE increase from the volcanic front toward the Japan Sea. Fujitani and Masuda (1981) demonstrated clearly this relation in the northeast Japan. They defined "inclination" which is the rate of change of common logarithm of chondrite-normalized value per increase of atomic number by unity. Their results are shown in Fig. 1D. It is clear that the inclination becomes larger with increasing distance from the volcanic front.

$^{87}Sr/^{86}Sr$ ratio

A number of measurements have been made on the $^{87}Sr/^{86}Sr$ ratio of volcanic rocks in the Japanese Islands (e.g., Hedge and Knight, 1969; Kurasawa, 1972 and 1979; Kaneoka et al., 1978; Nohda and Wasserburg, 1981). Recently, Notsu (1983) made more extensive studies on the variation of $^{87}Sr/^{86}Sr$ of basalts and andesites in the northeast Japan. He confirmed previous results showing that this ratio in both basalts and andesites decreases with distance from the volcanic front, as shown in Fig. 1E, although the data of basalts are few. He also found an anomalous high ratios in the southern part of the northeast Japan, which is excluded in Fig. 1E.

FeO/MgO ratio in basalts

The FeO/MgO ratio in basalts reflects the degree of fractionation of primary magmas which are believed to have relatively uniform FeO/MgO values. Fig. 1C shows the FeO^*/MgO in mole (total iron as FeO^*) for relatively magnesian basalts in the northeast and west Japan by Kushiro (1983). The FeO^*/MgO ratio of most magnesian basalt in the respective regions decreases away from the volcanic front. There are a number of analyses of basalts near the volcanic front, but there is no basalt with this ratio smaller than 0.7, whereas many basalts with this ratio smaller than 0.7 near the Japan Sea, many of which are alkali basalts.

DISCUSSION ON THE GENESIS OF MAGMAS IN THE JAPANESE ISLANDS

Kuno (1959, 1960) suggested that tholeiite, high-alumina basalt and alkali basalt magmas are produced by partial melting of the upper mantle peridotites at depths shallower than 150 km, between 150 and 250 km, and deeper than 250 km, respectively, on the assumption that the deep seismic zone is the site of generation of basalt magmas. Later experimental studies (e.g., Green and Ringwood, 1967; Kushiro, 1968), however, have revealed that the depth of generation of these three basalt magmas would be much shallower than those suggested by Kuno. A number of attempts have been made to determine the compositions of melts formed by partial melting of the upper mantle peridotite at high pressures; however, reliable data have not been obtained. Recently, Jaques and Green (1981) and Takahashi and Kushiro (1983) redetermined the major element compositions of melts equilibrated with peridotites at high pressures. The latter authors used a relatively undepleted lherzolite and a "peridotite capsule" technique with a careful examination for equilibrium. Although the abundance of incompatible elements cannot be determined with this technique, those of major elemets which are buffered by lherzolite minerals can be determined with a fair accuracy. Fig. 2 shows the P-T ranges for generation of melts with different basaltic compositions. The results of these experiments are, in general, consistent with those of the previous experiments (e.g., Green and Ringwood, 1967; Kushiro, 1973). As shown in Fig. 2, pressure for generation (or equilibrium with peridotite) of quartz tholeiitic melt is less than 8 kbar (<25 km),

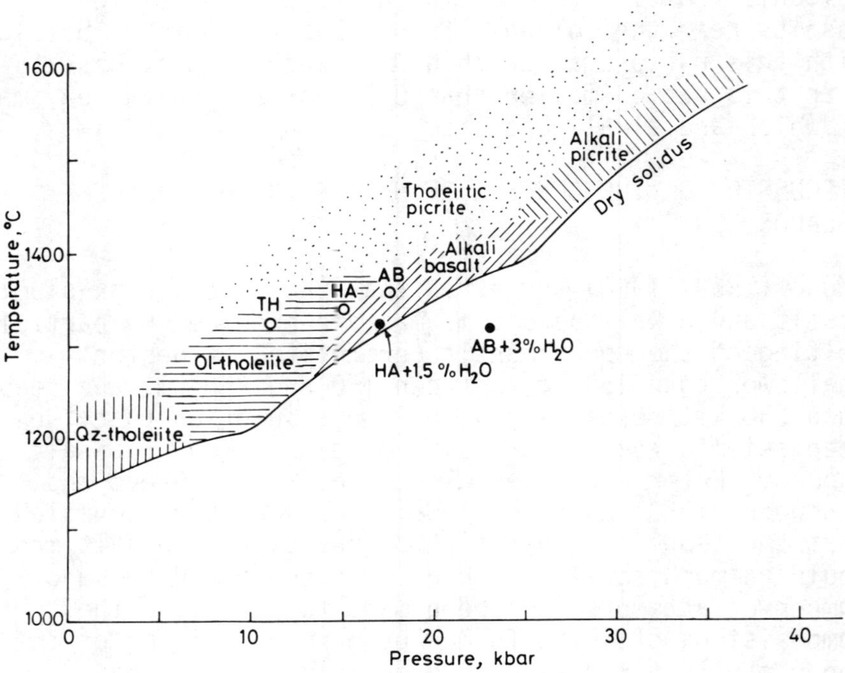

FIGURE 2. Conditions for equilibration of three primary basalt magmas with the upper mantle peridotites. TH, Primary tholeiite (in equilibrium with harzburgite); HA, Primary high-alumina basalt (in equilibrium with lherzolite), and AB, Primary alkali basalt (in equilibrium with lherzolite). Data from Tatsumi et al. (1983). The regions for generation of melts of basaltic and picritic compositions from a relatively undepleted lherzolite under anhydrous conditions are based on Takahashi and Kushiro (1983). Abbreviations: Qz, quartz; and Ol, olivine.

olivine tholeiitic melt up to about 15 kbar (<45 km), alkali olivine basaltic melt between about 15 and 25 kbar (45 - 70 km), and alkali picritic melt at pressures higher than about 25 kbar (>70 km) near but above the solidus of anhydrous lherzolite. Of course, the compositions of the erupted basalts cannot be compared directly with those of these partial melts unless the basalts keep the primary magma compositions. In the presence of H_2O, the depth ranges given above become deeper; however, they are still much shallower than those suggested by Kuno, as shown below.

Another experiments recently made are the determination of the liquidus phase relations of primary basalt magma compositions in island arcs to estimate the depth of their generation or segregation more specifically (Tatsumi et al., 1983). In these experimental studies, the compositions of the primary magmas of basalts in the northeast Japan were first estimated. The magnesian basalts containing olivine phenocrysts of tholeiite, high-alumina basalt, and alkali basalt series in the northeast Japan were selected. Assuming that these basalts are products of fractional crystallization involving separation of olivine, the compositions of three primary magmas were calculated by adding olivine to the basalts with step by step, keeping equilibrium between olivine and liquid in each step, until the liquid becomes in equilibrium with olivine of Fo_{89} which is the most magnesian olivine phenocryst found so far in the Quaternary basalts. This is a reverse process of maximum olivine fractionation.

The melting experiments were carried out on these primary basalt magma compositions under anhydrous conditions and hydrous conditions with H_2O up to 3 wt %. The amounts of H_2O in primary basalt magmas in the arcs are not known. In these experiments it is assumed that the H_2O contents in tholeiite, high-alumina basalt, and alkali basalt magmas are nearly 0, 1.5, and 3.0 wt % respectively, based on the estimation by Sakuyama (1979, 1983), though small amounts (<1 wt%) of H_2O may be present in the tholeiite magmas. Thus 1.5 and 3.0 wt % H_2O was added to primary high-alumina and alkali basalt compositions, respectively, and no H_2O was added to primary tholeiite composition.

The results indicate that olivine and orthopyroxene crystallize on the liquidus (or liquid is in equilibrium with harzburgite) at about 11 kbar for the primary tholeiite composition under anhydrous conditions. For the high-alumina basalt composition, olivine, orthopyroxene, and clinopyroxene are on the liquidus (or liquid is in equilibrium with lherzolite) at about 15 kbar under anhydrous conditions and about 17 kbar in the presence of 1.5 wt % H_2O. For the primary alkali basalt composition, olivine, orthopyroxene, and clinopyroxene are also on the liquidus at about 18 kbar under anhydrous conditions and at about 23 kbar in the presence of 3 wt % H_2O. These results are shown in Fig. 2. The experimental results indicate that the depth of generation or segregation of primary basaltic magmas becomes deeper from the volcanic front toward the Japan Sea even under anhydrous conditions. The results of these experiments are also consistent with those of the partial melting of a peridotite shown above, although the latter experiments give more specific P-T conditions than the former. Furthermore, the latter experiments give more information on the residues of partial melting; that is, the mineral assemblage of the residues is harzburgite after generation of primary tholeiite magma, whereas it is lherzolite after generation of primary high-alumina and alkali basalt magmas. It means that the degree of partial melting of the upper mantle peridotite would be higher for the generation of primary tholeiite magma than for generation of primary high-alumina basalt and alkali basalt magmas provided that the upper mantle before generation of the magmas were essentially the same.

Decrease in the degree of partial melting is consistent with the increase in K_2O and other incompatible elements of the magnesian basalts from the volcanic front toward the Japan Sea across the Japanese arcs, on the assumption that the abundances of K_2O and other incompatible elements in the source region of primary magmas are nearly the same aross the arcs. Not only the variations in K_2O content of basalts, but also those in the alkalis/alumina ratios are important for understanding the degree of melting and the melting phase relations.

Miyashiro (1974) pointed out based on the Sugimura's (1960) data, that $(Na_2O + K_2O)/Al_2O_3$ ratios of the Quaternay volcanic rocks with the similar SiO_2 contents increases from the volcanic front toward the Japan Sea. Magnesian basalts as well as the estimated primary bsalt magmas also show the same variations. This ratio in the melt formed by partial melting of peridotite does not change as long as feldspar, mica, and clinopyroxene containing jadeite component melt congruently. Increase of this ratio means that minerals with this ratio >1 melt more abundantly or those with this ratio <1 remain in the residue more abundantly. The former possibility is unlikely because there would be no minerals with this ratio >1 in the upper mantle peridotite. K-richterite and acmite are possible minerals, but the former mineral breaks down in the presence of aluminous phases under the upper mantle conditions (Kushiro and Erlank, 1970) and the latter mineral has not been reported in the upper mantle rocks. The latter possibility is, therefore, likely. There are several minerals with this ratio less than 1 in the upper mantle rocks; they are spinel, garnet and clinopyroxene containing $CaAl_2SiO_6$ component. Increase in the amount of these minerals in the residue after partial melting is quite possible by the decrease in the degree of partial melting, because aluminous phases are melted in the initial stage of partial melting as demonstrated by experiments (e.g., Ito and Kennedy, 1967; Kushiro, 1969; Mysen and Kushiro, 1977). Increase in H_2O and F away from the volcanic front can also be explained by the decrease of degree of partial melting of the source peridotite. Increase in the inclination of REE patterns may also be related to the decrease in the deree of partial melting of the source peridotite (Fujimaki and Kurasawa, 1980).

There is a relation between the depth of generation and the degree of partial melting as discussed by Kushiro (1980). This is due to the increase with increasing pressure of the solubility of alkali-bearing components and H_2O in melts coexisting with olivine and pyroxene; the melts in equilibrium with olivine and orthopyroxene at higher pressures can dissolve more alkalies and H_2O, so that if the composition of the source peridotite is the same, the amount of an initial melt is smaller at higher pressures.

Another important aspects of the experiments by Tatsumi et al. (1983) is that the temperature of generation of the three primary basalt magmas is significantly high even in the presence of 3 % H_2O; it is between 1320° and 1360°C depending on the composition. Such high temperature would not be spread extensively beneath the Japanese arcs, especially near the volcanic front where probable depth of generation or segregation of the primary tholeiitic magma is only about 40 km. It is suggested that such high temperature is restricted to a relatively small area, which we think is a diapir having ascended from the melt-rich zone just above the subducted slab through the high-temperature region of the mantle wedge.

The melting experiments also give information on the upper mantle materials beneath the island arcs. As shown above, primary tholeiite magma coexist with harzburgite at about 11 kbar. This means that the residue after segregation of primary tholeiite magma is harzburgite. Harzburgite defined here consists of magnesian olivine and orthopyroxene without clinopyroxene at or above the solidus. Orthopyroxene can dissolve appreciable amounts of CaO (up to 2.5 wt %) at such high temperatures ($\geq$1200°C) and may exsolve clinopyroxene upon cooling. Since tholeiites occur near the volcanic front, it is likely that harzburgite exists widely in the uppermost mantle along the volcanic front of the island arcs. Magnesian andesite and boninite which occur in some island arcs (e.g., Bonin Islands and southwest Japan) also have magnesian olivine and orthopyroxene as liquidus phases under hydrous conditions at pressures between a few kbar and 17 kbar, depending on the composition and H_2O content (Kushiro and Sato, 1978; Tatsumi, 1982; Kushiro, unpublished data). Near the Bonin Islands, a large amount of harzburgite has been dredged from the slope of a topographic high (Ishii et al., 1981). It is likely that the upper mantle near the Bonin Islands where boninites and magnesian andesites occur consists largely of harzburgite. In Papua both magnesian andesitic rocks (clinoenstatite rocks) and harzburgite also occur closely together in time and space (Dallwitz et al., 1966; England and Davies, 1973). Some calc-alkalic andesites might have been formed by fractional crystallization of magnesian andesite magmas. It is suggested that significant amounts of harzburgite may exist in the upper mantle beneath island arcs. This possibility can be examined by peridotite inclusions derived from the upper mantle. Very few breccia pipes or nodule-bearing localities have been found in island arcs. In the northeast Japan, Ichinomegata is the most well known locality for the upper mantle nodules. Examining the bulk chemical compositions of peridotite nodules (Kuno and Aoki, 1971) as well as their thin sections, it was found that harzburgite occupies about 30 % of the peridotite nodules. Other peridotites are mostly lherzolite. In other nodule localities in southwest Japan, harzburgite is also relatively abundant; for example, it is more than 30 % in Arato-yama as shown by Fujiwara and Arai (1982). These ratios are much larger than those obtained from the peridotite nudules from Hawaii (White, 1966; Kuno, 1968b) and Tahiti (Tracy, 1980). Peridotites dredged near the mid-oceanic ridges are mostly lherzolite, and a small amount of harzburgite has been found (e.g., Hamlyn and Bonatti, 1980). Although the data are not adequate to generalize at present, the upper mantle materials beneath island arcs seem to be more enriched in harzburgite than those in oceanic regions.

Basaltic magmas have much smaller densities than the upper mantle lherzolite at depths shallower than 60 km (Fujii add Kushiro, 1977; Stolper et al., 1981), and when the magmas segregate from the source, they would begin to ascend by buoyancy. If the densities of the magmas are smaller than those of the crustal rocks, the magma can ascend further through the crust to the surface, whereas if the former is larger than the latter, the magmas cannot ascend by buoyancy and must stop ascending at a certain level within the crust. It is important, therefore, to determine the densities of magmas in the pressure range within the island arc crust (i.e., up to 12 kbar). Such measurements have been made for melts with compositions of primary and fractionated tholeiites

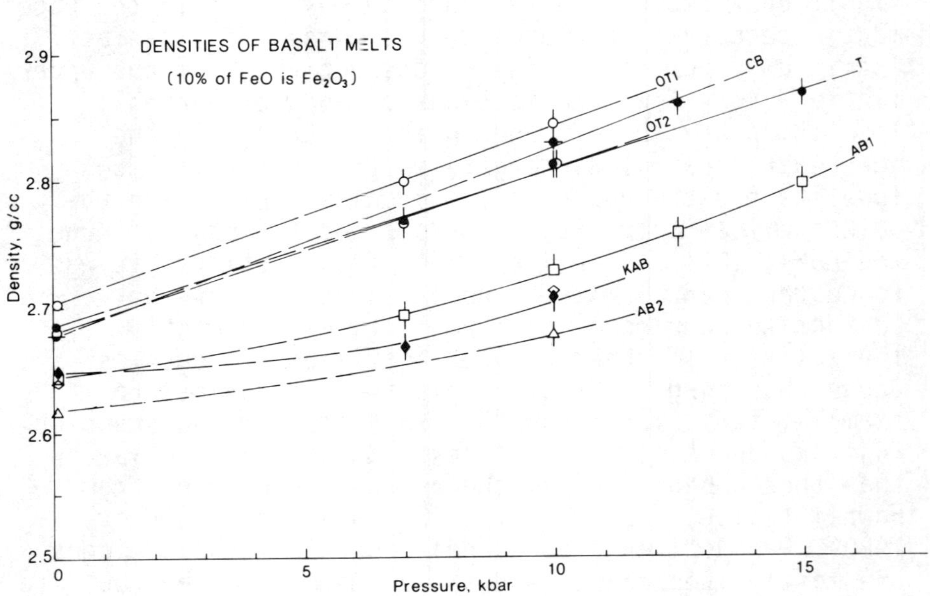

FIGURE 3. Densities of melts of seven different basaltic compositions. OT1 and OT2, Primary tholeiites; CB, Columbia River basalt; T, Tholeiite from Izu; AB1, Alkali basalt from Oki-Dogo, west Japan; KAB, K-rich alkali basalt from west Japan; AB2, Alkali basalt from Kuznetsk, U.S.S.R. (provided by L. L. Perchuk). Data except for AB2 are from Kushiro (1982 and 1983). The densities are recalculated to those for FeO/Fe_2O_3 (wt.) equal to 9, using partial molar volumes of FeO and Fe_2O_3.

and alkali basalts (Kushiro, 1982, 1983). The density variations of these melts in the previous figures are not corrected for the Fe_2O_3/FeO ratios. Fig. 3 shows the densities of several basaltic melts with Fe_2O_3/FeO ratio 0.11, which is the values for the least oxidized basalts in the Japanese Islands. The correction of the densities was made using the partial molar volumes of FeO and Fe_2O_3 given by Bottinga and Weill (1970) and assuming that the rate of change of the partial molar volume with pressure is the same for both FeO and Fe_2O_3 at pressures up to 15 kbar. The densities of melts of primary tholeiite compositions (OT1 and OT2), a magnesian tholeiite in the Izu region (T) and Columbia River tholeiitic basalt (CB) are distinctly higher than those of three alkali basalts. The density of melt of the Izu tholeiite with 1.1 wt % H_2O was also measured at 10 kbar. It is about 0.06 g/cc smaller than that under anhydrous conditions. The densities of these melts are compared with those of the crustal rocks. The densities of gabbro, basic granulite and amphibolite are all larger than those of the primary basalt magmas (i.e., $>$ 3.0 g/cc at pressures greater than 5 kbar and at temperatures higher than 1200°C). Both the primary tholeiite and alkali basalt magmas, therefore, can ascend through the lower crust by buoyancy.

The densities of primary tholeiite magmas are, however, larger than those of granites (average density is 2.67 g/cc at 1 atm and 2.69 at 6 kbar), whereas the densities of the primary alkali basalt magmas even without volatiles are smaller than those of granites. The alkali basalt magmas, therefore, can ascend through the upper crust consisting mainly of granitic rocks by buoyancy even in the absence of volatiles. On the other hand, the primary tholeiite magma cannot ascend through the granitic upper crust and would fractionate near its bottom. Frequent occurrence of peridotite inclusions in alkali basalts and their absence in tholeiitic basalts can be explained by the above mechanism including density filtering for primary tholeiite magma. Melt of the fractionated tholeiite, though most magnesian in the Izu region, is still larger than that of the average granite. Eruption of such a tholeiite magma in the Izu region may be due to the presence of some H_2O ($\geq$1.1 wt %) in the magma, or the upper crustal material which is denser than granite (e.g., granodiorite), or both.

Decrease in the FeO/MgO ratio of magnesian basalts with increasing distance from the volcanic front may also be explained by the increase in the density of primary magmas which would result in the increase of degree of fractionation of magmas within the crust. It may be partly due to the change in the stress regime within the crust across the arcs (Kushiro, 1983).

The higher $^{87}Sr/^{86}Sr$ ratios in basalts and andesite near the volcanic front than those near the Japan Sea could be due to the interaction of magmas with the surrounding crustal rocks during the process of density filtering. Magmas which erupted near the Japan Sea had less chance for such an interaction than those which erupted near the volcanic front. However, another possibility that the source regions for magmas which erupted closer to the volcanic front are more enriched in $^{87}Sr/^{86}Sr$ due to fluids supplied from the subducted slab (Notsu, 1983) cannot be discarded at present. Heterogeneity of the source region of arc magmas may have to be considered also for some trace elements, as discussed by Gill (1981).

The density of the Columbia River basalt magma is also larger than that of granite. It is unlikely that this plateau basalt magma contained an appreciable amount of H_2O. The eruption of such a dense magma through the upper crust, therefore, suggests that the fissure (or feeder) extends to a magma reservoir in the lower crust, or the upper crust has a larger density than granite. The former possibility is likely in the case of eruption of plateau basalts. If the magma reservoir exists within the lower crust, and a fissure extends from the surface to the magma reservoir, the bulk density of the rock column including both the upper crust and a part of the lower crust would become larger than the bulk density of the column of plateau basalt magma. The magma, therefore, can be extruded on the surface from the magma reservoir. The depth of the magma reservoir from which plateau basalt magmas can extrude depends on the densities of the crustal materials and the magmas.

The amount of volcanic materials erupted from the Quaternary volcanoes decreases sharply from the volcanic front toward the Japan Sea, as shown in Fig. 1A. The average volume of the erupted materials per one volcano is also largest near the volcanic front (about 36 km^3 for the first 50 km zone and 14-26 km^3 for other three zones). As discussed above, the diapirs containing partial melts may ascend from the lower part of the mantle wedge above the subducted slab to shallow levels where primary magmas segregate. If each volcano corresponds to one diapir, number of the diapirs is largest near the volcanic front. The size of diapirs is not necessarily larger near the volcanic front than in other regions, if the amount of melt or the degree of partial melting in the diapirs becomes smaller from the volcanic front toward the Japan Sea. This is probably the case, as shown by the experimental studies as well as by the trace element data.

Acknowledgements

The experimental studies reviewed in the present article have been made by a group involving Drs. T. Fujii, H. Fukuyama, M. Sakuyama, E. Takahashi, and Y. Tatsumi. The author thanks them for their excellent collaboration and stimulative discussion on various problems on the genesis of arc magmas. He also thanks Dr.H.Nagahara for editing the manuscript.

REFERENCES

Aramaki, S. and T. Ui (1983) Alkali mapping of the Japanese Quaternary volcanic rocks. J. Volcanol. Geotherm. Res., 18, 549-560.

Bottinga, Y. and D. F. Weill (1970) Densities of liquid silicate systems calculated from partial molar volumes of oxide components. Amer. J. Sci., 269, 169-182.

Dallwitz, W. B., D. H. Green, and J. E. Thompson (1966) Clinoenstatite in a volcanic rock from the Cape Vogel area, Papua. J. Petrol., 7, 375-403.

Dickinson, W. R. (1968) Circum-Pacific andesite types. J. Geophys. Res., 73, 2261-2269.

England, R. N. and H. L. Davies (1973) Mineralogy of ultramafic cumulates and tectonites from eastern Papua. Earth Planet. Sci. Letters, 73, 416-425.

Fujii, T. and I. Kushiro (1977) Density, viscosity, and compressibility of basaltic liquid at high pressures. Carnegie Inst. Washington, Year Book, 76, 419-424.

Fujimaki, H. and H. Kurasawa (1980) Lateral variation of REE pattern of basaltic magma across the Japan arc. J. Japan Assoc. Miner. Petrol. Econ. Geol., 75, 313-322.

Fujitani, T. and A. Masuda (1981) Light REE inclination and distance from volcanic front: a case of volcanic rocks in Northeastern Japan. Geochem. J., 15, 269-281.

Fujiwara, M. and S. Arai (1982) Ultramafic xenoliths in Arato-yama alkali basalt, Okayama Prefecture, as upper mantle materials beneath a typical island arc. J. Japan Assoc. Miner. Petrol. Econ. Geol., Spec. Issue No. 3, 219-227.

Gill, J. (1981) Orogenic Andesites and Plate Tectonics, Springer-Verlag, Berlin Heidelberg New York, pp. 390.

Green, D. H. and A. E. Ringwood (1967) The genesis of basaltic magmas. Contr. Miner. Petrol., 15, 103-190.

Hamlyn, P.R. and E. Bonatti (1980) Petrology of mantle-derived ultramafics from the Owen fracture zone, northwest Indian Ocean: Implications for the nature of the oceanic upper mantle. Earth Planet. Sci. Letters, 48, 65-79.

Hatherton, T. and W. R. Dickinson (1968) Andesitic volcanism and seismicity in New Zealand. J. Geophys. Res., 73, 4615-4619.

------------ and --------------- (1969) The relationship between andesite volcanism and seismicity in Indonesia, the Lesser Antilles, and other island arcs. J. Geophys. Res., 74, 5301-5310.

Hedge, C. E. and R. J. Knight (1969) Lead and strontium isotopes in volcanic rocks from northern Honshu, Japan. Geochem. J., 3, 15-24.

Ishii, T., K. Konishi and A. Omura (1981) Description of samples. In Preliminary Report of Hakuho Maru Cruise KH 80-3, K. Kobayashi (ed.). Ocean Res. Inst., Univ. of Tokyo, pp. 105-163.

Ishikawa, K., S. Kanisawa and K, Aoki (1980) Content and behavior of flourine in Japanese Quaternary volcanic rocks and petrogenetic application. J. Volcanol. Geotherm. Res., 8, 161-175.

Ito, K. and G. C. Kennedy (1967) Melting and phase relations in a natural peridotite to 40 kilobars. Amer. J. Sci., 265, 519-538.

Jaques, A. L. and D. H. Green (1980) Anhydrous melting of peridotite at 0-15 kb pressure and the genesis of tholeiitic basalts. Contr. Miner. Petrol., 73, 287-310.

Kaneoka, I., J. Matsuda, J. Zashu, E. Takahashi, and K. Aoki (1978) Ar and Sr isotopes of mantle-derived rocks from the Japanese islands. Bull. Volcanol., 41, 424-433.

Katsui, Y., Y. Oba, S. Ando, S. Nishimura, Y. Masuda, H. Kurasawa and H. Fujimaki (1978) Petrochemistry of the Quaternary volcanic rocks of Hokkaido, north Japan. J. Fac. Sci. Hokkaido Univ., Ser. IV, 18, 449-484.

Kawano, Y., K. Yagi and K. Aoki (1961) Petrography and petrochemistry of volcanic rocks of Quaternary volcanoes of northeast Japan. Sci. Rep., Tohoku Univ., Ser. III, 7, 1-46.

Kuno, H. (1959) Origin of Cenozoic petrographic provinces of Japan and surrounding areas. Bull. Volcanol., 20, 37-76.

-------- (1960) High-alumina basalt. J. Petrol., 1, 121-145.

-------- (1968a) Origin of andesite and its bearing on the island arc structure. Bull. Volcanol., 32, 141-176.

-------- (1968b) Mafic and ultramafic nodules in basaltic rocks of Hawaii. Geol. Soc. Amer. Mem., 115, 189-234.

Kuno, H. and K. Aoki (1970) Chemistry of ultramafic nodules and their bearing on the origin of basaltic magmas. Phys. Earth Planet. Inter., 3, 273-301.

Kurasawa, H. (1979) Isotopic composition of strontium in volcanic rocks from Fuji, Hakone and Izu area, central Japan. Bull. Volcanol. Soc. Japan, 24, 135-152.

Kushiro, I. (1968) Compositions of magmas formed by partial zone melting of the earth's upper mantle. J. Geophys. Res., 73, 619-634.

----------- (1969) Discussion of the paper "the origin of basaltic and nephelinitic magmas in the earth's mantle" by D. H. Green. Tectonophys., 7, 427-436.

----------- (1973) Origin of some magmas in oceanic and circumoceanic regions. Tectonophys., 17, 211-222.

----------- (1980) Changes with pressure of degree of partial melting and K_2O content of liquid in the system Mg_2SiO_4-$KAlSiO_4$-SiO_2. Carnegie Inst. Washington, Year Book, 79, 267-271.

----------- (1982) Density of tholeiite and alkali basalt magmas at high pressures. Carnegie Inst. Washington, Year Book, 81, 305-309.

----------- (1983) On the lateral variations in chemical composition and volume of Quaternary volcanic rocks across Japanese arcs. J. Volcanol. Geotherm. Res., 18, 435-447.

Kushiro, I. and A. J. Erlank (1970) Stability of potassic richterite. Carnegie Inst. Washington, Year Book, 68, 231-233.

Kushiro, I. and H. Sato (1978) Origin of some calc-alkalic andesites in the Japanese Islands. Bull. Volcanol., 41-4, 576-585.

Masuda, Y. (1979) Lateral variation of trace element contents in Quaternary volcanic rocks across northeast Japan. Bull. Univ. Osaka, Ser. A, 28, 105-125.

Masuda, Y., S. Nishimura, T. Ikeda and Y. Katsui (1975) Rare-earth and trace elements in the Quaternary volcanic rocks of Hokkaido, Japan. Chem. Geol., 15, 251-271.

Miyashiro, A. (1974) Volcanic rock series in island arc and active continental margins. Amer. J. Sci., 274, 321-355.

Mysen, B. O. and I. Kushiro (1977) Compositional variations of coexisting phases with degree of melting of peridotite in the upper mantle. Amer. Miner., 62, 843-865.

Nohda, S. and G. J. Wasserburg (1981) Nd and Sr isotopic study of volcanic rocks from Japan. Earth Planet. Sci. Letters, 52, 264-276.

Notsu, K. (1983) Strontium isotope composition in volcanic rocks from the northeast Japan arc. J. Volcanol. Geotherm. Res., 18, 531-548.

Robertson, J. K. and P. J. Wyllie (1971) Rock-water systems, with special reference to the water-deficient region. Amer. J. Sci., 271, 252-277.

Sakuyama, M. (1979) Lateral variations of H_2O content in Quaternary magmas of northeastern Japan. Earth Planet. Sci. Letters, 43, 103-111.

Sakuyama, M. (1983) Petrology of arc volcanic rocks and their origin by mantle diapir. J. Volcanol. Geotherm. Res., 18, 297-320.

Stolper, E., D. Walker, B. Hager and J. F. Hays (1981) Melt segregation from partially molten source regions: the importance of melt density and source region size. J. Geophys. Res., 68, 6261-6271.

Sugimura, A. (1960) Zonal arrangements of some geophysical and petrological features in Japan and its environs. J. Fac. Sci. Univ. Tokyo, Ser. II, 12, 133-153.

Sugimura, A., T. Masuda, K. Chinzei and K. Nakamura (1963) Quantitative distribution of late Cenozoic volcanic materials in Japan. Bull. Volcanol., 26, 125-140.

Takahashi, E. and I. Kushiro (1983) Melting of a dry peridotite at high pressures and basalt magma genesis. Amer. Miner., 68, 859-879.

Tatsumi, Y., M. Sakuyama, H. Fukuyama and I. Kushiro (1983) Generation of arc basalt magmas and thermal structure of the mantle wedge in subduction zones. J. Geophys. Res., 88, 5815-5825.

Tracy, R. J. (1980) Petrology and genetic significance of an ultramafic xenolith suit from Tahiti. Earth Planet. Sci. Letters, 48, 80-96.

White, R. W. (1966) Ultramafic inclusions in basaltic rocks from Hawaii. Contr. Miner. Petrol., 12, 245-314.

Wyllie, P. J. (1977) Crystal anatexis: an experimental review. Tectonophys., 43, 41-71.

GEOCHEMICAL AND MINERALOGICAL CRITERIA FOR THE IDENTIFICATION OF MINERALIZED KOMATIITES IN ARCHAEAN GREENSTONE BELTS OF AUSTRALIA

C.M. LESHER[1,2] AND D.I. GROVES[1]

[1]Geology Department, University of Western Australia, Nedlands, Western Australia, 6009.

[2]Present Address: J. Tuzo Wilson Research Laboratories, Earth and Planetary Sciences, University of Toronto - Erindale, Mississauga, Ontario L5L 1C6, Canada.

ABSTRACT

Komatiite-associated nickel deposits form an important component of a major class of magmatic Ni-Cu sulphide deposits hosted by rocks of komatiitic affinity in Archean greenstone belts. There are several geochemical and mineralogical criteria which identify the mineralized komatiite sequences and the specific host units.

i) On a regional scale, chromites from komatiites throughout mineralized ultramafic sequences are significantly enriched in Zn (>0.6 at% Zn) relative to unmineralized sequences and hosts to komatiitic dunite-associated nickel deposits. These anomalous chromite compositions may relate to separation of immiscible sulphides from parental magmas prior to emplacement; contamination by Zn-rich sulphidic sediments may also be important in lower units.

ii) At Kambalda, spinifex-textured zones of all komatiite flows exhibit significant chalcophile-element depletion, consistent with sulphide separation from the parental magma. Preliminary data suggest that only komatiite sequences associated with major (Kambalda-size) nickel deposits will show detectable chalcophile-element depletion.

iii) Data from Kambalda suggest that basal host units are structurally and geochemically distinctive components of the ultramafic sequence: they are thicker (up to 100m), better differentiated (texturally and compositionally), more magnesian (up to 45% MgO) and depleted in Ti, Al, Cr,

Proceedings of the 27th International
Geological Congress, Volume 9, pp. 283–302
PETROLOGY (IGNEOUS AND METAMORPHIC ROCKS)

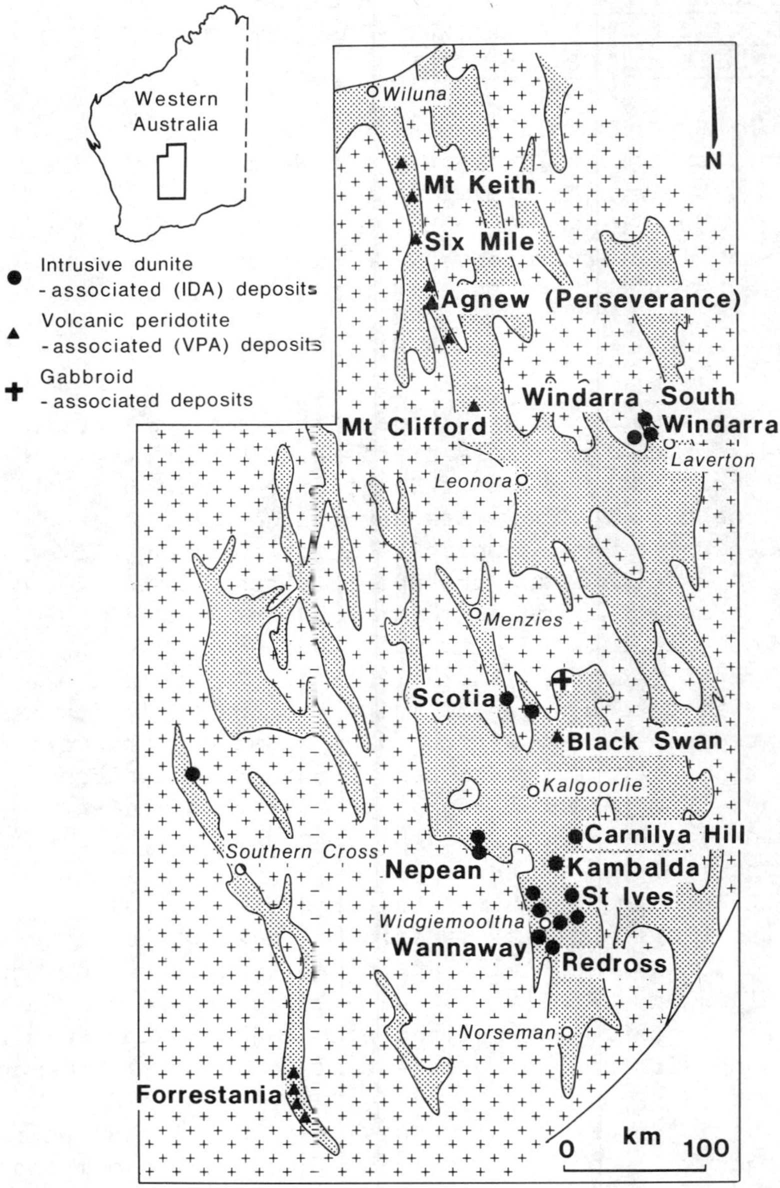

FIGURE 1. Schematic map of the eastern Yilgarn Block, Western Australia, showing distribution of the major komatiite-associated and komatiitic dunite-associated nickel deposits in Archaean greenstone belts.

Fe and Zn relative to adjacent barren flanking basal units. They were derived from more magnesian liquids (28-31% MgO) and comprise more forsteritic olivine (Fo94-91) than adjacent units and overlying flows (<26% MgO and Fo91-89). These geochemical and mineralogical data provide potentially useful criteria for mineral exploration as well as important constraints on ore genetic models.

INTRODUCTION

Following the discovery of the Kambalda deposits in Western Australia, there was rapid recognition of an important metallogenetic association between Ni-Cu sulphide deposits and ultramafic rocks of komatiitic affinity in Archean greenstone belts. The major deposits of this type occur in Western Australia, although examples of the association are also represented in Canada and Zimbabwe (Ross and Travis, 1981). Research has shown that there are two major groups within this metallogenetic association.

One group is associated with thick, largely serpentinized dunite lenses and pods which occupy extensive linear zones (Fig. 1). These deposits, here termed komatiitic dunite-associated nickel deposits, are reviewed by Marston et al. (1981) and Lesher et al. (1982), who consider that they most likely represent tectonically modified subvolcanic ultramafic sills (magma chambers) which, in part, fed overlying unmineralized komatiite lava piles.

The second, and currently most economically important group, here termed komatiite-associated nickel deposits, occur in thick, cumulate komatiites at, or near, the base of thick ultramafic lava flow sequences (Groves and Hudson, 1981; Marston et al., 1981; Lesher et al., 1982); the type examples at Kambalda (Fig. 1) are well-described by Gresham and Loftus-Hills (1981). The geometry of the ore zone for the Kambalda deposits is shown schematically in Figure 2.

There is general consensus that both groups of nickel deposits represent metamorphically modified magmatic deposits, and that the nickel ores crystallized from immiscible sulphide liquids. Only the timing of sulphur

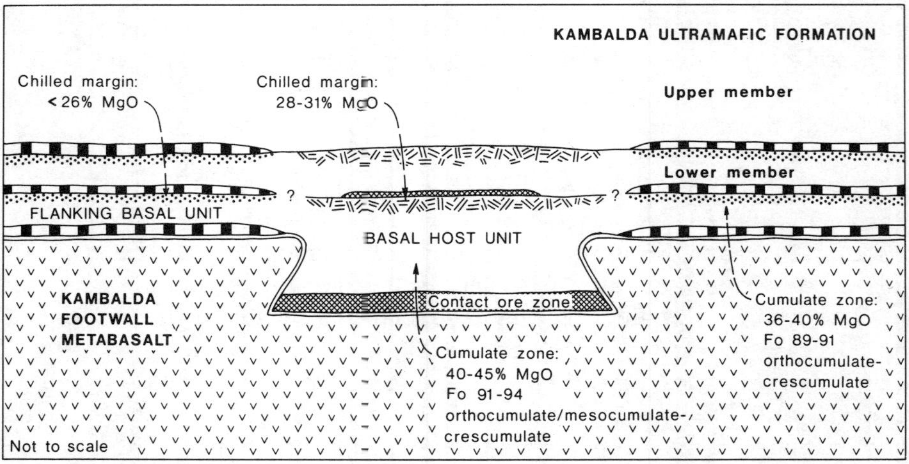

FIGURE 2. Schematic cross-section through a "typical" Kambalda ore shoot after allowance for superimposed deformation. The section shows the situation of sulphide ores in trough structures and contrasts the stratigraphic situations of ore and non-ore positions. It highlights the structural and geochemical differences between basal host units and barren flanking basal units. The former are strongly differentiated with relatively thin (5-25 per cent of total thickness) upper spinifex-textured divisions and anomalously thick lower cumulate divisions; barren flanking basal units are generally relatively poorly differentiated. Spinifex-textured zones in basal host units are strongly differentiated from fine random spinifex texture (31-28% MgO) through coarse platy spinifex texture (27-21% MgO) to pyroxene spinifex texture (20-16% MgO). The lower member comprises thicker, more magnesian flows (with intercalated metasediments) than the upper member (see Gresham and Loftus-Hills, 1981) for a more detailed geological comparison.

saturation and source of sulphur are contentious: compare, for example, Marston et al. (1981) and Lesher et al. (in press).

CHROMITE IN ULTRAMAFIC ROCKS

Chrome-spinel is a well-established petrogenetic indicator (e.g. Irvine, 1967) whose composition is sensitive to changes in silicate magma composition and to crystallization conditions such as fO_2 and fS_2 (e.g. Snethlage and Klemm, 1978). Thus, chromite has the potential to fingerprint magmas from which immiscible sulphide liquids have separated.

Chromite is a ubiquitous phase in komatiitic dunites and in ultramafic flows throughout thick komatiite sequences hosting komatiite-associated deposits. Chromites in adcumulate dunite bodies vary from subhedral cumulus crystals to lobate intercumulus grains up to 2 mm long. In komatiites, subhedral grains of about 0.1 mm diameter in ortho- and meso-cumulate zones contrast with smaller skeletal grains in spinifex-textured zones; chromite-rich zones may be developed between upper spinifex-textured and lower cumulate divisions in some strongly differentiated flows (Lesher, 1984).

Chromites in unserpentinized, albeit recrystallized, ultramafic hosts exhibit narrow Cr-magnetite rims; rare unzoned chromite in relict igneous olivine suggests that the rims are of metamorphic origin. The magnetite rim/chromite core ratio in altered ultramafic rocks increases with increasing metamorphic grade, and the outer rim becomes less chrome-rich. In addition, the chromite core shows a decrease in MgO and increase in FeO which Groves et al. (1977) attributed to metamorphic modification under relatively low fO_2 conditions. These and other changes produce considerable scatter in chromite compositions (Fig. 3). Nevertheless, the chromites overlap the field of most magmatic chromites in terms of their RCr_2O_4-RAl_2O_4-RFe_2O_4 end-member components (Groves et al., 1977).

In order to compare mineralized and unmineralized sequences, the following were analysed: 1) 64 chromites from units within both the upper and lower members of mineralized komatiite sequences, 2) 55 chromites from

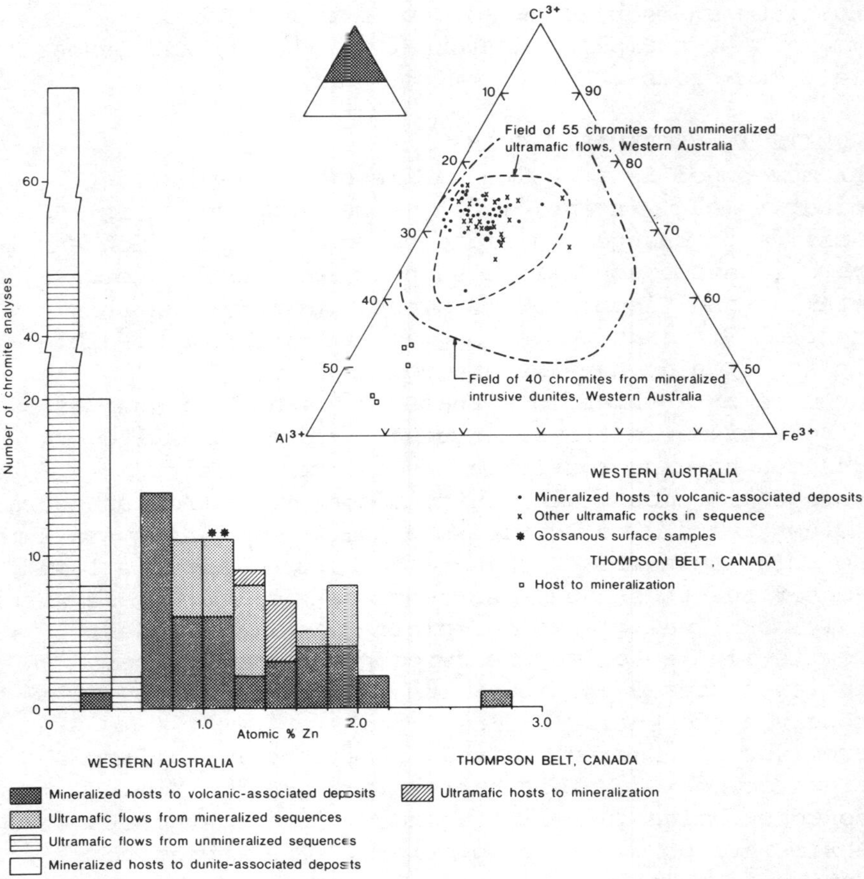

FIGURE 3. Composition of chromites from mineralized and unmineralized ultramafic rocks in Western Australia with emphasis on Zn content. Basal host units to komatiite-associated ores include Lunnon and Durkin shoots, Kambalda; Windarra; South Windarra; Carnilya Hill; Wannaway; Scotia. Unmineralized ultramafic volcanic units include Mt Clifford; Fly Bore; Mt Margaret; Hannan Lake; St Ives; Cranium; Hayes Hill; Forrestania. Komatiitic dunite-associated deposits include Mt Hope, Digger Rocks and Cosmic Boy, Forrestania; Mt Keith; Agnew; Yakabindie; Black Swan; Bullfinch. Data for Pipe No. 2, Thompson belt are shown for comparison. Chromites from two gossanous surface samples from Kambalda are also shown.

unmineralized komatiite sequences showing a similar range of alteration and metamorphic grade to the mineralized sequences, and 3) 40 chromites from komatiitic dunite bodies (Groves et al., in press); analytical procedures are described by Groves et al. (1977). The chromites from the three groups have essentially similar compositions with respect to most elements, but differ significantly in Zn context. All but two of the analysed chromites from mineralized komatiite sequences contain between 0.6 and 2.2 at% Zn (Fig. 3). They have a high Zn content, irrespective of the stratigraphic position of the ultramafic flows in which they occur, although there is evidence that Zn content of chromites decreases with stratigraphic height in the komatiite pile (M.J. Donaldson, pers. comm., 1981). In contrast, chromites from unmineralized ultramafic flow sequences and mineralized komatiitic dunites analysed in this study have lower Zn contents; there is no significant overlap between these and analysed chromites from mineralized komatiite sequences (Fig. 3).

It is most likely that the different Zn contents of chromites from mineralized and unmineralized komatiite sequences reflects the removal of immiscible sulphides from the former; Shimazaki and MacLean (1976) report that the partition coefficient for Zn between sulphide and silicate liquids in the system FeS-FeO-SiO_2 is 0.1-0.5, that is, that Zn is concentrated by a factor of 2-10 times in the silicate liquid, consistent with the low Zn contents of the nickel ores (<.02%: Groves and Hudson, 1981). There are no experimental or relevant thermodynamic data on Zn partitioning between chromite, olivine and silicate melt, but chromite/silicate partition coefficients should be in excess of 10 (using 100 ppm average Zn content of Kambalda spinifex-textured komatiites from Lesher et al., 1981 and minimum 1% Zn content of chromite from Groves et al., 1977 and Lesher, 1984) and olivine/silicate partition coefficients less than 2 (<190 ppm Zn(3σ) in Kambalda olivine: Lesher, 1984). Thus, early removal of sulphides would result in higher Zn contents in later crystallized chromites. If removal of immiscible sulphides is responsible for the contrast in Zn content, the data strongly suggest that the

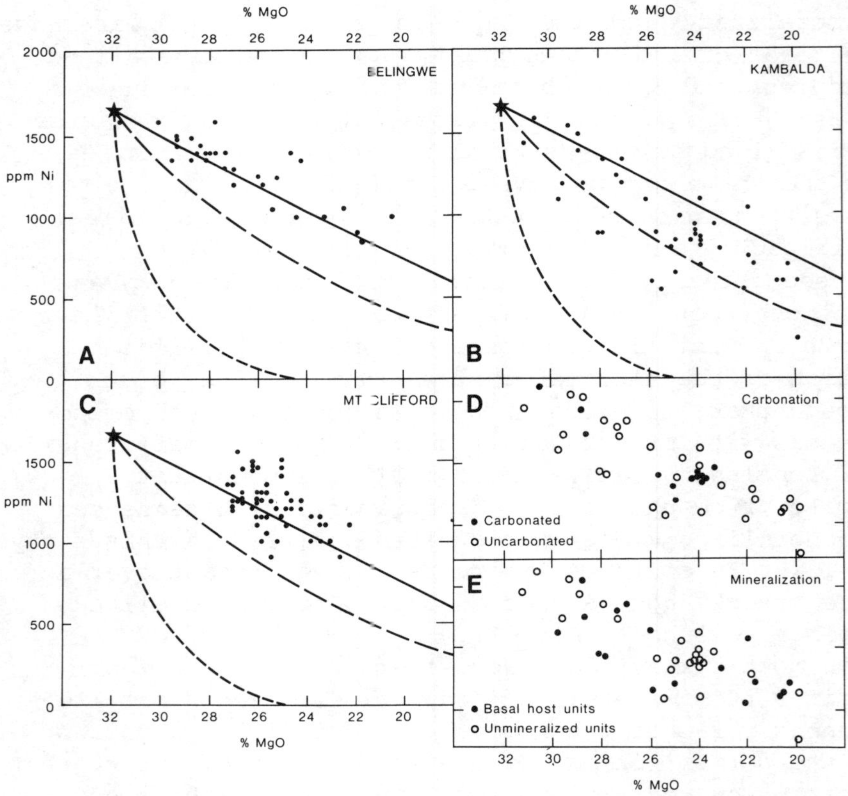

FIGURE 4. Ni in komatiites plotted against MgO content for A. Belingwe, Zimbabwe (data from Nisbet et al., 1977), B. Kambalda, and C. Mt Clifford, showing the chalcophile-element depletion of Kambalda komatiites. Superimposed are the model komatiite curves of Duke (1979) for olivine-only fractionation (solid line) and sulphide-saturated fractionation with olivine: sulphide ratios of 100:1 (long dashed line) and 10:1 (short dashed line). D and E show Ni plotted against MgO for Kambalda komatiites illustrating differences in alteration and sample location. D. Carbonated samples contain greater than 5 modal per cent dolomite. E. Basal host units contain basal nickel sulphide ores, whereas unmineralized units represent flanking or overlying flows. Data taken from Lesher et al. (1981) unless otherwise stated.

entire komatiite sequence hosting the nickel ores was sulphur-saturated at some stage in its evolution. The higher Zn content of chromites from lower units in the pile suggests the possibility that assimilation of Zn-rich sulphidic sediments (e.g. Lesher et al., in press) has increased the Zn content of the magma, but this mechanism does not explain Zn enrichment in chromites from upper units. The low Zn contents of chromites in mineralized dunites are surprising, in view of the data from komatiite sequences. Possible explanations include different magma conditions (e.g. fO_2, fS_2) or different timing of chromite crystallization relative to sulphide separation; the unsegregated, disseminated nature of the sulphides in most mineralized komatiitic dunites suggests relatively late sulphur saturation compared to the komatiite-associated deposits (see below).

CHALCOPHILE-ELEMENT DEPLETION IN KAMBALDA KOMATIITES

Despite problems with extrapolation of experimentally determined partition coefficients from mafic to ultramafic systems, there has been considerable success in numerically modelling the separation of sulphides from ultramafic magmas (e.g. Duke, 1979; Campbell and Naldrett, 1979). Such models predict that chalcophile-element abundances in derivative silicate liquids will be strongly influenced by the proportions of sulphide and silicate liquids equilibrating in the parental magma (see Fig. 4). The upper, spinifex-textured divisions of differentiated ultramafic lava flows are interpreted to result from rapid crystallization of undercooled, nuclei-free MgO-rich liquids (e.g. Arndt and Brooks, 1980). If the uppermost, finest-grained aphyric and random spinifex-textured zones represent derivative liquids from the parental magma, as seems most likely, they provide the opportunity to compare theoretical models with variations shown by mineralized komatiite suites.

Lesher et al. (1981) present the results of such a study at Kambalda (Fig. 1) which involved careful sampling, stratigraphic and petrographic documentation, and high-precision chemical analyses of aphyric and fine-grained random spinifex-textured zones (chilled margins) of flows from throughout the komatiite pile. A synthesis is presented below, and relevant geochemical data are summarized in Figure 4.

The Kambalda komatiites, irrespective of whether they are from near-ore or flanking positions (Gresham and Loftus-Hills, 1981; see Fig. 2), or from mineralized (normally basal) or unmineralized (stratigraphically higher) flows, are significantly depleted in chalcophile elements (Ni and Co) relative to komatiites from unmineralized sequences (e.g. Belingwe, Zimbabwe; Barberton, South Africa; Munro Township, Canada) and to a numerically modelled sulphide-free komatiite system; data for Ni vs MgO are shown in Figure 4. Comparison of komatiites subdivided on the basis of proximity to ore zones, and in terms of alteration mineralogy (Fig. 4), suggests that processes such as metal exchange with sulphide ores and modification by carbonation were not primarily responsible for this depletion. However, leaching during alteration processes reduced sulphur contents to well below saturation levels in the komatiites (Lesher et al., 1981).

The depletion is best explained as the result of partition of chalcophile elements into magmatic sulphide liquids which separated from the silicate magma. Importantly, the depletion in Ni (and Co) in komatiites from throughout the Kambalda ultramafic sequence indicates that all the silicate magma was at some stage in equilibrium with sulphide liquid; this is consistent with the chromite data. The magnitude of Ni depletion relative to the MgO content of the komatiites varies significantly from sample to sample, probably due to superimposition of additional processes on initial magmatic depletion. Possible effects include those related to centimetre-scale, non-systematic redistribution of metals during alteration, re-equilibration between derived silicate and sulphide liquids during eruption and emplacement, and low-pressure crystallization and fractionation of olivine. The possibility of post-eruption contamination by sulphides from interflow sediments (e.g. Lesher et al., in press) further complicates the picture.

The scatter of the data, combined with uncertainties as to its cause, preclude rigorous modelling of the system. The magnitude of the depletion suggests a mean olivine:

sulphide ratio for the system of *ca* 1500-2000:1 (Lesher et al., 1981). Recent evidence that the Kambalda komatiite sequence and associated nickel deposits may represent the distal segment of a much more extensive ultramafic flow sequence (e.g. Lesher et al., in press) require that this ratio is a minimum. These data thus suggest that sulphur saturation and sulphide segregation occurred prior to eruption and that a significant volume of sulphide may have segregated from the silicate magma at depth and been lost to the ore-forming system. Interpretation of chalcophile-element depletion of the basal host unit to mineralization (Fig. 2) is equivocal. Lesher et al. (1981) argued that the level of depletion was too low for the sulphides to have exsolved *in situ* from the volume of magma represented by the basal host units, and therefore that the initial magmas were erupted carrying sulphide droplets; they discuss the problems of ascent of sulphide-rich magmas inherent in such a model. However, studies of olivine morphology and composition in rare examples of these units where relict olivine is preserved suggest that the basal units formed by continued replenishment of magma through actively flowing lava conduits (Lesher et al., in press). Thus, the volume of magma from which sulphides exsolved could have been greatly in excess of that suggested by calculations based on vertical sections through the ore environments, and post-eruption exsolution of sulphides from the host magma cannot be discounted.

The chalcophile-element depletion trend observed for the Kambalda komatiites raises the possibility that similar trends may occur in other ultramafic volcanic sequences that host similar mineralization. However, preliminary examination of such sequences that host smaller, or less abundant, sulphide ores than the Kambalda sequence indicate considerable scatter with no detectable chalcophile-element depletion (Lesher et al., 1981); data for Mt Clifford are shown in Figure 4. It is suspected that very high silicate magma:sulphide liquid ratios for these magmatic systems relative to Kambalda inhibit recognition of any depletion (see Fig. 4).

These data suggest that chalcophile-element depletion of komatiites may only be detectable in ultramafic sequences

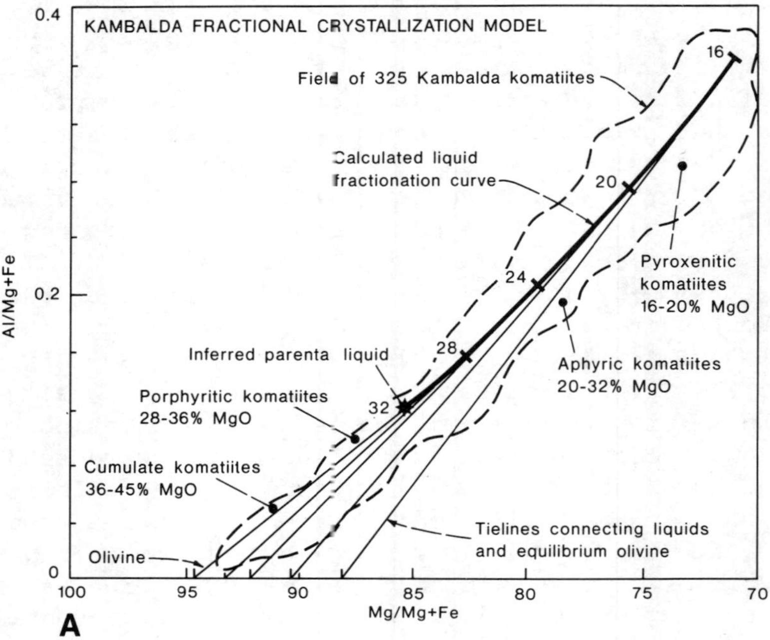
KAMBALDA FRACTIONAL CRYSTALLIZATION MODEL
Field of 325 Kambalda komatiites
Calculated liquid fractionation curve
Pyroxenitic komatiites 16-20% MgO
Aphyric komatiites 20-32% MgO
Inferred parental liquid
Porphyritic komatiites 28-36% MgO
Cumulate komatiites 36-45% MgO
Olivine
Tielines connecting liquids and equilibrium olivine
16
20
24
28
32
Al/Mg+Fe
Mg/Mg+Fe
0.4
0.2
0
100
95
90
85
80
75
70
A

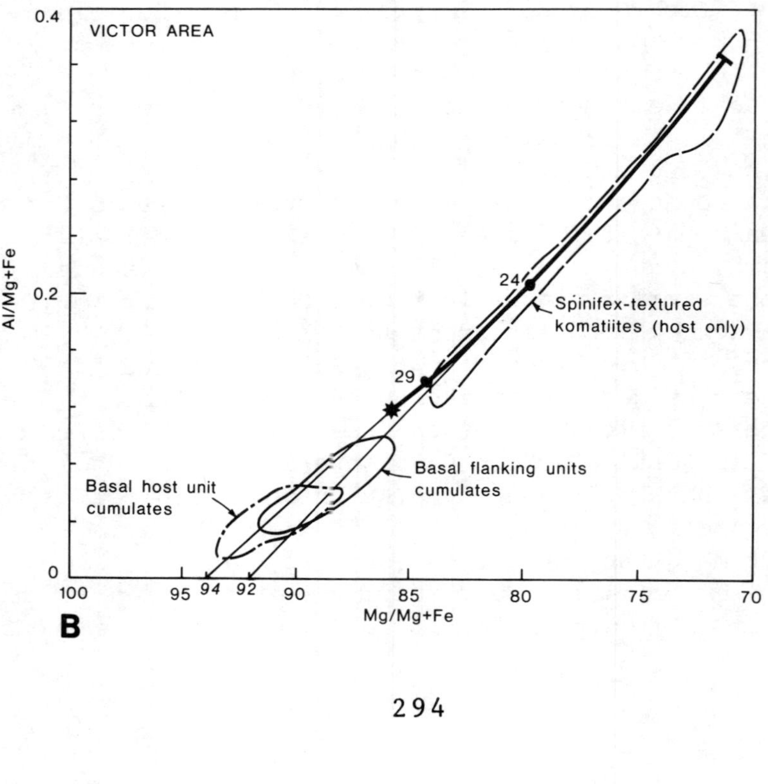
VICTOR AREA
24
29
Spinifex-textured komatiites (host only)
Basal flanking units cumulates
Basal host unit cumulates
Al/Mg+Fe
Mg/Mg+Fe
0.4
0.2
0
100
95
94
92
90
85
80
75
70
B

FIGURE 5 (Facing page). Summary of the major-element geochemistry of the Kambalda komatiite sequence illustrating differences between basal host units and flanking basal units: see Figure 2 for terminology. A. Al/Mg+Fe vs Mg/Mg+Fe plot for Kambalda komatiites. The dashed field encloses 325 Kambalda komatiites. The solid line is the computed liquid trend for olivine (plus 0.5 per cent chromite) fractional crystallization. Cumulate compositions should lie on tie lines (as shown) tangential to the liquid fractionation curve, in equilibrium with olivine, whose composition is determined by the Fe-Mg distribution coefficient ($K_D = 0.33$); olivine contains negligible Al_2O_3 and plots on the horizontal axis. The position of the cumulate composition along the tie line is defined by the relative proportions of liquid and olivine. For example, a suite showing equal degrees of olivine enrichment will define a cumulate trend at a low angle to the horizontal axis continuous with, but oblique to, the liquid trend. Such curvature is evident in the plotted data. B. Geochemical data from Victor Shoot, Kambalda. The field of spinifex-texture komatiites is in excellent agreement with the computed liquid fractionation curve (as in A), and the field of cumulate rock compositions is oblique to the liquid trend. These data support a fractional crystallization model and preclude simple mixing of a single less-magnesian liquid and an equilibrium olivine, as suggested by previous workers. The compositions of basal host unit cumulates overlap those of the flanking basal units, but lie on a locus of tie lines representing more magnesian liquids (*ca* 32-24% MgO: ave 29% MgO) and more magnesian olivine (Fo95-92: ave Fo94) than the flanking basal units (29-21% MgO: ave 24% MgO and Fo94-91: ave Fo92) broadly consistent with other geochemical data (Table 1).

hosting deposits of equivalent size and frequency to those at Kambalda. Even if such depletion is evident, there may be no sulphide ores present as current models are not capable of distinguishing pre-eruption separation from post-eruption *in situ* separation of sulphide liquids; in the former case, sulphides may be lost, due to gravity segregation from ascending magmas prior to eruption (Lesher et al., in press).

KAMBALDA BASAL HOST UNITS VS BARREN FLANKING BASAL UNITS

The geometry of the ore environment at Kambalda is shown schematically in Figure 2. The basal host units are virtually confined to distinctive trough structures and in most cases appear to be laterally continuous with adjacent barren flanking basal units (Fig. 2). The precise stratigraphic relationships between host units and flanking units are not known. However, in most areas the spinifex-textured top of the basal unit is above the projected level of the trough margins (Fig. 2) and the basal host unit is interpreted to laterally overlap the trough. There have been numerous models suggested for the genesis of the troughs, but detailed studies in areas of least strain (Lesher, 1984) suggest that they represent somewhat modified syn-volcanic stratigraphic depressions between non-overlapping footwall basalt flows (cf. Mauna Loa, Hawaii: Macdonald and Finch, 1950); their linear trends are interpreted to reflect topographic pitchlines in the volcanic environment (Lesher et al., in press).

The geochemistry of the ultramafic sequence at Kambalda is consistent with that of a continuum of komatiite flows whose compositional variations are related to evolution of parent liquids by olivine fractionation and to variable degrees of olivine enrichment and differentiation *in situ* (Fig. 5A). There are corresponding variations in the internal structure of the flows (Table 1).

The basal host units represent the most magnesian and internally differentiated end-member of this continuum; their major features are compared with those of barren flanking basal units in Table 1. Despite overlap of the geochemistry of these units (Table 1; Fig. 5B), data from Victor Shoot (where relict olivine is preserved in

TABLE 1. Comparison of structure and geochemistry of basal units from Kambalda.

	Basal Host units	Barren Flanking Basal Units
True Thickness	Range 15-100m Average *ca* 50m	Range 5-30m Average *ca* 20m
Differentiation	Well-differentiated Well-developed spinifex-textured zone	Poorly differentiated Poorly developed spinifex-textured zone
MgO content (Bulk composition)	Range *ca* 35-45% MgO Depleted in Ti, Al, Cr, Fe and Zn	Range *ca* 35-40% MgO Enriched in Ti, Al, Cr, Fe and Zn
Liquid Composition	Range *ca* 28-31% MgO	Range *ca* 16-26% MgO
Olivine Composition	Fo94-91 (Victor)	Fo91-89 (Victor)
Olivine Textures	Ortho-, Meso- and Crescumulate	Orthocumulate and Crescumulate
Olivine Enrichment	Range *ca* 15-60%	Range *ca* 15-60%
Interpretation	Central lava conduits	Flanking marginal areas

cumulate zones) confirm that the basal host units crystallized from more magnesian liquids (28-31% MgO upper chilled margins) than barren flanking basal units (16-25% MgO) and that they comprise correspondingly more forsteritic olivine (Fo94-91 vs Fo91-89) in cumulate zones. The basal host units therefore have a distinctive geochemical signature, and are not related to the flanking basal units by simple equilibrium olivine accumulation (cf. Marston et al., 1981; Gresham and Loftus-Hills, 1981).

Lesher et al. (in press) have integrated all relevant data on the ore environment and geometry of the lower member of the Kambalda ultramafic sequence. They conclude that the thick, linear basal host units represent actively flowing lava conduits, replenished from a source at some distance from Kambalda, which were largely confined to the troughs defined by non-overlapping foot-wall basalt flows. The thinner, flanking portions of the basal unit probably represent more slowly moving parts of the flow, in which case there would be less lava flow-through and therefore more *in situ* fractionation. Alternatively, they may represent lateral overflow from the central conduit as it periodically ruptured, thus allowing opportunity for fractionation of the lava prior to emplacement. Both models account for the more fractionated composition of the flanking basal units.

SUMMARY

Komatiite-associated nickel deposits represent an important group of magmatic deposits of komatiitic affinity in the Archean greenstone belts of Western Australia. The komatiite flow sequences which host these deposits have distinctive geochemical and mineralogical characteristics on a variety of scales:

i) Metamorphically modified, but originally magmatic, chromites from komatiites throughout mineralized ultra-mafic sequences are significantly enriched in Zn (>0.6 at% Zn) relative to rocks of similar composition from unmineralized sequences. The occurrence of Zn-rich chromites throughout the mineralized komatiite pile probably indicates sulphur saturation and immiscible sulphide liquid separation from the parent magma, but relevant experimental or thermodynamic studies are required to confirm this. Data from Finland, Canada and U.S.A. suggest that Zn-rich chromites may be more general indicators of proximity to nickel sulphide deposits (Groves et al., in press). Preservation of such chromites during weathering means that Zn-rich chromites provide a potential low-cost exploration tool for komatiite-associated nickel deposits.

ii) At Kambalda, there is significant chalcophile-element depletion in spinifex-textured flow tops interpreted to represent a series of liquids derived from the parental komatiitic magma by olivine fractionation. Such depletion is consistent with sulphur saturation and sulphide liquid separation from the parental magma prior to eruption of the komatiites. Rigorous modelling of the data is not possible, due to superimposed alteration effects on the komatiites and uncertainties as to the degree of magma flow-through during the evolution of individual flows. In particular, the recognition that the basal host units were formed by continued replenishment of magma though actively flowing lava conduits does not allow unequivocal interpretation of the timing of sulphide exsolution from these units. Preliminary data suggest that only komatiite sequences with deposits of similar size and frequency to those at Kambalda will show detectable chalcophile-element depletion, and this limits its usefulness as a general geochemical prospecting tool.

iii) On an even more restricted scale, specific host units can be distinguished from adjacent barren units (and overlying barren units) at Kambalda. Although the komatiite flows show a continuum of compositions, the basal host units represent differentiated, olivine-enriched lava conduits which were derived from more magnesian liquids (28-31% MgO) and more forsteritic olivines (Fo94-91) than any other units in the ultramafic pile. This leads to a distinctive structural (thicker and more differentiated) and geochemical (enriched in Mg and Ni, depleted in Ti, Al, Cr, Fe and Zn) signature for the basal host units. Flanking basal units may represent more slowly moving parts of the flow than the basal host units or represent periodic lateral overflow from the central conduit represented by the basal unit. Both processes could explain the contrasting structure and geochemistry. These geochemical data have direct application in exploration for komatiite-associated nickel deposits, and also provide important constraints on ore genetic models, particularly when used in conjunction with field and petrological data (e.g. Lesher et al., in press).

ACKNOWLEDGEMENTS

The authors are indebted to numerous mining companies, in particular Western Mining Corporation (Kambalda Nickel Operations), for access to deposits and support of research. We thank our co-workers, F.M. Barrett, M.J. Bickle, R.H. Brotherton, M.J. Donaldson and R.F. Lee for data contributed to this synthesis. C.M.L. was supported by a University of Western Australia Research Scholarship, and the early stages of the research were funded by the Australian Mineral Industries Research Association and the Australian Research Grants Committee. This paper represents a contribution to I.G.C.P. Projects 91, "Metallogeny of the Precambrian" and 161, "Sulphide Deposits in Mafic and Ultramafic Rocks".

REFERENCES

Arndt, N.T. and Brooks, C. (1980). Komatiites: Penrose Conference Report. Geology 8, 155-156.

Campbell, I.H. and Naldrett, A.H. (1979). The influence of silicate: sulfide ratios on the geochemistry of magmatic sulfides. Econ. Geol. 74, 1503-1506.

Duke, M.J. (1979). Computer simulation of the fractionation of olivine and sulfide from mafic and ultramafic magmas. Can. Mineral. 17, 507-514.

Gresham, J.J. and Loftus-Hills, G.D. (1981). The geology of the Kambalda nickel field, Western Australia. Econ. Geol. 76, 1373-1416.

Groves, D.I., Barrett, F.M., Binns, R.A. and McQueen, K.G. (1977). Spinel phases associated with metamorphosed volcanic-type iron-nickel sulfide ores from Western Australia. Econ. Geol. 72, 1224-1244.

Groves, D.I., Barrett, F.M. and Brotherton, R.H. (in press). Exploration significance of chrome-spinels in mineralized ultramafic rocks and Ni-Cu ores. In: ICAM 81 - Proceedings of the First International Congress on Applied Mineralogy, J.R.R. de Villiers and P.A. Cawthorn (eds). Geol. Soc. South Africa Spec. Publ. No. 7, in press.
Groves, D.I. and Hudson, D.R. (1981). The nature and origin of Archean stratabound volcanic-associated nickel-iron-copper sulphide deposits, In: Handbook of Strata-bound and Stratiform Ore Deposits, K.H. Wolf (ed.). Elsevier, Amsterdam, 9, 305-410.
Irvine, T.N. (1967). Chromian spinel as a petrogenetic indicator, Pt. 2. Petrological applications. Can. J. Earth Sci. 4, 71-103.
Lesher, C.M. (1984). Localization and genesis of komatiite-hosted Fe-Ni-Cu sulphide mineralization at Kambalda, Western Australia. Unpubl. Ph.D. thesis, Univ. West. Australia.
Lesher, C.M., Arndt, N.T. and Groves, D.I. (in press). Genesis of komatiite-associated nickel sulphide deposits at Kambalda, Western Australia: a distal volcanic-assimilation model. In: Sulphide Deposits in Mafic and Ultramafic Rocks, D.L. Buchanan and M.J. Jones (eds). Instn. Min. Metall., London, Spec. Publ., in press.
Lesher, C.M., Donaldson, M.J. and Groves, D.I. (1982). Nickel deposits and their host rocks in the Norseman-Wiluna Belt, In: Regional Geology and Nickel Deposits of the Norseman-Wiluna Belt, Western Australia, D.I. Groves and C.M. Lesher (eds). Univ. West Aust. Geol. Dept. and Extension Service Publ. No. 7, B1-B63.
Lesher, C.M., Lee, R.F., Groves, D.I. and Bickle, M.J. (1981). Geochemistry of komatiites from Kambalda, Western Australia: I. Chalcophile element depletion - a consequence of sulfide liquid separation from komatiitic magmas. Econ. Geol. 76, 1714-1728.
Macdonald, G.A. and Finch, R.H., (1950). The June 1950 eruption of Mauna Loa. Volcano Letter No. 509.
Marston, R.J., Groves, D.I., Hudson, D.R. and Ross, J.R. (1981). Nickel sulfide deposits in Western Australia: a review. Econ. Geol. 76, 1330-1363.
Nisbet, E.G., Bickle, M.J., and Martin, A. (1977). The

mafic and ultramafic lavas of the Belingwe greenstone belt, Rhodesia. J. Petrol. 18, 521-566.
Ross, J.R. and Travis, G.A., (1981). The nickel sulfide deposits of Western Australia in global perspective. Econ. Geol. 76, 1291-1329.
Shimazaki, H. and MacLean, W.H. (1976). An experimental study on the partition of zinc and lead between silicate and sulfide liquids. Mineralium Deposita 11, 125-132.
Snethlage, R. and Klemm, D.D. (1978). Intrinsic oxygen fugacity measurements on chromites from the Bushveld Complex and their petrogenetic significance. Contrib. Mineral. Petrol. 67, 127-138.

ARCHAEAN TONALITE-PLAGIOGRANITE SUITES OF KARELIA: GEOLOGICAL TYPES AND PETROGENESIS

LOBACH-ZHUCHENKO S.B.
Institute of Precambrian Geology and Geochronology, the USSR Academy of Sciences, Leningrad, USSR

Granitoids comprise over 90% of Karelian territory constituted by Archaean rocks. On geological grounds they can be divided into two main groups, i.e. granitoids showing spatial and genetic relationships with the greenstone belt development (GI) and gneiss-granites forming vast areals which separate greenstone belts (GII). Their recognition is based on the magnetic anomaly map compiled for the whole territory /I, Fig. 1/.

The earliest granitoids among those showing the spatial relation to greenstone belts are tonalites and plagiogranites which are followed by plagiomicrocline and microcline granites at later stages of the greenstone belt development /2,3:. The former are in turn subdivided into several age-groups, two of them being thoroughly studied in Central Karelia /2/.

The GIA tonalites generally intervene supracrustal greenstone sequences and basement rock complexes, less commonly they make up the domes occurring among minor greenstone structures. Leucocratic tonalites are found at the base of the Suna-River succession which thickens outwards the tonalites. Foliation in metabasalts is parallel to the contact. All the deformations observed in the metabasalts are also superimposed on the GIA tonalites /2/. Veins similar in composition to tonalites occur along the foliation in the direct contact in metabasalts, but in addiotion, the veins are reported from tonalites, being confined to an earlier foliation.

Thus, geological evidence suggests that the emplacement of a parent tonalitic magma of GI took place either before or synchronously with basalts, but it preceded the stage of early metamorphism and deformation of basalts. Tonalites were affected by K-metasomatism which gave rize to the local appearance of

Proceedings of the 27th International
Geological Congress, Volume 9, pp. 303–315
PETROLOGY (IGNEOUS AND METAMORPHIC ROCKS)

porphyroblastic microcline-plagioclase gneiss-granites /2,4/. The K-metasomatism is separated in time by a stage of early metamorphism and deformation of tonalites and volcanics, and by an intrusion of differentiated basic plutons.

Rarely, greenstone belts show direct conjugations with gneiss-granites, without a strip of leucocratic nonmagnetic granitoids of the GIA tonalite type. For example, foliated and migmatized biotite tonalites occur at the base of the Palaja Lamba greenstone belt. The substratum has been dated at 3.1 b.y. /5/, and vein material showed 2.8 b.y. Basalts and komatiites adjacent to the gneiss-granites are migmatized near the contact.

The GIB tonalites and plagiogranites form minor bosses, dykes and veins cross-cutting metamorphozed and deformed volcanics of a lower part of the greenstone succession. Their upper age boudary is determined by the presence of pebbles of conglomerates which are basal for the upper terrigenous-volcanogenous part of the greenstone succession.

Metamorphosed and deformed rocks of the latter part of the succession are cut by platioporhyry and plagiogranite-porphyry dykes.

The gneiss-granite areals /GII/ range in composition and structure due to the presence of remnants of basement rocks belonging to a presumably heterogenous basement, and also due to the nature and degree of its reworking during the formation and subsequent metamorphism of greenstones of a suprastructure. The rocks of tonalite-platiogranite composition occurring within the gneiss-granite areals belong to the basement comples (GIIA/ and to an younger group synchronous with the greenstone belt generation (GIIB). Tonalite-plagiogranite rocks forming a basement to greenstone belts are represented by "grey gneisses" as exampled by the Vodlozero gneisses older than 3.0. b.y. in age. A tonalite intrusion (GIIB) containing xenoliths of various rocks including "grey gneisses" is known in the Vodlozero gneiss-granite areal. Its emplacement occurred simultaneously with later stages of the greenstone belt development (2.65 b.y. ago) under conditions of granulite high-gradient metamorphism.

Gneiss-granitoids (GIIC), including tonalites and plagiogranites, which had been originated in situ by ultrametamorphism during metasomatism and metamorphic recrystallization of supracrustal rocks are exposed in South-West Karelia /6/. These processes are synchronous with the later high-gradient metamorphism of the supracomplex greenstones.

Thus, geological and geochronological data suggest that the generation of tonalites and plagiogranites took plaze both before that of greenstone belts and during the total prolong time-space of their formation.

In the wequence tonalites and granites are confined both to the basement complex and to greenstone structures (suprastructure).

Geological heterogeneity of rocks showing tonalite-plagiogranitic composition is reflected in chemistry which has been analysed both as total analysis by the method of main components, and for the geological groups in combination with trace element contents. The average composition of granitoids lacking K-feldspar (Tabl.I, N17) corresponds to a low-alkaline plagiogranite /7/. Great dispersion values for mafic elements, calcium and in particular trace element abundances indicate to significant variations in their composition. This is reflected in the diagram (Fig.1) where a more mafic portion of basement rocks is clearly separated from the granitoids synchronous to the formation of greenstone belts. Acid members of the basement complex exhibit, as seen in Fig.1, wide compositional ranges as opposed to younger tonalites and plagiogranites. Intrusive tonalites and plagiogranites of the greenstone belt time-level (GIIB, IA,IB) are characterized by compact, somewhat different fields. The 3 outlined fields-IB (Fig.1) correspond to 3 subgroups from various Karelian regions.

The GIA tonalites are composed of quartz, plagioclase, biotite (less 10%). Zircon is common accesory mineral ranging in size and U-content. Two zircon types are predominant, i.e. the first ($l \approx 1015$ c) with U-content about 1000.10^{-4}%, and the second, more abundant ($l \approx 1015$ c) with U-concent 80.10^{-4}%. Some

Some tonalites (Table I, N 1,2) show high Al_2O_3 (above 15%), Sr, Na_2O contents, low Fe+Mg, Y, Zr, U, abnormal high Th/U ratio, the absence of Ti-Y correlation. In our opinion /3,4,8/ rather high K/Rb ratios are due to a superimposed K-Rb metasomatism. The outlined chemical pattern (excluding Rb and U) seems to be determined by their mineral composition, i.e. by the abundant leucocratic minerals and by a high plagioclase/quartz ratio. Fig.2 shows that composition of the Suna tonalities falls within the plagioclase field, and they might have been originated under conditions of rather high pressures, both under "dry" melting conditions and assumed considerable H_2O-pressures. Tonalites contain minor REE concentrations, exhibit a high fractionation pattern of light and heavy REE (Ce^N/Yb^N=16.0), and have no Eu anomalies (Table 2, sample 222; Archaen tonalities /9, Fig.5 -14/, in particular, to tonalitic diapirs in Barberton, South Africa. The observed abundances of trace elements and REE are consistent with a model of melting of metamophosed basalts belonging to THI-type after Condie /9/. Oxygen isotopic composition (Table 1) suggests a relation of tonalities to basalts. However, the derivation of tonalites from the namely basalts seems to be unlikely as indicated by temporal link between the tonalites and metamorphism of the basalts making up greenstone belts in Karélia. To match the aveilable data in terms of the model it is necessary to assume the presence of older, geochemically similar mafic rocks at a lower crustal level. Nevertheless, the close spatial and temporal relationship between the generation of basaltic and tonalitic melts requires its explanation.

To construct an integral model it is necessary to take into account such features as extensive development of the GI tonalites, their uniform composition including that of different tonalites elsewhere in the world, and also the presence of two zircon types and the high Th/U ratio suggesting a complex crustal evolution.

The GIB plagiogranites (Tab.1,NN3-5) have a composition close to that of the GIA, but contain more SiO_2. Their composition points are somewhat shifted in Q-An-Ab diagram (Fig.2) towards quartz in comparison to those of the GIA tonalites. They commonly show lower Sr, Rb and Y contents. Their REE-distribution pattern (Table 2, Fig.3) is similar to that of the GIA tonalites. The facts that the GIB plagiogranites crosscut metamorphozed basalts and have a limited spread do not preclude a model of melt generation due to melting of metamorphosed basalts. Consistent with such an interpretation of plagiogranite genesis is oxygen isotopic composition (Table 1).

The GIIB tonalites are intrusibe rocks as evidenced by their relationships with xenoliths, composition and orientation fo the latter. The formation of the tonalites occurred under low-P granulite facies conditions. They are composed of quartz, antiperthitic plagioclase and biotite with TiO_2-content $>$ 4%. Hypersthene is found in heavy fractions. The xenoliths of melanocratic rocks commonly contain the $RP_y+MP_y \pm$ Hb association. As the GIA tonalites, they show higher Al_2O_3, Na_2O, Sr contents (Table 1, N 6,8), the absence of Ti-Y correlation have been found in the tonalites (Table I, N II). This uncommon composition is expressed by a combination of a high mafic index and high Ti, P, Zr and REE concentrations. Basic rocks with such chemical characteristics are absent both in greenstone belts of Karelia and in basement gneisses. These rocks are thought to represent restites formed during the generation of tonalitic melts.

In assuming the derivation of tonalites from melting of metamorphozed basic rocks similar to TH2 after Condie /9/ and typically occurring in the ancient gneiss complex of the Vodlozero block, they should correspond according to their major element contents to calculated compositions of restites produced by a melt of magmatic tonalites with 30%-fusion of the source. Calculation of a melting model based on trace element abundances and assumed restitic nature of two-pyroxene schist xenoliths seems to preclude a mechanism of partial equilibrium melting for the generation of these tonalites.

Si-metasomatism occurred within the tonalite fracture pattern to form more leucocratic rocks corresponding in composition to plagiogranites (Table 1, NN 7,9,10). Silica addition to tonalites results in the decrease of all trace element contents. The tonalites (Table 2, N 8) and plagiogranites (Table 3, N 18a and 20b) exhibit low REE-concentrations, a high degree of fractionation and positive Eu anomaly (Fig.4) is similar to that of GB of the Amîtsoq gneisses /10/.3 samples (Fig.4) are oriented paralelly each other showing a continued decrease in REE-concentration with SiO_2-increase. These data are in agreement with the field and mineralogical observations suggesting that hte plagiogranites (samples 18a and 20b) had been derived from tonalities (sample 18) during Si- metasomatism with a new-formed quartz. As evidenced by the composition of minerals and their associations, Si-metasomatism and also the emplacement of tonalites occurred under low-P granulite facies conditions. Such a sequence of events i.e. intrusion of a melt, metasomatism, is consistent with recent empirical data for the terranes affected by a high-gradient metamorphism /10/. Oxygen isotopic composition of plagiogranites (Table 1) appears to reflec a mixed juvenile - metahydrothermal genesis of fluids.

"Grey gneisses" of the Vodlozero block (GIIA) are representing by banded biotite and biotite-amphibole gneisses corresponding in composition to tonalites and plagiogranites (Table 1, NN 12-16); amphibolitic layers and rare conformable veins of gneissose leucoplagiogranites are found within them. These rocks have been repeatedly deformed, metamorphozed at high-T° amphibolite facies. In the Q-Ab-An diagram (Fig.2) the field of compositions of the Vodlozero gneisses is shifted, as compared to rocks of similar composition, towards anorthite. All the gneisses are characterized by high Sr-contents, and are divided based on Zr-content and a number of other features into two groups, the subdivision being supported by structural-metamorphic data suggesting a heterogeneity of the assemblage. Gneisses rich in lithophile elements, including Zr, made their appearance in relation to second fold generation and second schistosity. These gneisses exhibit an inconsistent REE-distribution pattern. The oldest member of assemblage, i.e. hornblende-biotite gneiss (sample 45, Tab. le2, Fig.5) shows low REE concentrations and a slight degree of their fractionation (Ce^N/Y^N=9.7). The extreme fractionation (Ce^N/Yb^N=273) is characteristic of a vien of leucocratic gneiss-plagiogranite (sample 46c). This vein according to geological data is younger than gneiss (sample 45), but has been emplaced before D_2-deformation and GII gneisses. Such a low concentration of heavy depletion of heavy REE at depth where liquidus had garnet as a stable phase. Biotite gneisses (sample 45e) formed during the second structutal-metamorphic vent show the REE-distribution pattern similar to that of sample 45; it is notable but a higher degree of fractionation due to the decrease of heavy REE-abundances (Ce^N/Y^N=19). Such a pattern is reported by Condie /9/ for acid volcanics of greenstone belts and is described by a model of thier derivation from partial melting of eclogite showing a TH 1-composition. The adoption of such a model finds some difficulties due to the absence of high-P metamorphism evidence and requirs more information to analyse the overall aturally occurring rock association.

Thus, rocks of tonalite-plagiogranite (trondhjemite) composition are important both in the framework of the oldest Karelian area, i.e. infrastructure and of the suprastructure formed in the Late Archean in connection to the greenstone belt development.

Such rocks can be derived under conditions of different geological processes, i.e. from metasomatic reworkint of ortho- and pararocks by fluids containing various components, from melting of rocks at different depths. Tonalites and plagiogranites generated by crystallization of a melt are predominant. The former exhibit various genetic relationships within the tonalite-plagiogranite suite. In many cases there are gradual passages between tonalites and plagiogranites suggesting their cogeneity. In some cases tonalites can be shown to have an igneous nature and to be transformed into plagiogranites by metasomatism.

Further information about crystallization temperatures of melts and fluid phase composition of granitoids will provide a well-founded basis for the elaboration of preferrential petrogenetic models.

REFERENCES

1. Krylov I.N., Levchenkov O.A., Lobach-Zhuchenko S.B., Chekulajev V.P. Heterogenous structure and evolution of Archean lithosphere of the Karelian granite-greenstone terrain. 27th IGC, vol.5, "Geology of the Precambrian" (in Russian).
2. Geology and petrology of Archean granite-greenstone sequence of Central Karelia. Leningrad, "Nauka", 1978, 262p. (in Russian).
3. Krylov I.N., Lobach-Zhuchenko S.B. Geochemistry of granotoids of the Suna complex (Central Karelia).- In "The oldes- granitoids of the USSR. Grey gneiss complex". Leningrad, "Nauka", 1981, p.68-81 (in Russian).
4. Lobach-Zhuchenko S.B., Chekululajev V.P., Baikova V.S. Episodes and types of granite formation in the Precambrian of the Baltic Shield. Leningrad, "Nauka", 1974, 208p. (in Russian).
5. Lobikov A.F., Lobach-Zhuchenko S.B. Isotopic age of granites of the Palaja Lamba greenstone belt in Karelia. Dokl. the USSR Academy of Sciences, vol.250, N 3, p.729-733 (in Russian).
6. Geology and petrology of granite-gneiss terrain of South-West Karelian. Leningrad, "Nauka", 1969, 226 p. (in Russian).
7. Classification and nomeclature of magmatic rocks. Moscowm "Nedra", 1981 (in Russian).
8. Lobach-Zhuchenko S.B., Krylov I.N. The geochemical evolution of the Karelian granites. In: Methods of the geochemical mapping and bouder prospecting in the eastern part of the Baltic Shield, Geol. Surv. of Finland, 1980, p.115-132.
9. Condie, K.C. Archean greenstone belts, 1981, 434 p.
10. O'Nions, R.K., Pankhurst, R.J. Early Archean rocks and geochemical evolution of the Earth's crust. Earth P.Sci.Lett., v.38, 1978, pp.211-236.
11. Sedova I.S., Glebovitsky V.A. Modes of development of ultrametamorphism under amphibolite facies conditions. Izv. of the USSR Academy of Sciences, ser. geol., 1983 (in Russian).

12. Yoder, H.S. Albite-anorthite-quartz-water at 5 kb. -Carnegie Inst. Washington Yearbook, 66, 1967, p.477-478.
13. Hoffman, C. Natural granitic rocks and granite system Qz-Or-Ab-An(H_2O) and Oz-Ab-An-(H_2O), Neues Jahrbuch für Mineralogie, Heft 7, 1976, pp.289-306.
14. Steward D.B. Four-Phase curve in the system $CaAl_2Si_2O_8$-SiO_2-H_2O between 1 and 10 kilobars. Schweizer. miner.petrogr. Mitt., B 47, H1, 1967, p.35-60.
15. Luth, W.C. The system $NaAlSi_3O_8$-SiO_2 and KAl Si_3O_8-SiO_2 to 20kb and relationship between H_2O content, P_{H_2O} and P_{total} in granite magmas.-Amer.J. Sci., A 267, Schairer vol. 1969, p.325-341.

Table 1. Chemical composition of tonalites and plagiogranites of Karelia*

	1	2	3	4	5	6	7	8	9
SiO_2	67.20	68.48	69.37	70.42	70.94	68.19	70.27	67.88	70.96
TiO_2	0.40	0.31	0.31	0.28	0.28	0.45	0.31	0.30	0.25
Al_2O_3	16.61	17.24	15.76	16.17	15.80	16.03	15.42	17.21	14.60
Fe_2O_3	1.29	0.59	0.63	0.52	0.48	0.88	0.81	0.73	0.65
FeO	1.95	1.52	2.59	1.80	1.14	2.72	2.01	2.27	2.38
MnO	0.05	0.03	0.03	0.02	0.02	0.06	0.04	0.03	0.03
MgO	1.23	0.63	1.23	0.81	0.90	1.43	1.17	1.03	1.52
CaO	3.11	3.25	3.29	3.39	3.85	3.14	3.11	3.37	3.03
Na_2O	4.28	5.34	4.60	4.81	4.85	4.31	4.31	4.84	4.08
K_2O	1.93	1.28	0.96	0.76	0.91	1.14	1.39	1.29	1.25
P_2O_5	0.12	0.07	0.26	0.09	0.07	0.12	0.07	0.10	0.08
H_2O	-	0.12	-	0.10	-	-	-	0.08	0.08
LOI	-	0.97	-	1.00	0.81	-	-	0.83	1.16
Rb	73	40	24	20	34	39	39	39	35
Sr	474	702	221	159	286	345	312	366	290
Zr	133	151	130	130	128	182	154	187	131
Y	8	8	5	5	5	10	9	5	5
SiO_2	76.08	45.38	68.62	61.20	62.32	68.24	70.58	68.95	11.29
TiO_2	0.12	2.50	0.31	0.84	0.47	0.38	0.19	0.37	0.05
Al_2O_3	13.60	12.84	15.99	15.67	15.84	16.13	16.54	15.51	1.34
Fe_2O_3	0.72	3.49	0.98	3.37	2.04	0.50	0.07	1.11	0.52
FeO	0.58	11.15	1.65	3.42	3.30	2.49	1.37	2.04	1.28
MnO	-	0.20	0.04	0.12	0.13	0.03	0.01	0.06	0.002
MgO	0.43	8.73	1.15	2.71	3.15	1.62	0.97	1.22	0.41

CaO	2.57	9.68	3.98	5.51	5.66	4.42	4.16	3.58	1.03
Na_2O	3.99	2.29	4.72	4.16	4.00	4.33	4.34	4.50	0.52
K_2O	1.14	104	1.22	1. 6	1.06	1.13	0.83	1.27	0.34
P_2O_5	0.04	0.50	0.09	0.26	0.14	0.11	0.06	-	-
H_2O	0.02	-	-	-	-	0.06	-	-	-
LOI	0.44	1.60	0.47	0.62	0.63	0.68	0.43	-	-
Rb	17	19	45	43	35	28	17	44	735
Sr	266	113	500	461	431	475	510	385	20511
Zr	90	450	88	245	113	240	214	151	5268
Y	6	32	8	17	14	11	7	11	83
Pb	9	5	10	12	11	10	6	-	-
n	1	1	[illegible]	3	1	1	1	142	-

*1 - average composition of the Suna River tonalites (GIA); 2 - sample 222 - the Suna River tonalite; 3 - average composition of plagiogranites from Central Karelia (GIB); 4 - sample A20/3 and 5 - sample 1506/II - GIB plagiogranites; 6 - average composition of GIIB tonalite; 7 - average composition of GIIB pagiogranite; 8 - sample 18 - GIIB tonalite; 9 - sample 20b, and 10 - sample 18a - GIIB plagiogranites; 11 - two-pyroxene schist of xenolith in tonalite, sample 20/3 (restite); average composition of grey gneisses of the Vodlozero complex; 12 - dacitic, 13 - andesitic compositions; 14 - sample 45; 15 - sample 45e; 16 - sample 46c: gneisses of the Vodlozero complex; 17 - average composition and 18 - dispersion of tonalites and granites form Karelia.

Table 2. REE concentrations and oxygen isotopic composition in Archaean tonalites and plagiogranites from Karelia [1)]

	222	A20/3	1506/II	18	20b	18a	45	45d	46v
Ce	30	41	27	17.4	15.0	10.1	34	38.6	111
Nd	16	14	7.9	9.2	8.2	6.0	16.1	14.9	31.2
Sm	2.5	2.5	1.8	1.37	1.07	0.94	3.29	2.25	2.89
Eu	0.61	0.62	0.42	0.60	0.64	0.46	0.94	0.84	0.63
Tb	0.16	0.15	0.09	0.16	0.13	0.09	0.40	0.18	0.14
Yb	0.48	0.36	0.20	0.30	0.29	0.18	0.89	0.52	0.10
Ce^N/Yb^N	16.0	29.6	33.3	14.9	13.3	14.2	9.7	19.0	273.0
$\delta^{18}O$,‰	7.3	6.73	-	7.9	8.3	8.3	-	-	-

[1)] See footnote to Table 1.

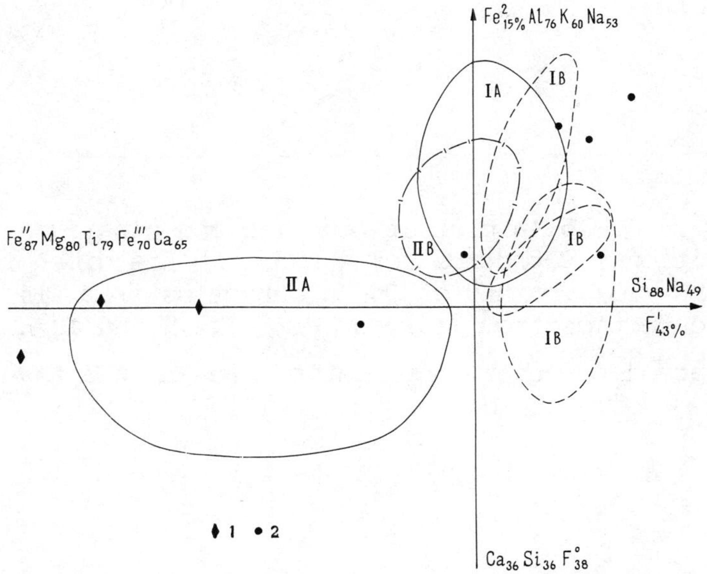

FIGURE 1. Component diagram for plagiogranites and tonalites of the Karelia. The Vodlosero complex: 1, plagiogneisses of tonalitic composition, 2, more melanocratic gneisses.

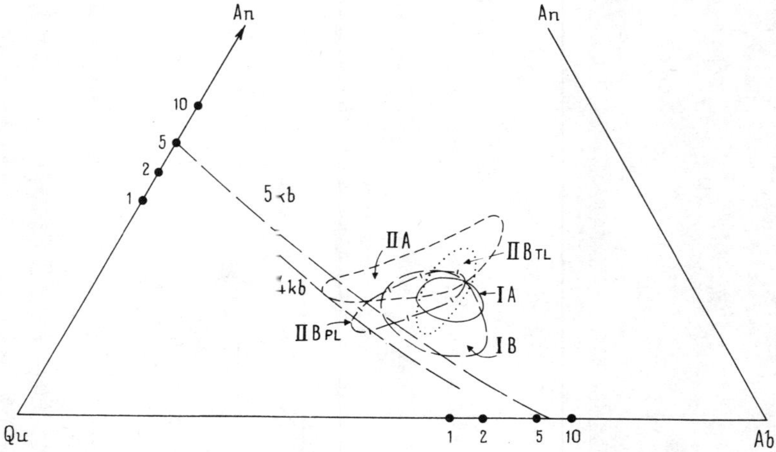

FIGURE 2. Qz-Ab-An plot of tonalite and plagiogranite compositions. Cotectics for 4 and 5 kb are giben from /12/ and /13/, points of Qz-An eutectics from /14/, Qz-Ab eutectics from /15/ for P_{H_2O}=1,2,5 and 10 kb. Explanation to other abbreviations see in the text.

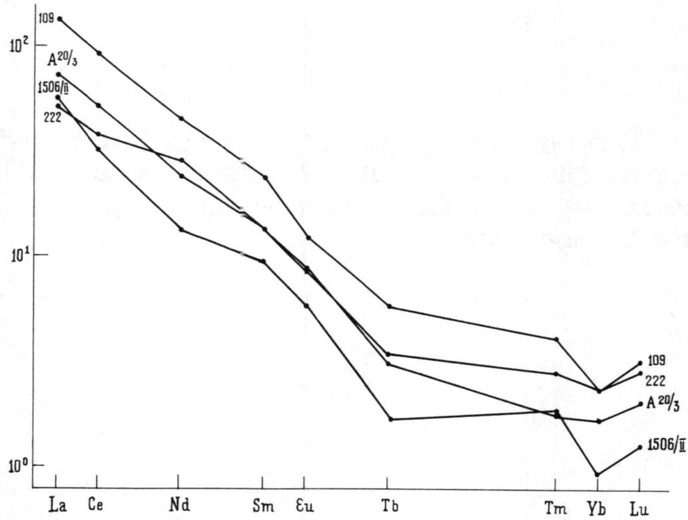

FIGURE 3. Chrondrite-normalized REE distribution in the G¼A tonalites / sample 222/ and plagiogranites / sample 109, A2 , A20/3 and 1506/11/.

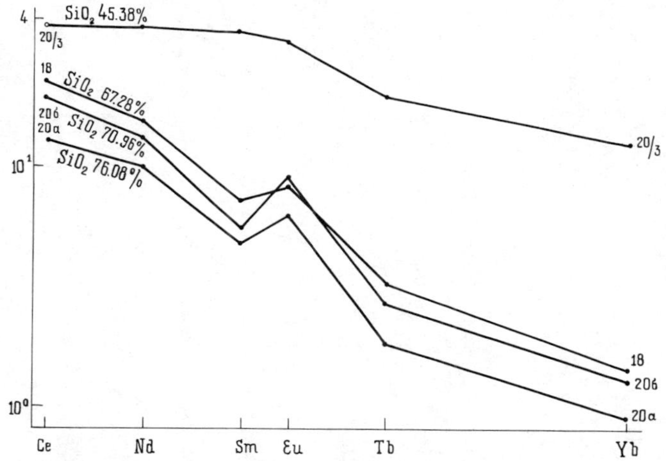

FIGURE 4. Chrondite-normalized REE distribution in the G11B tonalites /sample 18/ and plagiogranites / samples 20a and 20b/. Sample 20/3 - two-pyroxene schist/restite/.

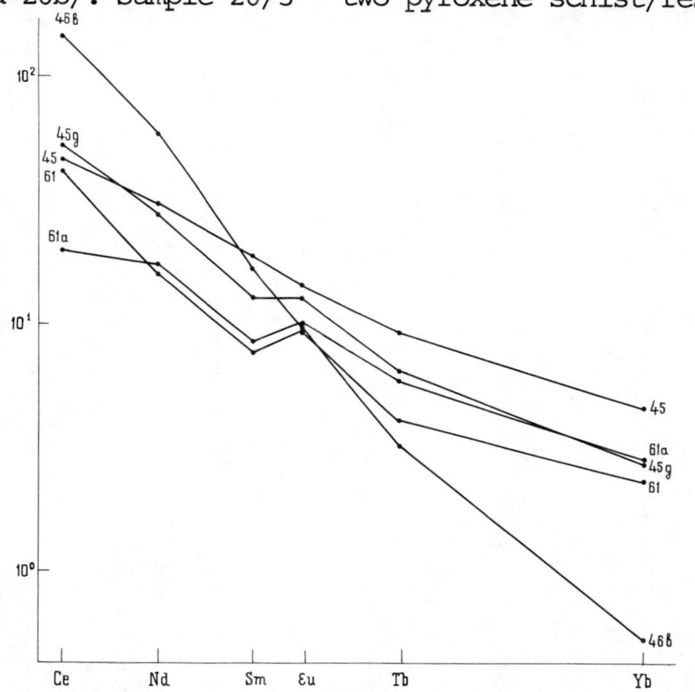

FIGURE 5. Chrondite-normalized REE distribution in the G11A rocks. Samples 45, 45e and 46c - the Vodlozero gneisses; samples 61 and 61a - substratum and vein material of migmatites from the Palaja Lamba complex.

ACID MAGMATIC ROCKS OF MID-OCEANIC UPLIFTS

LUCHITSKY I.V.
Institute of the Lithosphere. The USSR Academy of Sciences Moscow.

It has been known for long and stated by many scientists (1-6) that acid magmatic rocks are widely spread in the oceans. These rocks are widelly found in the marginal oceanic zones (the Cape Verde Islands, the Seychelles, Revilla Gigedo), in the areas of mid-oceanic uplifts (Island, Kergelen, the Ascention island, the Easter island and others). Among them not only different products of volcanic erruptions, presented by rhyolite row rocks, but typical granites may be observed. Nowadays the question of distribution of not only rhyolites but granites as well in the oceans and mid-oceanic uplifts is solved positively and is no longer the subject to discussion. Both types of rocks are found on the margins as well as on the mid-oceanic uplifts.

Formation of rhyolites and granites on the mid-oceanic uplifts and their slopes in particular appears the main problem now, as there is no doubt that in the oceanic margins these rocks might belong (not obligatory) to the earth crust blocks, separated from the continents. In the meanwile rhyolites and granites of the mid-oceanic uplifts are located in the area of, as it is considered, typical development of oceanic crust. Formation of acid rocks under such conditions is possible, as supposed, only as a result of magma differentiation, close in its composition to low potassium content toleiites, composing oceanic crust. However, in this case one may expect that rocks, spread over the mid-oceanic uplifts, which have low potassium content belong to plagiogranite and rhyiolites containing considerable potassium fieldspar amount.

This is how modern ideas of the prevailing of plagiogranites and sodium rhyiolites in the mid-oceanic uplifts have been formed. Putting forward such ideas one usually draws a parallel between present day and ancient oceanic crust referring to plagiogranites and sodium rhyolites of the upper parts of ophiolite suite known by discriptions of many scientists (7-9). Nevertheless, the real

Proceedings of the 27th International
Geological Congress, Volume 9, pp. 317–326
PETROLOGY (IGNEOUS AND METAMORPHIC ROCKS)

estimate of the particularities of acid magmatic rocks of mid-oceanic uplifts not always coinsides with such ideas, what can be proved by the following examples.

Iceland. If in the beginning of our century it was supposed that in Iceland there are not many acid rocks, nowadays their wide distribution may be considered proved, thus allowing to state their significant role as they comprise 10% of the total volcanic formations of the island. Nowadays it is known that in Iceland, together with rhyolites, there are granites as well. We have at our disposal numerous chemical analysis of Islandic acid rocks (10-12) among them those carries out in the last century and included into the general work of Wolff (13).

Basing on the data of 52 previously made and 6 newly made analysis Walkes (4) compiled avarage chemical composition of rhyolites in Iceland. It corresponds norm ratio; for 100% of anhydrous matter (in %%) potassiumfeldspar - - 20%, albite - 46 and quartz - 34% what corresponds to a normal rhyolite row. Presence of normal quantity of kalispar soes not allow to refer these rocks to plagiogranites, as for average composition of typical plagiogranite, specified by Dixon and Rutherford (4) ratio kalispar-albite-quartz (per norm parameters) presents, correspondingly 2:54:44 (fig 1). Quantity of norm kalispar in the potassium-most-rich plagiogranites is less than 5%. Thus, rhyolites of Iceland differ greately from the plagiogranite row rocks by considerable potassium content.

The granites known in Iceland, also have kalispar-albite-quartz ration 23:60;17 what does not allow to refer them to plagiogranites. These rocks, by their nature, may be refered to granosienites with considerable potassium content, or normal granites.

Kergelen. The archipelago is considerably shifted southwards from the axis of the mid-oceanic uplift due, as supposed, to oceanic plate's shift and located nearly simmetrically regarding the undersea Brocken uplift (16). In 1960-65, on the task of French Commission on Arctic Studies Nougier investigated the archipelago and made up the geological scheme of the island based on the data of 1:200 000 scale survey. On the scheme in the area of Raye du Bati at the extreme SW of Kergelen a large massif of grain crystal rocks slightly stretched in Ne direction, 25 km long, about 15 km wide is shown. According to Nougier's data normal composition of this massif corresponds the composition of leucocratic syenites with varying quartz content reaching 30% and defining transfer of syenites into typical granites. The massif sects the host basalts, sometimes injection of aplitic veins are observed near contact zones. Thus, detailed research of Nougier proved already known fact of the spread over the Raye du Bati peninsula of granites which for the first time were mentioned in 1924 by Lacroix basing on the Poe samples (in work 2 the first chemical analysis from this area is given) later it was noted by Hobert-de-la Rue and other scientists.

Though in joined publications Nougier (16) gave average composition of the massif rocks on the data of 19 analysis corresponding alkali syenite (syenite poor in calcium is mentioned and stated that analogues analysis in tripple diagramm quartz-ortoclase-albite by quartz content do not reach the field of 500 continental granites as per Tuttle and Bowin (17) such statements not strictly reflect the results of chemical analysis of rocks, published by him earlier in his fundamental work (5). Specification based on this data of normal composition of granites and syenites of the massif located on the Raye du Bati peninsula shows that in 50% of chemical analysis mentioned by Nougier, quartz content exceeds considerably 25% level, reaching sometimes 44%. Almost one third part of analysis reveals quartz content equal to or consederably higher than 30%. Alkali microgranites of Loranche mentioned in the same work also contain about 30% of quartz. In general the spread of granites in Kergelen became absolutely evident mainly in a giant (about 300 km^2) syenitic massif of the Raye du Bati peninsula and small body of the Loranche peninsula.

In general, the idea of the distribution of acid rocks in Kergelen may be completed by the information about the role of rhyolites in the island. Though basalts are the main rock type in Kergelen, trachytes - only partually, nevertheless, there are numerous outdrops of rhyolites on the Loranche peninsula. On the diagramm quartz-ortoclase-albite given by Nougier chemical analysis of rhyolites as converted to norm composition show that they also tend to continental granites' field (fig.1). It should be specially emphasized that among the syenites and granites of Kergelen there are not more than 27% of rocks which contain more alkali than alumina; there are than 30% of such rocks among rhyolites. In other words typical alkali syenites, granites and rhyolites in Kergelen not prevail, but they combine only one third of their total quantity.

To evaluate peculiarities of mid-oceanic uplifts acid rocks' composition (for Kergelen as well) it is essential to take into account characteristic alkali ratio expressed by relative role of potassium and sodium field spars norm content. It is easily seen on fig.1, that for granites of Kergelen, containing over 20% of quartz 8 analysis from 12 show the presence of potassium field spars in the amount from 20 to 50%. Thus all these rocks are typical representatives of usual rocks of granitic and rhyolitic row and do not belong to plagiogranites or sodium rhyolites.

Apart from Iceland and Kergelen the distribution of acid magmatic rocks is typical for some other islands located on the mid-oceanic uplift.

Ascension Island. Here rhyolites and granitic inclusions in basalts, aglomerates, to be more exact, have been noted for long (18). Daly (3) has described obsidians of rhyolitic composition. Convertion into norm composition of chemical analysis of these rocks yeald ratio kalispar-albite-quartz equal to, correspondingly, 21:55:24 and 29:43:27, what witnesses that they belong to the row of normal, but not low potassium rhyolites. As far as granites are concerned they belong to the row of rocks, characterized by the excess of alkali refered to alumina, i.e. to alkali rocks. Their model per cent composition is the following: Quartz -30%, pertite -62%aegerine -1; arphedsonit -1; aenigmatite -1; Zicon, pyrochlorine -less than 1. Ration kalispar-albite-quartz in granites is 26:37:37. Thus, visually under the microscope and by calculations these rocks do not belong to plagiogranites; they belong to the row of alkali granites with high potassium content.

The Easter Island. The island located on the East

Pacific mid-oceanic uplift for long time is known for its rhyolitic obsidians, noted in 1922 by Tilley and later - by Lacroix, Labbat, Bandy and others. Average composition of these rocks has been defined by Gorshkov (20) basing on 7 analysis. Here also there are no prof or similarty of rhyolitic obsidians and plagiogranites. Ratio kalispar-albite-quartz 22:48;30, characteristic for obsidians stresses comparatively high content of kalispar and quartz, typical for normal granites and thyolites. Higher alkali content in these rocks has not been extablished.

St. Pauls Island. in the Indian Ocean is also known for its acid magmatic rocks. According to Lacroix, volcanic activity on the island started with the effusion of acid lavas, later named rhyolotoids. He specified their composition basing on the presence in the basic mass of orthoclase needles, sometimes of spherolitic structure and crystalls of trydimith.

The given data of mid-oceanic uplifts acid magmatic rocks may be enriched by references to gneiss inclusions, noted in the lavas of the Tristan da Cunha Island (21,22) and granite cobbles on the Azores (13). One should be carefull about their gracial origin and transfer with icebergs from the north.

In general. the data considered above quite definitely specifies the significant role of granites in formation of mid-oceanic uplifts. These granites have typical grain-crystalline structure and belong, most often to the row of normal, not alkali rocks, containing considerable quantities of kalispar. Thus, the idea of the domination of plagiogranite-type acid row rocks on the mid-oceanic uplifts is groundless. The same may be refered to rhyolites among which granits of normal, not alkali or plagiogranitic row revail. In all the granites and rhyolites relative role of kalispar is defined by their presence in quantities exceeding 20%; among the considered ones they make up 2/3 of all the acid rocks.

Experiments proved the possibility of obtaining the residual plagiogranite melt by means of partual (10%) melting of ocanic basalts with low (0,006%) potassium oxide content (23). The same experiments revealed that in order to get 10% melt with K_2O content of 2,39%, i.e. more or less close to normal granites by composition, it is necessary to have Kilauea -type basalts with 0.44% K_2O content as initial ones (fig.2). This melt, experimentally obtained from basalts is close to granitic one by its composition and has rather low K_2O content even if compared to average parameters, characterizing typical granites and rhyolites of Iceland, where K_2O content is equal to 3% and sometimes is even higher (see fig.2), with only one exclussion - average composition of rhyolitic erruption of Askja in 1985 - 2,53%. As far as Kergelen and the Ascension island are concerned, for the acid rocks typical for them, deviations from the average parameters are even more significant. The general picture of K_2O content variations in granites and rhyolites is very impressive, as it includes variations, reaching 7% K_2O. On the Easter Island acid rocks contain, in average, over 3,5% K_2O, as per 7 analysis.

As it is evident on Fig.2, even 10% melt of Kilauea basalts which are differenciates of oceanic toleiitic magma, does not reach K_2O content, typical for enormous amount of different rhyolites and granites, located in different areas of mid oceanic uplifts. Thus, basing on the above material of the distribution of acid rocks in the mentioned islands and taking into account experimental data, one should consider origination of rhyolites and granites of mid-oceanic uplifts by means of 10% melting from oceanic toleiitic basalts or Kilauea type basalts hardly possible. Probably one cannot exclude the opporunity of the formation due to assimilation of sialitic basement (21) submerged blocks or as a result of partual melt of silica enriched remnant, similar to sialitic crust (4), if such a remnant could have ever been obtained. Its quite possible, that distribution of acid rocks over mid-oceanic uplifts can correspond the inhomogenity of the Earth's upper mantle (6). It is evident, that to explain the origin of acid rocks within the area of mid-oceanic uplifts different alternatives micht be suggested. Among them, the hypothesis, suggesting that these rocks had been formed as a result of partual melting of oceanic toleiitic low potassium basalts. The present review of data concerning distribution and compositional particularities of mid-oceanic uplift acid magmatic rocks shows that experimental basis of this hypothesis does not take into account really excisting wide spector of variations.

REFERENCES

1. Backstrom H. Beitrage zur Kentniss der islandischen Liparite. - Geologiska Foreningens, 1981, s. 637-678.
2. Lacroix M.A. Les roches eruptives grenues de l'Archipel de Kuergelen. Comptes Rendus de l'Academie des Sciences, t.179, 1924, p. 113-119.
3. Daly R. The Geology of Ascension Island. Proc.Am. Acad.Arts a Sciences, v.60, N1, 1925, p.71.
4. Walker G.P.L. Acid Volcanic Rocks in Iceland. Bull Volcanol, t.29, 1966, p. 375-406.
5. Nougier j. Contribution a l'etude geologique et geomorphologique des isles Kuergelen. Com.Nat. Franc. Rech. Atract., N27, t1, 171p., t.2, 255p. 1969-1970.
6. Luchitsky I.V. Onacid magmatic rocks of the oceans (In Russian) Geotectonica, N5, 1973, c.22-34.
7. Thayer T.P. Himmelberg G.R. Rock succession in the alpine-type mafic complex at Canyon-Mountain, Oregon. 23-rd Int Geol, Congr., v.1, 1968, p.175-186.
8. Moores F.M., Vine F.J. Troodos Massif, Cypur and other ophiolites as oceanic crust: evaluation and implications Phil. Trans.R.Soc. Land., v.A268, 1971. p.433-466.
9. Coleman R.G., Peterman L.E. Oceanic Plagiogranite. J. Geophys. Res., V.80, 1975, p. 315-318.
10. Sparks S.R. Sigurdson H. Wilson L. Magma mising: a mechanism for triggering acid explosive eruptions. Nature, 267, 1977, p.315-318.
11. Sigvaldson G.E. The petrology of Hecla and origin of silicic rocks in Iceland. In: the eruption of Hecla 1947-48. V.1 Sci Icelandica, 1974, 44p.
12. Carmichael J.S.E. The petrology of Thingmuli a tertiary volcane in Eastern Iceland. J. Petrol., v.5, N3, 1964, p. 435-460.
13. Wolff F. Der Vulkanismus Bd.2. 1931, p. 829-1112.
14. Dixon S. Rutherford M.L. Plagiogranites as late stages immiscible liquids in ophiolite and mid-oceanic

ridge wuites: an experimental study. Earth a Planet Sci. Litt. v.45, N1, p.45-60.
15. Blake Dutt The net - veined complex of the Austerharn intrusions south-eastern Iceland. Journ of Geology, v.74, N6, 1966, p.891-907.
16. Watkins K.D., Gunn V.M., Nougier J., Baksi A.K. Kergelen: continental fragment of Oceanic Iceland? Geol.Soc. Amer.Bull., V.85, 1974, p.201-212.
17. Tuttle O.F., Bowin N.C. Origin of granite in the light of experimental studies in the system Na AP Si_3O_8--K AP SiO_3O_8-H_2O America Mem., 74, 1958, 153 p.
8. Roedder E., Coombs D.S. Immiscibility in granite Melts, indicated by fluid inclusions in ejected granite blocks from Ascension Island. Jorn. Petrol. v.8. pt.3. 1967, p.417-451.
19. Tilley C. Density, refractivity and composition relations of some naturale grasses. Mineralogy Mag.v.19, 1922, p.276-294.
20. Gorshkov G.S. Volcanism of the Kuril Island Arc (In Russian). M. "Nayka", 1967, 268 p.
21. Gilluly J. Geological contract between continents and oceanic basins. Geol. Soc. Amer. Spec. Pap., N62, p.7-18.
22. Barth T.F. Mineralogical petrography of Pacific lavas, part 2. Rocks, amer.Journ., SCi, v.221, p.492-530.
23. Spulber S.D., Rutherford J. Origin of silicic rocks in oceanic crust. J. Petrol., V. 24, N1, 1983, p.1-25.

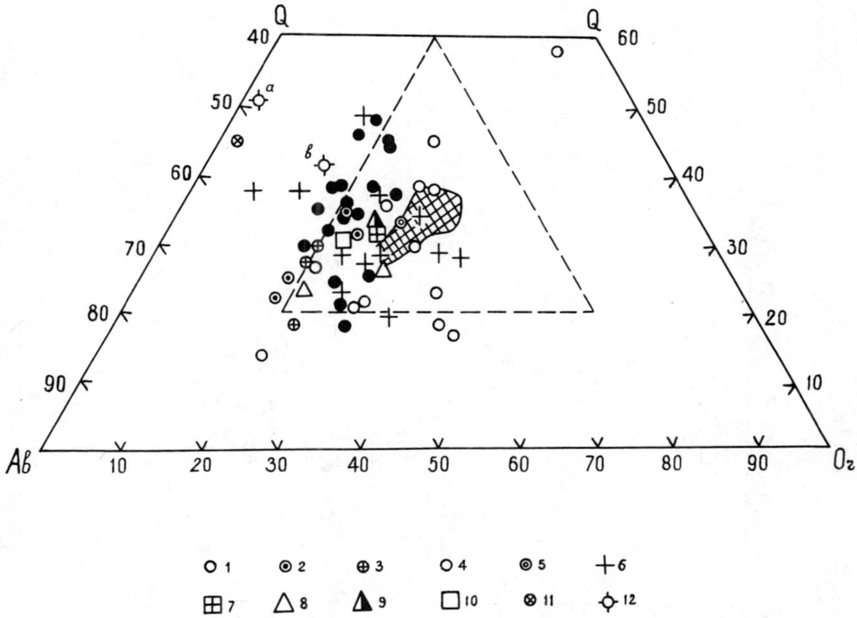

FIGURE 1. Tripple diagramm quartz-albite-ortoclase for rhyolites and granites of mid-oceanic uplifts. Legen: Iceland -1- rhyolites, 2- their average composition, 3- granites; Kergelen -4- rhyolites, 5- their average composition, 6- granites, 7- their average composition; the Ascension Island -8- rhyolitic obsidians, 9- granitic inclusions in aglomerates; the Easter Island - 10- rhyolitic obsidians (average composition) 11- average composition of plagiogranites as per (14), 12- - experimental melts (10%) of toleiitic basalts (a) and Kilauea type basalts (b) as per (23). All the analysis are calculated for 100% adhydrous composition. Shaded is the area of 90% of 500 continental basalts as per (17).

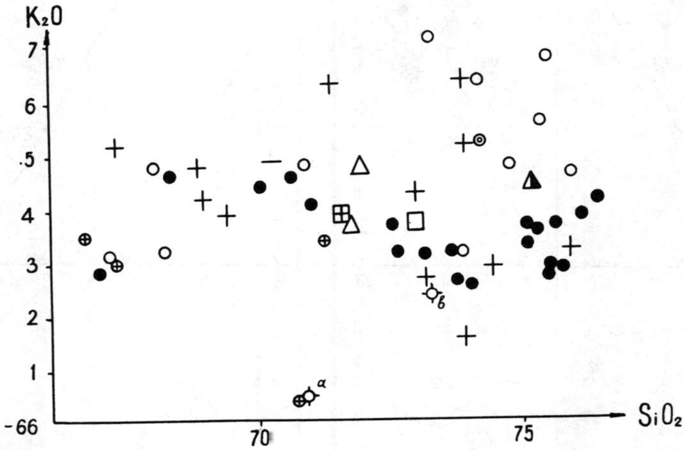

FIGURE 2. K_2O content on the rocks of rhyolite-granite row. Legend as above.

PERIDOTITE NODULES IN KIMBERLITES AND BASALTS AS INDICATORS FOR LITHOSPHERE DEEP-SEATED STRUCTURE

A.A. MARAKUSHEV
Department of Geology, Moscow University, Moscow, USSR

The paper summarizes the evidence for peridotites found as nodules in kimberlites and alkali basalts to show that they are intrusive in nature and belong to the magnesian-alumina and iron-titanum magmatic series. Contrary to the accepted view it is inferred that they are layered intrusions in the crust rather than wold-wide layers of the upper mantle. Periodites contain mineral generations that occur in the order of decreasing pressure of formation, being the result of magma's stepwise crystallization during emplacement and in intermediate hearts at different horizons in the mantle and crust. The earliest mineral assemblages have recorded the maximum depths of the hearts (150-200 km) beneath the ancient platforms. They are associated with the emplacement of diamondiferous and exlogite magmas. On transition to oceanic regions, magma hearts occur at depths as shallow as 50 km. This agrees with the contrasting thermal regimes in the mantle under the ocean and continents, as shown in fig.1. The scheme is based on Mac Donalds's computed model (3), in which the mantle beneath the continental crust is depleted in radioactive elements (^{238}U, ^{235}U, ^{232}Th, ^{40}K and others) and therefore gives off far less heat than the mantle under the oceans. However, this difference is offset by the radioactive input from the continental crust which is of greater thickness and by far more abundant in radioactive elements than the oceanic crust. As the result, the heat flux on the continents and in the oceans is nearly equal (3). For all reasons, the continental mantle, beiing more depleted in the radioactive elements (heat-generators), is colder than the oceanic mantle. The alternative explanation for the differences in the mantle thermal regimes in various geological environments (fig.1) is based on the convection hypotheses. In this hypothesis, currents of the material occuring in the mantle beneath the oceans

Proceedings of the 27th International
Geological Congress, Volume 9, pp. 327–342
PETROLOGY (IGNEOUS AND METAMORPHIC ROCKS)

rise along the rift zones of the midoceanic ridges and, moving away from the ridges and under the continents, lose heat brought from the deepest parts of the mantle.

This hypothesis adequately explains the observed regeneration of the oceanic crust along the mid-oceanic and continental regions. These zones fell into two major types - atlantic and pacific (3). The atlantic type is characterized by the progressive stage in the development of the mantle convection currents which are responsible for the steady expansion of the Atlantic and Indian oceans whose margins (shelf, continental slope, continental foot) exhibit only moderate seismicity. By contrast, the pacific-type zones in the oceanic areas steadily deminishes, with thick folded crust of the continental type being developed on the margins. These tectonically active transitional zones are represented by folded mountain ranges on the margins of the continents or by island arcs bounded by deep-water trenches in the oceans.

A composite picture of temperature distribution in the mantle beneath the oceans and continents as sketched in fig.1 is supported by geophysical evidence according to which at depth the mantle material undergoes partial melting (the wave guide zone). The boundary of this zone is defined by the periodotite to yield small amounts (close to 1%) of basaltic melt. Depending on the thermal regime in deep-seated areas, the partial melting of the peridotite mantle beneath the continents is not very distinct at depth as great as 250 km while beneath the oceans it is by far more pronounced at much shallower depth (50 km) (3). The wave-guide surface marks the lower boundary of the lithosphere which at the margins of the oceans dips sharply under the continents as is shown in fig.1 for the pacific-type transitions. At the Atlantic margins, where the thermal regime is more uniform, the wave-guide immersion zone seems to be displaced into the continental region. The final stage in the evolution of tectonically active (geosyncline) pacific-type transitional zones involves the contraction regime which leads to the development of folded orogenic systems with thick continental crust. The adjoining thin oceanic lithosphere plates tend to slide down under the continent in the direction of the isotherm trend (fig.1) thereby procucing earthquakes whose epicentres also shift downwards in the Benioff zones. The maximum depth of magma hearths from which magmas intrude the upper horizons in the crust or erupt into the surface, increases in the same direction (from the ocean towards under the continent), This conclusion transpires from studies on mineral parageneses formed in deeplseated nodules (peridotites, pyroxenites, eclogites, babbros) in basalt and kimberlites. Mineral crystallization stages recorded in these paragenesis relate to different depth facies (I, II, III) whose location in fig.1 is based on the contrasting temperature regimes in continental and oceanic mantles. Current ideas on actual rocks consistituting the worldwide upper mantle zones (fig.1) are purely hypothetical. "No geologist of geophysicist can lay claim to having ever held a rock fragment in his hand that was derived from depth below the Mohorovicic discontinuity, be it on the continents or in the oceans" (1,p.339). True, many geologists tend to regard peridotite or eclogite nodules from alkali basalts and kimberlites as xenoliths detached from the relevant layers of approach, volcanoes are sort of naturals drill holes which supply rock specimens onto the earth's surface from different horizons in the upper mantle and thus make it possible to reconstruct mantle's regional cross-sections. However, this hypothesis is at variance with the many known features of these rocks. Normally, they form regular magmatic series controlled by magmatic differentation, their structural characteristics and mineral assemblages appearing in the order of decreasing pressure of crystallization (7). It thus seems that they are product of some complex evolution which starts in deep-seated magma hearths and envolving primary (in hearths) crystallization and subsequent consolidation of magmas as they intrude higher levels in the mantle and crust to form layered or multiphase intrusive and subvolcanic complexes (lopoliths, sills, dykes). Although pipe-like volcanic vents (diatremes) with which the emplacement of kimberlites, kimberlite-like rocks and alkali basalts is associated formed later than these preceding intrusions, there can be traced genetic links between them. Fig.2 presents comprehensive diagrams based on the petrochemical data for deep-seated nodules from kimberlites and alkali basalts in continental areas and from alkali basalts in oceanic regions. The represented by two magmatic series (I and II) whose analogues widely occur in layered multiphase intrusives of the basite hyperbasite type. The differences between porphyritic and granular peridotite from modules are in a way similar to those commonly found between subvolcanic and plutonic rocks. Granular peridotites contain accessory phlogopite, graphite and chromite normally not found in porpyritic rocks. Bulk chemical analyses show that hte granular peridotites are depleted in sodium, iron, calcium, titanum and aluminium as compared with porhyritic peridotites such as those found in a Lesotho pipe (4):

porphyritic lherzolite - $Na_{0.15}Ca_{0,67}Mg_{19,24}Ni_{0,06}$ x $Mn_{0,04}Fe^{2+}_{1,61}Fe^{3+}_{0,41}Cr_{0,04}Al_{0,66}Ti_{0,04}Si_{13,28}O_{50}$;

granular lherzolite - $K_{0,04}Na_{0,04}Ca_{0,16}Mg_{21,04}Ni_{0,05}$ x $Mn_{0,02}Fe^{2+}_{1,18}Fe^{3+}_{0,40}Cr_{0,04}Al_{0,22}Si_{13,26}O_{50}$.

These elements fractionate into the residual fluid phase during final, slower crystallization of granular periodotites and this depletion seems to due to the fluid migration. The two groups of peridotite differ by their $Fe_2O_3/(FeO+Fe_2O_3)$ which is higher in granular peridotites whose crystallization temperatures are lower (800-1000°) relative to porphyritic peridotites (1100-1300°C). The pyroxenes and garnets from granular periodotites are depleted in titanum and enriched in chromium because aluminium gets lost from rocks through migration when the intergranular liquid starts to separate (4).

The porphyritic structure of peridotites is a feature of subvolcanic rocks and reflects at least two stages in their crystallization: deep in the upper mantle and within the earth's crust following their intrusion into pipe-like structures. There are many indications to the upward intrusion of diamondiferous ultramafic magmas, such as zoning in phenocrysts from peridotites. Many authors noted the presence of zoned crystals in high-pressure minerals from peridotites. The calcium content of the garnet in the core of such crystals is generally lower than in the rims. This alteration reflects the main trend in the garnet evolution from the dunite to harzburgite, lherzolite and wehrlite types. The porphyritic peridotites testify most convincingly to the upward intrusion of peridotite magmas containing intratelluric phenocrysts. A specimen of these rocks from the Udachnaya pipe was found to contain violet and orange garnet phenocrysts of variable composition (13). This suggests that the telescoping of phenocrysts, common for volcanic formations, had taken place as they were settling in the uprising ultrabasic magma.

The origin of kimberlites is associated with some unusual deep-seated process of peridotite fluidization (14). The resulting kimberlite magmas entrap the remaining peridotites (nodules) and other rocks (fig.2) together with their desintegrated grains and high-pressure mineral intergrowths (including diamond). Composition of kimberlites as recalculated to the anhydrous material for major rock-forming acides closely approximates that of the peridotites from which kimberlite magmas are formed on substitution.
Below are given the data for the Mir pipe in Yakutia:

peridotite - $K_{0,08}Na_{0.34}Ca_{1,57}Mg_{14,39}Ni_{0,04}Mn_{0,04}$ x

$Fe^{2+}_{2,05}Fe^{3+}_{0,97}Al_{1,43}Ti_{0,15}Si_{13,85}P_{0,04}O_{50}$;

kimberlite - $K_{0,18}Na_{0,13}Ca_{2,15}Mg_{15,89}Ni_{0,05}Mn_{0,02}$ x

$Fe^{2+}_{0,87}Fe^{3+}_{1,65}Cr_{0,04}Al_{1,07}Ti_{0,38}Si_{12,87}P_{0,09}O_{50}$.

Kimberlite only differs in that it contains iron in a higher oxidation state and is richer in potassium, phosphorus, calcium and other elements (rare earths, alkalis, zirconium, fluorine etc.) which tend to accumulate in residual magmas and subsequently pass into fluids giving rise to the kimberlitization of peridotites. The fluid flows which produce the replacement of peridotites by kimberlite magmas must have been derived from the same deep-seated hearths as peridotites and eclogites, though there is a considerable age difference between them. Extensive studies of diamondiferous pipes have revealed close links between these rocks and also the presence of intermediate varieties - phlogopite peridotites, biotite eclogites, glimmerites and other rocks bearing primary calcite, apatite, zircon, serpentine etc. Biotite, phlogopite and zircon have been identified as inclusions in diamonds. Magma hearths of this type must have taken a long time to develop-starting in precambrian times through the paleozoic to the mesozoic, the kimberlite volcanism being the final stage in their evolution. The same duration is typical of the development of continental rift structures (12).

Basalt magmas entrapped nodules as they were replacing intrusive peridotites and associated rocks (fig.2). In this process, the alkalinity of magmas was increasing owing to the acid-base interaction of components. For this reason, peridotite nodules only occur in the alkali-type basalt volcanic series. Rocks found in nodules from kimberlites and basalts are similar petrochemically but they commonly differ from one another in their mineral composition and mineral parageneses, representing intrusion magmatic series associated with magma hearths at different depths. The early mineral assemblages from these rocks reflect the crystallization conditions at the greatest depths. They give an idea of the maximum depths "sampled" by magma hearths: those associated with kimberlite pipes may attain the horizon of the diamondiferous mantle (III. fig.1), while under basalt volcanoes they reach but seldom depths great enough for pyrope peridotites to develop (II. fig.1), being confined commonly to facies I (fig.1). However, the maximum depths of magma hearths will not be determined if no crystallization occurred in them (the temperature remained above the liquidus) or if the early crystallization products were completely altered by reactions with residual fluid melts during their upward intrusion or in intermediate hearths (in the mantle and crust). The reaction relations of minerals from peridotites have fixed the stages of magmatic crystallization that occurred in these hearths or on the way of magma to the surface. In this process, mineral parageneses of peridotites were consecutively confined to facies III-II-I (fig.1).

Later mineral assemblages in the groundmass or in the interstices of the earlier-formed mineral grains which are often altered or have kelyphite rims reflect the conditions at which gradual crystallization and final consolidation of magmas occurred at different levels in the mantle and crust. The depth at which ultrabasic and basic rocks were formed in layered and other intrusions (before they were trapped as nodules in kimberlites or alkali basalt magmas) can be deduced from mineral assemblages. The spinel-enstatite assemblage is a reliable indicator for ultrabasic rocks (fig.3). At pressures of the order of 1500 MPa (close on 50 km), these minerals react to yield olivine-garnet (pyrope) assemblages. However, with spinel becoming richer in chrome (i.e. aluminium decreases), the pressure of the reaction steadily increases which is reflected in the diagram as a series of isolines for the Al_2O_3 content in the spinel from garnet peridotites of facies II and III in fig.3. Within the diamondiferous facies III, the only stable spinels in garnet peridotites are those with Al_2O_3 less than 10%.

The diagram also shows the compositional variations in the peridotite enstatites with temperature and pressure, after Newton's estimates (10). Similar relationships were obtaines by Perchuk et al. (11) and by other authors. If peridotites contain no garnet, the Al_2O_3 of the enstatite depends little on pressure, being mainly the function of the temperature (facies I, fig.3). In the facies II peridotites, the alumina content of enstatite decreases rapidly with increasing pressure (that is, with increasing depth of crystallization of ultrabasic magmas), to expand the garnet stability field in these rocks. Enstatites with less than 1% Al_2O_3 are the only stable phase in garnet peridotites from facies III.

Fig.4 presents the isothermal sections of the diagram in fig.3 (for 800^0, 1000^0 and 1200^0C) and shows that the equilibrium chrome-spinel and enstatite from garnet peridotites exhibit the same pattern of compositional variations: with increasing pressure, both minerals become depleted in alumina which passes into the garnet composition. In the diagram (fig.4) the pressure increases in the general case along the isotherms towards the origin. The compositional fields of these minerals are shaded in this diagram according to (2) where the enstatite-chromespinel assemblages in peridotites are grouped with the inclusions in basalts (4, fig.4) and in kimberlites. The latter contain both early and late mineral assemblages (6 and 5 in fig.4, respectively). The early mineral assemblage in peridotite nodules from kimberlites is represented by mineral inclusions in diamonds and porphyritic minerals in peridotites, with spinel occurring as octahedral crystals 0,1-1,0 mm in size (2). The later-generation minerals in the garnet peridotites from diamondiferous pipes include the chromespinel - enstatite assemblages that have crystallized from the residual magmatic melts between the early mineral grains (granular peridotites) or in the groundmass of porphyritic rocks, often in the form of symplektic eutectoid intergrowths. In the same group are found pyroxene-spinel assemblages with an exsolution structure, the position of the spinel being controlled by the crystallography of host enstatite grains. "The spinels in the groundmass of the same rocks have almost identical composition with the exsolved spinels" (2,p.933). The consistent compositional changes that occurred in the enstatites and spinels on transition from early to late mineral assemblages (6 ⟶ 5, fig.4) reflect a marked drop in the formation pressure of garnet peridotites in kimberlite pipes, supporting their intrusive origin. According to (2), mineral assemblages in peridotites form a continuous compositional series (shaded areas 6 - 5 - 4, fig.4) ranging from the nodules in peridotites to the nodules in basalts.

Additional data are summarized in the diagrams in figs.3 and 4. In general, the peridotites from kimberlite pipes vary greatly in the Al_2O_3 content of the enstatite of garnet peridotites from a pipe in Lesotho (6) ranges from 0,69 to 3,58%, the variability within one sample being 1,48-3,58%; the Al_2O_3 content in the spinel varies accordingly (43,56--49,91%). The kelyphite rims produced at the final stages contain orthopyroxenes highly enriched in Al_2O_3 (7,52-8,72%). All these compositional variations occur on transition from facies III associated with the earliest mineral assemblages in the peridotites from kimberlite pipes (including those in diamond) to facies II and I associated with typical peridotite mineral assemblages that form nodules in alkali basalts.

Diamondiferous peridotite magmas were so rich in carbon gases (CH_4, CO etc.) that the diamond along with the earliest intratelluric minerals (phenocrysts) were the first to have crystallized. At later stages, during the upward movement of peridotite magmas or in intermediate chambers in the mantle and crust these minerals underwent magmatic resorption and reaction replacement by later-generation minerals. The minerals found as inclusions in diamonds differ from those in peridotite nodules in that they are richer in MgO and Cr_2O_3 and more depleted in CaO, Al_2O_3, FeO, Na_2O and K_2O.

The compositions of the enstatite-spinel assemblages in the peridotites from kimberlite pipes bear some indications that during the peridotite crystallization they were shifting from the deepseated diamondiferous to a shallower facies, transitional to the one associated with the graphite-bearing garnet and enstatite-spinel peridotites that occur in the subalkali and alkali basalt nodules. Even the earliest mineral assemblages found in nodules, exhibit a high degree of compositional correlation with the rocks constituting the magmatic series (fig.2). This suggests the liquid (pre-crystallization) layering to have occurred in magma hearths, primary and intermediate alike, at different horizons in the mantle and crust. The basite-hyperbasite layering predominated throughout the early differentiation of magma hearths to have produced stable assemblages of garnet peridotites with eclogites or spinel peridotites with spinel-pyroxenites and gabbros in nodules. As with peridotites, diamond-bearing eclogites contain various-pressure mineral assemblages (8), associated with the deep-seated and shallower crystallization conditions. The pressure range is particularly wide in the porphyritic-type diamond-bearing eclogites. Diamond crystallizes out of ultrabasic (peridotitic) and basic (eclogitic) magmas saturated in hydrocarbon gases. Analyses of gas-liquid inclusions in crystals having different habitus 9 suggest that the peridotite and eclogite types of the diamondiferous petrogenesis had considerably different fluid regimes. In ultrabasic, highly reducing environments, the diamond crystallized by the reaction $CH_4 + 2\ CO = 3\ C$ (diamond) $+ 2\ H_2O$. In basic magmas crystallization of diamond proceeded by other scheme: $CH_4 + CO_2 = 2\ C$ (diamond) $+ 2\ H_2O$.

The gases in the clearly eclogite-type diamonds (cubic crystals) are more strongly oxidized (virtually no CO is present). In the course of magmatic layering, there was a distinct division of fluid components into the reduced and oxidized groups.

Consequently, the study on peridotites and associated rocks which occur in kimberlites and basalts as nodules though they do not represent the mantle material has provided valuable information on the deep-seated structure of continental and oceanic regions of the earth, and on the thermal regime of their development.

No direct evidence is available on the deep-seated (subcrustal) matter of the lithosphere, and the current conceptions are based on computations and analogies with meteorites. Models suggested for lithosphere structure under the continents and oceans differ to reflect the differences in thermal regimes in these areas. The upper mantle is subdivided into the graphite-bearing and diamondiferous zones, the latter being the nearest the surface under the continents (occurs at the depth of 150 km). Diamonds in mantle rocks are believed to fill interstices between olivine grains in mixture with the nickel-iron and sulphide phases, cohenite and wustite. Nodules of peridotites and eclogites from kimberlites and basalts, including the diamondiferous ones, do not represent mantle rocks, contrary to the widely held belief. These are intrusive rocks which form regular magmatic series associated genetically with the differentiation of fluid magmas in chambers, the deepest of which must have originated within the diamondiferous mantle under the continents. Several mineral generations can be traced in rocks, with the earliest (diamon among them) having crystallized in the deep-seated zones, and subsequent generations - within the higher levels in the mantle and crust. The mantle crystalline matter could be partially preserved in magmas to spike the diamond crystallization as is evidenced by parageneses of uncommon minerals observed by Bulanova in diamond crystal cores (nickel iron, cobalt-nickel troilite, iron carbide, copper wustite and others) and totally different from mineral inclusions in diamond in that they represent mineral assemblages of intrusive peridotites and eclogites. Kimberlite magmas are formed by fluidization and replacement of diamondiferous peridotites and therefore contain remnants (nodules) of these rocks and their desintegrated minerals. Kimberlites are a member of the rock series: kimberlites and carbonatites - wolgidite and cedricite - alnoite - olivine melilite basalts.

REFERENCES

1. Afanasjev G.D. Magmatic formations and general problems in geologic petrology (Selected papers). - M.: Nauka, 1981, 514 p.

2. Basu A.R., MacGregor I.D. Chromite spinels from ultramafic xenoliths. - Geochim. et Cosmochim. Acta, 1975, vol.39, p.937-945.

3. Bott M. Interior of the Earth. - M. : Mir, 1974, 374 p.

4. Boyd F.R., Nixon P.H. Structure of the upper mantle beneath Lesotho.- Carnegie Inst. Year Book 72, 1973, p.431-445.

5. Deep-seated xenoliths and the upper mantle./ Ed. V.S. Sobolev. - Novosibirsk: Nauka, 1975, 269 p.
6. Lock N.P., Dawson J.B. Garnet-olivine reaction in the upper mantle: evidence from peridotite xenolith in the Letseng-La-Terae kimberlites, Lesotho. - In: Transactions of the Royal Soc. of Edinburgh: Earth Sci., 1980, vol.71, p.47-53.
7. Marakushev A.A. On the volcanic origin of diamondiferous rocks from kimberlite pipes. - Geol. and Geophys., 1982, N 8, p.16-17.
8. Marakushev A.A., Botkunov A.I., Shcheka S.A. On the problem of the genesis of diamondiferous plagioclase-cearing rocks. - Mineralogicheskii Zhournal, 1982, N 4, p.14-22.
9. Melton C.E., Giardini A.A. The composition and significance of gas released from natural diamonds from Africa and Brasil. - Amer. Mineral., 1974, vol.59, N 7, P.775-782.
10. Newton A.C. Thermodynamics in the geothermometry and geobarometry of peridotites. - In: Phys.-chem. problems of petrol., vol. 1. M.: Nauka, 1979, p.103-117.
11. Perchuk L.L., Aranovich L.Ya., Kosiakova N.A. Thermodynamic models of generation and evolution of basalt magmas. - Vestnik Moskovskogo Universiteta. Ser.4, Geology, 1982, N 2, p.3-25.
12. Phipotts A.R. Rift magmatism in the Eastern North America. - In.: Continental Rifts. M.: Mir, 1981, p.100-117.
13. Pokhilenko N.P., Sobolev N.V. Garnets of variable composition in a sheared lherzolite sample from the Udachnaya kimberlite pipe. - In.: XI IMA Meeting, vol.2. Novosibirsk: Nauka, 1978, p.9-21.
14. Ryabchikov I.D. The nature of kimberlite "magmas". - Geol. rudnykh mestorozhd., 1980, N 6, p.18-26.
15. Sobolev N.V. Deep-seated inclusions in kimberlites and composition of the upper mantle. - Novosibirsk: Nauka, 1974, 264 p.

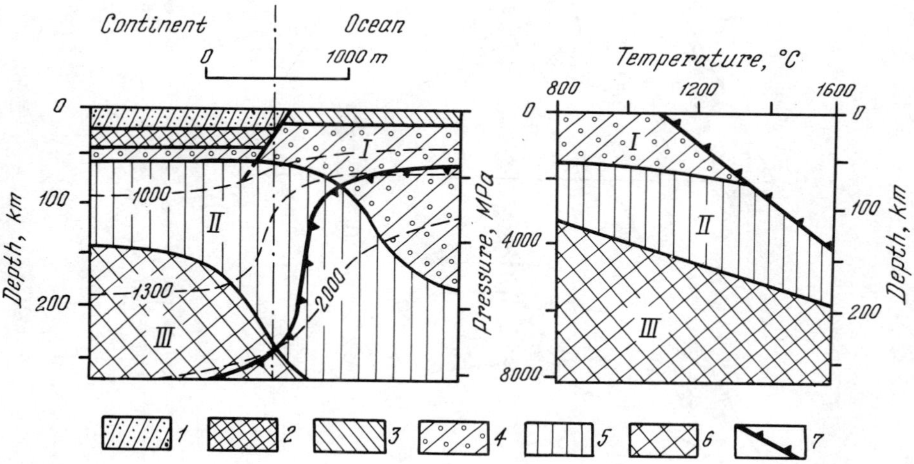

FIGURE 1. Schematic distribution of temperatures and rock mineral facies in deep-seated continental and oceanic zones. 1-2 - continental crust (1- granite-gneiss layer, 2- basic granulite layer); 3- oceanic basalt crust; 4- spinel peridotites with graphite; 5- garnet peridotites with graphite; 6- garnet peridotites with diamond; 7- lithosphere boundary (determined from the peridotite solidus). The temperature isotherms (T^0C) are based on MacDonald's calculated model (3). I-III - depth mineral facies: enstatite-spinel (I), pyrope-enstatite-chromespinel graphite-bearing (II) and diamondiferous (III)

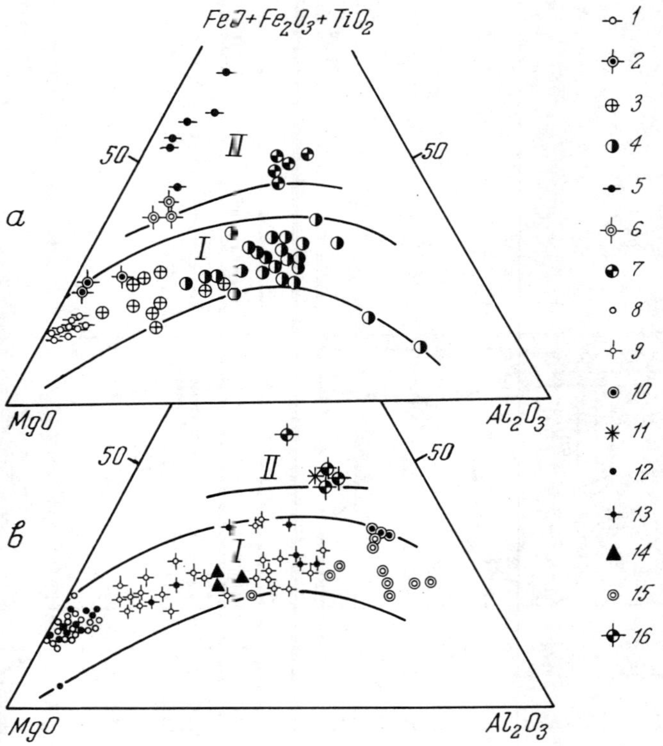

FIGURE 2. Petrochemical diagrams (mass.%) for ultrabasic and basic rocks forming nodules in kimberlites (a) and in alkali basalts (b). 1-7 - nodules in kimberlites: 1- garnet dunites and peridotites; 2- their varieties with ilmenite and phlogopite; 3- garnet pyroxenites and their varieties with ilmenite and phlogopite; 4- eclogites and kyanitic eclogites; 5- phlogopite-ilmenite peridotites; 6- phlogopite-ilmenite pyroxenites; 7- rutile eclogites; 8-11- nodules in alkali basalts and gabbro-dolerites on platforms: 8-spinel dunites and peridotites; 9-spinel and garnet pyroxenites; 10-camptonites; 11-ilmenite pyroxenites; 12-16- nodules in alkali basalts of islands and island arcs: 12-spinel dunites and peridotites; 13-spinel pyroxenites; 14-garnet pyroxenites? 15- gabbro; 16- ore titanum-augite gabbro. I-II - magmatic series for magnesian-alumina (I) and iron-titanum (II) rocks

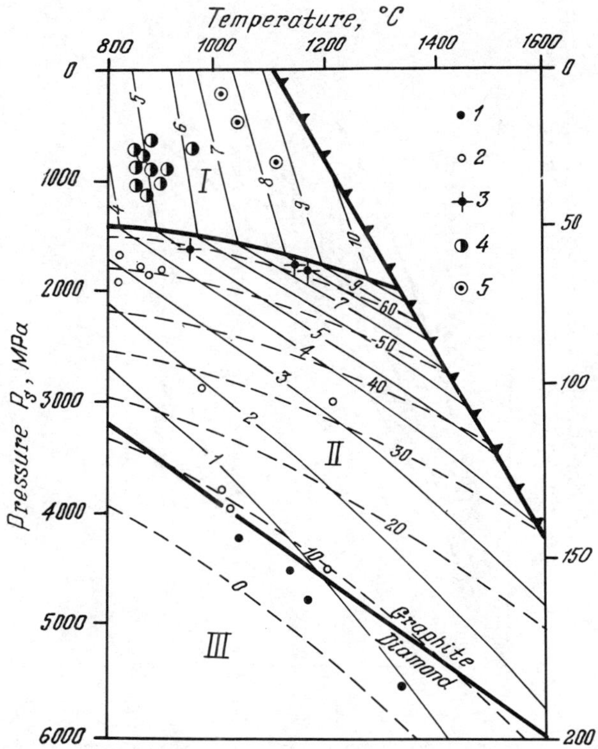

FIGURE 3. The equilibrium garnet (Gr)+forsterite (01)= orthopyroxene (Opx)+clinopyroxene (Cpx)+spinel (Spl) and the Al_2O_3 content (mass.%) in orthopyroxene (the solid isolines 1-12 are from (10)) and in chrome-spinel (the dashed isolines 0-60). I-III - mineral ficies: spinel peridotites (I), garnet peridotites with graphite (II) and garnet peridotites with diamond (III) as in fig.1. 1- crystalline inclusions in diamond; 2-nodules of garnet peridotites in komberlites; 3-nodules of garnet peridotites in basalts; 4-nodules of spinel peridotites in basalts; 5- kelyphite rims in nodules of garnet peridotites from kimberlites

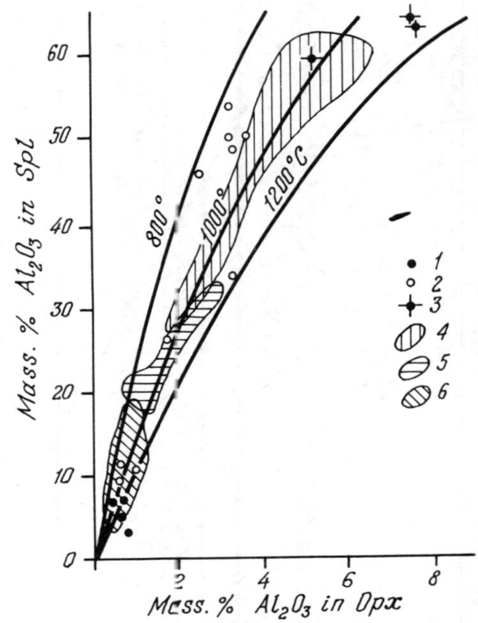

FIGURE 4. Concomitant changes in the chromespinel (Spl) and enstatite (Opx) compositions in garnet peridotites with temperature (800, 1000 and 1200°C - equilibrium isotherms are as in fig.3). 1-2 - in kimberlites: 1- crystalline inclusions in diamonds; 2- minerals from peridotite nodules; 3- nodules in alkali basalts; shaded areas are for the compositions of enstatite and chrome-spinel in ultrabasic nodules from basalts (4) and kimberlites (5- later-generation minerals, 6- early-generation minerals) from (2)

PRINCIPLES OF IDENTIFICATION OF POTENTIAL ORE-BEARING BASIC AND ULTRABASIC IGNEOUS ROCKS

MIKHAILOV N.P., MOSKALEVA V.N.
A.P. Karpinsky All-Union Geological Research Institute, Leningrad, USSR.

One of the central problems in studies of magmatic geology and metallogeny is the relation of endogenic ore deposits and magmatism. It has been examined since long time in different aspects and many scientific schools made their contributions into it. Soviet geologists proposed a formational analysis of magmatic associations and for the last two decades this method acquired an exceptional importance in the study of regularities of relation of endogenic ore mineralization and magmatic rocks. It is attempted to study associations of magmatic rocks, their relations with tectonic evolution of large blocks of the Earth's crust, regularities in their space-time occurrences on the scale of geologic history (Scientific schools of All-Union Geological Research Institute, Leningrad; Institute of geology of ore deposits, petrography, mineralogy and geochemistry, Moscow, Institute of geology and geophysic, Novosibirsk).

The studies in nature of relations of ore formations with different associations of magmatic rocks (magmatic formations - the term used by the Soviet geologists) is the principal practical aspect in advances of petrographic-metallogenic investigations.

Synthesis of abundant materual dealing with characteristics of certain associations of magmatic rocks (magmatic complexes), correlation of magmatic complexes of the same type that occur in different geologic epochs and in different regions and retaining principal peculiarities of their petrographic composition, structure and relation to the environment, permit to state that every type of magmatic formations is characterized with a specific potential ore concentration (the metallogenic specialization) and this ore concentration may be represented either by one or by several types of ore mineralization that generated in the process of formation of magmatic associations and their subsequent transformation / 5, 6 /.

Proceedings of the 27th International
Geological Congress, Volume 9, pp. 343–351
PETROLOGY (IGNEOUS AND METAMORPHIC ROCKS)

Specially important in correlation of magmatic formations and endogenic mineralization types is the convergence of the latter, that means ore paragenesis of the same type may be registered in different magmatic formations. This convergence of ore paragenesis is explained when we correlate the ore mineralization types and ore-bearing (ore generating) magmatic rocks, that are heterogenous to a certain extent being of different genesis and of different occurrence in natural associations (formations). This fact, on the hand, makes more difficult the recognition of distinct petrological criteria in potential and real ore concentration in magmatic associations and, on the other hand, it emphasizes the significance of correlation of ore mineralization types and composition of the associated magmatic rocks.

The convergent aspects of heterogenous magmatic formations become appreciated when a complex of petrographic methods are applied, petrochemical studies mostly useful in volcanic rocks analysis inclusive. The mineral composition, however, is the most important in correlation of intrusive rocks of the same types but of different magmatic formations. And the modern classification and nomenclature of intrusive rocks is based on the mineral composition / 2 /. It is quite important for the associations of basic and ultrabasic rocks in which occur ore deposits of mainly magmatic origin. A direct and distinct correlation of ore components and enclosing magmatic rocks is fairly well known for these deposits.

Associations of basic and ultrabasic rocks are formed within a wide range of deep-facies of magmatism, the latter reflecting certain levels in the Earth's crust with definite pT-conditions and dominant forms of magmatic activity.

Commonly associations of basic and ultrabasic rocks of normal alkalinity (alpin-type dunite-peridotite, pyroxenite-peridotite, peridotite-orthopyroxenite-norite, dunite-clino-pyroxenite-gabbro, gabbro-anorthosite, syenite-gabbro formations) generate in mesoabyssal facies (5-10 km) where pT-conditions are favourable for the processes of magma generation (in the zones of high heat flows) and for the emplacement and crystallization of abyssal magmatic melts. Under these conditions the outer lithostatic pressure of surrounding medium exceeds the inner pressure in intrusive bodies and favous intra-intrusive magmatic differentiation, generation of camera and vein pegmatite, interaction of melts and fluids in the magmatic phase and formation of ore concentrations inside the intrusive massives.

The hypabyssal facies displays a depth range of 1,5-5 km where are lacking the processes of magma generation and regional metamorphism; it is a level of solidification of allochthonous magmatic mass with its significant enrichment in volatile components. Here the intra-intrusive pressure exceeds the outer lithostatic pressure and it favours formation of hydrothermal and hydrothermal-metasomatic ore concentrations in the zone above intrusions during evolution of post-magmatic process. Fluid regime develops towards increase of alkalinity and that is why appear widely distributed alkaline magmatic associations which are not typical of facies of a greater depth. This facies does not contain intrusive ultramafic formations of normal alkalinity; the ultramafic rocks are the extreme members of essentially gabbroic (gabbro-diabase, gabbro-wehrlite), alkaline-ultramafite and alkaline-gabbroid formations / 3 /. All of the intrusive formations are typified with spatial and genetic relations with volcanic formations of similar composition and commonly they together form volcano-intrusive assoviations.

The perographic and metallogenic aspects of intrusive associations of basic and ultrabasic rocks depend on the depth (and on the pT-conditions correspondingly) of their generation. Within each intrusive formation are distinctly recognized the typomorphic (indicator) ore-bearing magmatic rocks for definite ore components. And the ore-bearing rocks that make up components of different magmatic formations generated under different depth facies in the crust reveal common ore components but locally differ in genetic types of concentrations of the latter.

All basic and ultrabasic intrusive formations of normal alkalinity are divided into two groups according to their chemical and mineral composition and conformably these differ in metallogenic specialization.

In accordance with petrochemical properties a group of essentially magnesian formations is established, it is characterized with ore mineralization of chrome and platinoids, and other group of magnesian-ferriferous formations mainly typified with iron-titanium mineralization with subordinate platinoid and chromite mineralization of a wide compositional range of ore minerals. The differences of these groups of intrusive formations are more clearly seen through their petrographic-mineralogical properties. Diagram-model of mineral composition of main rock types of these formations (Fig. 1) uses as its basic the petrographic-mineralogic characteristics of many intrusive complexes and it shows clearly the significance of mineral composition of magmatic rocks in particular the significance of a pyroxene component in determining their potential ore content. On the diagram it is distinctly plotted two groups of associations of basic and ultrabasic rocks according to their mineral composition: a) essentially orthopyroxe-bearing (on the left and upper parts of diagram) built up of the alpin-type dunite-peridotite, peridotite-orthopyroxenite-norite and partially pyroxenite-peridotite formations; b) essentially clinopyroxene-bearing (on the richt and lower parts of diagram) made up of dunite-clinopyroxenite-gabbro, gabbro-anorthosite, wehrlite-gabbro, syenite-gabbro, gabbro-diabase formations.

These two groups of magmatic formations are of different depth of generation and of different potential ore content (Table). Thus, chromite and platinum mineralization is characteristic only of orthopyroxe-bearing group and not typical of the clinopyroxe-bearing one. And the occurrence of ores of different quality and chemism is related to definite types of ultramafites. In the alpin-type dunite-peridotite formation the high-chrome magnochromites associate with dunite, lower chromic ore - with harzburgite, rarely with orthopyroxenite and lherzolite; the latter contains magnetite parallel with chrome-spinel when abundance of clinopyroxene (diallage) grows. The pyroxenite-peridotite formation is typified with the same types of ore-bearing rocks but they have higher content of iron and titan; they contain ores of low chrome and higher iron concentration that are regarded as transitional to titan-bearing chrome-magnetite. In the stratiform intrusions of the peridotite-orthopyroxenite-norite formation the major potentially chromite-bearing rocks are also high magnesian dunites and harzburgites; besides, in the stratiform intrusions of this type the cumulative beds of chromitite are known from orthopyroxenite (bronzitite), from norite and even anorthosite / 1 / and that is not recognized about the chromite ores from any other magmatic formations.

The regular attachment of chromite deposits only to the orthopyroxene-bearing rocks and their absence in clinopyroxene associations can be explained by the fact that lower temperature and low pressure mafic and ultramafic associations generating in the hypabyssal facies are characterized by a higher alkalinity of their fluid regime. It results in crystallization of clinopyroxene mainly and to simultaneous supplanting of chrome-spinellid by magnetite (even in accessories).

The magmatic deposits of iron-titan ores usually associate with ultramafites and gabbroids of clinopyroxebearing formation. And there exists a definite correlation of ore aspects with clino-pyroxene composition in orebearing rocks. Magnetite low titan ore deposite (commonly vanadium-bearing) associate with the clinopyroxene ultramafite and gabbroid that contain diopside (diallage) or diopside-hedenbergite as a typomorphic mineral. High titanic titanomagnetite and ilmenite ores (with apatite) occur in gabbroid formations of normal and higher alkalinity (gabbro-diabase, wehrlite-gabbro, syenite-gabbro, gabbro-anorthosite, alkaline-gabbroic) and the orebearin rocks are wehrlite, clinopyroxenites and gabbro with augite or titanaugite as a typomorphic rock-forming mineral. It also pints to a higher alkalinity in the fluid regime and to a progressive solution of iron and titanium in silicates / 4 /.

Uncertain is the correlation with typomorphism of rockforming pyroxene in ore-bearing rocks enclosing sulfide coppernickel deposits, these latter occur in mafite-ultramafite associations with orthopyroxene as well as with essentially clinopyroxene-bearing associations and it becomes clear because of absence of correlation between Cu-Ni and Ca. Nevertheless, even in this case the ore composition depends on the nature of wall rocks. The relations of Ni:Cu ration with basicity and maficity of ore-bearing rocks are well known: the ores with dominance of Ni over Cu (2:1 and higher) occur in ultramafites enriched in olivine; and the ores associated with gabbroids demonstrate the inverse ratio. The environment of generation, that is the saturation with sulphur, is of great importance in genesis of sulfide copper-nickel mineralization, and the realtion of saturation with the alkalinity grade is reverse in this case: high alkaline medium with high pH is not favourable for reaction of sulphur. That is why we have no great sulfide copper-nickel deposits in alkaline-ultramafite and alkaline-gabbroid formations. The genetic types of mineralization in these formations are added due to pegmatite formation, to phlogopitization, carbonatite generation. The contrast in composition of alkaline-ultramafite and alkaline-gabbroid formations (from ultramafite to salic foid syenite) that were generated in hypabyssal facies, affects concentration of ore components in them: the titanomagnetite ore with perovskite accumulates in ultramafic rocks (olivinite, dunite with clinopyroxene, clinopyroxenite); as the salic foid-syenite rocks are mainly typified with apatite and rare-earth mineralization; apatite and iron, rare metal and rare-earth mineralization is located in carbonatites and the zones of phlogopitization enclose vermiculite deposits.

Thus we have a correlation of composition of ultrabasic and basic rocks with qualitative-quantitative composition of the associated ore paragenesis for the magmatic (and partially for permatite and carbonatite) ore concentrations of Cr, Pt, Ni, Co, Cu, Fe, Ti, V, P, REE, TR. This correllation is recognized not only at the level of species but also at varieties of rocks (that is when composition of rock-forming minerals varies). This accordance in composition of ore-bearing rocks and ore paragenesis is a convincing evidence of genetical relations of the above mentioned mineralizations and the processes of generation of magmatic rocks that enclose these deposits. The correlations of petrogenetic and ore-forming factors are reflected in objectively established features of composition and structure of magmatic formation and these become criteria of its potencial presence of ore.

The regularities in relations of ore paragenesis and

magmatic formations examined above are the base for the elaboration of methods of identification of potentially ore-bearing basic and ultrabasic magmatic rocks.

Convergence of ore concentrations and convergence of ore-bearing rocks in different magmatic formations gives preference to petrological criteria in evaluation of perspective ore presence in a given magmatic complex.

When modern methods of formational analysis are applied in exploration of mineral deposits they give us a reliable determination of magmatic associations from the point of view of their belonging to a certain magmatic formation, then they permit to establish species and varieties of typomorphic (indicator) rocks in terms of ore-generating or ore-enclosing for a given type of mineralization and all this helps us in progress studies of substantially based and specific forecasting of endogenic mineral deposits.

REFERENCES

1. Cameron E.N., Desborough G.A. - Occurrence and characteristics of chromite deposits - Eastern Bushveld complex. In: Magmatic ore deporits. Economic Geology, Monograph 4, 1969, pp. 23-41.
2. Classification and nomenclature of magmatic rocks. Ed. by O.A. Bogatikov, N.P. Mikhailov, V.I. Gon'shakova, M.,Nedra, 1981, 159 pp.
3. Magmatic formation of the USSR. V.1-2. Ed. by V.L. Masaitis, V.N. Moskaleva, N.A. Rumyantseva, L., Nedra, 1979, 318 pp, 279 pp.
4. Marakushev A.A. Petrogenetis and ore formation (Geochemical aspects). M., Nauka, 1979, 263 pp.
5. Principles and methods in evaluation of in geological formations. Issue 2. Magmatic formations. Ed. by V.L. Masaytis, V.N. Moskaleva, N.A. Rumyantseva. L., Nedra, 1983, 259 pp.
6. Ore content and geological formations of the Earth's crust structures. Edited by D.V. Rundquist, Leningrad, Nedra, 1981, 421 pp.

TABLE 1. Correlation of ore-components and mineral composition of ore-bearing rocks of ultramafic and mafic formations

Deep facies of magmatism	Magmatic formations and their ore-bearing rocks	Typomorphic rockforming minerals	OLIVINE		ORTHOPYROXENE		CLINOPYROXENE			
			fo	ol,hort	enst	br,hyp	di,hed	di-aug	aug	ti-aug
HYPABYSSAL	Gabbro-diabase	clinopyroxenites							Fe,Ti,Zn, Pb,Cu,Ni	
		gabbro-diabases							Fe,Ti,Zn, Pb,Cu,Ni	
	Wehrlite-gabbro	wehrlites						Cu,Ni		
		gabbro							Fe,Ti	
MEZOABYSSAL	Seynite-gabbro	clinopyroxenites								Fe,Ti,ap
		gabbro,leucogabbro							Fe,Ti,ap	
	Gabbro- -anorthosite	clinopyroxenites							Fe,Ti,ap	
		gabbronotites							Fe,Ti,ap	
	Dunite- -clinopyroxenite- -gabbro	dunites	Pt,Pd	Pt,Pd						
		clinopyroxenites					Fe(Ti)			
		gabbro					Fe(Ti)			
MESOABYSSAL	Peridotite- -orthopyroxenite- -norite	olivinites, dunites		Cu,Ni,Pt, Cr,Pt						
		harzburgites				Cu,Ni				
		orthopyroxenites				Pt,Cu Ni,Cr				
		norites, gabbronorites				Cu,Ni,Cr	Fe(Ti)			
	Pyroxenite- -peridotite	dunites		Cr						
		olivinites		Cu,Ni						
		orthopyroxenites				Cu,Ni,Cr				
	Dunite-peridotite (alpine type)	dunites	Cr,Ir,Os							
		harzburgites			Cr					

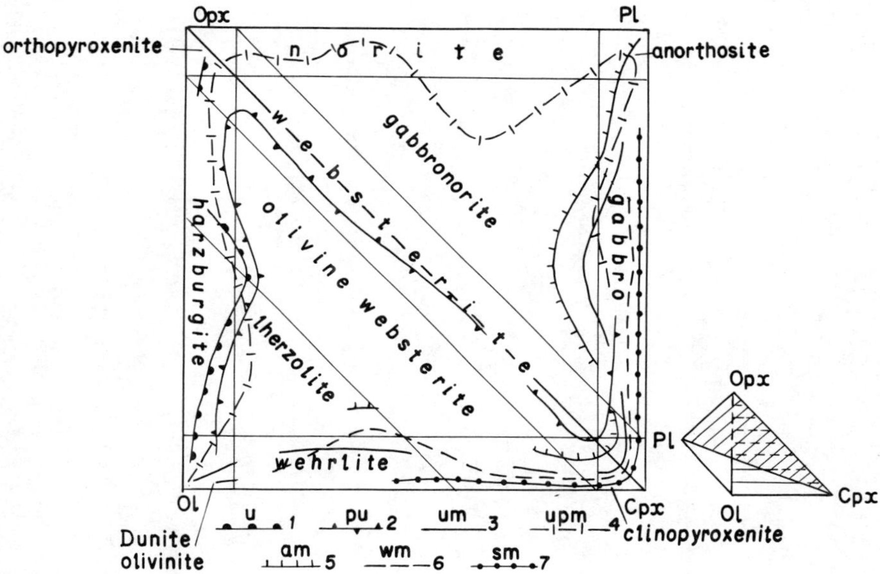

FIGURE 1. Diagram-model of petrographo-mineralogical composition of main intrusive formation of basic and ultrabasic rocks (projection of unfolded tetrahedron Ol-Opx-Cpx-Pl on the plan). Intrusive formations: 1 - alpintype dunite-peridotite; 2 - pyroxenite-peridotite; 3 - dunite-clinopyroxenite-gabbro (concentric complexes); 4 - peridotite-orthopyroxenite-norite (stratiform or layered complexes); 5 - gabbro-anorthosite; 6 - wehrlite-gabbro; 7 - syenite-gabbro.

EXPERIMENTAL STUDIES OF THE ROLE OF FLUORINE AND BORON IN THE FORMATION OF LATE-STAGE GRANITIC ROCKS AND ASSOCIATED MINERALISATION

D.A.C. MANNING
Department of Geology, The University,
Newcastle upon Tyne, NE1 7RU, U.K.
M. PICHAVANT
Centre de Recherches Pétrographiques et Géochimiques,
CNRS, BP 20, 54501 Vandoeuvre-lès-Nancy, France.

ABSTRACT

Fluorine and boron are believed to play a particularly important role in the petrogenesis of volatile-enriched late-stage granitic rocks and their associated ore deposits, especially tin and tungsten. Recent experimental studies have extended earlier work using natural materials to determine the effect of fluorine and boron separately on phase relationships in the hydrous 'synthetic granite' system, and can be used to model the behaviour of F and B-enriched residual magmas. Liquidus and solidus temperatures are greatly reduced (at 1 kbar water pressure) to 630°C with 4 wt% F and to below 660°C with 4,5 wt% B_2O_3. The minimum composition is displaced in both cases towards the albite apex, but this effect is less marked for the boron-bearing system than for the fluorine-bearing system. Magmatic differentiation producing F or B-enriched granitic magmas may consequently lead to normative albite-enriched residual melts, particularly in the case of F-rich systems. Melt-fluid and crystal-fluid interaction processes are strongly dependent on the volatile species involved (F,B,Cl), as illustrated by alkali exchange equilibria between feldspars and aqueous solutions with fluorine or borate anions. Similarly, the partitioning of tungsten between melt and vapour shows that fluorine-rich melts are able to transport and concentrate metals, whereas chlorine-bearing fluids may effectively extract metals from magmas.

Proceedings of the 27th International
Geological Congress, Volume 9, pp. 353–372
PETROLOGY (IGNEOUS AND METAMORPHIC ROCKS)

INTRODUCTION

Of all the 'volatile' elements which may be present during the formation of late-stage granitic rocks and associated mineralisation (especially Sn and W) fluorine and boron are believed to play a particularly important role in petrogenetic and ore genetic processes. Both elements are commonly (and usually separately) enriched in late-stage granitic rocks associated with pegmatitic and high temperature Sn-W mineralisation, and are expressed mineralogically as topaz, lithium micas and tourmaline. Fluorine-rich granites commonly contain 1-2 wt% F and exceptionally reach 3,5 wt% F (as in ongonites), while boron-rich granites average 0,2-0,3 wt% B_2O_3 with maximum values up to 1 wt% B_2O_3 (Pichavant and Manning, 1984). The role of boron and fluorine in post-magmatic processes is called upon in greisening and tourmalinisation, with the formation of secondary topaz, lithium micas and tourmaline, and, with the introduction of calcium, the development of fluorite. The combination of both mineralogical and experimental studies is essential in order to more fully understand the behaviour of both elements in rock forming processes. Experimental studies follow two approaches - 1) studies of the effect of fluorine and boron on phase relationships in the granite system and how these might control the behaviour of F and B-enriched residual magmas; and 2) studies of the effect of fluorine and boron on element redistribution during silicate-fluid interaction. In view of the complexity of the natural systems of interest, the amount of experimental work required is considerable, and many studies have already been carried out. With fluorine, most of these have involved determination of the melting behaviour of natural granitic rocks in F-bearing systems (e.g. Anfilogov, et al., 1973; Glyuk and Anfilogov, 1973a,b; Kovalenko, 1979; Wyllie and Tuttle, 1961) with only a limited amount of work involving relevant synthetic compositions (Koster van Groos and Wyllie, 1968; Manning, 1981). Many of these studies are reviewed by Bailey (1977). The effect of boron on phase relationships in the granite system has attracted rather less attention, but for synthetic compositions includes that of Chorlton and Martin (1978) and Pichavant (1981). Studies of the effect of fluorine and boron on metal transport and deposition are very limited (e.g. Hards, 1976; Manning and Henderson, 1984; Stemprok, 1982).

Rather than providing a review of the existing literature, the purpose of this paper is to draw attention to some aspects of the authors' recent work which extends previous experimental studies. For reasons of space, the application of experimental results to natural fluorine and boron-rich granites is not discussed in detail, and is in any case considered more fully by the authors elsewhere (Manning and Pichavant, 1983; Pichavant and Manning, 1984).

EXPERIMENTAL STUDIES OF SYNTHETIC FLUORINE AND BORON-BEARING SYSTEMS

1) Some general points

As in all experimental studies, great care has to be taken to design the experiments so that the interpretation of the results is made as simple as possible, and so that reproducible equilibrium results can be obtained. The choice of starting material is important - the use of synthetic materials rather than natural starting materials allows the compositions involved to be closely controlled, and the form of the starting materials (gel, crystals or glass) can be chosen to assess the degree of equilibrium. In detail, the preparation of starting materials and experimental procedure depend on the aim of the experiment. For liquidus studies in the synthetic system Qz-Ab-Or-H_2O-F it has been found preferable to add fluorine as NaF to appropriate Na-deficient gels to give compositions in Qz-Ab-Or-F, rather than adding F as a fluoride solution (Manning, 1981). Similarly, for liquidus studies in the system Qz-Ab-Or-H_2O-B_2O_3 boron is also added as a component of the gel rather than in solution (Pichavant, in preparation). Again for liquidus studies, the fluid:silicate ratio in the charge should be such as to limit the proportion of free fluid (taking into account the solubility of water in F and B-bearing melts) to only a slight excess under run conditions, so that the composition of the melt is not significantly affected by incongruent dissolution into the aqueous phase. This problem is particularly serious if large excess volumes of fluoride solutions are present (see discussion in Manning, 1981). Experiments dealing with fluorine and boron also need to take into account the different partitioning behaviour of each element between granite melt and fluid phases, and the consequences for assessing melt fluorine or boron contents. On the one hand, fluorine partitions very strongly in favour of the melt (K_D = molality of F in vapour/molality of F in melt = 0,18 (850°C), 0,12 (680°C); Hards, 1976) and so this problem is relatively insignificant, but boron partitions in favour of the fluid (K_D = 3; Pichavant, 1981). Although not affecting liquidus results (as long as the proportion of fluid is kept small) these effects will affect the interpretation of solidus and partitioning results, and are taken into account in the subsequent presentation of experimental data. Finally, it should be noted that the experimental studies reported here do not take into account the stability relationships of some important F and B-bearing mineral phases, such as micas, topaz and tourmaline. The crystallisation of these will remove F or B from the melt (or fluid) and so reduce the effects of either element on phase relationships. Both topaz and tourmaline may be early to crystallise - they are both known to be stable at magmatic temperatures, and tourmaline appears to be a relatively refractory host for boron (Rosenberg, 1972; Barton, 1982; Benard and Pichavant, 1982).

2) Liquidus phase relationships

a) the system Qz-Ab-Or-H_2O-F at 1 kbar pressure: Liquidus phase relationships for the synthetic system Qz-Ab-Or-H_2O with 1,2 and 4 wt% added fluorine at 1 kbar pressure are summarised in Figure 1 (Manning, 1981). Minimum compositions and temperatures are given in Table 1. It is important to note that with up to 4 wt% F in the melt

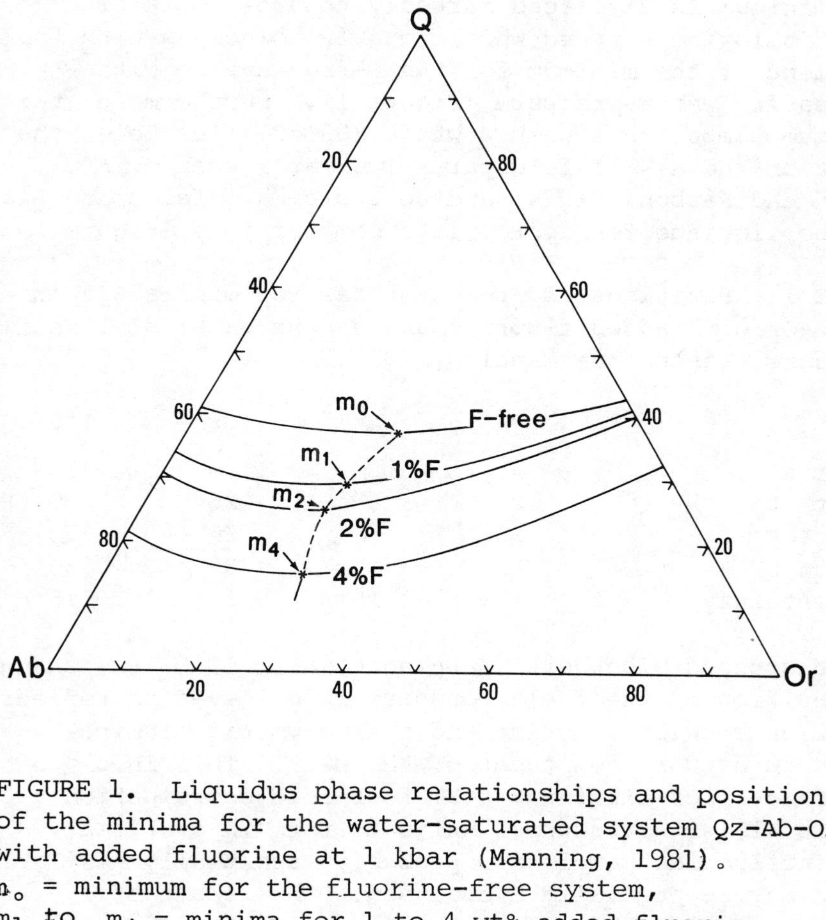

FIGURE 1. Liquidus phase relationships and positions of the minima for the water-saturated system Qz-Ab-Or with added fluorine at 1 kbar (Manning, 1981). m_o = minimum for the fluorine-free system, m_1 to m_4 = minima for 1 to 4 wt% added fluorine respectively. The solid lines show the phase boundaries for a given fluorine content.

quartz and alkali feldspar are stable under the experimental conditions - no F-bearing crystalline phases were encountered. No evidence was found to suggest that fluoride-silicate liquid immiscibility exists for these compositions at near-liquidus conditions. The main feature of the experimental results are that in addition to a reduction in minimum temperature, the position of the minimum is displaced markedly towards the albite apex following a trend which roughly coincides with the movement of the minimum for the F-free system with increasing water pressure (Figure 1). Furthermore, the minimum temperature with 4 wt% F (630°C) lies below the crest of the alkali feldspar solvus at 1 kbar (657 °C; Smith and Parsons, 1974) and so two alkali feldspars lie on the liquidus for appropriate compositions (Figure 1).

TABLE 1. Positions and temperatures for minima within Qz-Ab-Or with added fluorine and excess water at 1 kbar pressure, taken from Manning, 1981.

	Qz	Ab	Or	T°C
F-free	37	34	29	730
1 wt% F	29	45	26	690
2 wt% F	25	50	25	670
4 wt% F	15	58	27	630
uncertainty	±1	±2	±2	±5

As discussed elsewhere (Manning et al., 1980) the change in position of the field boundary is believed to reflect reaction between fluorine and the polymeric aluminosilicate network component of the melt. This involves the removal of aluminium from four-fold coordination within $AlSi_3O_8$ structural units to form interstitial aluminofluoride complexes, probably essentially AlF_6^{3-}, which may be partially hydrolysed. The result of interactions of this type is to increase the activity of the silica component of the aluminosilicate network at the expense of the feldspar component, with a concomitant increase in the size of the quartz stability field.

b) the system Qz-Ab-Or-H_2O-B_2O_3 at 1 kbar pressure: The results of preliminary experiments to determine the liquidus for the synthetic system Qz-Ab-Or-H_2O-B_2O_3 with 4,5 wt% added B_2O_3 at 1 kbar are summarised in Figure 2. A detailed report of these experiments will be published later (Pichavant, in preparation). The starting materials were glasses prepared for several compositions in Qz-Ab-Or with 4,5 wt% B_2O_3. Loss of boron from the glass to the fluid was minimised in the experiments by working with very little H_2O in the charge (10 wt% H_2O) while maintaining water saturation. Microprobe analysis of the liquidus feldspars has not revealed any anomalous Al concentrations suggesting that boron is kept within the melt under the experimental conditions. Only 'forwards' experiments (those using glass as starting material) have been used in the construction of Figure 2 and this probably results in slight underestimation of the true liquidus temperatures.

The results show that the liquidus temperatures are markedly reduced for all the compositions investigated. For some compositions reductions in liquidus temperature of more than 100°C are found, and there is a slight shift of the minimum composition towards the albite apex (N.B. the precise composition of the minimum has not yet been determined; the temperature of the minimum lies below 660°C). However, the change in minimum composition for the boron-bearing system is much less significant than the change observed for the corresponding fluorine-bearing system (Figure 1). It is considered that for geologically reasonable boron concentrations (<1 wt% B_2O_3) only very small changes in minimum composition can be expected.

In view of the limited changes in liquidus phase relationships for the boron-bearing system compared with the fluorine-bearing system (Figure 1) it is considered that the incorporation of boron has a relatively limited effect on melt structure. It is probable that the same type of structural units which exist in the hydrous, boron-free, aluminosilicate melt are essentially retained; boron may be present as triangular BO_3 units dispersed interstitially through the aluminosilicate network. However, the shift of the minimum for the

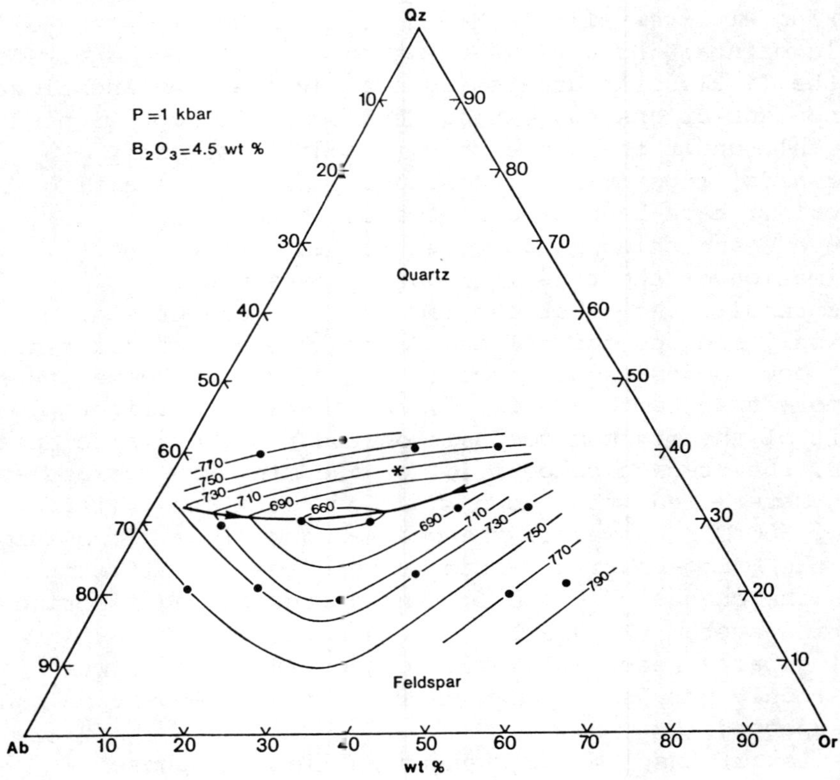

FIGURE 2. Liquidus phase relationships for the water saturated system Qz-Ab-Or with added B_2O_3 (4,5 wt%) at 1 kbar (Pichavant, in preparation). o = compositions investigated; * = minimum at P_{H_2O} = 1 kbar for the boron-free system (Tuttle and Bowen, 1958).

system Qz-Ab-Or-H_2O-B_2O_3 (B_2O_3= 4,5 wt%) at 1 kbar closely follows the shift in the minimum for boron-free compositions with increasing water pressure and water solubility (Tuttle and Bowen; Luth et al., 1964). It appears from this that the changes in aluminosilicate melt structure with increasing boron content at 1 kbar pressure and with increasing water pressure in the absence of boron may be similar. However, this may reflect the effect of boron on water solubility, which increases with increasing melt boron content (Pichavant, 1981; 1983a).

3) Solidus phase relationships

Solidus phase relationships in the system Qz-Ab-Or-H_2O with added F or B may be deduced from the liquidus phase relationships (Figures 1 and 2) because liquidus and solidus temperatures are reduced to the same value for the minimum composition for a given F or B content in the melt. Large reductions in solidus temperatures are therefore expected, and are detailed below.

a) the system Qz-Ab-Or-H_2O-F at 1 kbar pressure: Although many experimental studies have shown that the solidus temperature for the system granite-H_2O-F is reduced with increasing fluorine content (e.g. Bailey, 1977) detailed studies for the synthetic system have yet to be carried out. As a corollary to the liquidus results, the solidus temperature must be reduced by at least as much as the reduction in minimum temperature, but as shown by exploratory experiments (Manning, 1981) there appears to be a large melting interval for fluorine-bearing compositions. This may in part be due to the increase in melt fluorine content as the degree of crystallisation of F-free phases (quartz and alkali feldspar) increases. The results of exploratory experiments below the liquidus for near-minimum compositions indicate that with 2 wt% F melts exist down to temperatures below 600°C at 1 kbar, and with 4 wt% F to below 550°C. The fluorine contents of the residual melts in these experiments are not known, but are sufficiently high to give rise to textures possibly due to quench fluoride-silicate liquid immiscibility.

b) the system Qz-Ab-Or-H_2O-B_2O_3 at 1 kbar pressure: Solidus studies in the system Qz-Ab-Or-H_2O-B_2O_3 at 1 kbar have been carried out by Chorlton and Martin (1978) and Pichavant (1981). Both introduced boron to the system as H_3BO_3 aqueous solutions. There is some uncertainty in the B_2O_3 concentrations of the starting solutions used by Chorlton and Martin (1978), who also used glass rather than crystalline starting materials. Consequently, it is considered that their experimental results are less accurate than those of Pichavant (1981) who used crystalline starting materials. The solidus relations are summarised in Figure 3, taken from Pichavant (1981). With B_2O_3 concentrations expressed as wt% in solution, the solidus curve is theoretically independent of the fluid:silicate ratio used in the experiment (Wyllie and Tuttle, 1961). However, changes in the silicate composition due to interaction with the fluid phase (particularly if large fluid:silicate ratios are used) may affect the observed solidus temperatures (discussed in Pichavant, 1981). As shown in Figure 3, there is a large reduction in solidus temperature with increasing solution boron content. The B_2O_3 content of the near solidus melt (calculated from the melt-vapour partition coefficient for boron, the experimental fluid:silicate ratio and the estimated melt fraction) concomitantly increases along the solidus curve to reach about 4 wt% B_2O_3 for 12 wt% B_2O_3 in the starting solution (solidus temperature 610°C). Two alkali feldspars are stable together with quartz, fluid and melt along the solidus curve below the crest of the alkali feldspar solvus (657°C; see before).

4) Melt-fluid interaction

In order to extend the liquidus studies, a series of experiments has recently been carried out to assess the role of fluorine and other volatiles on metal partitioning between melts in Qz-Ab-Or and fluid at 800°C and 1 kbar pressure (Manning and Henderson, 1984). For experimental reasons, tungsten was chosen as the metal of interest, with fluoride, chloride and phosphate solutions as well as water alone as the fluid phase.

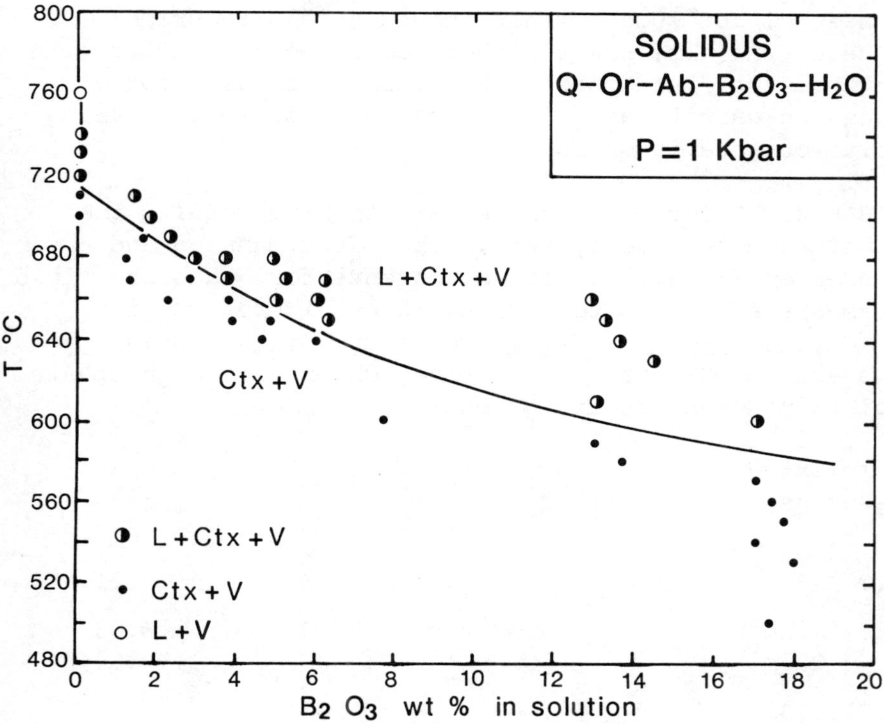

FIGURE 3. Solidus temperature versus B_2O_3 concentration in solution for a synthetic crystalline composition in Qz-Ab-Or (close to the minimum at P_{H_2O} = 1 kbar) at 1 kbar (Pichavant, 1981).
L = liquid (silicate melt); ctx = crystals; V = vapour.

The results obtained are summarised in Table 2. Tungsten essentially shows two types of behaviour. With water alone and fluoride solutions, tungsten partitions into the melt, whereas with chloride and phosphate solutions tungsten partitions strongly into the fluid. These partitioning experiments provide little direct evidence of complex speciation, and can be satisfactorily explained assuming that W-Cl complexes and polymeric tungsten oxyanions (isopolytungstates and heteropoly-tungstates) are present (Manning and Henderson, 1984) as the dominant species. The results demonstrate that fluorine-enriched melts may concentrate metals such as tungsten, but suggest that should a chloride-rich aqueous phase be present these could be removed from the melt to the fluid phase. Preliminary results for tungsten partitioning in the presence of boron-bearing solutions are inconclusive.

TABLE 2. Selected tungsten partitioning results for granite melt-fluid systems, taken from Manning and Henderson (1984). Tungsten contents for glass and fluid phases are rounded to 2 significant figures; K_D is defined as molality W_{fluid}/molality W_{glass} and is subject to 30% error. Fluoride, chloride and phosphate solution strengths are given as molalities.

complexing species	W_{glass} ppm	W_{fluid} ppm	K_D
H_2O	4400	490	0.11
NaF, 1m	5100	2700	0.53
0.5m	5000	4200	0.84
NaCl, 3.4m	1000	6500	6.7
3m	2100	8700	4.1
2m	2500	6300	2.6
1m	2700	5100	1.9
0.5m	3900	5700	1.5
$Na(PO_4)_{1/3}$, 2m	2700	6400	2.4
1m	2700	7100	2.6
0.5m	2900	5900	2.0

Some constraints on the composition of aqueous phases in coexistence with F or B-rich magmas can be deduced from compositional data for the fluid phase in equilibrium with F or B-bearing melts in the synthetic system Qz-Ab-Or-H_2O. Increasing the B_2O_3 or F content of the aqueous phase leads to an increase in the silicate solute content (Pichavant, 1981; Wyllie and Tuttle, 1961) and a decrease in the K/(K+Na) ratio of the fluid in comparison with Cl-bearing fluids under the same PT conditions (Burnham, 1967 and next section). Boron-bearing fluid phases in equilibrium with granite magmas are believed to consist mainly of Si, B and Na in solution, with the solute corresponding to about 90% normative quartz and 10% normative feldspar.

5) Crystal-fluid interaction

Postmagmatic (subsolidus) alteration associated with fluorine and boron-rich granites often involves metasomatic processes such as albitisation or microclinisation, as outlined by Scherba (1970). These reactions may be modelled by the experimental determination of alkali exchange between feldspars and boron or fluorine-bearing aqueous solutions as well as chloride solutions. The results of two (Na,K) ion exchange isotherm determinations (T=500°C, P = 1 kbar) for alkali feldspars and boron and fluorine-bearing aqueous solutions (1 and 0,7 molal respectively) are shown in Figure 4 (Pichavant, 1983b) where they are compared with similar ion exchange data for chloride solutions (Orville, 1963; Lagache and Weisbrod, 1977). Whereas no significant changes occur in the composition of the feldspar, the K/(K+Na) ratio for the solution differs considerably for different anionic species.

Solutions in equilibrium with two feldspars become more sodic with Cl^-, borate and F^- anions respectively, with fluoride-bearing solutions exceptionally rich in Na. Experimental results of this type demonstrate that the effect of a two feldspar assemblage on buffering the fluid alkali ratio is very much a function of the anionic species involved. As an example, it can be seen that percolation of F or B-rich hydrothermal solutions would lead to potassic alteration of the wall rock because Na from wall rock feldspars would be exchanged for K from the fluid, enriching the fluid in Na. Such an effect would be more marked for fluoride-bearing solutions than boron-rich solutions. On the other hand, a drop in the F or B concentrations of the fluid (through crystallisation of F or B-bearing phases, or dilution with F or B-poor fluids) would produce the opposite effect, as a result of the change in fluid composition towards more K-rich values (Figure 4).

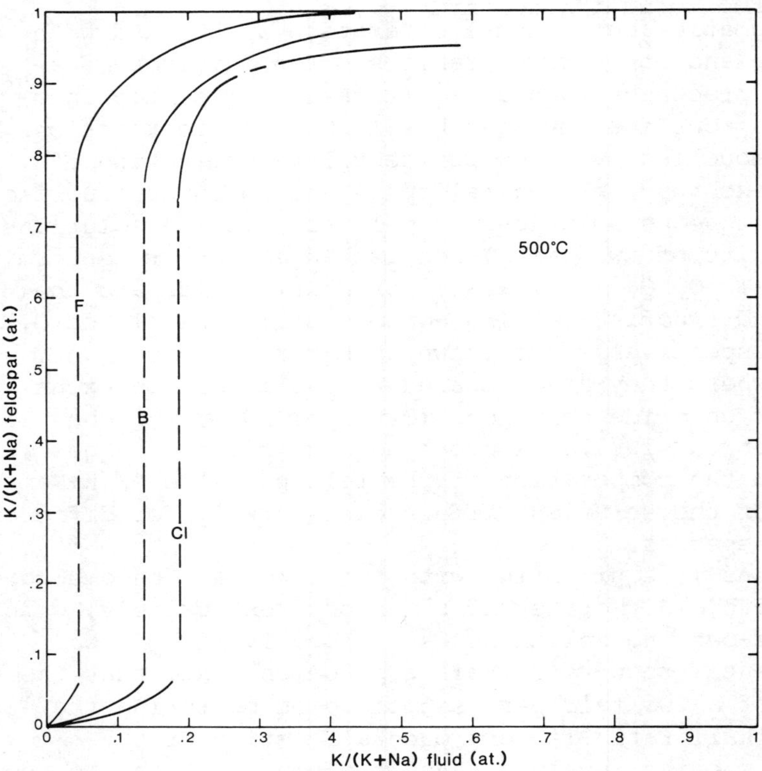

FIGURE 4. Alkali distribution curves between feldspars and aqueous solutions in the presence of chloride, borate and fluoride anions at 500°C, P = 1 kbar. From Pichavant, 1983b.

DISCUSSION AND CONCLUSIONS

One of the most important results of the detailed phase equilibria studies concerns the temperatures down to which fluorine and boron-enriched granitic magmas can exist. These results are summarised in Figure 5, in which the minimum melting curves for fluorine and boron-bearing compositions are given together with selected mineral equilibria. It is apparent from this that muscovite may become a stable liquidus phase, rather than aluminium silicate (or topaz) in peraluminous (and fluorine or boron-rich) compositions at lower pressures than would be necessary for equivalent fluorine and boron-free compositions. As already mentioned, the effect of fluorine in particular also enables two alkali feldspars to crystallise on the liquidus. The effect of fluorine and/or boron enrichment consequently can have an important effect on the major (as well as accessory) mineral content of a late-stage granitic rock. Secondly, the observed change in minimum composition with increasing fluorine content implies that F-enriched late stage granites will, in addition to carrying high fluorine contents, be characteristically richer in normative albite than corresponding fluorine-poor granites, with plots displaced towards the albite apex in Qz-Ab-Or. This has frequently been observed, and can in the light of these experimental results be considered to be a magmatic feature in many cases rather than a consequence of post-magmatic albitisation. In the case of boron-enriched residual magmas, the experimentally observed change in minimum composition is less marked. Furthermore, the boron concentrations of such magmas are not likely to exceed 1 wt% B_2O_3, which would correspond to a small shift in minimum position in Qz-Ab-Or. This marks a difference between B-rich minimum melts, which are not significantly different in composition from B-free minimum melts at a given pressure, and F-rich minimum melts, which are significantly displaced towards the albite apex. The normative Qz-Ab-Or composition of some volatile-rich natural granitic glasses such as the Macusani glass (Barnes et al., 1970) suggests that the effects of the combined addition of fluorine and boron on minimum compositions are additive.

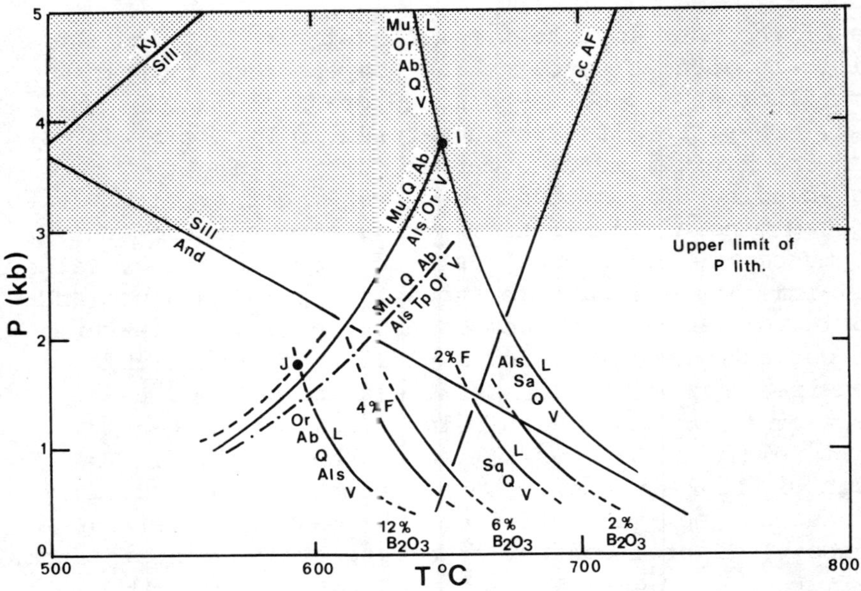

FIGURE 5. PT diagram showing selected equilibria for the crystallisation of F and B-rich granitic compositions. Symbols used: Q = quartz; Sa = alkali feldspar solid solution; Ab = Na-rich alkali feldspar solid solution; Or = K-rich alkali feldspar solid solution; Als = Al_2SiO_5 polymorphs - Ky = kyanite, Sill = sillimanite, And = andalusite; Mu = muscovite; Tp = topaz; L = silicate melt; V = vapour phase.
Sources of data; Al_2SiO_5 equilibria - Holdaway, 1971; alkali feldspar critical curve (ccaF) - Thompson and Waldbaum, 1969; Mu + Ab + Q = Or + Als + V - Chatterjee and Froese, 1975; same equilibrium with F buffered by andalusite-topaz-H_2O taken from Barton, 1982; Q + Sa + Als + V = L and Q + Or + Ab + Mu + V = L (peraluminous F and B-free minimum melt) from Luth, 1976 and Petö,1976. F concentrations (wt% F) are concentrations in the melt; B_2O_3 concentrations (wt% B_2O_3) are concentrations in the aqueous starting material solution.
Minimum melting curves for F and B-bearing compositions are extrapolated either side of the value at P = 1 kbar. The dashed curve is the muscovite stability curve

schematically drawn for 12 wt% B_2O_3 in the vapour phase. The dash-dotted curve is the muscovite stability curve with F buffered by andalusite-topaz-H_2O. Point I is the minimum pressure for the coexistence of muscovite and melt under water-saturated conditions, and point J is similar but with 12 wt% B_2O_3 in addition to water. The upper limit for the lithostatic pressure of emplacement of the natural F and B-rich granites is taken to be 3 kbar.

Melt-fluid and crystal-fluid interaction processes have been shown to depend strongly on the nature of the volatile species involved. This concerns major elements (notably the alkalis) and also metals such as tungsten. As indicated by the tungsten partitioning experiments, fluorine-rich residual melts can affectively concentrate and transport metals. However, should a chlorine-rich fluid phase coexist with such magmas then metals may be removed to enter the fluid phase. Such a mechanism suggests that the genesis of ore-bearing granitoids is not simply related to the effect of volatiles in modifying melt structure, so permitting the incorporation of incompatible elements, but also depends on the effect of different volatile species on the removal of metals from the melt to a coexisting fluid phase.

REFERENCES

Anfilogov, V.N., Glyuk, D.S. and Trufanova, L.G. 1973. Phase relations in interaction between granite and sodium fluoride at a water pressure of 1000 kg/cm^2. Geochem. Int. 10, 30-33.

Bailey, J.C. 1977. Fluorine in granitic rocks and melts: a review. Chem. Geol. 19, 1-42.

Barnes, V.E., Edwards, G., McLaughlin, W.G., Friedman, I. and Joensuu, O. 1970. Macusanite occurrence, age and composition, Macusani, Peru. Bull. geol. Soc. Am. 81, 1539-1546.

Barton, M.D. 1982. The thermodynamic properties of topaz solid solutions and some petrologic applications. Am. Mineral. 67, 956-974.

Benard, F. and Pichavant, M. 1982. Fusion experimentale des leucogranites a tourmaline du massif granitique de la Margeride (France). 9e RAST, Soc. geol. Fr., 42.

Burnham, C.W. 1967. Hydrothermal fluids at the magmatic stage. In: Barnes, H.L. (ed) Geochemistry of hydrothermal ore deposits, Holt, Rinehart and Wilson, New York, 34-76.

Chatterjee, N.D. and Froese, E. 1975. A thermodynamic study of the pseudobinary join muscovite-paragonite in the system $KAlSi_3O_8$-$NaAlSi_3O_8$-Al_2O_3-SiO_2-H_2O. Am. Mineral. 60, 985-993.

Chorlton, L.B. and Martin, R.F. 1978. The effect of boron on the granite solidus. Can. Mineral. 16, 239-244.

Glyuk, D.S. and Anfilogov, V.N. 1973a. Phase equilibria in the system granite-H_2O-HF at a pressure of 1000 kg/cm^2. Geochem. Int. 10, 321-325.

Glyuk, D.S. and Anfilogov, V.N. 1973b. Phase equilibria in the system granite-water-potassium fluoride at a water vapour pressure of 1000 kg/cm^2. Dokl. Akad. Sci. USSR, Earth Sci. Sect. 210, 237-238.

Hards, N.J. 1976. Distribution of elements between the fluid phase and silicate melt phase of granites and nepheline syenites. NERC Rpt. Progress in experimental petrology, 3, 88-90.

Holdaway, M.J. 1971. Stability of andalusite and the aluminium silicate phase diagram. Am. J. Sci. 271, 97-131.

Koster van Groos, A.F. and Wyllie, P.J. 1968. Melting relationships in the system $NaAlSi_3O_8$-NaF-H_2O to 4 kilobars pressure. J. Geol. 76, 50-70.

Kovalenko, N.I. 1979. Experimental studies of the formation of rare-metal Li-F granitoids. Nauka, Moscow, 152 pp (in Russian).

Lagache, M. and Weisbrod, A. 1977. The system two alkali feldspars-KCl-NaCl-H_2O at moderate to high temperatures and low pressures. Contrib. Mineral. Petrol. 62, 77-102.

Luth, W.C. 1976. Granitic rocks. In: Bailey, D.K. and MacDonald, R. (eds) The evolution of the crystalline rocks, Academic Press, London, 335-417.

Luth, W.C., Jahns, R.H. and Tuttle, O.F. 1964. The granite system at pressures of 4 to 10 kilobars. J. Geophys. Res. 69, 759-773.

Manning, D.A.C. 1981. The effect of fluorine on liquidus phase relationships in the system Qz-Ab-Or with excess water at 1 kbar. Contrib. Mineral. Petrol. 76, 206-215.

Manning, D.A.C., Hamilton, D.L., Henderson, C.M.B. and Dempsey, M.J. 1980. The probable occurrence of interstitial Al in hydrous, F-bearing and F-free aluminosilicate melts. Contrib. Mineral. Petrol. 75, 257-262.

Manning, D.A.C. and Henderson, P. 1984. The behaviour of tungsten in granitic melt-vapour systems. Contrib. Mineral. Petrol. in press.

Manning, D.A.C. and Pichavant, M. 1983. The role of fluorine and boron in the generation of granitic melts. In: Atherton, M.P. and Gribble, C.D. (eds) Migmatites, melting and metamorphism. Shiva, Nantwich, 94-109.

Orville, P.M. 1963. Alkali ion exchange between vapour and feldspar phases. Am. J. Sci., 261, 201-237.

Peto, P. 1976. An experimental investigation of melting relations involving muscovite and paragonite in the silica-saturated portion of the system K_2O-Na_2O-Al_2O_3-SiO_2- H_2O to 15 kb pressure. NERC Rpt. Progress in experimental petrology, 3, 41-45.

Pichavant, M. 1981. An experimental study of the effect of boron on a water saturated haplogranite at 1 kbar pressure. Geological applications. Contrib. Mineral. Petrol. 76, 430-439.

Pichavant, M. 1983a. Melt-fluid interaction deduced from studies of silicate-B_2O_3-H_2O systems at 1 kbar. Bull. Mineral. 106, 201-211.

Pichavant, M. 1983b. (Na, K) exchange between alkali feldspars and aqueous solutions containing borate and fluoride anions; experimental results at P = 1 kbar. (Abstract). 3rd NATO Advanced Study Institute on feldspars, felspathoids and their parageneses. Rennes, France.

Pichavant, M. and Manning, D.A.C. 1984. Petrogenesis of tourmaline granites and topaz granites: the contribution of experimental data. Phys. Earth. Planet. Int., in press.

Rosenberg, P.E. 1972. Compositional variation in synthetic topaz. Am. Mineral. 57, 169-187.

Scherba, G.N. 1970. Greisens. Int. Geol. Rev. 12, 114-150; 239-255.

Smith, P. and Parsons, I. 1974. The alkali feldspar solvus at 1 kbar water vapour pressure. Mineral. Mag. 39, 747-767.

Stemprok, M. 1982. Tin-fluorine relationships in ore-bearing assemblages. In: Evans, A.M. (ed) Metallization Associated With Acid Magmatism 6, John Wiley, Chichester, 321-337.

Thompson, J.B. jr. and Waldbaum, D.R. 1969. Mixing properties of sanidine solid solutions. III. Calculations based on two phase data. Am. Mineral. 54, 811-838.

Tuttle, O.F. and Bowen, N.L. 1958. Origin of granite in the light of experimental studies in the system $NaAlSi_3O_8$-$KAlSi_3O_8$-SiO_2-H_2O. Mem. geol. Soc. Am. 74, 153 pp.

Wyllie, P.J. and Tuttle, O.F. 1961. Experimental investigation of silicate systems containing two volatile components. Part II. The effects of NH_3 and HF, in addition to H_2O, on the melting temperatures of albite and granite. Am. J. Sci. 259, 128-143.

THE PETROLOGY OF THE UPPER PART OF THE CRITICAL ZONE OF THE BUSHVELD COMPLEX AND ITS BEARING ON THE ORIGIN OF THE MERENSKY REEF

NALDRETT, A.J.,
University of Toronto, Toronto, Canada
GASPARRINI, C.,
Minmet Scientific, Toronto, Canada
BARNES, S.J.,
University of Toronto, Canada
SHARPE, M.R., and VON GRUENEWALDT, G.,
University of Pretoria, South Africa

ABSTRACT

The PGE-bearing Merensky reef lies at the base of one of 5 cyclic units that occur at the top of the Critical Zone. Trace element data indicate that there is a stepwise decrease in the Sr content of plagioclase on passing upwards through the units. The MgNo of bronzite shows a marked upward decrease in each cyclic unit, increasing abruptly again at the base of the overlying unit. In many areas the MgNo of bronzites in the pegmatoidal reef, below the upper chromite layer, is significantly higher than in the non-pegmatoidal bronzite a few cm above the chromite.
The Ni/Cu ratio of the Merensky reef is 2.3:1, in contrast to the ratio of 1.2:1 characterizing the remainder of the Merensky unit. The Pt/Cu ratio in the Merensky reef is 5 to 12×10^{-3}, corresponding to a Pt tenor in the sulfides of 560 ppm, in contrast to a ratio of 0.1 to 0.25×10^{-3} and tenor of 18 ppm characteristic of the rest of the unit. Other units or parts of units also have distinctive Ni/Cu and Pt/Cu ratios.

Many of the new observations made in this study are incompatible with former hypotheses proposed to account for the PGE enrichment. The changes in MgNo of bronzite, Sr content of plagioclase, and variation in $(Sr^{87}/Sr^{86})_o$ ratio (reported by Kruger and Marsh, 1982) argue strongly for successive injection of

Proceedings of the 27th International
Geological Congress, Volume 9, pp. 373–389
PETROLOGY (IGNEOUS AND METAMORPHIC ROCKS)

pulses of hot, primitive new magma. It is proposed that these pulses are less dense than magma in the chamber and rise up as turbulent plumes. The high PGE tenor of the sulfides of the Merensky reef is attributed to efficient mixing of these sulfides and magma during ascent in the planes, and subsequent very convective overturn. The development of the Merensky reef and associated cyclic units are well explained by the subsequent 'roll-over' (Campbell et al, 1983) and mixing of the new pulses.

INTRODUCTION

The Bushveld complex (Figure 1) is the source of 75 percent of the world's platinum and much of its chromium and vanadium. All of these resources are found within the layered mafic and ultramafic rocks of the Rustenburg Layered Suite. The platinum group elements (PGE) are concentrated mainly at two horizons, the Merensky reef and the UG-2 chromitite, associated with weakly disseminated sulfides. The PGE tenors of these sulfides are very much higher than is the case for most deposits of Ni-Cu sulfides; the main focus of this paper is the reason for these high tenors, and for the localization of these unusual sulfides.

GEOLOGICAL SETTING

The Merensky reef is associated with a cyclic unit which is one of a number (5 or more) occurring at the top of the Critical Zone of the Complex (Figure 2). Each unit shows the same progression from bronzitite (Bc) to norite (PBc) to anorthosite (Pc). Some also have chromitite (Cc) overlain by harzburgite (Oc) underlying the bronzitite. The Merensky reef lies at the base of the Merensky unit, which is the second from the top of the Critical Zone: it is overlain everywhere by the Bastard unit and underlain in the southwestern (Rustenburg) area of the Western Bushveld by the Boulder unit, and in the northwestern part (Union and Amandelbult sections) by the Pseudo

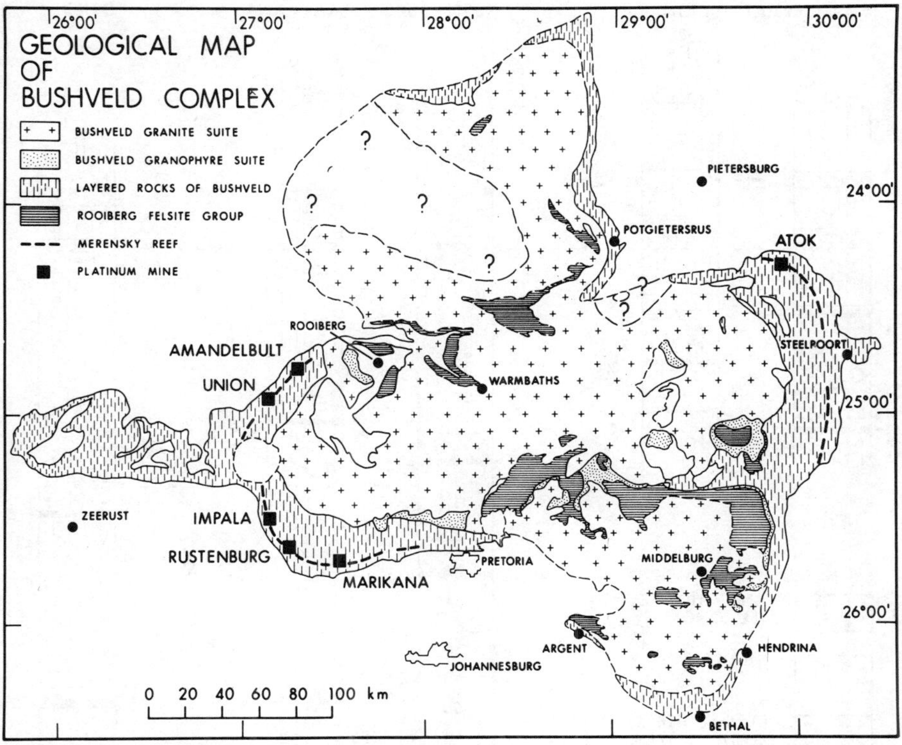

FIGURE 1. Geological map of the Bushveld Complex (modified after Willemse, 1969). The Rustenburg Layered Suite is labelled 'Layered Rocks of Bushveld'.

unit.

The Boulder unit is characterized by ellipsoidal, partly pegmatoidal masses of bronzitite, partially rimmed by chromite, set in anorthosite. The Pseudo unit has a pegmatoidal bronzitite at its base, overlain by harzburgite, bronzitite and anorthosite. Both the Merensky and Bastard units consist of bronzitite, norite, and anorthosite. The base of the Merensky unit is marked in most areas by a 30 cm to 1 m thick zone of pegmatoidal bronzitite: this contains a somewhat higher than normal (1 to 5 percent) proportion of sulfide, which carries the PGE and thus constitutes the ore, and is referred to as the Merensky

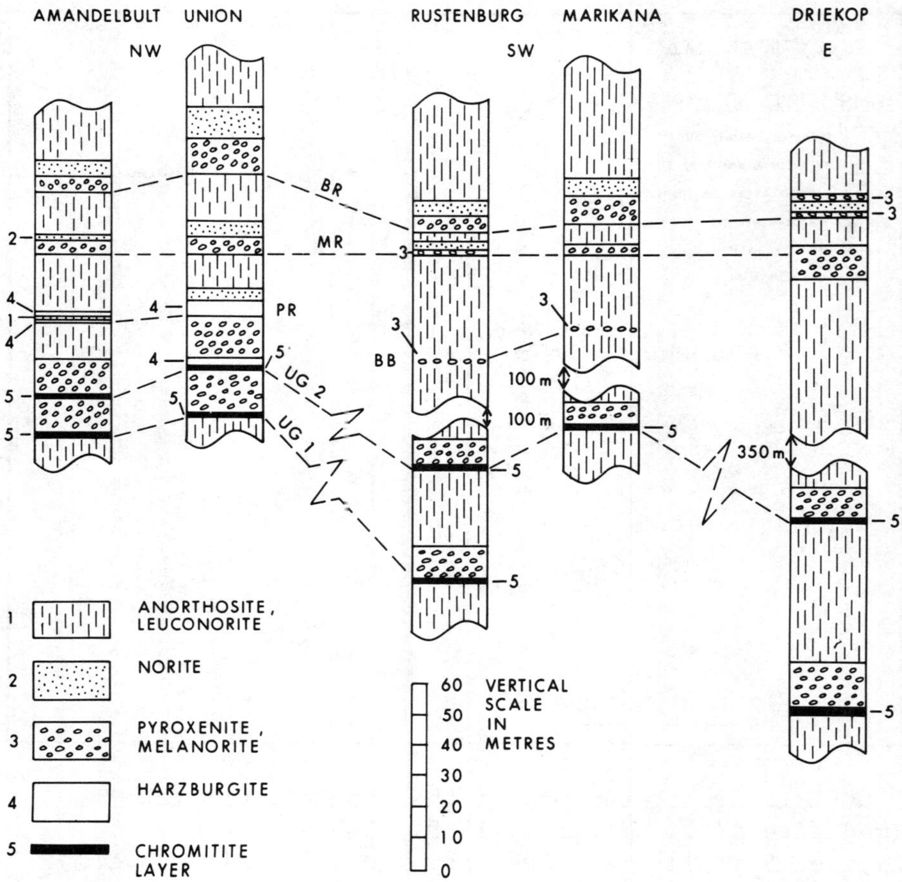

FIGURE 2. The sequence of cyclic units in the vicinity of the Merensky Reef. BR = Bastard Reef; MR = Merensky Reef; PR = Pseudo Reef; BB = Boulder Bed. (after Naldrett, 1981; data from Vermaak (1976), Farquhar (1984) and Gain (1980).

reef. Irregular patches of pegmatoid characterize the base of the Bastard unit. Thin (3 to 30 mm) seams of chromite occur at the base of both the Merensky and Bastard units and also at the top of the Merensky reef.

WHOLE ROCK GEOCHEMISTRY

Chemical profiles through the Bastard, Merensky and Pseudo-Boulder units at Union and Rustenburg reflect the mineralogy of the units for the most part (Figure 3): since so many of the rocks contain only 2 major minerals, bronzite and plagioclase, MgO serves as a nearly perfect index of the relative proportions of the two. As would be expected, trace elements such as Cr (concentrated mainly in bronzite) and Sr (concentrated mainly in plagioclase) correlate linearly with MgO, the correlation being positive and negative respectively. The Sr correlation (Figure 4) indicates that there is a generally stepwise decrease in the Sr content of plagioclase on passing upwards from the Pseudo-Boulder unit to the Merensky to the Bastard. While the decrease is abrupt at the Pseudo-Boulder unit-Merensky unit contact, it occurs within the Merensky unit rather than at the Merensky-Bastard contact.

Trace element abundances indicate that the bronzitites at the base of each unit contain a much higher proportion of trapped intercumulus liquid than the norites or anothosites. The areas showing partial or complete pegmatoidal texture are those with the highest concentrations of incompatible elements. Significantly, within such an area, there is no difference in the trace or major element content of pegmatoid as compared to non-pegmatoid; thus while pegmatoid tends to occur in areas of high incompatible elements, within these areas it is not marked by a distinctive composition.

MINERAL COMPOSITIONS

A detailed programme of microprobe analysis has shown that the MgNo of bronzite shows a marked upward decrease in most cyclic units, increasing abruptly again at the base of the overlying unit (Figure 5). In many areas the MgNo of bronzites in the pegmatoidal reef, below the upper chromite seam, is significantly higher than in the non-pegmatoidal bronzite a few cm above the chromite. Plagioclase shows a gradual upwards decrease

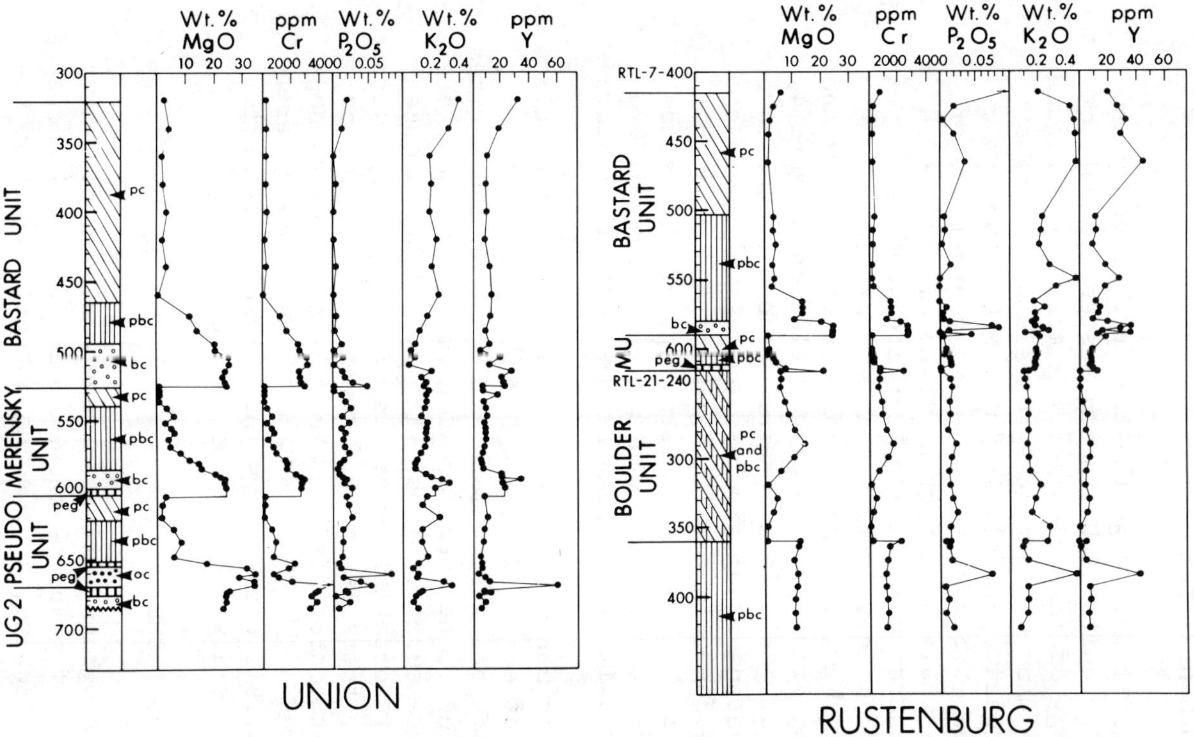

FIGURE 3. Profiles of the concentration of some major and trace elements in the Pseudo-Boulder, Merensky and Bastard units over the Union and Rustenburg sections.

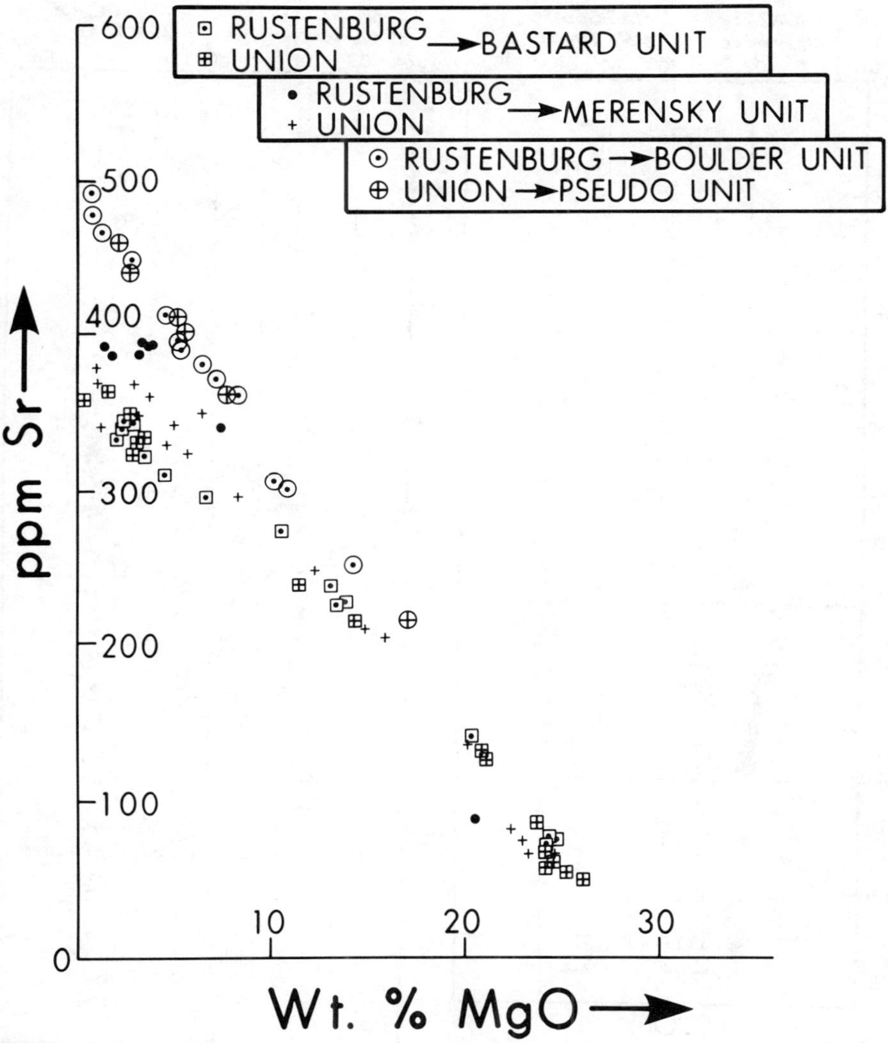

FIGURE 4. Variation of Sr with MgO in cyclic units at the top of the Critical Zone. Note that rocks containing more than 18 percent Mgo contain no cumulus plagioclase and thus do not fall, and would not be expected to fall, on the trends exhibited by those with cumulus plagioclase.

in anorthite content. Zoning in plagioclase is very much greater in the intercumulus material of the bronzitites than in rocks with cumulus plagioclase, supporting the

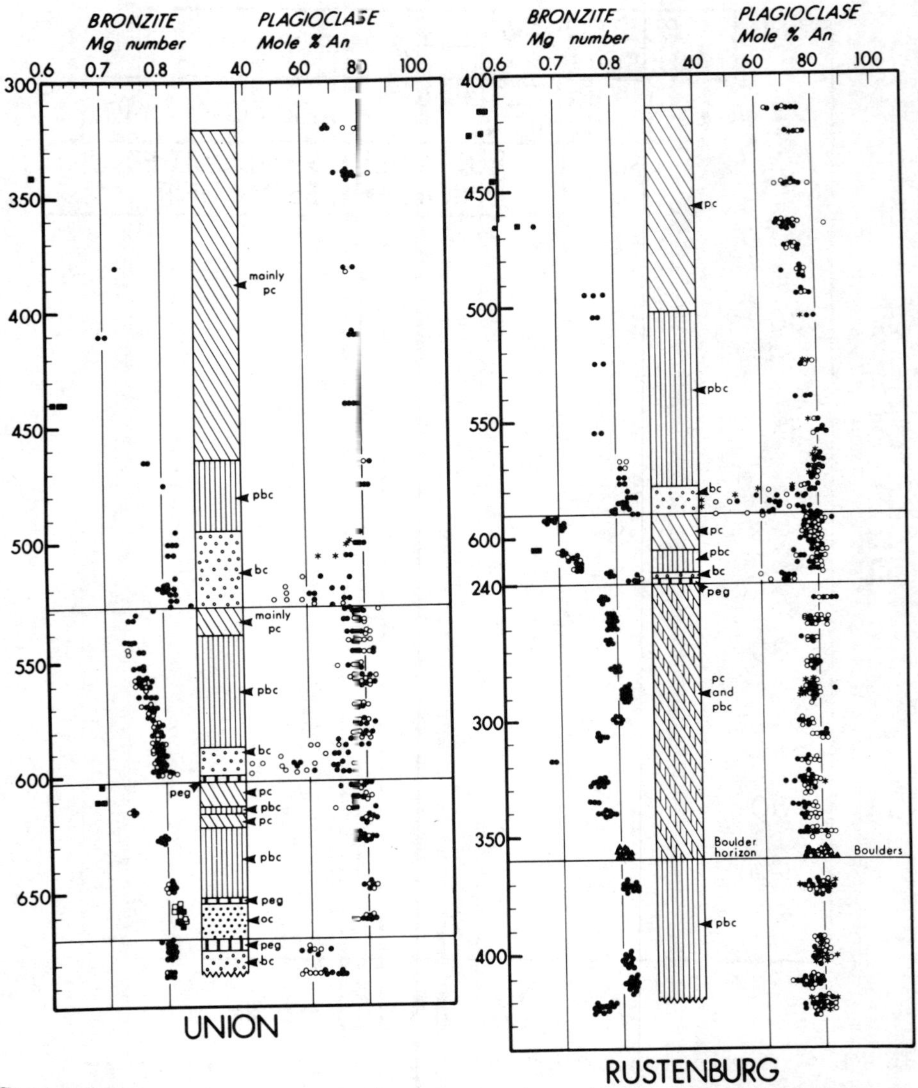

FIGURE 5. Variation in the MgNo of bronzite and An content of plagioclase, as determined with the microprobe, in the Pseudo-Boulder,Merensky and Bastard cyclic units at Union and Rustenburg. Hollow symbols = margins of grains; solid symbols = cores of grains; circles = cumulus grains; squares = intercumulus grains; stars = small plagioclase grains enclosed within cumulus bronzite; triangles = pyroxene and plagioclase within the boulders of the Boulder Bed.

conclusion that the bronzitites, particularly those near the base of each unit, are orthocumulates in contrast with the adcumulus nature of the norites and anorthosites.

COMPOSITION OF SULFIDES

Nickel also shows a positive correlation with MgO, as would be expected as a result of its ability to substitute in bronzite. Some rocks define a straight line correlation compatible with pure bronzite containing 700 ppm Ni (Figure 6). Many rocks fall above this straight line, containing more Ni than can be accounted for on the basis of bronzite. When this "excess" Ni is plotted against Cu for an individual unit, a strong positive, linear correlation is obtained, indicating that the excess Ni is in sulfide, and giving the Ni/Cu ratio of this sulfide (Figure 7). Different units and different parts of units have distinctive Ni/Cu ratios. That of the Merensky reef is 2.3:1, very different to the Ni/Cu ratio of 1.2:1 which characterizes the remainder of the Merensky unit. The Pt/Cu ratio is also characteristic of a given part of a unit: that in the Merensky reef is 5 to 12×10^{-3}, corresponding to a Pt tenor in the sulfides of 560 ppm, in contrast to the ratio of 0.1 to 0.25×10^{-3} and tenor of 18 ppm that characterizes the rest of the Merensky unit (Figure 8). The Bastard unit has a Ni/Cu ratio of 1.2:1 and Pt/Cu ratios of 0.45×10^{-3}. The division between these zones of high and low Ni/Cu and Pt/Cu ratio is extremely sharp, occurring across the upper chromite seam in the case of the Merensky unit.

DISCUSSION

Many of the new observations made in this study are incompatible with former hypotheses proposed to account for the PGE enrichment, for example those calling upon hydrothermal fluids to introduce the metals, or upon the upward percolation of highly fractionated and PGE rich residual silicate fluids expelled by compaction of the underlying crystal pile. The petrologic

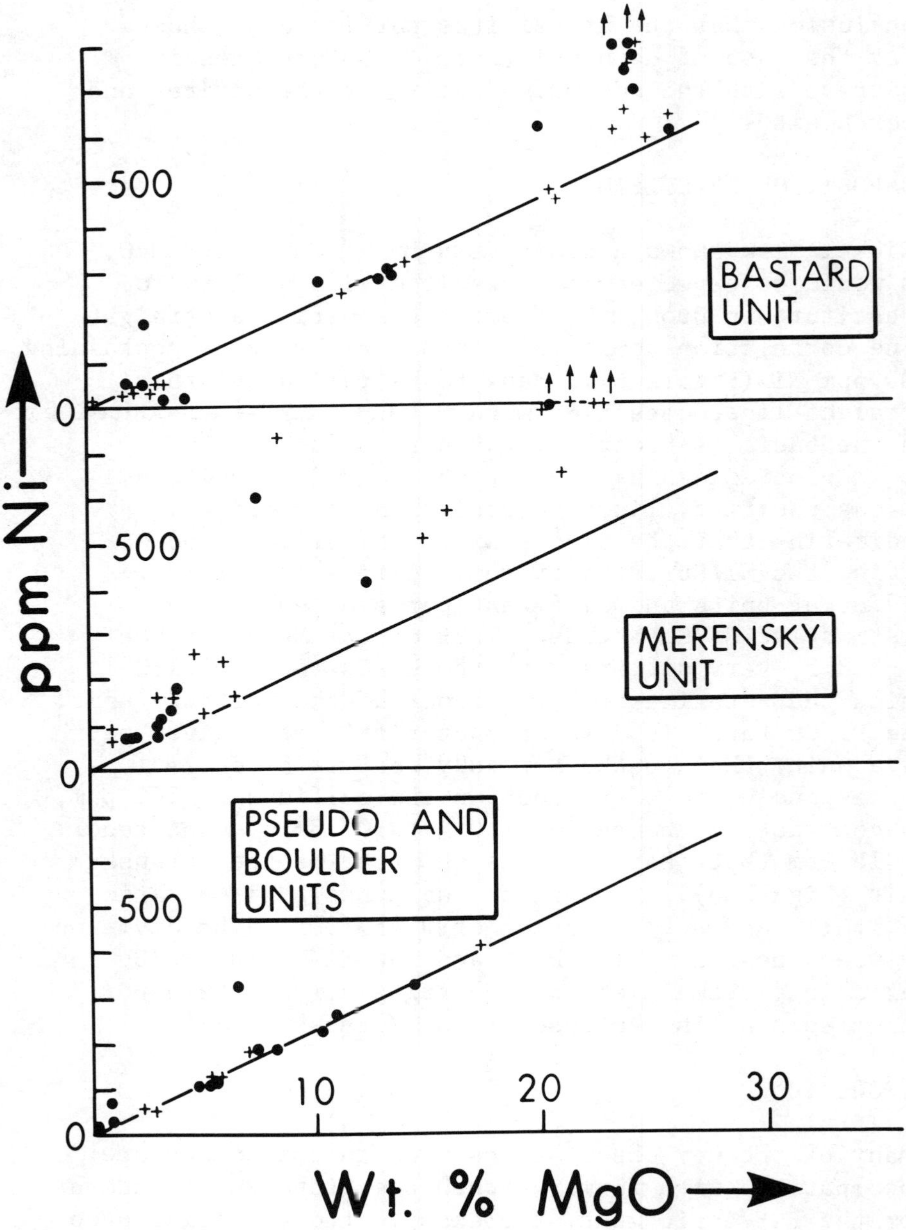

FIGURE 6. Variation of Ni with MgO. Dots = samples from Union: crosses = samples from Rustenburg.

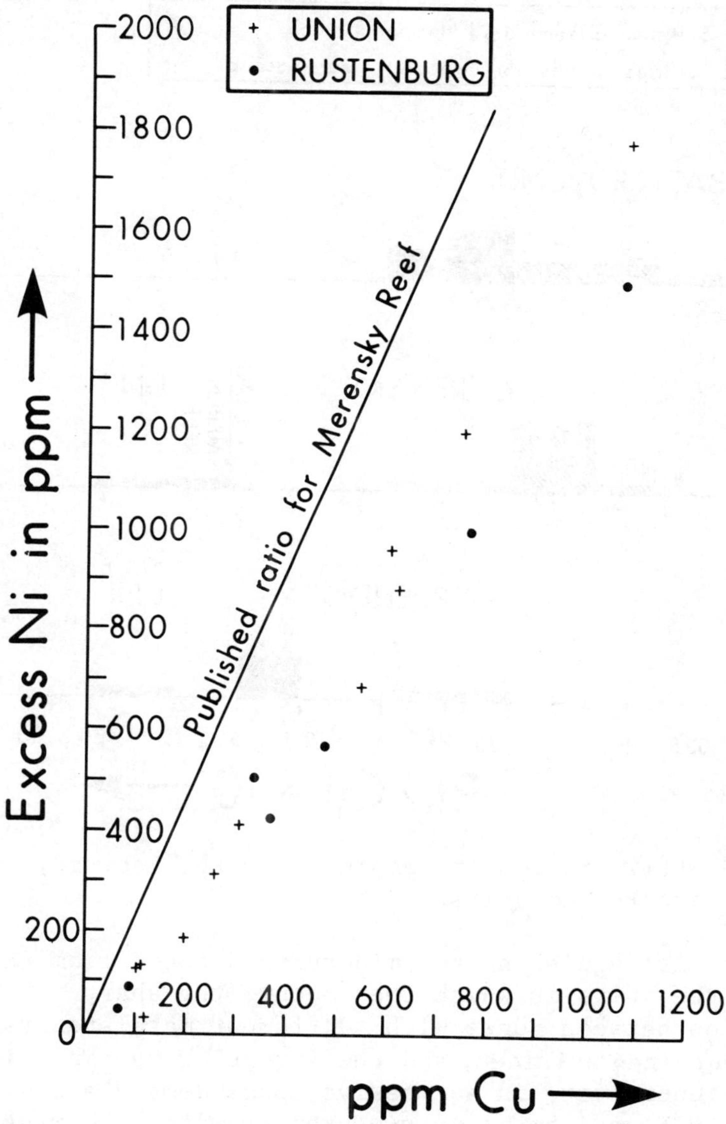

FIGURE 7. Plot of 'excess' Ni versus Cu for samples from the Merensky cyclic unit. Excess Ni is defined as that in excess of the amount defined on the basis of whole rock MgO content by the solid lines in Fig. 6.

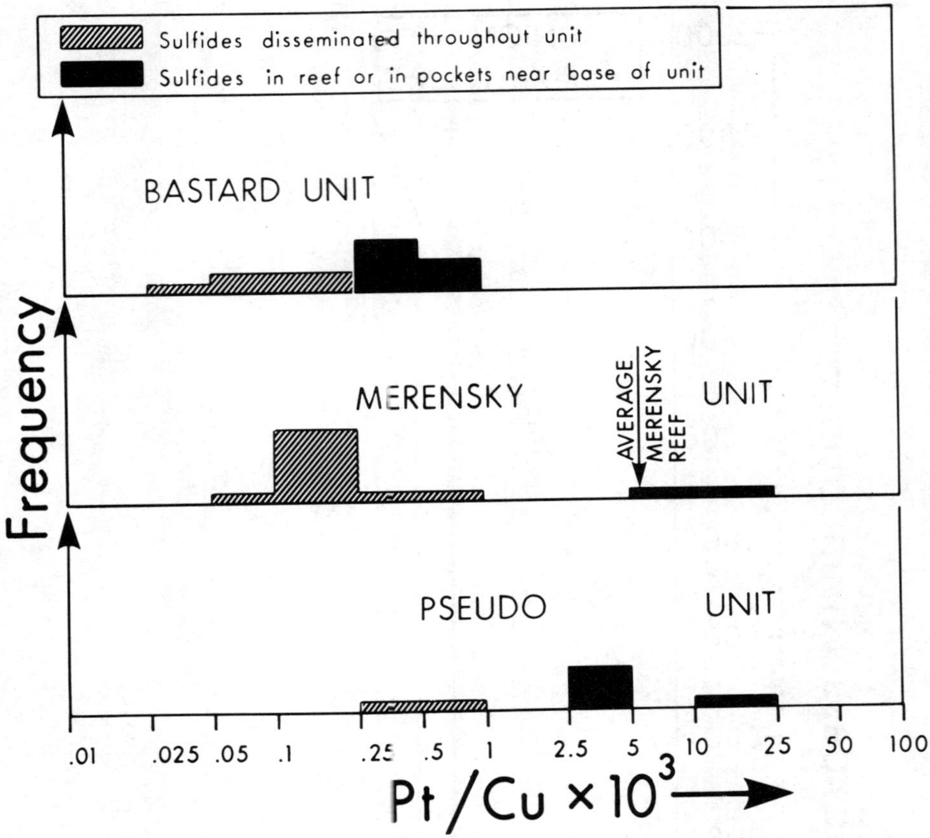

FIGURE 8. Pt/Cu ratios in samples from the Bastard, Merensky and Pseudo units.

features distinguishing the mineralized zones from the rest of the units in which they occur, the sharp boundaries between zones with different metal tenors in their contained sulfides, and the irregular way in which sulfide tenors vary in successive zones demand a model in which PGE enrichment is synchronous with the crystallization of the rocks themselves.

A model, modified after that of Campbell et al (1983), has been developed that can explain both new and old observations. The changes in MgNo of bronzite, Sr content of plagioclase, and variation in $(Sr^{87}/Sr^{86})_o$ ratio (reported by Kruger and Marsh, 1982) argue strongly for successive injection of pulses of new magma. As illustrated in Figure 9, it is proposed that the magma in the chamber of the intrusion was density stratified into a series of doubly diffusive, convecting cells (McBirney and Noyes, 1979). New pulses entered the chamber of the intrusion with a density that was lower than that of the magma immediately overlying the crystal cumulates, but greater than that at the roof of the chamber. Consequently, the new magma rose rapidly as a turbulent plume, drawing within itself and mixing with some of the magma that it was rising through, and spreading out at an intermediate level within the chamber.

The high PGE tenor of the sulfides of the Merensky reef is attributed to their segregating early, possibly as a result of hybridisation in the plume, and thus their mixing very efficiently with their host magma, both in the plume and during the subsequent very vigorous convective overturn of the hot new magma after it had spread out as a horizontal layer. The Merensky reef formed as a result of'roll-over', once cooling and crystallization caused the density of the sulfide-bearing new magma in the layer to exceed that of the underlying magma. Roll-over involved the formation of 'down-spouts' of the dense magma-crystal mixture which then spread out over the chamber floor as a horizontal orthocumulate layer; this subsequently became the reef.

The remainder of the new magma (that which did not participate in the initial roll-over) then mixed completely with underlying magma to give rise to a magma from which the bronzitite and norite crystallized. The replacement of the cotectic assemblage plagioclase-bronzite of the norite by the monomineralic assemblage plagioclase within a given cyclic unit is evidence that the crystallization of yet another magma, with plagioclase alone on its liquidus, was involved; this magma is thought to be that which overlay the plane at which the new injection spread out. If the thickness of magma overlying the Merensky reef horizon was significantly less than the present 6 km of rock above this horizon, so that the convective cell above the plane of injection

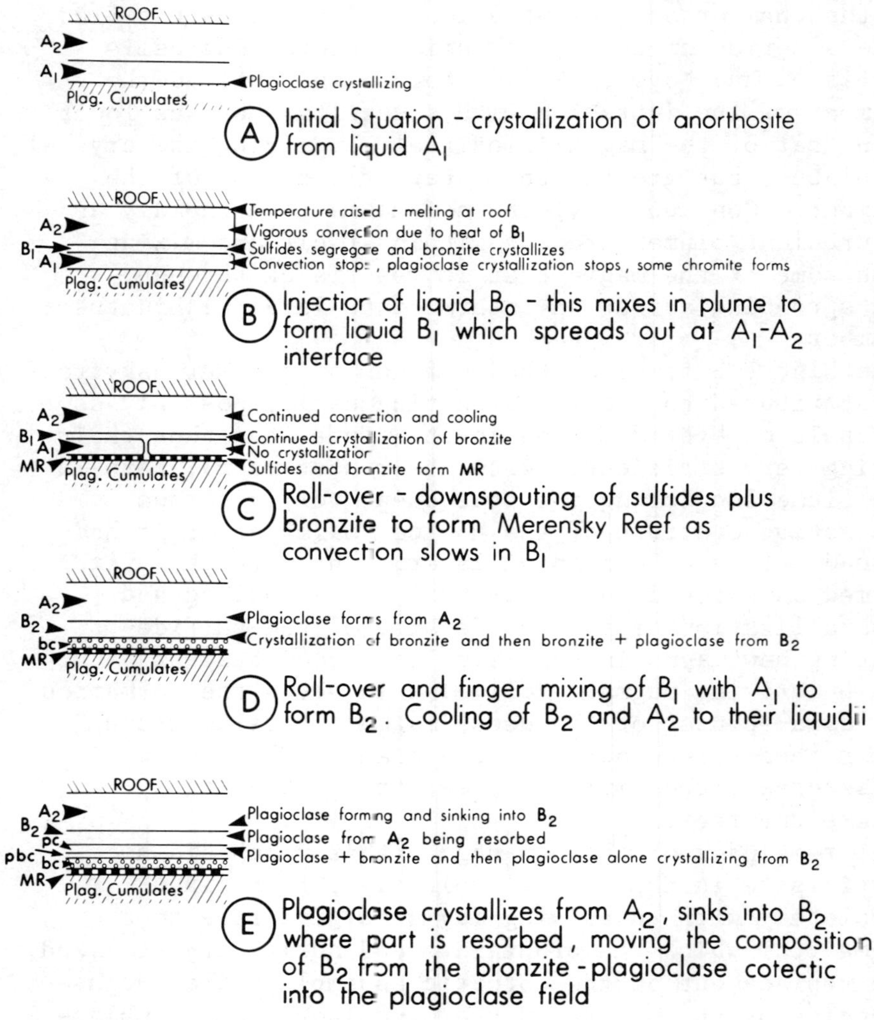

FIGURE 9. Diagramatic representation of the model for the genesis of the Merensky unit.

of new magma was in contact with the roof rocks, it is possible that the magma of this uppermost cell became contaminated with Sr from the roof; this would have had the result that the moderate $^{87}Sr/^{86}Sr$ initial ratio characteristic of Bushveld rocks beneath the Merensky horizon (about 0.705) was increased by the abundance of radiogenic Sr that characterizes the roof. The mixing of progressively greater amounts of the plagioclase crystallizing from the upper cell with that crystallizing co-tectically from the doubly saturated underlying magma accounts for the progressive change observed in Sr content and Sr initial ratio in the plagioclase cumulates of the Merensky unit.

Chromite seams at the base of the cyclic units, at the top of the Merensky pegmatoid, and around the peripheries of boulders are reaction products forming at the contacts between two dissimilar magmas. The zones and patches of pegmatoid are the result of volatiles becoming concentrated in the trapped interstitial fluid of orthocumulate-textured layers within the intrusion.

ACKNOWLEDGEMENTS

We wish to thank Johannesburg Consolidated Investments Ltd. for their advice with this project and for providing access to their mines and core farm on numerous occasions. All of their geologists were most helpful, but in particular we would like to thank Mr. Frank Vermaak, Dr. Morris Viljoen, Dr. Christopher Lee, Mr. Piet Coetze and Mr. David Kennedy. We also wish to thank Dr. C. Frick for access to the microprobe of the Geological Survey of South Africa, Mr. Cleland Wallace of the Geological Survey for microprobe analyses, Mr. John Thompson of the University of Toronto for microprobe analyses, and Dr. Ronald Hancock of the University of Toronto and Dr. Joy Sowa of Erindale College, University of Toronto for help with the neutron activation analysis for PGE. Much of the microprobe work and all of the gamma ray counting was undertaken in laboratories under the supervision of Dr. Michael Gorton of the University of Toronto and we thank him for his advice and help. Professor Naldrett was supported by a CSIR fellowship while in Pretoria during the early stages of this project. Subsequent expenses of the project have been supported by the Bushveld Research Institute of the University of Pretoria, a grant from the Cooperative Scientific Programmes of the CSIR to Professors Naldrett and Von Gruenewaldt, and NSERC grant no. A4244 to Professor Naldrett.

REFERENCES

Campbell, I.H., Naldrett, A.J., and Barnes, S.J.,(1983). A model for the origin of the platinum-rich sulfide horizons in the Bushveld and Stillwater complexes. Jour. Petrol., 1983.

Farquhar, J., (1984). The Geology of Western Platinum Mine, in "Some ore deposits in Southern Africa:Bushveld volum", ed. E. Vermaak. Geol.Soc. Southern Africa (in press).

Gain, S.B.,(1980). The geology and PGE distribution in the Upper Group Chromitite layers at Maandagshoek 254, KT, Eastern Bushveld Complex. Inst. for Geological Research on the Bushveld, Research Report, No.22, 24 pp.

Irvine, T.N., Keith, D.W., and Todd, S.G., (1983). The J-M Platinum Palladium reef: II, Origin by double-diffusive convective magma mixing and implications for the Bushveld Complex. Econ.Geol., 78, (in press).

Kruger, F.J. and Marsh, J.S., (1982). Significance of Sr^{87}/Sr^{86} ratios in the Merensky cyclic unit of the Bushveld complex. Nature, 298, 53-55.

McBirney, A.R. and Noyres, R.M., (1979). Crystallization and layering of the Skaergaard intrusion. Jour.Petrology 20, 487-554.

Naldrett, A.J., (1981). Pt group element deposits. In platinum group elements: Mineralogy,Geology,Geochemistry. (ed. L.C. Cabri). Can.Inst.Min. Met. Special Volume, 23, 199-231.

Stolper, Edward, (1980). A phase diagram for mid-ocean ridge basalts: Preliminary results and implications for petrogenesis. Contrib. Mineral. Petrol., 74, 13-27.

Vermaak, C.F., (1976). The Merensky Reef - Thoughts on its environment and genesis. Econ. Geol. 71, 1270-1298.

Walker, D., Shibata, T., and Delong, S.E., (1979). Abyssal tholeiites from the Oceanographer Fracture Zone. II. Phase equilibria and mixing. Contrib. Mineral Petrol., 70, 111-125.

GENERATION OF PRIMARY MAGMAS IN PRIMITIVE AND ALTERED MANTLE

I.D. RYABCHIKOV
Institute of the Geology of Ore Deposits, Petrography, Mineralogy and Geochemistry of USSR Academy of Sciences

ABSTRACT

P-T diagrams and the composition of liquid phases produced at various degrees of partial melting of primitive mantle lherzolite have been calculated. It has been demonstrated that primary magmas of komatiitic series were generated at great depths (at pressures 30-50 kbar) and at the degree of melting c.a. 50%. Complementary crystalline restite has harzburgite mineral composition. Highly magnesian basalts which may be primary melts with respect to MORB are generated at pressures ca 20 kbar and degree of melting 15-20% in the presence of olivine and two pyroxenes. At still lower degrees of melting primary magmas corresponding to high-alumina oceanic tholeiitic are produced. The available experimental data on fluid mass transfer under the conditions of upper mantle show that two subcrustal zones favourable for fixation of alkalis from ascending fluids may exist: in the uppermost part of the mantle where sodium may be predominantly accumulated and more deep seated zone enriched in potassium. It implies that potassic alkaline magmatism must have deeper source as compared with sodic alkaline basaltic magmas.

Introduction

A substantial progress has been achieved during the recent years in deciphering the composition of the earth's primitive mantle. It has been suggested by A.E. Ringwood (1975) that this composition could be bracketed by the most fertile peridotite nodules brought to the surface by alkaline basalts and kimberlites on the one hand, and by the most magnesian komatiites corres-

Proceedings of the 27th International
Geological Congress, Volume 9, pp. 391–407
PETROLOGY (IGNEOUS AND METAMORPHIC ROCKS)

ponding to the largest degrees of the partial melting of the mantle material, on the other. Afterwards the validity of this approach was demonstrated by the detailed studies. The nodules of the certain type of spinel lherzolites have been discovered which were proved by isotope geochemical investigations to be unaffected by any superimposed processes since the earliest stages earth's evolution, and which are characterized by ratios among refractory lithophile elements similar to chondritic and whose compositions are projected close to intersection of terrestrial and cosmochemical variation trends (Jagoutz et al., (1979). On the other hand, the majority of komatiites exhibit approximately chondritic ratios for those refractory lithophile elements which were not substantially captured by olivines and orthopyroxenes (Sun, 1982). The comparison of these facts shows that the bulk composition of early Archean upper mantle is fairly reliably established.

Nevertheless, the geochemistry of trace elements and isotopic unvestigations show that present day mantle sources of basaltic magmas are characterized by the long lived geochemical inhomogenity resulted from the depletion in incompatible elements due to the extraction of partial melts and/or enrichment in certain incompatible elements caused by mantle metasomatism.

The quantitative evaluation of these processes, changing with time the compositions of magma generating reservoir, is the principal aim of a present paper.

The calculations of subliquidus equilibria in dry primitive mantle

In order to find out the composition of the melt coexisting with the sesidual minerals of partially molten mantle material we used experimental results published in a number of works. These data involve compositions of clinopyroxenes and equilibrium melts from experiments performed with natural basalts at various pressures (Bender et al., 1978; Thompson, 1975), and also compositions of silicate liquids coexisting with two pyroxenes or two pyroxenes plus olivine (Bender et al., Ito and Kennedy, 1974; Mysen and Kushiro, 1977; Scarfe et al., 1979). We also utilized a value 0.3 ± 0.03

obtained by a number of authors for Mg/Fe distribution coefficient for coexisting olivine and melt. This value is rather insensitive to the variations in temperature, pressure and liquid phase composition. The same values of Mg/Fe distribution coefficients were accepted for phase assemblage clinopyroxene+melt and orthopyroxene+melt.

The distribution of estatitic component between coexisting clino- and orthopyroxenes was estimated from empirical formula derived by P. Wells (1977). Mole fraction of Ca in orthopyroxene coexisting with clinopyroxene was calculated using Margules parameters and free energy values for phase transitions in the systems $CaMgSi_2O_6$-$Mg_2Si_2O_6$ and $CaSi_2O_6$-$Fe_2Si_2O_6$ (Lindsley, 1981; Lindsley et al., 1981). Alumina contents in orhopyroxenes associated with garnet was estimated using the data by D. Perkins and R. Newton (1980) and M. Akaogi and S. Akimoto (1978). In order to find out the distriburion of Al and Ti between bwo coexisting pyroxenes we used experimental results published by I. Ryabchikov and D. Green (1978).

Methods permitting to relate concentrations of rock-forming components in melts and solid phases for peridotitic and basaltic bulk compositions were described earlier (Ryabchikov et al., 1982). These methods are based on the solution of simultaneous equitions including the expressions for apparent equilibrium constants for exchange reactions between clinopyroxenes and melts estimated using the above sources of experimental information, and also expressions describing constraints imposed by soichiometry of the present phases and electroneutrality principle.

Using this method we calculated the compositions of melts which may be in equilibrium with residual minerals of primitive mantle lherzolite (Jagoutz et al., 1979) as the functions of degree of melting (α) for 10, 25 and 50 kbar (Fig.1). These data were utilized for the construction of α-P diagram (Fig.2). Using the latter and taking into account the data for liquidus temperatures of the rocks belonging to komatiitic series with various MgO contents (Arndt, 1976; Bender et al., 1978; Bickle, 1978; Bickle et al., 1978) we obtained approximate P-T diagramme for the composition of primitive mantle (Fig.3).

The calculated diagrammes may be used for the evaluations of the compositions of generation of primary magmas characterized by diverse composition. The rough estimate of pressure and degree of melting is carried out using Fig.2 localizing the intersection of the appopriate MgO isopleth and line for the constant Mg-quent refinement of these results is reached on the basis of the comparison of calculated and observed concentrations for all major components.

We used this approach for the evaluation of the formations conditions for the rrimordial magma of komatiitic series described in South-East Karelia (Soviet part of Baltic shield) (Kulikova and Kulikov, 1981). These rocks are characterized by the relationships among Ca, Al and Ti similar to chondritic ratios, and their composition in general is typical for Archean komatiites. Teble 1 gives the comparison of the composition of spinifex textured peridotitic komatiites containing ca 28% MgO from Vetreny Poyas (Windy Belt) (Kulikova and Kulikov, 1981) with the calculated composition of the melt which should be produced at approximately 50% melting of primitive mantle at 35 kbar (Table 1, Nos 2, 3). Good consistency between calculated and observed concentrations for all major oxides except alkalis can be seen from these data. Alkalis were sufficiently mobile during greenstone alterations of ultrabasic effusives.

The estimated pressure of the generation for the magmas of more magnesian komatiites with ca 33% MgO is approximately 50 kbar at the degree of melting ca 60%. In both cases residual minerals of partially molten material are represented by olivine and orthopyroxene.

Melts close in coposition to the primary magmas of MORB series (Irvine, 11979) may be derived from primitive mantle at pressures close to 20 kbar and degrees of melting ca 15-20% are represented by lherzolitic mineral assemblage (olivine+two pyroxenes). In this field ratios between refractory lithophile elements are distorted as compared to approximately chrondritic characteristics of primitive mantle and the majority of komatiites. It results from the intense capturing of Ca, Ti and some other cations by clinopyroxene. Thus, for primitive mantle lherzolites and the majority of komatiites we have Al_2O_3/CaO=1.1-1.2 while for the calculated composition of partial melt at 21 kbar and $\alpha = 17\%$ Al_2O_3/CaO= 1.28 (Table 1, Nos 5). Primary magmas of MORB series are characterized by even higher values of this rato (Al_2O_3/CaO=1.35 - Irvine, 1979). It is likely to be caused by the fact that MORB magmas are derived not from primitive magma but from subcrustal material whose comosition had been modified during the foregoing episodes of the processes of magma generation.

It can be seen from Fig.1 that at even lower degrees of partial melting of mantle lherzolite melts with very high Al_2O_3 content should arise, which could be primary magmas for high alumina varieties of tholeiites in oceanic regions (Ryabchikov et al., 1982).

Possible composition of the MORB source material

Although it has been shown above that the melts with the concentrations of rock-forming oxides similar to MORB primary magmas may be derived by the partial melting of primitive mantle, isotope geochemistry convincingly shows that the mantle source of suboceanic basalts lost part of less refractory components (e.g. Rb as compared to Sr adn Nd in relation to Sm) billions years before present. The most likely explanation of this fact is the loss of a part of liquid phase during certain episode of anatexis in the past. Depth, at which such processes coud have taken place, may be estimated by comparing K/Rb raros in the basalts of various types. This ratio is close to 1000 in MORB, while in the basalts of oceanic islands, which, as it is usually assumed, are derived from less depleted mantl, K/Rb ratio is ca 300. In as much at at relatively shallow depth and in the absence of micas and samphiboles both K and Rb are stongly incompatible elements whose separation during partial melting appears inexplicable, we may hypothesize that these processes proceeded at sufficiently high pressures wher potassium ceased to be incompatible element, because it would be noticiably captured by clinopyroxene in the form of the potassium analogue of jadeite. According to experiments conducted by N. Shimizu

(1974) such conditions may be realized at pressurs above ca 50 kbar. Therefore, we have provisionally accepted 50 kbar as a pressure at which the depletion of mantle in less refractory components by the loss of a fraction of liquid phase took place, and which resulted in the formation of reservoir whose partial melting during the later epochs gave rise to MORB. Comparison of element ratios evaluated from the data of isotope geochemistry for primitive mantle with approximately chondritic ratios of lithophile elements (Rb/Sr=0.029; Sm/Nd=0.3) and for the source of modern MORB (Rb/Sr=0.008; Sm/Nd=0.38) permits to estimate the average degree of melting during early depletion as ca 10% and the fraction of liquid trapped in interstices between residual silid phases as 0.5%. The estimated composition of sillicate liquid generated at 10% melting of primitive mantle at 50 kbar and bulk composition of residual crystalline material are given in Table 1 (Nos 6 and 7).

Because the postulated parameters characterising this process of partial melting correspond to stability of garnet in quite extensive temperature range above solidus, liquid phase has Al_2O_3/CaO significantly less than 1 and may be represented by meimechitic or possibly kimberlitic magmas, while coplementary solid residuum would be correspondingly enriched in Al with respect to Ca as compared to the composition of primitive mantle. This conclusion is consistent with the composition of suboceanic upper mantle estimated by S. Maaloe and K. Aoki (1977) who give Al_2O_3/CaO ratio for xenoliths of spinel lherzolites from oceanic regions which is significantly higher than in primitive mantle (Al_2O_3/CaO=1.78 - Maaloe and Aoki, 1977).

If depleted mantle reservoir is indeed characterized by higher Al_2O_3/CaO ratio than primitive mantle, then the generation of primary highly magnesian primary magmas giving rise to MORB and which, as it has been pointed out above, also posses high Al_2O_3/CaO ratios, may proceed in the presence of much smaller amounts of residual clinopyroxene than in the case of the generation by partial melting of a liquid similar in composition from the material of primitive mantle. Table 1 shows the calculated composition of a melt produced by 16% melting of depleped mantle (Table 1, No 7) at 20 kbar. This composition is sufficiently similar to the estimated composition of the primary magmas for MORB series (Table 1, No 4) given by T.Irvine (1979). The degree of melting (16%) is close in this case bo the boundary between fields olivine+clinopyroxene+orthopyroxene+melt and olivine+orthopyroxene+melt (18%) and therefore the residual mineral assemblage should be represented in a given case by harzburgite with a small admixture of clinopyroxene.

Thus, the results of our calculations show that the melt fairly similar with respect to the contents of major oxides may be derived from primitive mantle and from the partly depleted mantle material. It implies that the depleted mantle of oceanic regions should not necessarily be characterized by certain fixed composition. It is much mbr likely that there is a whole spectrum of the compositions of rocks which lost in Precambrian various amounts of less refractory components. During the generation of the magmas of modern abyssal tholeiites crysalline restite may contain variable amount of clinopyroxene not only due to the variations in the degree of melting but also as a result of inhomogeneity of source material.

The variations of masses of clinopyroxene which were in equilibrium with the primary tholeiitic melts in the zones of magma generation are corroborated by the data on Sr isotope composition and Sr/Ca ratios in the part of Atlantic studied by FAMOUS project (White, 1979): with the constant values of $^{87}Sr/^{86}Sr$ ratios significant variations in Sr/Ca values are observed which testifies to the participation in magmatic equilibria of a phas of a phase rich in Ca, i.e. clinopyroxene.

The compositions of primary magmas estimated for tholeiites of oceanic islands are characterized by lower Al_2O_3/CaO ratios (e.g. for Kilauea Al_2O_3/CaO=1.19 - Irvine, 1979) as compared with MORB. It may be due to the number of reasons: different depths of magma generation, different degrees of melting during the generation of primary magmas and lower degree of depletion of source mantle material by less refractory components in case of oceanic islands. The last factor is confirmed by the data on isotope geochemistry of Sr and Nd. However, Ti/Ca and Ti/Al ratios in the basalts from oceanic islands are significantly higher (3 times and more) than it could be estimated from partial melting models of partly depleted or even primitive mantle.

Some investigators explained high concentrations of titanium in certain parts of upper mantle by mantle metasomatism. The transport of material by fluid phase in subcrustal horizons of the earth will be discussed in the next section of present paper.

Fluids and redistribution of chemical elements in mantle

Mineralogical and geochemical investigation of basaltic rocks and mantle derived nodules shows that under certain circumstances fluids were present and were chemically active in some parts of eartch's upper mantle. It is proved by the studies of fluid inclusions in the minerals of mantle origin, by the presence of micas and amphiboles in lherzolitic nodules and also by the data on geochemistry of isotopes and minor elements.

Experimental studies of equilibria among fluids and rock forming minerals at the parameters prevailing in upper mantle also confirm the possibility of intense mass transfer by compressed gases in subcrustal zones of the earth. E.g., according to our data total concentration of nonvolatile oxides in aqueous fluid equilibrated with forsterite and enstatite at 1100°C and 30 kbar is close to 35 wt% (Ryabchikov et al., 1982), while for the solutions coexisting with forterite and phlogopite under the same conditions this value exceeds 50 wt% (Ryabchikov and Boettcher, 1980). The solubility of sodium containing minerals is also fairly high. E.g., the total content of silicate components in aqueous fluid saturated with respect to jadeite at 650°C and 20 kbar is equal to 35 wt%, and with the further incrase in temperature this value grows rather quickly.

Therefore the solubilities of rock-forming mineral assemblages at mantle parameters is close to the solubilities of readily soluble salts at ambient conditions.

Extrapolating the available expermental data we may conclude that in the pressure range 30-40 kbar we may expect the presence of the critical end point in the system lherzolite-water, above which the gradual transition between aqueous-silicate fluids and hydrated sillicat melt takes place, i.e. sharp boundary between magmatic processes and fluid mass transfer should disappear.

Extensive variations of the solubilities of various compounds influids ensure profound chemical differentiation during the processes which include mantle fluids. This differentiation may be even more efficient due to the presence of maxima on solubility-pressure curves for the fluids in equilibrium with mantle lherzolite. The reality of such maxima is proved by the comparison of our experimental data for the solubility of jadeite in aqueous fluids, showing that this value decreases with rising pressure with the data for sodium aluminosilicates having low densities (albite, nepheline) whose solubilities increase with pressure.

Depth ranges corresponding to these maxima represent zones in the mantle where the precipitation of respective components from the ascending fluids shoud be most efficient.

As it follows from our experimental data, for sodium such conditions (that is distinct maximum of solubility) are realized at the boundary between spinel lherzolites, on the one hand, and plagioclase lherzolites or amphibole lherzolites on the other. In the first case sodium is present in the form of dense jadeitic component in clinopyroxene, while at the lower depths this element is present in the form of albitic component in plagioclase of sodium minal in amphibole with the far larger partial volumes than volume of jadeite. The decrease in partial molar volume of sodium in crystalline phases is the reason for the appearance of maximum in the concentration of this component in equilibrium fluid as function of pressure.

These data imply that the intense fixation of sodium from ascending fluids shoud proceed in pressure range 15-20 kbar. Ther are strong arguments in favour of the presence of amphibolized lherzolites in the respective depth rnage, in as much as their xenoliths are in many cases carried to the surface by alkali basalts.

A maximum of solubility in fluid as a function of pressure should also exist for potassium, although at much higher pressures (70-100 kbars) than for sodium. It is due to the fact that at shallower levels traces of potassium present in the mantle peridotite yield microscopic (and sometimes megascopic) grains of potassium minerals such as phlogopite or sanidine, while at higher pressures all thid amount of potassium may enter clinopyroxene in the form of potassic analogue of jadeite.

From this one may conclude that there should exist in the mantle two spatially sperated zones: the shallower one enriched in sodium and the second at greater depths with the elevated concentrations of potassium.

The involvement of these zones in the process of magma generation should give rise to sodic alkaline basalts in the case of shallower levels, and it should produce highly potassic basic melts when anatexis takes place at much deeper zones. Available petrological data indeed show, that the primary highly potassic magmas are of deeper origin as xompared to sodic alkaline basalts. It follows, for instance from the fact that the latter bring to the surface xenoliths of spinel lherzolites only, while highly potassic rocks in many cases contain nodules of garnet peridotites. All this is consistent with the conclusions from our experimental data. The enrichment of many kimberlites with potassium also testifies to the very deep source of this element. In this relation recent discoveries of diamondiferous highly potassic rocks lamproites in Australia seems to be very noteworthy.

REFERENCES

AKAOGI M.,AKIMOTO S. High pressure phase equilibria in a garnet lherzolite, with special references to Mg^{2+}-Fe^{2+} paritioning among constituent minerals. Phys. Earth Plan. Int., 1979, vol.19, pp.31-51.

ARNDT N.T. Melting relations of ultramafic lavas (komatiites) at 1 atm and high pressure. Carnegie Ist. Wash.,

1976, Ybk 75, pp.555-561.
BENDER J.F., HODGES P.N., BENCE A.E. Petrogenesis of basalts from the project FAMOUS area: experimental study from 0 to 15 kbars. Earth Plan. Sci. Lett., 1978, vol.41, pp. 277-302.
BICKLE M.J. Melting experiments on peridotitic komatiites. Progr. Exp. Petrol., 1978, vol.4, pp. 187-195.
BICKLE M.J., FORD C.E., NISBET E.G. The petrogenesis of peridotitic komatiites: evidence from high pressure melting experiments. Earth Plan. Sci. Lett., 1978, vol.37, pp. 97-106.
ELTHON D., SCARFE C.M. High-pressure phase equilibria of a high-magnesia basalt: implications for the origin of mid-ocean ridge basalts. Carnegie Inst. Washington, Y.B. 79, 1980, pp. 277-281.
GREEN D.H., NICHOLLS I.A., VILJOEN M.J., VILJOEN R.P. xperimental demonstration of the existence of peridotite liquid in earliest Archean magmatism. Geology, 1975, vol.3, pp. 11-14.
IRVINE T.N. Rocks whose composition is determined by crystal accumulation and sorting. In: The evolution of the igneous rocks (figtieth anniversary perspectives). 1979, pp. 245-306, Princeton Univ. Press, Princeton.
ITO K., KENNEDY G.C. The composition of liquids formed by partial melting of eclogites at high temperatures and pressures. J. Geology, 1974, vol.82, pp. 383-392.
JAGOUTZ E., PALME H., BADDENHOUSEN H., BLUM K., CENDALES M., DREIBUS G., SPETTEL B., LORENTZ V., WÄNKE H. The abundances of major minor, and trace elements in the Earth's mantle, as derived from primitive ultramafic nodules. Proc. Lunar Planet. Sci. Conf. 10th, 1979, pp. 2031-2050.
JAQUES A.L., GREEN D.H. Anhydrous melting of peridotite at 0-15 kb pressure and the genesis of tholeiitic basalts. Contrib. Mineral. Petrol., 1980, vol.73, pp.287-310.
KULIKOVA V.V., and KULIKOV V.S. New data on Archean peridotitic komatiites in eastern Karelia (in Russian). Doklady AN SSSR 1981, vol.259, p. 693-697.
LINDSLEY D.H. The formation of pigeonite on the join hedenbergite-ferrosilite at 11.5 and 15 kbar: experiments and solution model, Amer. Mineral., 1981 vol.66, pp. 1175-1182.

LINDSLEY D.H., GROVER J.E., DAVIDSON P.M. The thermodynamics of the $Mg_2Si_2O_6$–$CaMgSi_2O_6$ join: A review and an improved model. In: Advances in Physical Geochemistry, p. 146-172. Springer-Verlag, N.Y., 1981.

MAALOE S., AOKI K. The major element composition of the upper mantle estimated from the composition of lherzolites. Contrib. Mineral. Petrol., 1977, vol.63, pp.161-173.

MYSEN B.O., KUSHIRO I. Compositional variations of coexisting phases with degree of melting of peridotite in the upper mantle. Am. Mineralogist, 1977, vol. 62, pp. 843-865.

PERKINS D., NEWTON R.C. The compositions of coexisting pyroxenes and garnet in the system CaO-MgO-Al_2O_3-SiO_2 at 100-1100°C and high pressures. Contrib. Mineral. Petrol., 1980, vol.75, pp. 291-300.

RINGWOOD A.E. Composition and petrology of the earth's mantle McGraw Hill, N.Y., 1975, 584p.

RYABCHIKOV I.D, BOETTCHER A.L. Experimental evidence at high pressure for potassic metasomatissm in the mantle of the earth. Amer. Mineral., 1980, vol.65, pp.915-919.

RYABCHIKOV I.D., BOGATIKOV O.A., PILOYAN G.O., BABANSKY A.D. Primary magmas of series of oceanic tholeiites. Internat. Geol. Rev., 1982, vol.24, N 2, pp.143-156.

RYABCHIKOV I.D., and GREEN D.H. Role of carbon dioxide in petrogenesis of highly potassic magmas. In book: Problems of petrology of earth's crust and upper mantle (in Russian). Nauka Publishers, Novosibirsk, 1978, pp.49-64.

RYABCHIKOV I.D., SCHREYER W., ABRAHAM K. Aqueous fluids in equilibrium with pyroxenes and olivines at 20-30 kb and 900-1100°C. Contrib. Mineral. Petrol., 1982, vol. 79, pp 80-84.

SCARFE C.M., MYSEN B.O., RAI C.S. Invariant melting behaviour of mantle material: partial melting of two lherzolite nodules. Carnegie Inst. Washington, Y.B. 78, 1979, pp. 498-501.

SHIMIZU N. An experimental study of the partitioning of K, Rb, Cs, Sr, and Ba between clinopyroxene and silicate liquid at high oressures. Geochim. Cosmochim. Acta, 1974, vol.38, pp. 1789-1798.

SUN S. Chemical composition and origin of the earth's primitive mantle. Geoch. Cosm. Acta, 1982, vol.46, pp. 179-192.

THOMPSON R.N. Primary basalts and magma genesis. II Snake river plain, Idaho, USA. Contrib. Mineral. Petrol., 1975, vol.52, pp.213-232.
WELLS P.R.A. Pyroxene thermometry in simple and complex systems. Contrib. Mineral. Petrol., 1977, vol.61, pp. 129-139.
WHITE W.M. Geochemistry of basalts from the FAMOUS area: a reexamination. Carnegie Inst. Washington, Year Book 78, 1979, pp. 325-331.

TABLE 1. Estimated compositions of upper mantle and primary magmas (all compositions are recasted to 100 wt% of shown oxides)

	1	2	3	4	5	6	7	8
SiO_2	45.47	46.83	46.76	47.88	48.49	43.46	45.64	47.96
TiO_2	0.20	0.33	0.36	0.64	0.85	1.17	0.09	0.42
Al_2O_3	3.99	7.14	7.12	13.77	13.50	5.89	3.78	14.48
FeO*	7.88	10.55	10.20	9.48	9.06	13.81	7.22	8.39
MgO	38.58	28.48	28.33	15.86	15.42	23.96	40.20	16.26
CaO	3.52	6.39	6.54	10.18	10.57	9.95	2.81	10.95
Na_2O	0.33	0.02	0.64	2.08	1.93	0.96	0.26	1.73
K_2O	0.03	0.01	0.06	0.12	0.18	0.30	0.002	0.01
m**	0.897	0.83	0.83	0.75	0.75	0.76	0.91	0.78

* Total amount of iron oxides recalculated as FeO

** $m = Mg/(Mg+Fe)_{atomic}$

1. = primitive mantle according to Jagoutz et al (1979)
2. = average spinifex textured peridotitic komatiite from Kenozero--Sumozero greenstone belt, south-east Karelia (Kulikova and Kulikov, 1981)
3. = calculated composition of liquid phase produced during partial melting of (1) at 35 kbar and degree of melting equal to 52%
4. = primary magma of abyssal tholeiites estimated by Irvine (1979)
5. = calculated composition of liquid phase produced at 17% partial melting from bulk composition (1) at 21 kbar
6. = calculated composition of liquid phase produced at 10% partial melting from bulk composition (1) at 50 kbar (presence of potassium in clinopyroxene was neglected, though it may be important, see text).
7. = estimated composition of depleted mantle produced by subtraction of 10% of (6) from (1)
8. = calculated composition of liquid phase produced at 16% partial melting from bulk composition (7) at 20 kbar

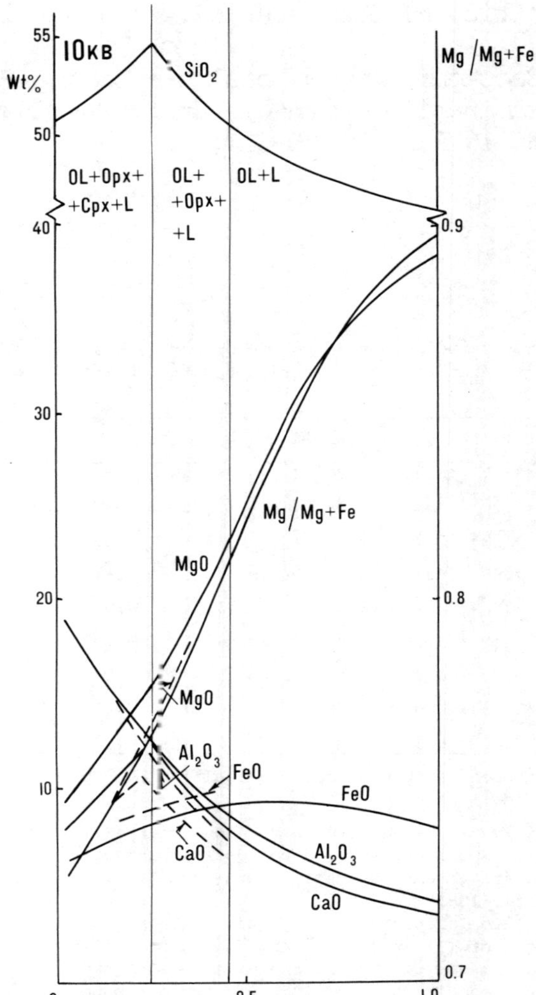

10KB
Wt%
SiO2
Mg/Mg+Fe
OL+Opx+
+Cpx+L
OL+
+Opx+
+L
OL+L
MgO
MgO
Al2O3
FeO
FeO
CaO
Al2O3
CaO
α
55
50
40
30
20
10
0
0.5
1.0
0.9
0.8
0.7

FIGURE 1(a).

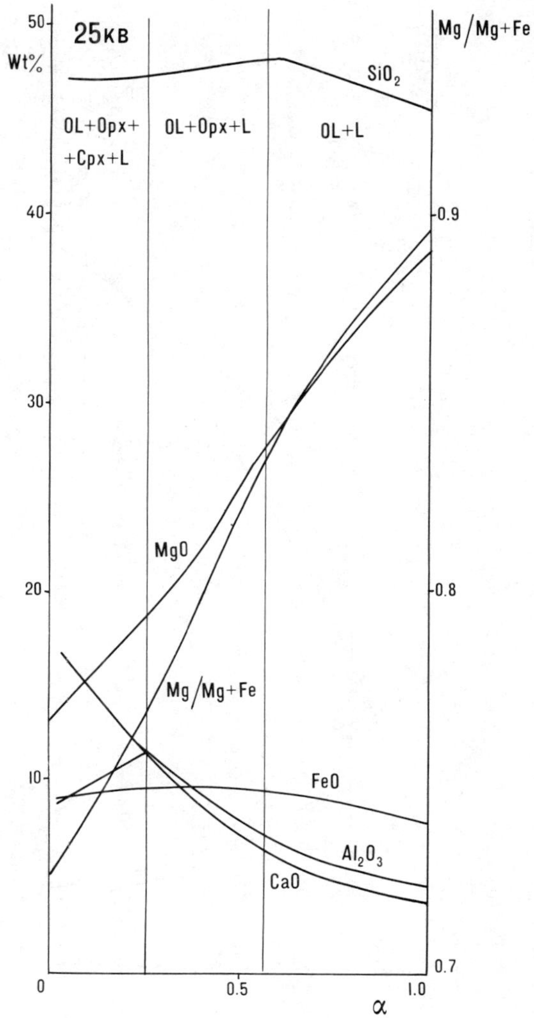

FIGURE 1(b).

FIGURE 1. Composition of liquid phases generated during partial melting of primitive mantle lherzolite (Table 1, No 1) as the functions of degree of melting (α) at 10 kbar (α), 25 kbar (b) and 50 kbar (c). At the diagramme for 10 kbar dashed lines show data by A.L. Jaques and D.H. Green (1980) for pyrolite bulk composition. L=liquid; OL=olivine; Opx=orthopyroxene; Cpx=clinopyroxene; Ga=garnet.

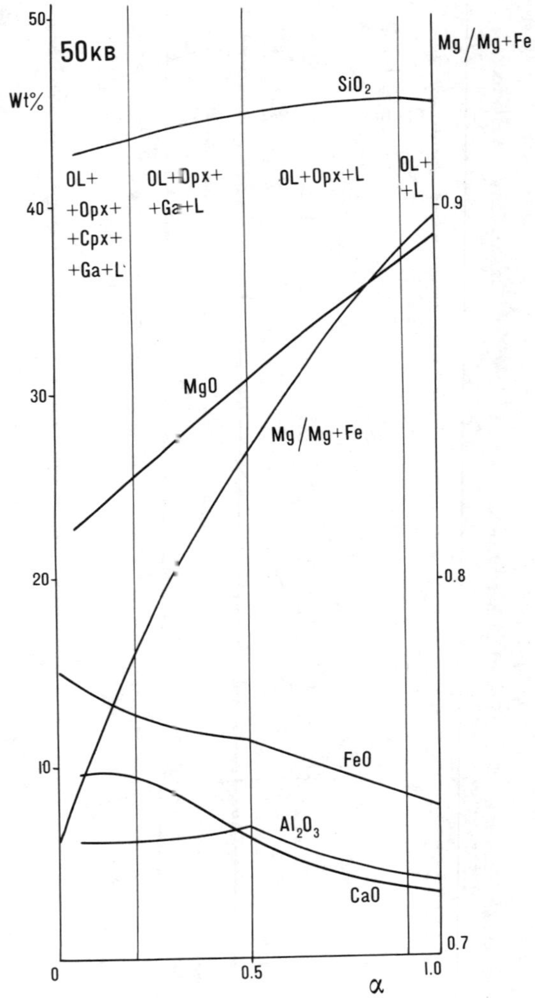

FIGURE 2. Stability fields of various phase assemblages in the melting range of primitive mantle lherzolite in coordinates degree of melting (α) - pressures (P). Symbols for the phases are the same as at Fig.1. Dashed lines are MgO isopleths for the melts; dotted lines correspond to equal values of Mg/(Mg+Fe) atomic ratios. Area covered by small dots corresponds to the melts with $CaO/Al_2O_3>1$.

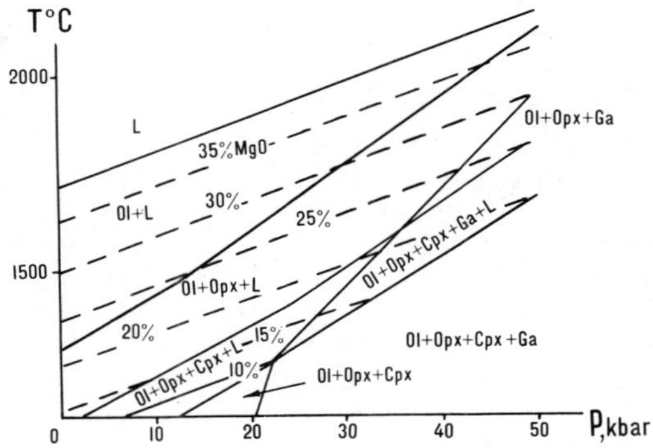

FIGURE 3. P-T diagram for primitive mantle lherzolite. symbols are the same as for Fig.1.

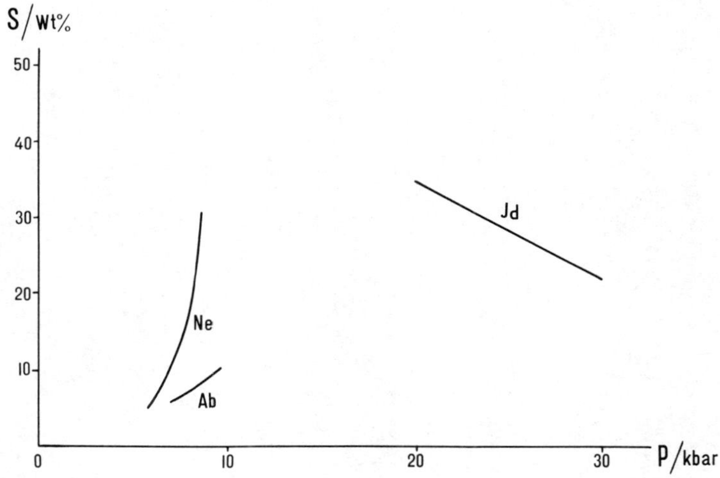

FIGURE 4. Solubilities (S) of albite (at 700-760°C), nepheline (at 750-900°C) and jadeite (at 650°C) in aqueous fluid as the funtion of pressure.

COMPARISON OF THE ANCIENT GREENSTONE BELTS' MAGMATITES WITH THE ROCKS OF THE MOON

SIMON A.K., GONJSHAKOVA V.I., FRIKH-KHAR D.I.
Institute of Geology Ore Deposits, Petrography, Mineralogy and Geochemistry of the USSR Adacemy of Sciences, Moscow, USSR

ABSTRACT

The magmatites of the nuclear stage greenstone belts are first subject in the Earth's history adequate to be compared with the rocks of the Moon, whose basic structural and compositional features were formed 4.6 - 3.0 billion years ago. The comparisons performed revealed substantial qualitative and quantitative differences between the two subjects. Sedimentary rocks played a significant role in the Earth's records. The combinations of volcanic and plutonic formations are different, too. The lunar magmatites differ dramatically from their terrestrial analogues in the contents of petrogenetic elements. It all shows that at the age of 2.5 - 3.5 b.y. ago the Earth witnessed the mechanisms and the processes of matter differentiation, different from those which determined the major structural and compositional properties of the continental, oceanic surface of the Moon.

Following E.V. Pavlovskyi /1982/, K. Kondi /1983/, E.V. Gaposhina, B.D. Dvorkina, V.A. Rudnik /1984, in press/ and a team of other investigators, the authors of this report acknowledge the age-diversity of the Archean greenstone belts and the irregularity of their origin in the Protogene, similarly to the geosynclinal foldbelts in Neochrone.

The natural periodicity of the Earth' record on the historical-geological basis suggests the existence in the Protogene of several stages of the lithospheric development within the platform segments thereof, despite the diverging views on the subject /Pavlovskyi, 1982; Bogatikov *et al*, 1983/. The stages which have been described best are presented in Table I.

The subject of the present report are the greenstone

Proceedings of the 27th International
Geological Congress, Volume 9, pp. 409–422
PETROLOGY (IGNEOUS AND METAMORPHIC ROCKS)

belts, primarily of the nuclear stage /granite-greenstone areas, according to Leonov/, parts of which started to be formed, possibly, at an earlier, "lunar" stage. The stage of granite-greenstone areas or the nuclear stag is, according to some investigators, the stage of the lithospheric development, during which the more or less "rigid" blocks of the sialic crust were being formed. Those are "protocontinents, according to J.Engel /1974/ and O.A.Bogatikov and co-workers /1983/ or "protoplatforms", according to E.V. Pavlovsky /1979, 1982/ and G.P.Leonov /1980/. In addition to peculair tectonics, characterized by the conception and growth of cu-olate plagiogranitic and gneissic "nuclei" and of the irrelgularly shaped "greenstone nuclei" in the negative intercupola intervals, the stage was characterized by a specific magmatism. Characteristic therof are rocks of granite-gneissic, migmatic and charnockitic ultrametagenic and magmagene formations and the initial granites of the rapakivi type in positive structures, whilst in negative "greenstone belts" - characteristic is a complex association of tholeiitic and ultrabasic rocks in volcanic and plutonic facies woth a strongly subdued quantity of andesites and of the sedimentary and tuffogene-sedimentary formations.

The stage under examination involves most of presently identified greenstone "belts" of the Canadian, Rhodesian-Transvaalian, Western-Australian, Baltic and Ukranian shields

Stratigraphic sections and vertical sequences of the nuclear stage greenstone belts

It is possible that some chrononlogically insignificant belts covered at their outset an earlier, lunar stage of the Earth's formation /the age of the Barberton belt's early deposits has been for example estimated by some investigators to 3,5 b.y./. However, in the historical and geological terms, all the belts, which have preserved the supercrustal formations and structures, belong to the nuclear, or to a younger stage. Meanwhile, the lunar stage covers only some relicts of the greenstone rocks, encountered within certain regions stratigraphically lower, than the greenstone belts of the nuclear stage.

It is known that within the supercrustal pile sections of the considered tectonic structures, volcanites are prevailing, occasionally saturated with interstratal near-surface intrusions. A significant role in the total mineral mass is played by the mezpabyssal and abyssal intrusives, primarlily of hte basic and acidic composition.

All investigators note a strong prevailence within the greenstone belts sections of the igneaous rocks, primarily those pertaining to the colcanic facies /70-85%/ over the sedimentary rocks /30-15%/, thus pointing to a deterministic role of the first-mentioned for the indentification and classification of the said structures. Using this guide, A.Glyckson /1980/ defined two essential belt types, one of which is characterized exclusively by a bimodal association of volcanites, the other - by either bimodal and calc-alkaline or by the latter alone.

It is possible that both belts are also age-diversed, but this way or another, both fit into the age interval of the nuclear stage and /which is important/ do not differ in the composition of the common magmatite varieties, though differing in rocks' combinations in section.

In the most general form, the formational sequences of the greenstone belts include the following magmagene and sedimentary formations:

a/ supercrustal:
- komatiite-basaltic;
- tholeiit-basaltic;
-dacite-riolitic;
greywacke-shale /including turbidites/;
arenitic;
ferruginous-silliceous;

b/ intercrustal:
- dunite-harzburgitic;
-peridotite-pyroxenite-gebbro-noritic;
- diorite-plagiogranitic.

A schematized view of the stratigraphic position and on the succesion of the above-listed formations in combined sections of various belts and/or individual structural units thereof, is shown in Fig.1.

Regarding the igneous rocks' sequences, which give an idea magmas' genesis and of the magmatic process development, many investigators hold the opinion that abundant here are rocks of the tholeiitic, calc-alkaline and alkaline sequences. Besides, a number of investigators: K.Kondi /1983/, V.S. and V.V. Kulikov /1982/, et al. use the greenstone belts to indentify another sequence - the komatiitic sequence, consisting mainly of ultrabasic effusives /normally referred to as peridotitic komatiites/ and of high-magnesium basalts /basaltic komatiites/.

According to K.Kondi /1983/, prevailing /from 50 to 75% in various belts/ are to tholeiitic sequence rocks; the komatiitic rocks' occurrences sonstitute quantities to 10%; exceptionally /Barberton/ they can amount to 25% of the total rock' volume; the calc-alkaline rocks can constitute from 5 to 40%; finally, the alkaline sequence rocks are a rarity, occurring, however, in insignificant quantities, e.g. trachytes of the Timiscoming group in the Abitibi belt.

A comparative analysis of the composition of the Earth's greenstone belts' magmatites with the continental-oceanic rocks of the moon

It is known /e.g. from Schmitt, 1975, Wood, 1975/ that the basic surface-forming processes of the Moon took place 4.6 - 4.4 - 4.1 - 3.9 - 3.0 b.y. ago. The resulting rocks experienced occasional meteoric bombardments and a superficial impact of the solar wind, i.e. were practically preserved. In the contemporary structure of the lunar crust these rocks, which are undeniably of magmatic origin, form huge structural storeys. The most ancient rocks make up continents. They formed during the initial period of the Moon's development at the time of its shell differentation /4.6 b.y./ and in the course of its successive magmatic activity /4.4 - 4.1 b.y./. The rocks are represented by plutonic associations:

- troctolite-spinel troctolite-olivinic gebbro;
- highly aluminiferous olivinic gabbra- olivinic

norite-leucogabbro-anorthosite; and by a volcanic association of bipyroxenic basalts.

The lunar oceans are made up by rocks, whose age is 3 b.y. Prevailing here are the associations of volcanic rocks: picro-basalt-olivinic basalt; basalt-leucobasalt; olivinic basalt-ilmenite = olivinic basalt - ilmenitic basalt. Distinguished are rocks of plutonic associations: peridotite-olivinic melanogabbro; olivinic high aluminiferous gabbro-leucogabbro-anorthite; olivinic high aluminiferous gabbro-leucogabbro-anorthosite; pyroxene-ilmenite = olivinic gabbro.

Naturally, it has been suggested that the lunar magmatic associations broadly feature analogues of the synchronous or terrestrial magmatites of the nuclear stage greenstone belts, i.e. the most ancient terrestrial rocks which preserve features of the original composition and structure.

This offers new opportunities for looking into the composition of the original terrestrial crust, as well as into the process of its formation and into some other problems of the "pre-Archean" and "early-Archean" geology.

We have endeavoured to compare the essential data on the composition and chemistry of the lunar rocks with similar data on the greenstone belts taken from K. Kondi /1983/, A.P. Vinogradov /1975/, V.L. Barsukov *et al*. /1979/. We have also drawn petrochemical diagrams in order to distinguish the characteristic compositional features of the lunar and nuclear rocks/see, for example, Fig. 2a, 2b/ and have taken account of the statistical data on the occurrences of the major varieties in sections and territorially.

In contrast to the assumption about the similarity between the lunar and nuclear rocks, they exhibit substantial differences which amount to:

- the differences of the chemistry of the common groups and affinities of rocks, similar in mineral composition and in structural and petrographic features. Normally, the lunar rocks are poorer in silica and alkalies /esp. sodium/, but are richer in magnesium and especially titanium. The Fenner's evolution tendency in the lunar rocks is generally more prominent than in the nuclear stage rocks. The ferruginity of various lunar basalts is by 10-20 and more units higher, relative to their nuclear terrestrial counterparts /Table 2/, e.g. even at lower silicity thereof;

- the difference in the combinations of magmagene rocks and, especially, in their quantitative ratios in sections of the nuclear greenstone belts and generally of the granite-greenstone areas on the one hand and in the composite "section" of the continental and oceanic areas of the visible side of the Moon - on the other hand. Among the volcanic belt formations the ultrabasic - basic/komatiitic/ formation takes up the thickness from 3 to 16% /rarely, less than 1% and more than 20%/, the tholeiite-basaltic formation - from 3 to 5%, the andesitic formation from 2 to 25% /rarely from 0 to 40%/, the dacite-riolitic formation from 3 to 15%. Among the plutonic formaions inside the belts a minor part /according to the occurrence area/ is taken up by the dunite-harzburgitic formation; a wider abundance is enjoyed by the formation of the interbedded peridotite - pyroxenite - norite - gabbro intrusions and by the formaion of the tonalite-granodiorites. From 15 to 30% of the greenstone sections' thickness is taken up by sedimentary formations. In the composite "section", i.e. the combination of the continental and oceanic rocks of the Moon, the total ratio of the basic plutonic and volcanic rocks, including their high-aluminiferous, high-titanium and KREEP-varieties, practically approaches 100%. The ultramafites, granophyres and ferrogabbro represent exotic findings, though the latter are believed to still stronger emphasize the high ferruginity of the lunar formations; meanwhile the granites have not been encountered;

the existence within the granite-greenstone areas of huge fields /ocupying more than a half of the total area/ of gneissoid granodiorite - trondhjemite - tonalitic rocks, which formed either synchronically or prior to the conception of the nuclear greenstone belts /basaltic conglomerates, etc./.

Conclusions

The above-listed and still some other differences

between the magmagene rocks of the Moon and the nuclear stage of the Earth's development lead to the conclusion that at the age of 2.5 - 3.5 billion years ago, the mechanisms and the scale of matter differentiation on the Earth differed from those which determined the major structural and compositional features of the continental-oceanic surface of the Moon.

Thus, the stage of the granite-greenstone areas' formation has set back from our view all the abovesaid processes, deterministic for the Earth's development, as well as the formations produced thereby and has substantially obscured them from clear understanding.

Undoubtedly, the basic features of the initial sialic crust had already formed by the beginning of the said stage and a certain mechanism of the lithospheric matter blending had been activated. On the one hand that was expressed through the involvement into the magmatic process of the sialic matter, through the appearance of substantial acid rocks and through the general reduction of the common varieties of the basic magmatites, relative to the Moon; on the other hand - through the appearance of a great number of ultrabasic abyssal formations in the upper horizons of the lithosphere.

The sialic protocrust, whose existence has been clearly observed from the outset of the nuclear stage, presumably differed dramatically from the composition of the magmatic rocks, formed as a result of the Moon's magmatic processes and making up the crust thereupon.

Thus, it can be suggested that the magmatism of the initial Earth's development was either altogether different from the lunar magmatism, or its products were substantially modified in to the pre-nuclar stage of the Earth's evolution. In the latter case, according to the absolute age determination, at the time of the active magmatism on the Moon, the green magmatites, similar to the lunar counterparts, would not only be formed, but also be modified through both the endogenic and exogenic processes.

At this moment we arrive at the conclusion that for both the first and the second evolution variants the initial stage /stages?/ of the Earth's magmatism and evolution differed from the evolution of the Moon within the analysed age interval of the latter /4.6 - 3.0 b.y. ago/.

REFERENCES

Barsukov V.L., Dmitriev L.V., Garanin A.V. The basic geochemistry of the lunar rocks. In.: The Soil of the Continental Region of the Moon., Moscow, Nauka, 1979.

Bogatikov O.A., Borsuk A.M., Dmitriev Yu.I., Kovalenko V.I., Pyabchikov I.D. Magmatic formations in the evolution of the lithosphere. Izvestia AN USSR, Geol. series., 1983, No 1.

Bogatikov O.A., Dmitriev Yu.I. Basalts and anorthosites of the Moon. In.: The Problems of Petrology., Moscow, Nauka, 1976.

Vinogradov A.P. The differentiation of the matter of the Moon. In.: Cosmochemistry of the Moon and Planets., Moscow, Nauka, 1975.

Wood J.A. Survey of the lunar rocks types and comparison of the lunar and the terrestrial crust. In.: Cosmochemistry of the Moon and Planets. Moscow, Nauka, 1975.

Gaposhina E.V., Dvorkina B.D., Rudnik V.A. The petrochemical evolution of the Archean greenstone belts' volcanism. In.: The Role of Magmatism in the Evolution of the Lithosphere., Moscow, Nauka, 1984 /in press/.

Glykson A. The stratigraphy and evolution of the primary and secondary greenstone complexes; data from the shields of the Southern semisphere. In.: The Early History of the Earth., Moscow, Mir, 1980.

Goodwin A.M. Ghe gigantic meteoric bombardment and the development of the continental terrestrial crust. In.: The Early History of the Earth., Moscow, Mir, 1980.

Kondi K. The Archean greenstone belts., Moscow, Mir, 1983.

Kulikov V.S., Kulikova V.V. et al. The Sumozerskyi-Kenozerskyi greenstone belts. The Southern-Vagozerskyi greenstone belts. In.: The Volcanism of the Archean Greenstone Belts of Karelia., Leningrad, Nauka, 1981.

Kulikov V.S., Kulikova V.V. The ultrabasic effusives

in the Pre-Cambrian of the south-eastern Baltic shield. In.: Mantle Xenolites and the Problem of the Ultrabasic Magmas., Novosibirsk, Nauka, 1983.
Leonov G.P. Historical Geology. MGU Publication, 1980.
Nestoyanova O.A. The greenstone complexes of the early Pre-Cambrian of werstern Karelia. In.: The Greenstone Belts of the Ancient Shields., Moscow, Nauka, 1982.
Pavlovskyi E.V. The problem of the "greensone belts" of the early Pre-Cambrian., Izvestia An SSSR. Geol. Series., 1979, No4.
Pavlovskyi E.V. The evolution of the volcanism and of the tectonics of the early Pre-Cambrian. In.: The Greenstone Belts of the Ancient Shields., Moscow, Nauka, 1982.
Rybakov S.I., Svetova A.K., Robonen V.I., Raevskyay M.B. The formational composition and major petrochemical evolution features of the volcanism. In.: The Volcanism of the Archean Greenstone Belts of Karelia., Leningrad, Nauka, 1981.
Savochkina E.N. The ultramafites of the greenstone belts of the Western Karelian Early Pre-Cambrain., In.: The Greenstone Belts of the Ancient Shields., Moscow, Nauka, 1982.
Sivoronov A.A., Berzenin B.Z., Maljuk B.I. et al. Metamorphized volcanogenous formations of the early Pre-Cambrian greenstone belts of the Ukranian shield. Article 1. Structure and Composition. Geol.J., 1981, vol.41, issue 5.
Scmitt G.G. The evolution of the Moon: Model of 1974. In.: Cosmochemistry of the Moon and Planets., Moscow, Nauka, 1975.

FIGURE 1. The typical sections and formations of the World ancient shields' nuclear stage greenstone belts
Names of the shields and belts:
1. Canadian shield, the Abitibi belt. 2. Rhodesian-Transvaalian shield, the Barberton belt. 3. Western-Australian shield, the Kalgruli-Norsman belt a/ Kulgardi-Karravang b/ region to the east of Kalgurli.
4. Hindustan shield, the Dharvar belt. 5. Baltic shield, Gimolsko-Kostomukshskyi belt. 6. Ukranian shield, Verkhotsevskyi trough. 7. Schematic composite section of the visible surface of the Moon.
The formations of magmatic and sedimentary rocks:
1 - tholeiite-basaltic. 2 - komatiite-basaltic.
3 - greywacke-shale. 4 - of polymictic conglomerates.
5 - dacite-riolitic. 6 - of ferruginous quartzites.
7 - arenitic. 8 - arkose-shale. 9 - basit-ultrabasitic rocks presumably /according to S. Nakvi/ of a lunar type. 10 - of gneissoid tonalites / the gneisses of Dharvar/. 11 - carbonate-amphibolitic /modified tholeiites and carbonates/. 12 - of the grey banded gneisses. 13 - andesitic. 14 - ultramafites of the Moon. 15 - complex /continental plutonic rocks of the Moon/. 16 - of the cryptic basalts of the Moon.
17 - of the peridotites, olivinic gabbro of the Moon.
18 - of the oceanic basalts of the Moon.

Note: thickness is given in km; most of the sections have been grossly schematized, the names of the formations pick up the names of the dominating rock varieties.

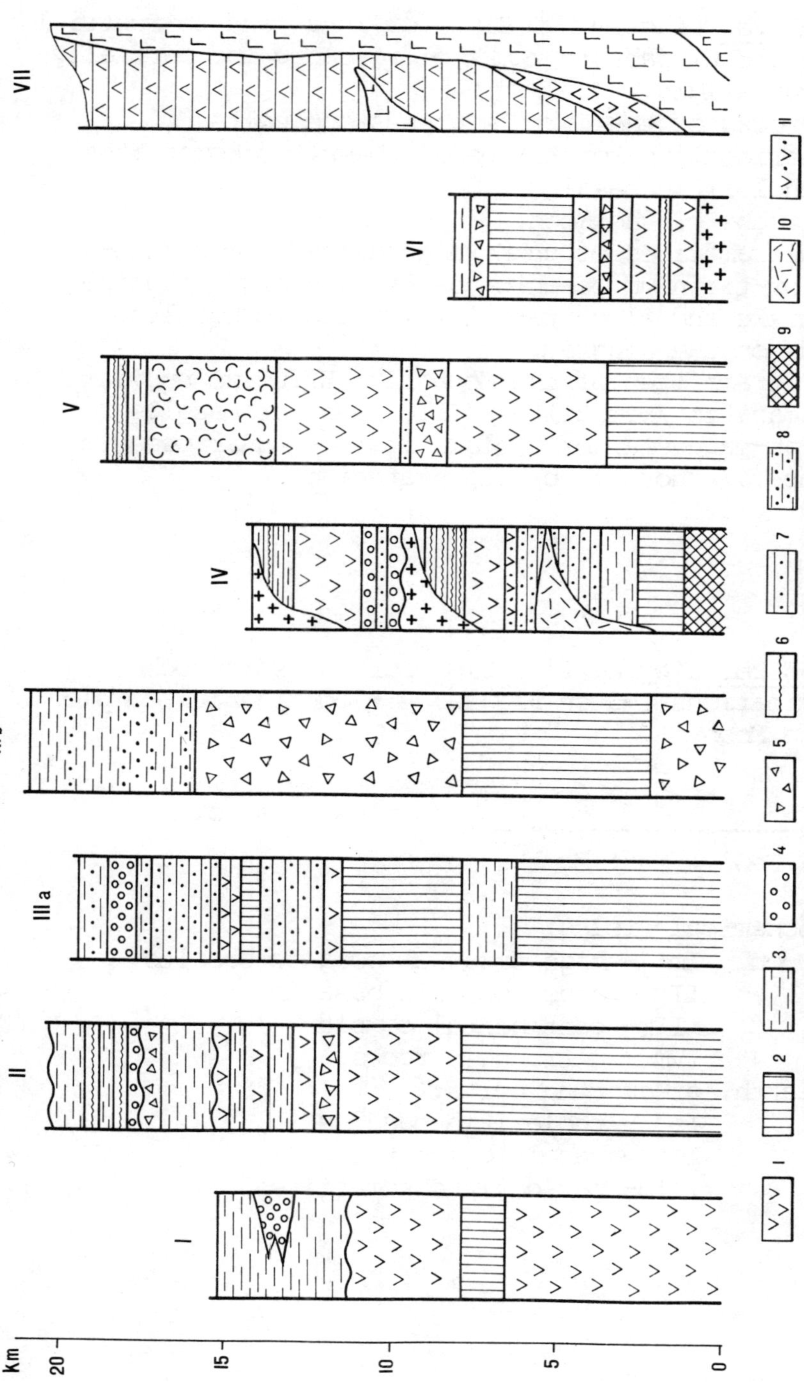
Km
20
15
10
5
0
I
II
III a
III b
IV
V
VI
VII
1
2
3
4
5
6
7
8
9
10
11
12
13
14
15
16
17
18

FIGURE 2a. Diagram $MgO-SiO_2$ for lunar and terrestrial basic and ultrabasic rocks of volcanic and partially unknown arigin
E. - rocks of the Earth's greenstone belts:
1 - peridotitic komatiites 2 - basaltic komatiites
3 - tholeiitic basalts
M. - rocks of the Moon:
1 - ultramafites of unknown genesis 2 - moderate- and low-titanium oceanic basalts 3 - high-titanium/ ilmenitic and ilmenite-olivinic oceanic basalts.
Fields on the diagram:
I - Ultramafites of the Moon II - Effusive nuclear ultramafites /komatiites/ III - Nuclear basalts a/ high-magnesial b/ tholeiitic IV - Lunar oceanic basalts a/ "normal" b/ high-tatanium

FIGURE 2b. Diagram $f' - SiO_2$ for the same rocks. Denominations 1-3 on Fig.2a; 4 - continental complex of the Moon

$$f = \frac{FeO'}{FeO' + Mn + Mg} \quad \text{mass \% ratio.}$$

Petrochemical fields:
The Moon - Im - high titanium oceanic basalts
IIm - other oceanic basalts
IIIm- continental complex
IVm - ultrabasic rocks
The Earth, greenstone belts:
Ie - tholeiitic basalts
IIe - basaltic komatiites
IIIe- peridotitic komatiites

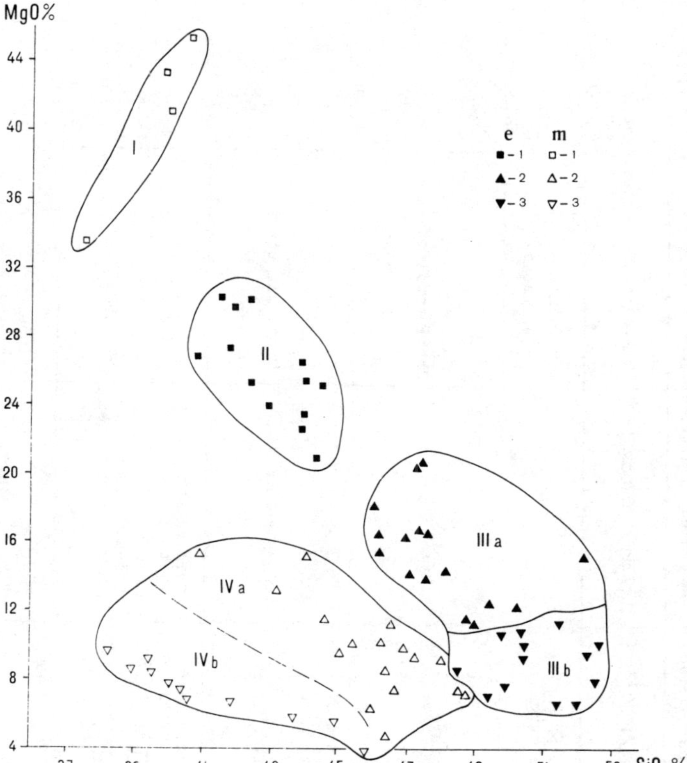
MgO%
SiO2%
I
II
III a
III b
IV a
IV b
e
m
■ - 1
□ - 1
▲ - 2
△ - 2
▼ - 3
▽ - 3

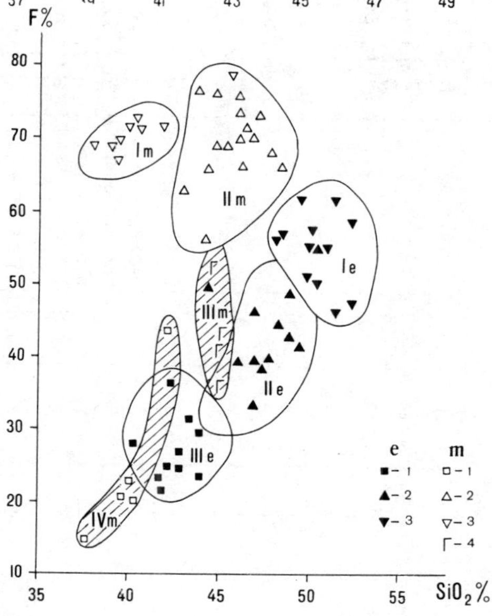
F%
SiO2%
I m
II m
III m
IV m
I e
II e
III e
e
m
■ - 1
□ - 1
▲ - 2
△ - 2
▼ - 3
▽ - 3
Γ - 4

Table 1. The natural periodization of the Earth's early history, according to different investigators

Isotopic age /b.y./	Geochron	Eon	A.M.Goodwin /1980/	G.P.Leonov /1980/*			E.V.Pavlovskyi/1982/	O.A.Bogatikov et al. /1983/	Isotopic age /b.y./
0.5	Neogene	Phanerozoic	Platform stage	Platform interval	Plate stage	Transition s.l.		Continental-oceanic stage	0.5
		Proterozoic			Pre-plate stage			Continental stage	
1.5	Protogene			Period of transition /S str /			Protoplatform stage	Cratonnal stage	1.5
				Pre-platform interval	Protoplatform stage				
2.5		Archean	aggregation of the protocontinental nuclei		Granite-greenstone areas' development stage /3.2-3.3 b.y./		/termination from 2.4 to 3.0 b.y./ Nuclear stage / Protogeosynclinal	Nuclear stage	2.5
3.5			Nuclear stage		Initial stage		"Lunar" stage	"Lunar" /of the initial crust/ stage	3.5

* G.P.Leonov and A.M.Goodwin provide periodization only for the platform sectors of the Lithosphere.

ON THE TYPIFICATION OF MAGMATITES AND THEIR POTENTIAL ORE-CONTENT

TAUSON L.V.
Vinogradov geochemistry Institute, Ac.Sc., Irkutsk, USSR

The potential ore-bearing of magmatic rocks is determined by the character of initial magmas and the conditions of processes of the crystallization emanation differentiation at their intrusion, the drystallization and cooling. Available data on the distribution of ore and trace elements in magmatites show that in magmatic processes dominates the crystallo-chemical dispersion of these elements. From this in the earth-crust the basic mass of ore and trace elements is in a state of dispersion. For instance, in the first kilometre of the continental crust the correlation of the concentrated (predicted resources) and dispersed forms of such different, upon geochemical relations, metals as copper and niobium makes 1:10,000. After V.I. Vernadsky (2) we can maintain, that the dispersion 'must be considered as a characteristic property of the matter of our planet'.

Thus, the basic atom mass of trace elements is dispersed in the forming magmatic rocks and does practically not take part in the following postmagmatic processes.

Against the ore and trace elements the volatile magma components (H_2O, Cl, CO_2, S etc.) do generally leave the magmas at their intrusion and crystallization. Thereby they do in vaporgas phases involve a part of the ore and trace elements. Several of them (water, fluorine) act as modifiers of silicate melts able to substantially change the structure and thermophysical properties of the latter. An important petrogenic consequence of this modifying role of volatiles is the possibility of formation of 'low-temperature melts' (ongonite, elvan) with a lowered viscosity.

Solution processes of volatile components in magma, their separation from the magmatic melts as vapo-gas mixtures and the migration of the latter within the magmatic chamber are united by the emanation process of

Proceedings of the 27th International
Geological Congress, Volume 9, pp. 423–436
PETROLOGY (IGNEOUS AND METAMORPHIC ROCKS)

magma differentiation. After the scale and consequence this process is like the process of crystallochemical matter dispersion and does finally determine the potential ore-bearing of magma.

When the initial volatile concentration in magmatic melts is defined by the character of initial magmas then the further geochemical history of volatiles and their ore and trace elements is largely determined by the particularities of the emanation differentiation in the magmatic chambers. Naturally, the character of this differentiation does greatly depend on the composition of the initial magma, the size of the magmatic chambers and the depth of the intrusion standing.

Upon the latter two parameters there are distinguished four morphological groups of intrusives: the abyssal and mesoabyssal batholiths, hypabyssal and subvolcanic intrusions.

In the abyssal batholiths the distribution of ore and trace elements is regulated largely by the processes of crystallization differentiation leading to a gradual dispersion of trace elements and especially in rocks of the major intrusive phases. The volatile magma components and their elements migrate into the lower parts of batholiths enriching the late acid differentiates of the magmatic chambers. The processes of emanation differentiation are here rather limited. In mesoabyssal batholiths, being usually displaced magmatic bodies, the processes of crystallization and emanation differentiation are weakened. They are characteristic of a plurality of facies varieties in the major intrusive phases, the unevenness of trace-element distribution and also of a usual degasedness of acid differentiates.

In hypabyssal and subvolcanc intrusives is characteristic a rare weakening of the crystallization and a heightening of the emanation differentiation. The influx into the apical parts of such intrusions of a considerable quantity of volatiles leads here to the formation of local chambers of 'low-temperature' melts differing by a rare enrichment of trace elements. In the intrusions of these morphological types the observed unevenness of the trace-element distribution is defined almost by emanation differentiation processes. However, as mentioned above, the potential ore-bearing of the magmatic rocks is not only determined by the conditions of the emanation differentiation processes in the magmatic chambers but also by the particularities of initial magma composition. Such a significance of the initial magma composition has predetermined the working-out of principles of geochemical typification of magmatic rocks.

The study of the magmatite geochemistry showed that even in the case of petrochemically close rocks the content of trace elements there can substantially differ as to the geochemical type. For instance, in the case of granitoids the most changeable contents are K, F, Li, Rb, Sr and Ba. The ore element concentrations in granites are not so characteristic and do slightly change from type to type.

Advantageous for the geochemical typification are leucocratic granites (Table 1).

The data show that within this analogous, on mineralogical and petrochemical features, granite group the ratio $F.(\frac{Li+Rb}{Sr+Ba})$ changes within four orders.

It seems that such a considerable difference in the trace-element composition of like rocks can be taken as a strong proof of the effective use of the priciples of the geochemical typification of magmatic rocks.

As a geochemical type of magmatic rock should be understood a rock group characteristic of the same conditions and mode of formation which are reflected by the confinement to a definite geodynamic situation, similarity of chemical, low-element and mineral compositions and also by the formation in similar geological conditions of ore-magmatic systems close in the potential ore-bearing.

First we have done the geochemical typification of granitoids (6).

It is known that at present the majority of investigators do agree on the possibility of granite formation by three different ways.

The basic granitoid mass cropping out, apparently, did originate as a result of the palingenic melting of the continental crust matter. This is especially true of the Phanerozoic granitoids. Their smaller part could form as a result of the magma differentiation of the basic and median compositions. Finally, a granitoid part, probably a considerable one in the precambrian stages of the Earth's geological history, could come up by anatexis, ultrametamorphism and rock granitisation of the crystalline basement of the continental crust. But, within these large genetic granitoid groups can considerably differ upon the formation conditions, potential ore-bearing and the content level of the indicator trace-elements.

Our investigations came up with twelve geochemical granitoid types.

The first four ones form by magma differentiation of the basic and median compositions. Among them there are distinguished plagiogranites of the tholeiitic series, granites of the andesitic series, granitoids of the latitic series and agpaitic rare-metal granites. The first three ones are acid differentiates of the tholeiitic, andesitic and latitic (shoshonitic) magmas, but the latter are understood as derivatives of the alkaline-basalt magmas.

Palingenic granitoids are also divided into four types: granitoids of the calcareous-alkaline series, plumasitic raremetal leucogranites, granitoids of the alkaline series and alkaline raremetal granites.

Palingenic calcareouslalkaline granites form in the region of a low metamorphism whereas granitoids of the alkaline series do form in regions of a deep metamorphism. Plumasitic raremetal leucogranites are spatially and apparently also genetically related to the batholiths of the calcareous-alkaline granitoids but alkaline raremetal granites are related to granitoids of the alkaline series. The last four geochemical types belong to the group of ultrametamorphic granitoids. Among them are distinguished enderbites, charnockites, rapakivi and proper ultrametamorphic granitoids.

Geochemical studies of granulite-gneiss complexes (5) show that enderbites are to take as a product of ultrametamorphism and melting of plagiomagmatites, charnockites can be related to the shadow K-spathic migmatites. The rapakivi granites do form a proterozoic granitoid group separate in time and conditions of formation. The ultrametamorphic granitoids are but a result of a deep metasomatic protocrust transformation.

The raremetal characteristic of the distinguished geochemical granitoid types is given in Table 2. The data show that the greatest changeability of concentration lies with rubidium (from 3 to 440 ppm) and barium (from 40 to 2300 ppm). The content of ore elements in granites changes only slightly. In comparison with the median granite the higher tin content is characteristic only for plumasitic and agpaitic raremetal granites. And when with plumasitic raremetal leucogranites are known rather large tin deposits then with the agpaitic ones there are no large deposits. The geochemical typification of basites (4,7) following that of granitoids made it possible to distinguish eight of their basic types: anorthosites, komatiites, basalts, oceanic and continental tholeiitic basalts, Na- and K-alkaline basalts, andesitic and latitic series. In the last two the most widespread are not rocks of the basic composition (andesite-basalts and shoshonites) but rocks of the median composition (andesites and latites). Besides, in these associations are widespread also acid rocks (granitoids of andesitic and latitic series).

Table 3 shows that in basites the greatest concentration changeability goes with Ti, K, Rb, Sr, Ba, Cr and Ni. Therefore for them are especially informative the ratios $\frac{K}{Ti}(\frac{Sr+Ba}{Cr+Ni})$, changing from 0.002 in komatiite basalts to 500 in latites, that is more than by five orders. The established geochemical types of granitoid and basite are confined to a definite geodynamic situation.

Anorthosites and komatiites and also all granitoids of the ultrametamorphic group are mainly widespread within large lithospheric plates with the continental crust and formed on its early development stage.

Oceanic tholeiitic basalts are mainly composed of lithospheric plates with the oceanic crust and generate within the median oceanic ridges. Besides, they are constantly observed in zones of continent-ocean joint. In the latter they are in the frontial parts of the Banioff-Zavaritsky zone, neighbouring with Na- and K-alkaline basaltoids. Continental tholeiites are widespread within large lithospheric plates with the continental crust.

Rocks of andesitic series occupy the median part of the Benioff-Zavaritsky zone and the latitic ones their rear part, the continental one. The mentioned displacement regularities occur clearly in the Andes. There on the eastern slope of the Peruan-Chilian trough are delоped tholeiites, on the Pacific coast of Peru and Chile are developed rocks of the andesitic series and in the Central Andes (Bolivia) show rocks of the latitic series. In the development regions of these series are often met intermediate varieties which go under the name of latiandesites. Besides, in the median and especially in the rear parts of these zones develops a palingenic granitic magmatism, magmatism, mainly calcareous-alkaline and plumasitic raremetal geochemical types. Rocks of the latitic and andesitic series are widespread also in the intracontinental mobile regions characteristic of an ancient continental crust broken up into small plates. A typical example for such regions is the Mongol-Okhotsky belt.

In such regions along with intrusions of the latitic and andesitic series are widespread all four types of palingenic granitoids and especially large are granitoids of the calcareous-alkaline series. Often they do form gigantic abyssal batholiths. For instance, the Susamirsky batholith in Middle Asia stretches more than 1,500 km with a breadth of 400 km and a volume of around six million cubic kilometres.

The disposition of palingenic granitoids in the intracontinental mobile belts is regular: the granitoids of the alkaline series and raremetal granites of the alkaline series do form in regions of ancient folding with a great metamorphism degree of the sedimentary-metamorphic sheath, granitoids of the calcareous-alkaline series and plumasitic raremetal leucogranites in regions with a lesser metamorphism of the substrate of the palingenic melting.

For the granitoids of the calcareous-alkaline and the

alkaline series are more characteristic abyssal and raremetal granites of the alkaline series do often form hypabyssal intrusions and rarer mesoabyssal batholiths.

The hypabyssal and subvolcanic form of intrusives are characteristic also of rocks of the andesitic and latitic series. In riftogenic zones of the intracontinental mobile belts there are developed intrusions of agpaitic raremetal granites and their possible basaltoid forerunners (K-alkaline basalts). The assesment of the content level of ore, trace and volatile xomponents in granitoids and basites of different geochemicla types and also the analysis of process conditions of emanation produced that the maximal potential ore-bearing is with hypabyssal and subvolcanic intrusions of basaltoids of the latitic and andesitic geochemical series and also with hypabyssal intrusions of the plumasitic raremetal leucogranites, agpaitic raremetal granites, palingenic granitoids of the calcareous-alkaline series and alkaline raremetal granites. With the assessment of the potential ore-bearing of magmatites the most important criteria are hte high concentration level of their volatile components (water, fluorine, chlorine, boron, sulphur and carbonic acid) and also the accumulation of a series of trace elements and some volatiles (fluorine, chlorine) in the minerals of the late crystallization stage. In this respect biotite is of a special interest. Analysing the concentration level of volatiles it goes about those which in magmatic rocks can form own mineral phases and in considerable mass can remain there. This is true for fluorine, boron and sulphur. For instance, for rocks of the latitic series is characteristic the augmented concentration level of fluorine and boron whereas in plumasitic raremetal leucogranites occur only high fluorine contents. Analysing the accumulation conditions of volatiles in magmatites is not only important for the concentration level assessment but also for the determination of fugacity of oxygen, which along with temperature and pressure controls the mineral balance in the enogenic processes. Recently has been shown (8) that a method of assessment of oxygen fugacity can be the analysis of its content in the position (OH, F) in hydrate minerals where oxygen takes the part of a microelement subsituting the major ions OH^- and F^-. Thereby the ratios of ions of hydroxyl, fluorine, chlorine and oxygen in biotites of magmatic rocks are important. Investigations show that the ratios of these ions in biotites can be used as a criterium at the assessment of the potential ore-bearing and the ore productivity of magmatitic intrusions of various geochemical types. The diagram (Fig.1) shows that intrusions able to generate gold mineralization must differ by a maximal oxygen fugacity and an augmented concentration level of chlorine. Only in this case the mobilization and migration of gold is possible as an ion $AuCl_4^-$ which is the most probable form of its transition in conditions of magmatic temperatures. In the case of tin as established earlier (1), the most probable transition form is the ion $(Sn(OH,F))^{2-}$. Naturally, the augmented fluorine concentration in biotites in the position (OH,F) should be taken as an important criterium of the intrusion's tin-bearing. It is substantial that in biotites of tin-bearing intrusions the tin content is usually in comparison with tinless ones several times higher. Considering the potential ore-bearing of geochemical magmatite types first of all should be paid attention to the hypabyssal intrusions and granitoids of the latitic series. Now it is clear that intrusions and granitoids of the latitic series do differ by the potential ore-bearing of a large number of elements (Sn, W, Pb, Zn, Mo, Cu, Au, Ag, etc) the feature of magmas of this series is a considerable enrichment of all volatiles (water, fluorine, chlorine, boron, sulphur, carbonic acid) and also of incoherent for basaltoids rare lithophile elements and particularly barium and strontium. The mentioned geochemical particularities are not only characteristic of latites (monzodiorites), but also of their basic forerunners (shoshonite-monzogabbro) and acid deribatives (granites of the latitic series). The high content level of water and fluorine in magmatic melts of hupabyssal intrusions of this series predominates the intensive crystallization differentiation of shoshonite-latite melts including the formation of acid differentiaties, the rhyolites and granites of the latitic series. The development in such intrusions of intensive processes of emanation differentiations predetermines their high potential ore-bearing. It is understood that in many world regions the deposits of many metals are related to large volcano-come ring structures characteristic of an intensive development of latitic magmatism in intrusive (small intrusions), effusive and dike facies. Within these structures with a diameter of 30-40 km weakened zones forming a complex network of radial and concentric faults are most favourable for the formation of strong mineralized zones with vein, veinlet-impregnated and stockwork mineralization.

The commonness of the structural position, the periodically close development of the latitic magmatism and the postmagmatic processes make these ring structures large ore-magmatic systems which are metallogenically very important.

Presently in Eastern Transbaikalia there are distinguished more than twenty such ring structures, the majority of which is confined to zones of deep faults or their sections. It is important that almost to each of these structures are confined tens of deposits and ore occurrences of raremetal and polymetal mineralization.

The presence of latitic magmatism has been established in many regions of the mediterranean intracontinental mobile zone (USSR, Bulgaria, England, Czechoslovakia, etc.). In all of these regions there are related hundreds of ore deposits to the latitic complexes, among which are such giants as Climax, Bingham, Potosi, Cornwall, etc.

The wide development of the latitic rock series in ultracontinental mobile zones and in those of transition from the continent to the ocean and also the duration of development and the high productivity of the formed oremagmatic sestems stresses the necessity of a wider investigation of the geology, petrology, geochemistry and metallogeny of the magmatites of this geochemical series.

Among other granitoids of a high potential ore-bearing there are distinguished hypabyssal intrusions of plumasitic raremetal leucogranites. They are with a high content level of fluorine and water but a low one of boron and chlorine.

With the granitoid intrusions of this geochemical type are related large deposits of tin, wolfram, niobium, tantalum and other trace elements.

The intrusions of this type and the related raremetal mineralization are known in South-East Asia, Central Europe, Western USA, Eastern Australia and in the Mongol-Okhotsky belt. In these regions the intrusions of plumasitic raremetal leucogranites are usually neighbouring with batholiths of palingenic granites of the calcareous-alkaline series. With a high potential ore-bearing go hypabyssal intrusions of the andesitic series. These magmas differing by a high concentration level of water and chlorine have usual a poor fluorine. Therefore they are mainly related to deposits of copper molybdenum, lead and zinc. Against the magmatites of the andesitic series agpaitic raremetal granitoids are particular with a high fluorine concentration and a low one of water. This predetermines their alkaline character and the specific set of related ore concentrations (zirconium, rare earths, niobium, etc).

The proposed typification system of magmatites makes a better mapping of endogenic geochemical dispersion fields possible and helps reliably restore the past geodynamic situations.

Moreover, the study of the geochemical history of volatiles in magmatic processes and the use of geochemical criteria of assessing the potential ore-bearing of magmatites makes it possible on the base of petrologo-geochemical data to realize a regional and local metallogenic prediction.

REFERENCES

1. Barsukov V.L. K geokhimii olova.-Geokhimya, 1957, N 1, p.36-45.

2. Vernadsky V.I. O rasse janii khimicheskikh elementov. Otchjet dejatelnosty Akad.Nauk SSSR za 1926 god. Lenin-

grad, 1927, p.1-15.
3. Vinogradov A.P. Srednee soderzhanie khimicheskikh elementov v glavnykh tipakh izverzhennykh gornykh porod.-Geokhimya, 1962, N 7, p.555-571.
4. Kuzmin M.I., Popolitov E.I. Geokhimicheskie osobennosti magmatizma ostrovnykh dug i aktivnykh kontinentalnykh okrain i nekotorye problemy petrogenezisa. -Geokhimya, 1978, N 5, p.691-699.
5. Petrova Z.I., Levitsky V.I. Petrologya, geokhimya i posledovatelnost formirovanya porod sharyzhalgayskoy serii.-In: Geologya granulitov. Irkutsk, 1981, p.13-49.
6. Tauson L.V. Geokhimicheskie tipy n potentsialnaya rudonosnost.-Moskva, Nauka, 1977.-279 pp.
7. Tauson L.V. Magmatism i rudoobrazovanie. Moskva, Nauka, 1979.-36pp.
8. Troshin Yu.P. Geokhimicheskoe znachenie variatsiy soderzhanya kisloroda v biotite.-Geokhimya, 1983, N 4, p.514-521.

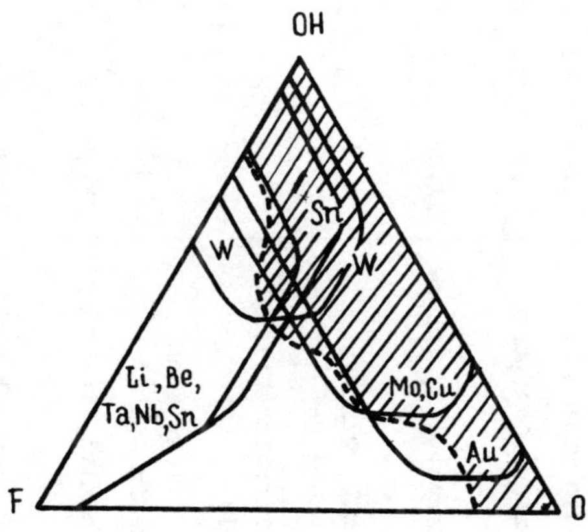

FIGURE 1. Composition fields, positions (OH,P) of biotite from intrusions of various orebearing on the diagram F-OH-H (8). Shaded region: content Cl - 0.3 at% in position (OH,F).

Table 1. Trace-element characteristics of leucocratic granites of rare geochemical types

Complexes	SiO_2	Na	K	F	Li	Rb	Sr	Ba	$F(\frac{Li+Rb}{Sr+Ba})$	Geochemical type
	weight percent				ppm					
Alaskite Aldan granites	72.49	2.3	5.4	0.015	7	156	360	2940	7	Ultrametamorphic
Additional intrusion phase, Kirensky complex	74.37	2.8	3.9	0.03	42	185	50	310	190	Palingenic calcareous -alkaline series
Kharalginsky complex, Trans-baikalia	74.02	2.8	4.1	0.28	72	360	90	260	3.460	Plumasitic rare-metal leucogranites
Plagiogranites, Mainsky complex, Western Sayan	73.16	3.1	0.5	0.015	2	4	187	169	2.5	Plagiogranites, tholeiitic series
Rhyolites, latitic series, Transbaikalia	73.11	2.4	3.9	0.03	31	185	280	1060	4.8	Granites, latitic series
Alkaline granites, Khan-Bogor, Mongolia	75.15	3.5	3.6	0.14	75	180	8	20	12.750	Agpaitic raremetal granites
Granites rapakivi, Kolarsky massif	72.36	2.3	4.2	0.17	26	220	100	1100	350	Rapakivi
Ongonites of Mongolia and Transbaikalia	72.96	2.9	3.8	0.61	100	980	17	44	108.00	Li-F facies, plumasitic raremetal granites

Table 2. Rare metal characteristics of geochemical granitoid types

	Elements	Derivatives of basic and median magmas				Palingenic granitoids				Ultrametamorphic granitoids				Median granite	
		Plagio-granites, tholeiitic series	Granitoids, andesitic series	Granites, latitic series	Agpaitic raremetal granites	Granites, calca-reous-alkaline series	Plumasitic rare-metal leuco-granites	Granites, alkaline series	Rare-metal granites, alkaline series	Ender-bites	Charno-ckites	Rapakivi	Ultra-meta-morphic gran-itoids	Our data	After Vino-gradov (3)
per-cent	K	0.26	2.0	3.5	3.6	3.3	3.9	4.1	3.8	1.1	3.3	4.5	5.0	3.3	3.3
	Na	3.1	3.0	3.4	4.0	2.8	2.8	3.5	3.1	2.4	2.4	2.4	2.0	2.9	2.8
	F	0.015	0.07	0.08	0.2	0.08	0.27	0.05	0.09	0.05	0.10	0.17	0.02	0.08	0.08
ppm	Li	5	18	21	105	50	180	27	52	12	15	30	10	42	40
	Rb	4	100	125	270	175	440	140	270	22	90	250	140	160	200
	Br	140	260	700	12	330	70	650	170	280	300	110	260	340	300
	Ba	60	550	1700	40	830	175	1550	500	450	1000	1400	2300	1020	830
	Sn	3	2	5	18	6	22	4	6	2	2.5	7	3	6	3
	Pb	3	10	23	46	27	28	19	20	3	10	40	12	22	20
	Zn	75	43	70	390	51	40	64	43	50	50	90	35	70	60
	K:Rb	650	200	280	130	190	90	290	140	500	370	180	360	200	160
	$F(\frac{Li+Rb}{Sr+Ba})$	6	100	50	15000	155	6800	40	430	25	80	300	12	120	170
	Relative distribution*	2	8	5	2	40	5	15	3	4	3	5	8	-	-

* All granites are 100%

Table 3. Rare metal characteristics of geochemical basite types

	Elements	Anorthosites	Komatiitic basalts	Tholeiitic basalts		Na-alkaline basalts	K-alkaline basalts	Calcareous-alkaline (andesitic)		Latitic series	
				oceanic	continental			andesitic basalts	andesites	shoshonites	latites
%	Ti	0.12	0.3	0.7	1.1	1.6	1.65	m 0.65	0.4	1.2	0.8
	K	0.5	0.03	0.17	0.6	1.1	2.9	0.9	1.3	2.3	3.1
	Na	3.0	0.8	1.8	1.9	2.5	2.5	2.4	2.7	2.8	2.7
ppm	Rb	2.5	2	3	16	25	85	15	30	65	110
	Sr	440	30	135	270	580	1300	360	380	1270	1220
	Ba	220	35	50	290	380	970	600	270	1920	1470
	Ni	10	640	145	90	130	100	60	20	50	10
	Cr	25	2390	320	140	200	310	100	55	140	15
	V	40	180	250	320	300	270	240	175	130	45
	K:Rb	2000	150	570	420	440	340	600	430	350	280
	$\frac{K}{Ti}\left(\frac{Sr+Ba}{Cr+Ni}\right)$	80	0.002	0.1	1.3	2	10	8	29	32	500

MAGMATIC ASSOCIATIONS OF RECENT ISLAND ARCS AND OROGENIC BELTS OF CONTINENTS

FROLOVA T.I.
Department of Geology Moscow University Moscow USSR

ABSTRACT

Magmatic associations of Island arc and orogenic belts of continents developed above Benioff zones differ in essence from magmatic associations connected with other Earth's structure by a great role of calc-alkaline rocks. This is connected with the fact, that material of Earth's crust takes part in the process of magma generation. The generation of basaltic magmas rich by lithophylic elements take place in conditions of high fluid pressure. Its further evolution with formation of acidic and intermediate volcanic rocks is the result of interaction of basaltic magma with Earth's crust by means of magmatic substitution. The increase of role of calc-alkaline associations and of volumes of acidic and intermediate rocks in continental orogenic belt in comparison with island arcs is connected with the increase of continental Earth's crust thickness above orogenic belts and with the displacement of magma generation levels in the crust. Benioff zone is the supplier of fluids but not a place of maggeneration.

Recent island arcs and active continental margins, as well as orogenic belts of continents, both recent and ancient, are the major structures of the Earth where intermediate and acid magmas are generated. A common feature of these structures is their initiation above the Benioff zones, that can be defined in Recent structures and reliably reconstructed in ancient volcanic belts. Due to the fact that these belts are associated with various types of the Earth's crust, the evidence obtained from their comparative analytical studies is of particular importance, contributing to the knowledge of the role of the crust in the formation of andesite and acid volcanism.

Proceedings of the 27th International
Geological Congress, Volume 9, pp. 437–453
PETROLOGY (IGNEOUS AND METAMORPHIC ROCKS)

Volcanic belts conjugated to Benioff zones represent linear planetary structures, unconformably superimposed into older formations. They are restricted to troughs that in the process of their evolution experienced recompensation and upwarping. These troughs are developed on deepseated faults, adjoining tectonic elements which differ in age and in the thickness and structural framework of the crust. The belts under study, notwithstanding their high diversity, domprise a continuous series from island arcs to orogenic continental belts. Having many features in common, distinctions between them are tentative. Their distinguishing features are: regular position in the lateral series of tectonic structures, which includes the trench or foredeep, connected with the surface exposure of the Benioff zone, the amagmatic borderland, volcanic belt and the depression or marginal sea; absence of ophiolites; generally homodromic order of magmatic sequences, complicated by minor antidromic cycles; increase in alkalinity towards the back of the belts and end-stages of evolution; the local occurence of adic volcanic rock types and the general areal distribution of intermediate and basic rock associations.

Magmatic Associations of Island Arcs.

The variety of island arcs results from the differences in the preceding historical development of the region. Two types of island arcs have been recognized: arcs developed on oceanic crusts (Tonga-Kermadec, Izu-Bonin, Marianas, South Sandwich Islands); and arcs developed on continental or transitional types of crusts (Japan, Kamchatka, Kuriles, Lesser Antilles). This subdivision is tentative and largely reflects the predominant crustal type. Due to the absence of ophiolite assemblages and the wide distribution of andesites, the island arcs (like the active continental margins) have been treated as andesite geosynclines (1) or liminairs (7).

The close correlation in the chemical composition and mineral paragenetic associations of the rock types of different acidity (10), their close positional and time-relationship, the presence of relative inclusion of more basic rocks in acidic rocks, correspondence between the major rock-mass composition of the more basic and more acidic rock-types also the smooth serial curves, indicate the genetic relationship between the igneous rock associations of the island arcs. Three types of genetic series can be recognized: the tholeitic low-potash, calc-alkaline and high alkaline (shoshonite and others) series (23). The tholeitic series are most typical of oceanic arcs, where they are of predominant occurence (Tonga, South Sandwich Islands). They are also characteristic for continental arcs where during the early developmental stages as well as throughout the entire period of existence of the arcs they occur in the external zones (Kamchatka, Japan, Lesser Antilles). The tholeitic rockassociations are weakly differentiated, being represented, chiefly by basalts, and are formed under subaqueous and subaerial conditions. They are products of fissure eruptions, and frequently build up shield volcanoes and slag cones, more seldomly stratovolcanoes.

The calc-alkaline series, most widespread in the island arcs, are typical of the mature developmental stages of oceanic arcs, and also predominate in the internal zones of continental arcs. They are erupted under subaerial conditions, more seldom in coastal sea environments where they comptise large composite volcanoes, often with calderas. The volcanoes are commonly of homodromic development with the attendant formation of large andesite volumes. The volcanicity has an explosive nature, with concomitant exceptionally high-heat avalanches, pumice-stone outpourings, lahars and other pyroclastic accumulations. Extrusive domes and ignimbrite flows of acid rocks are restricted to calderas and volcano-tectonic ring depressions.

High alkaline rock associations are of scant occurence in the contonental back-arc zones. They comprise small stratovolcanoes, complicated by slag cones.

Intrusive rock types are scarce. They reveal a spatial and chemical relationship to coeval effusive rock associations, concluding the stages of igneous activities. A homodromic succesive order of manifestations is reflected in the multiphase intrusions (11).

A characteristic feature of the igneous activities is their shift in time towards the back-arc zones. This is either concomitant with the subsidence of the frontal zone, when the volcanic rocks are overlain by marine deposits, and its subsequent upwarping (Lesser Antilles, Tongan island arc), or the frontal zone persists in the form of an uplift (Japan, Kurils).

The lateral change from the tholeitic series to the calc-alkaline and then alkaline series, occuring form the frontal zone towards the back-arc zones, is the cause of the tranversal zonality of the island arcs, which was first recognized by H. Kuno (6). This zonality was initially interpreted in terms of the dependency ofthe rock composition upon the position of the Benioff zone, which is regarded as the locus of magma chambers (24). Further studies revealed that, despite the correlation between depth to the focal zone and certain chemical elements (K, Rb, Sr, Si) of the volcanic rocks, this fact cannot be regarded as proof of the relationship between the magma source and the Benioff zone (10, 19, etc.). The numerous inconsistencies in this relationship and the obvious regional distinctions in the rock types can be caused by various factors, determining the specific chemical properties of the igneous rocks, such as the heterogenity of the arc basement, the different rates of evolutionary development of the arcs, the hypogene fluid flow intensity, etc. In this way reliable inferences can be drawn concerning the dip of the Benioff zone in ancient arcs, though quantitative estimates should be made with great caution.

Magmatic associations are restricted to different geotectonic environments. Tholeiitic associations develop during the subsidence of retarded upwarping of the area under conditoins of tension (8). Evidence for this is found in the restriction of tholeiitic basalts of the early stages of arc-formation to subaqueous fissure-erupoin products (15) and to external island arc zones. In the latter zones the controlling factors in the conditions of tension are step-faults and grabens, dipping oceanward, or towards intra-arc thoughs. Under conditions of subsidence and tension high-alkaline rock series are also formed in the back-arc zones, being frequently restricted to the off-shore slopes of the marginal sea basin (Kuriles, Japan). Calc-alkaline rock associations occur in areas that are characterized by a different type of developmental conditions. During the early developmental stages of the island arcs they are associated with shallow throughs. Later the troughs undergo recompensation owing to the abundant igneous rock-mass supply, and the growth of geoanticlinical structures which throughout the following developmental stages are the realm of calc-alkaline volcanism. They are characterized by conditions of predominan conpression, supporting evidence for which is found in the geophysical investigations, carried out in Kamchatka and the Kurile Islands (13). The volcanic centres themselves are associated with compensatory depressions which complicate the geoanticlines (11). This tends to suggest that within the general pattern of regional compression, tension was a local manifestation.

Magmatic Associations of continental Orogenic Belts. These belts develop on crusts of the continental or transitional types, i.e., in contrast to the island arcs they are always continental, marking the juncture of geological structures varying in age and tectonic setting. The thickest Earth's crust in the basement of the belts is connected with the back-belt areas. Examples of orogenic continental beltss are: Okhotsk-Chukotsk (K_1al-K_2c), East Sikhote-Aline(K_2e-N_1), Honsu(K_2c-K_2m), Neoandes(N_1^3-Q). As a rule, these belts are formed on a geosynclinal basement. The older geosynclines at the base of the belts are commonly assigned to andesite geosynclines or liminairs, i.e., they are analogues are the Okhotsk-Chukotsk volcanic-plutonic belt which is superimposed on the Taigonos geosyncline, a buried island arc (1); also the Neoande belt which is superimposed on the Araucan geosyncline, assigned to liminair structure (7).

The igneous rock associations of the orogenic belts are comprised predominantly of calc-alkaline rock series with a subordinate occurence are high-alkaline rock series. Weakly diffentiated basaltic-andesite-basaltic associations of calc-alkaline type are of wide occurence predominantly in the frontla zone (Okhotsk-Chukotsk belt). A characteristic feature of the differentiated calc-alkaline series, most wide spread in the orogenic belts, is the extensive development of intermediate and acid rock types, with basalts playing a subordinate role. The former comprise indepent rock formations - andesite, rhyolitic, ignimbritic - or extensive, according to their silicic acid content, rock sequences. They characteristically occur in complicated structural forms - isometric structures of the caldera type and caldera-like depressions which are frequently infilled with comagmatic shallow intrusive rocks. Most of them are synvolcanic compensatory structures. The differentiated calc-alkaline rock associations increase in volume to wards the back-zones of the composite orogenic belts (1).

High-alkaline rock associations are of the Earth's crust, commonly to the baek and flanking zones, or the concluding developmental stages of the belts (the Okhotsk-Chukotsk belt), comprising weakly differentiated trachybasalt or rhyolite-trachyte rock formations.

Intrusibe rock bodies, in contrast to those of the island arcs, are widespread in orogenic belts and being of later time-formation, are comagmatic to the effusive rocks. They are rather large, myltiphase plutonic masses restricted to magmatogenic uplifts and exibit a homodromic successive order of development. As compared to the effusive rock types they are more acidic in composition, consisting predominantly of granitoids. The syenite-diorite-alkaligranite rock formation is associated with the late developmental stages of the belt (1).

In contrast to the island arcs, manifestations of the temporal migration of the igneous activities towards the back-belt part are practically lacking. Of all the orogenic belt reveals a shift in volcanicity, generally towards the back-belt zone, as compared to the late geosynclinal volcanicity. As to the belts of East Asia, there is no evidence of migration Honsu and East Sikhote-Aline belts; in the Okhotsk-Chukotsk belt, where volcanic outbursts were of simultaneous occurence throughout the territory during the Albian, in the back-belt zone these activities terminated earlier (1). The temporal evolution of the igneaous activity is more distinctly expressed in the basalts and andesites formed during the early developmental stages, the subsequent volumtric increase of rhyolitic rock formations and the termination of the volcanicity in more basic rock types. The lateral transversal zonality in the distribution of coeval ignous ricks, which is distinctly apparent in the island arcs, is not of regular occurence in the orogenic belts. In certain belts e.g. the Sikhote-Aline, it is utterly avsent (2). At the same time, variations in igneous activity are reflected everywhere along the strike of the belts, and resultsfrom the heterogeneity of the basement (1,2,7).

The difference in the types of igneous rock associations can be interpreted in terms of the different geodynamic conditions in the areas of their occurence. For instance, the weakly differentiated basalt rock associations are restricted to the fromtal belt zones where systems of graben-type troughs and horsts occur (1,7), created under conditions of tension. The general upwarping, concomittant with the formation of magma-feeding faults, occuring under conditions of prevalent compression, results in the creation of differentiated andesite-rhyolite and ignimbrite rick formations. The high-alkaline rock formations are associated with the stages of belt stabilization.

Petrogenesis.

Basalts are present in most of the volcanic rock associations of the island arcs, and in a large part of the orogenic volcanic-plutonic belts. Only in the rhyolite rock associations is the correlation weak or non-evident, although andesites are present here. The homodromic order of temporal development of the igneous activities over lenghtly periods, the general petrological similarity between the intermediate and acid rock types, associated with or without basalts, suggest that in most cases the source magmas of the genetic series are basaltic.

The basalts of all the genetic rock series in the igneous

belts conjugated to the Benioff zones, despite certain essential distinctions, exhibit many common features, implying their generation under largely similar conditions. In contrast to the basalts of other regions e.g. oceans which without doubt are of manitle origin, these basalts are characteristically of high alumina and ferric oxide content, high iron and low titanium content, poor in magnesium, chromium, and nickel, and rich in potassium, lytophyle elements and light lanta-enhanced in the tholeiitic basalts of the islands arcs, increases in the basalts of orogenic continental belts. The persistence of these distinctions tends to suggests that above the Benioff zones is a substratum, different from that in the oceans and close in composition to the basic rocks, that has a paramount role in the generation of the magma. Evidence for this is found also in predominantly crustal composition of the inclusions (8,11). From this it is conceivable that the "andessite line", which was traced already at the beginning of the century, not only delimits the island arcs from the ocean, but is also the boundary between fundamentally different conditions of magma generation.

Inferences concerning the origin of the basalts were derived from experimental studies of the olivine-plagioclase-clinopyroxene-silica system. Examination was made of the basalt projection onto the face of the fusibility diagramm of the olivine-plagioclase-clinopyroxene system (Fig.2); in estimates of the normative minerals, placed at the veritices of the triangle, the olivine content was reckoned together the olivine component of the hyperstene. The position of the pseudoinvariant data points (eutectics) and pseudomonovariant lines (cotectics), corresponding to two or three equilibrium phases of the magma melt composition, are given according to the data obtained from laboratory experiments carried out under different dry and water vapour pressures (17,22,24 and other). The basalt data points of the island arcs and orogenic belts, in contrast to the oceanic basalts, alre located near the "water eutectic region. The data points of island arcs exhibit a "tail" extending towards the "dry2 conditions, due to the presence of tholeiitic basalts, whereas the data points of orogenic basalts are narrowly delimited, forming a dense area that is hifted towards the "water" eutectics. This serves to suggest that they all undergo segregation under conditions of high fluid pressure, which in the orogenic belts is more prevalent.

Due to the fact that the ratio of load pressure to fluid pressure cannot be determined, appoximate estimates were made of the pressure under which the segregation of the basalts occurs. It was found that thpeiitic basalts segregate under pressures of 7-8 kbar, whereas basalts of the calc-alkaline series segregate under pressures of 10-12 kbar (Fig.2). These figures are confirmed by geophysical data. For instance, according to these data, the magma sourc on Kunashir Island is located probably at depths between 35-50 km, and intermediate poxkets are located even nearer to the surface (5). Evidnece for shallow depth of magma chambers is found in the thermal anomalies beneath the island arcs, where the temperature of induced melting of the basalts is reached at depth within 40-60 km (14). Thus, the level ofbasalticmagma accumulation situated above the Benioff zones, obviously within the limit of basaltic layer of Earth's crust. The controlling factors in the basalt composition of different genetic rock series are the changing conditions in the basalt magma segregation, invilving different melting depths, fluid pressure levels and different role of the crust. The major factors controlling basalt evolution are in causal relationship with their segregation conditions and thus differ essentially. The petrological distinctions between the major genetic rock series have been under discussion in recent publications (98, 11, 22, 25). The characteristic features of the Tholeiitic low-potash series, known only in the island arcs, are low alkali content, particularly of potassium, low content of minor litophylic elements (Rb, Sr, Ba, Zr), also low iron oxidation level, as compared to other island arc series. The tholeiitic rock series are of weak differentiation. The controlling factor in their evolution is crystallization differentiation of the magma. Theoretical estimates of this process are in good agreement with real series (4,12, etc.). The plutonic allivalite inclusions and gabbroids, that are encountered in the tholeiitic series, are manifestations of the early stages of the process occuring at crustal depth (11). The variational curves of such rock series are characterized by a gradual increase in the alkali content and the rapid rise of iron content.

The rock types of the most widespread calc-alkaline series are characterized by higher alkali (potassium) content, associated minor lithophyllic elements and light lanthanoids, also iron oxideds, and low titanium content. The higher oxidation level of the source mgma ant its evolutionary products is reflected in the appearance of amphiboles, most frequently in the volcanic rocks of orogenic belts, and in the early crystallization of magnetite (16). Another distinction of the orogenic volcanic series from the island arcs series is the wide occurence of acid rock types with limiting levels of silics content, and lower magnesium and calcium contents.

The process of crystallization differentiation is also the controlling mechanism in the formation of the weakly differentiated basaltic andesite-basalt series of the calc-alkaline rock types, which is reflected in the gradual increase of the alkali and iron contents. However this fractionation trend can be traced commonly only to the andesite composition, owing to which most of the basaltic andesites are regarded as basalt differentiates. During the more advance d developmental stages, the curves undergo levelling and alkali accumulation has a slower rate; at the same time the acceleration iron in content reveals a sharp drop, an integral feature of the calc-alkaline series. Attempts to explain this peculiarity by fracionation, involving the segregation of phases higher in iron content than the magma, were unsatisfactory. Such phases are amphiboles in the wet magmas pressures up to 27 Kbar, and garnet under higher pressures or in dry magmas. However, the liquidus phase of the calc-alkaline rock series is commonly plagioclase nor amphiboles which besides are rare occurence. Garnet is even more scarce (23). Amphibole fractionation fractionation probably occurs from exceptionally water-rich magmas of the orogenic belts, but is not a widespread process. The decrease in the rate of iron- and alkali accumulation during the evolution of the calc-a-kaline series can be interpreted in terms of the inclusion of granite-metamorphic layer materials of the Earth's crust in the process of magma generation. This process is similar in type to that of magmatic replacement under the influence of fluid flow, and lead to the acidification fo the magmas. The acidic nature of the medium, resulting from the effects of acid-basic interaction, causes the removal of iron from the melt, its being a weaker base than magnesium, and this is one of the reasons for the slower rate of the iron content increase (8); in addition the Osborne effect causes the early crystallisation of the iron oxide phases. The increase in the rate of alkali accumulation. The interaction between the magmas and the crust becomes more intensive, when a systemof intermediate magma pockets is present in the basal complex of the igneous belts; this can be feasibly explained by the existence of conditions of compression in the zones of the calc-alkaline series formation, and is confirmed by geophysical data (5, etc.). Supporting evidence for this is found in generally acknowledged petrological criteria - the presence of orthopyroxene and calcic plagioclase (8, 12, 16, etc.) and also in the preservation of the high alumina content in the extreme acidic members of the calc-alkaline series. The large volumes of igneous rocks with limiting silica-contents, showing no apparent relationship to the source basalts, typical for the orogenic belts, tends to indicate that the fluid flow was of long duration and high intensity, stimulating the interaction between the crustal materials and the magmas.

The interaction between the fluid and the melt is characteristically most strongly manifested in magmas that are stabilized near the granite e utectics, which are of most wide occurence in the orogenic volcanic belts. This is reflected in the decrease in the alkali content and concurrent sharp increase in the potassium to soda ration, and results from the increase of the fluid acidity in the process of magma evolution. This effect can be feasibly explained in terms of the distribution of potassium and sodium between the fluid and the melt; due to the co-existence of the acidic fluid, the melt becomes enriched in potassium, whereas the total alkalinity of the melt increases as a result of the removal of alkalies (leaching) during its segregation (9). The acidic nature of the fluid causes the early crystallisation of quartz phenocrysts in the rhyolites, and its segregation is marked by their resorption.

Thus, the igneous activities in the belts developing above the Benioff zones, are characterized specifically by the segregation and evolution of the magma under conditions of high fluid (predominantly water) pressure, involving the inclusion of materials of different levels of the Earth's crust during different stages of proces, and this is particularly significant during the formation of the calc-alkaline series. Evidence for the involvement of the crust in the process of magma generation is found in the following: the enrichment of the sourc magmas in lithophyllic elements compared to the mantle magmas (oceanic); the direct correlation between the volumes of calc-alkaline rock associations and the thickness of the crust (25); the increase in the volumes of intermediate and acid rock types in the orogenic volcanic belts, where crustal processes are of more intensive development; the increase in the initial strontium-isotope ratios in the process of colcanic-belt evolution (7); the heterogeneity in the igneous activities along the trend of the belts, reflecting the basement heterogeneity; the tholeiitic basalt specification in the oceanic arcs, as compared to the continental, which cannot be explained by the earlier development of the oceanic arcs (in the western part of the Pacific they all initiated earlier, than the continental arc systems: the time of inition of oceanic arc is Paleogene in contrast of initiation of continental arcs, which is Neogene).

The voluminous fluid flow can be explained by the intensive massheat transfer above the Benioff zones. The temporal evolution of the fluids results from the general changing conditions of magma generation during the formation of the volcanic belts (increase in depth of magma generation, increase in total alkalinity and, paricularly, of the potash content during this process), as well as from mixture with the crustal fluids. The lateral zonality observed on the island arcs, also the temporal shift of the igneous activity from the frontal zones towards the back-arc zones, are seemingly associated with the overthrust of the island arcs onto the ocean during the opening of the marginal seas, and the subsequent passage of various island arc-zones over the evolving zone of magma generation. The violation of this zonality in the orogenic continental belts results from the existence of intermediate long-duration crustal-source systems. Besides, due to the fact that the orogenic continental belts develop on thicker and more rigid continental crusts, the formation of associated more weakly than in island arcs; consequently, their horizontal shift is allso more limited.

REFERENCES

1. Belii V.F. Rock Formations and Tectonics of the Okhotsk-Chukotsk Belt, 212 pp. Formatsii i tectonika Okhotsko-Chukotskovo vulkanogennovo poyasa. Moskwa, Nauka, 1978, 212s.
2. Vetrennikov V.V. Specific Features of the Volcanicity, Tectonic Setting and Ore-Localization in Marginal-Continental Volcanic Belts. Osobennosti vulkanisma, tektoniki i orudenenia okrainno-materiko-vikh vulkanicheskikh poyasov. M., Nedra, 1976, 155s.
3. Geological History of Japan Islands. 717p. geologicheskoye razvitiye Yaponskikh ostrovov.M., Mir,1968,717s
4. Dmitriev L.V., Sharaskin A.J. The main features of oceanic bottom magmatism. In: Problems of Petrology.M. Osnovnye cherty magmatisma dna okeana. V kn. Problemy petrologii. M.,1976.c.173-189.
5. Zlobin T.K., Fedorchenko V.I., Petrov A.V., Nemchenko G.S. The lithosphere Structure of Kunashir Island (the Kurile Islands) inferred from Seismic Data. Structura litospheri o.Kunashir (kurilskie ostrova) po seismicheskim dannim. Tikhookeanskaje geologija,.N.1,1982,s.92-100.
6. Kuno H. Lateral basalt Magma Variations Across the Continental Margins and Island Arcs.pp.249-261.

Lateralnije variatsii basaltovoy magmi vkrest okrain kontinentov i ostrovnikh dug, Sb.:Okraini kontinentov i ostrovnije dugi. Mir, M., 1970,s.92-103.
7. Lomize M.G. Liminair Systems and their Relation to Island arcs. Liminarniye sistemi i ikh sootnosheniye s ostrovnimi dugami. Geotectonika, N.2,1983,s.92-103.
8. Marakushev A.A., Frolova T.I., Yakovleva E.B., Emelyanenko P.F. Magmatism and the Earth's Crust. Magmatism i zemnaya kora. Vestnik MGU, ser.Geol.,1977,N1,1977.
9. Perchuk L.L., Frolova T.I. Effects of Fluid Composition on the Evolution of Basal Series. Vliyaniye sostava fluidov na evolutsiya basaltovikh serii. DAN, SSSR, 1980,T.253,N6,s.1436-1439.
10. Piskunov B.N.,Abdurakhmanov A.I.,Kim Ch.U. The composition-Depth Relationship in the Kurile Island Arc Volvanoes and its Petrological Significance. Sootnosheniye sostav-glubina dlya vulkanov Kurilskoy ostrovnoy dugi i yego petrologicheskoje znacheniye. Vulkanologia i seismologia, 1979, N 4,s.57-67.
11. Popolitiv E.I., Volinets O.N. Geochemistry of Quarternary Volcanicity in the Kurile-Kamchatka Island Arcs and Certain Aspects of Petrogenesis. Geokhimicheskiye osobennosti chetvertchovo vulkanizma Kurilo-Kamchatskoy ostrovnoy dugi i nekototiye voprosi petrogenesisa. Novosibirsk, Nauka, 1980,180s.
12. Ryabchikov L.D., Bogatikov O.A., Babanskii A.D. Problems bearing on the Origin of Calc-Alkaline Magmas. Problemi proskhozhdenia izvestkovo-schelochnikh magm. Vulkanologia i seismologia,N 1979, N 1, s.28-40.
13. Simbireva I.G., Fedotov S.A., Fecfilaktov V.D. The Geodynamics of the Kurile-Kamchatka Arc, on the basis of Seismological Data. In: volcanicity and Geodynamics, pp.91-103. Geodinamika Kurilo-Kamchatskoy dugi po seismologicheskim dannim. V kn: Vulkanism i Geodinamika., 1977,s.91-103.
14. Smirnov Ya.B., Sugrobov V.M. The Earth's Heat Flow in the Kurile-Kamchatka and Aleutian Provinces. III. Estimates of deep Temperatures and the thickness of the lithosphere. Zemni teplovoy potok v Kurilo-Kamchatskoy i Aleutskoy provintsiyakh. Vulkanologia i seismologia, 1980,N2, s.3-13.
15. Frolova T.I. Comparative Petrological Studies of

Volcanic Rocks in the Lesser Antilles and South Sandvich Island Arcs. In: Tectonic and Geodynamics of the Carribean Region, p.78-96. Sravnitelnaya petrologia vulkanicheskikh porod Maloy Antilskoy i Yuzhno-Sandvichevoy ostrovnikh dug. Sb.; Tektonika i geodinamika Karibskovo regiona. M., 1979, s.78-96.
16. Frolova T.I., Perchuk L.L. Volcanism of Oceanic Islands and Islands Arcs. Vulkanism okheanicheskikh Ostrovov i ostrovnikh dug. Vestn. MGU, ser. geol., 1982, N6, s.27-39.
17. Khitarov N.I., Pugin V.A. Problems Bearing on the Origin of Basalt Magmas (in the light of Experimental Data). In: Geodynamics, Magmageneration and Volcanicity. Problema proiskhozhdenia basaltovikh magm (v svete ekspeimental-nikh dannikh). V kn.: Geodinamika, magmoobrasovaniye i vulkanism. Petr.-Kamch., 1974, s.250-271.
18. Bailey J.C. Geochemical Criteria for a Refined Tectonic Discrimination of Orogenic Andesites. Chemic. Geol., 1981, 32, pp.139-154.
19. Cawthorn R.G. Petrological Aspects of the Correlation Between Potash Content of Orogenic Magmas and Earthquake. Miner. Mag., v.41, 1977,pp.173-182.
20. Ewart A., Bryan W.B. Petrography and Geochemistry of the Igneous Rocks from Eua, Tongan Island. Geol. Soc. of Amer., Bull.v.83, 1972, pp. 3281-3298.
21. Green D.H., Ringvood A.E. Genesis of Basaltic Magmas. Contr. Miner. Petrol.,v.15,N 2, 1967,pp. 103-190.
22. Jakes P., White A.J.R. Composition of Island Arcs and Continental Growth. Earth and Planet. Science, Let.12, 1971, pp.224-230.
23. Yoder H.S.Yr., Tilley C.E. Origin of Basalt Magmas. Journal of Petrol. v.3, N3, 1962, p.247.
24. Hatherton T., Dickinson W.R. Relationship Between Andesit Volcanism and Seismicity in Indonesia, the lesser Antilles, and Other Island Arcs. J. Geophys. Res., v. 74, 1969, pp.5301-5310.
25. Miryashiro A. Volcanic Rock Series in Island Arcs and Active Continental Margins. Amer. J. Sci., v. 274, pp.321-355.

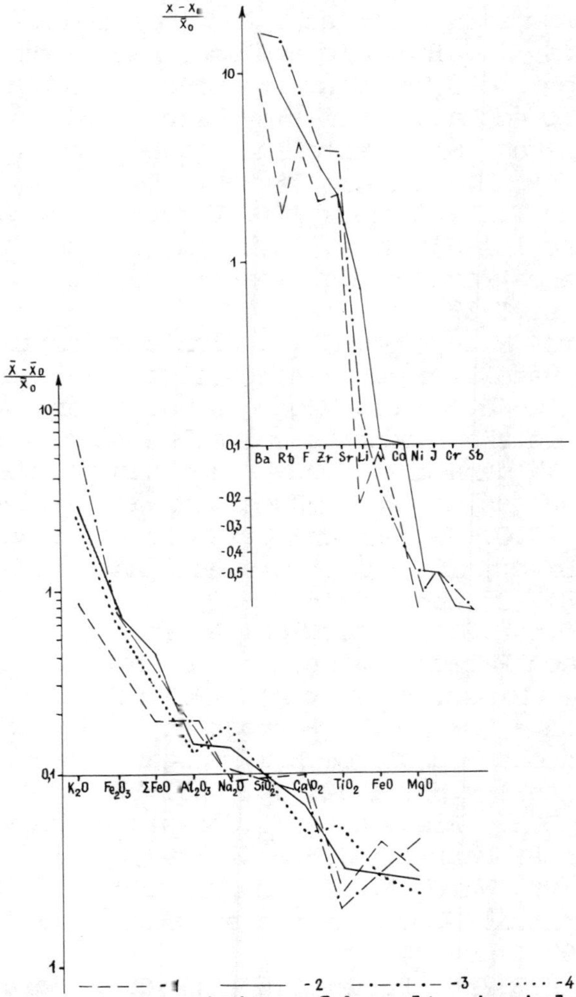

FIGURE 1. Average composition of basalts in island arcs and orogenic continental belts, standartized according to the oceanic basalt composition (4). East Kamchatka and the Kurile Islands basalts: low potash tholeiite, 2 - calc-alkaline series, 3 - high-alkaline series, 4 - basalts of Okhotsk-Chukotsk volcanic belt (1). On abscissa - the average content of oxides and elements in oceanic basalts ($\bar{X}_o$), adopted as unit. On ordinate axis - the ratio $\frac{\bar{X} - \bar{X}}{X_o}$, where $\bar{X}$ - the average content of oxides and element in the rock normalized.

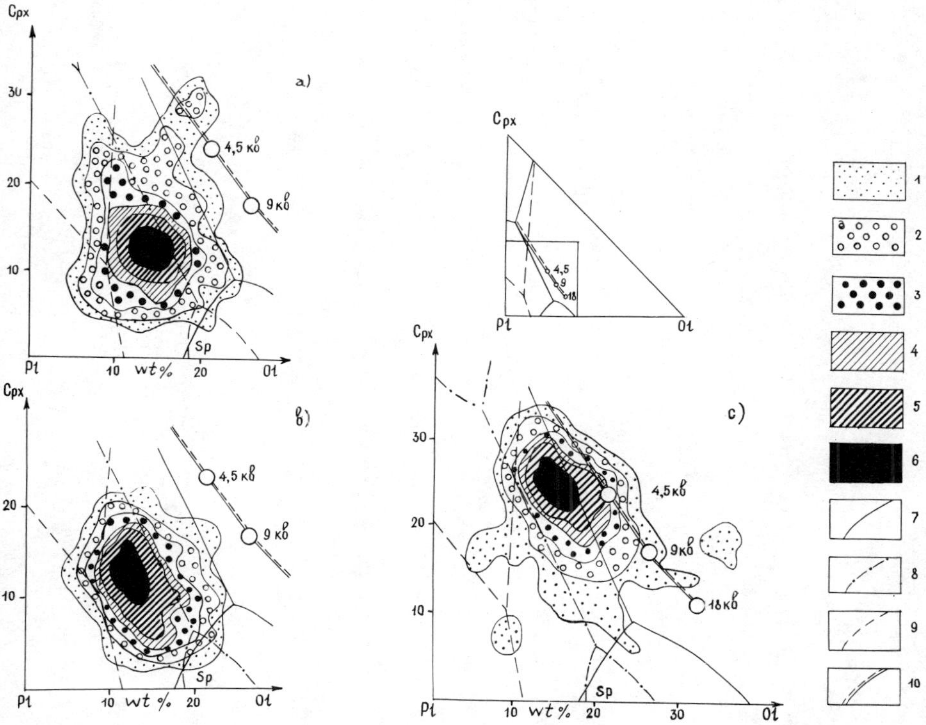

FIGURE 2. Distribution of basalt data points, plotted on the fusibility diagramm of the plagioclase-clinopyroxene-olivine system. l - island arcs (110 anal.), II - orogenic belts (116 anal.), III - mid-oceanic ridges (147 anal.).
1-6 - density of data points: 1 - 1 to 2.5%; 2 - 2.5 to 5%; 3 - 5 to 7.5%; 4 - 7.5 to 10%; 5 - 10 to 12.5%; 6 -12.5 to 15%. 7 -9 - cotectic lines: 7 - under pressure of 1 atm. for plagioclase with 70% anortite content and olivine of 20% fayalite content; 8 - under pressure of 1 atm., for plagioclase of 40% anortite content and olivine of 40% fayalite content; 9 - under 3000 atm. water pressure for plagioclase of 70% anortite content with forsterite and diopside; 10 - line of displacement of the triple-bond point (pseudoeutectic) of plagioclase N 70- clinopyroxene-olivine under pressure of 4,5,9 and 18 Kbar. In the upper right corner- the pseudotriple fusibiliyy diagramm of the plioclase-clinopyroxene-olivine system.

CARBONATION REACTIONS FOR MANTLE LHERZOLITE AND HARZBURGITE

WUU-LIANG HUANG and PETER J. WYLLIE
Exxon Production Research, P.O. Box 2189, Houston, Texas 77001 USA; Division of Geological and Planetary Sciences, California Institute of Technology, Pasadena, California 91125 USA

ABSTRACT

The synthetic system $CaO-MgO-SiO_2-CO_2$ contains assemblages corresponding to lherzolite and harzburgite: forsterite(Fo)+ orthopyroxene(Opx)+clinopyroxene(Cpx) and Fo+enstatite(En). Fully carbonated, these are transformed into siliceous limestones: dolomite(Do)+magnesite(Mc)+quartz/coesite(Qz) and MC+Qz, respectively, which can coexist with free CO_2. Partly carbonated, they form dolomite-lherzolite (Do+Fo+Opx+Cpx) and magnesite-harzburgite (MC+Fo+En) which cannot coexist with CO_2, but can coexist with (H_2O+CO_2), buffered to high H_2O/CO_2. Review of experimental and theoretical data for decarbonation reactions and pyroxene-carbonate exchange reactions in $CaO-MgO-SiO_2-CO_2$ shows one invariant point Q_1 for Opx+Cpx+Do+Mc+ Qz+CO_2 near 1090°C and 34 kb; and another Q_2 for Cc+Do+Cpx+Opx+Fo+CO_2 between the 7 kb and 13 kb, 800°C and 950°C. Dolomites in the high-pressure lherzolite reaction (6): Do/Cc+Opx= Fo+Cpx+CO_2 contain more than 70% $CaCO_3$, at 1200°C and 27 kb; below the crest of the calcite-dolomite solvus, the reaction for carbonate-lherzolite is represented by (6A)

Proceedings of the 27th International
Geological Congress, Volume 9, pp. 455–473
PETROLOGY (IGNEOUS AND METAMORPHIC ROCKS)

with magnesian calcite down to Q_2, and by (6B) with near-stoichiometric dolomite below Q2. Within limited depth intervals in the upper mantle with H_2O present, solid magnesite-harzburgite can coexist with partially melted lherzolite. Eruption of kimberlite could transport xenoliths of lherzolite and magnesite-harzburgite, with subsequent disruptive dissociation of magnesite and dispersal of the low-calcium garnet characteristic of depleted harzburgites from the deep roots of cratons. This result is consistent with the explanation of Boyd and Gurney for the correlation of such garnets with diamond-bearing kimberlites.

INTRODUCTION

The upper mantle is a great reservoir of peridotites. Peridotites may be transported from this reservoir into the crust, or they may be generated within the crust by two means. Crystallization differentiation of basic magma yields precipitates of ultrabasic rocks. The metamorphism or metasomatism of siliceous dolomites yields assemblages including the minerals of peridotites. In this contribution, we review recent experimental and theoretical data (mainly at high pressures) which demonstrate the effect of CO_2 on the mineralogy of peridotite, and the reactions which can yield peridotite from siliceous dolomite of suitable composition. The effect of H_2O and the involvement of serpentine is omitted, but the effect of H_2O on melting reactions in the mantle is introduced.

THE SYSTEM $CaO-MgO-SiO_2-CO_2$ AND "MODEL" PERIDOTITES

Bowen (1940) used the system $CaO-MgO-SiO_2-CO_2$ for representation of the minerals in siliceous dolomites, and this is also adequate for modelling the major minerals of many peridotites (Wyllie and Huang, 1976a); without Al_2O_3 as a component, the accessory aluminous minerals plagioclase, spinel and garnet are absent. The minerals of peridotites occurring in this system include forsterite (Fo), enstatite (En), orthopyroxene (Opx), diopside (Di), and clinopyroxene (Cpx); the minerals of limestones include calcite and its solid solution (CC, Cc), dolomite and its solid solution (Do), magnesite and its solid solution (MC, Mc), and quartz and other polymorphs of SiO_2 (Qz).

The available experimental and theoretical data for many reactions in the system $CaO-MgO-SiO_2-CO_2$ were recently reviewed and combined with new experimental data by Wyllie et al (1984), for compilation of a revised array of reactions. In this contribution, we will deal with particular reactions, and concentrate on the phase assemblages produced which correspond to peridotite bulk compositions. Peridotite mineral assemblages include model lherzolite (Fo+Opx+Cpx) and harzburgite (Fo+Opx, Fo+En). These may be carbonated by CO_2 at high pressures or low temperatures and, when fully carbonated, they are transformed into limestones which can coexist with CO_2, Do+Mc+Qz, and MC+Qz, respectively. When partly carbonated, the model lherzolite and harzburgite are converted to dolomite-lherzolite (Do+Fo+Opx+Cpx) and magnesite-harzburgite (MC+Fo+En), respectively, which cannot coexist with CO_2. Other model peridotites are wehrlite (Fo+Cpx) and websterite (Opx+Cpx).

Following the numbering system adopted by Wyllie and Huang (1975,1976b) and Wyllie et al (1984), the reactions of particular interest are: for model lherzolite at temperatures above the calcite-dolomite solvus:

$$Do/Cc + Opx = Cpx + Fo + CO_2 \quad (6)$$

for model harzburgite,

$$MC + En = Fo + CO_2 \quad (3)$$

for model websterite,

$$Mc + Cpx = Do + Opx \quad (0)$$

and

$$Do + Cpx = Cc + Opx \quad (01)$$

and for model wehrlite,

$$Do + Cpx = Fo + Cc + CO2 \quad (9)$$

At temperatures below the calcite-dolomite solvus, carbonate-lherzolite is represented by reactions:

$$Cc + Opx = Cpx + Fo + CO_2 \quad (6A)$$

$$Do + Opx = Cpx + Fo + CO_2 \quad (6B)$$

CARBONATION OF MODEL LHERZOLITE AND WEHRLITE

Figure 1 illustrates the complete tie-figure for reaction (6), and the phase assemblages occurring on either side of the reaction curve. The assemblages of interest are lherzolite with CO_2, and the partly carbonated dolomite-lherzolite without CO_2. Notice that for other bulk compositions, phase assemblages with CO_2 correspond to partly carbonated dolomite-harzburgite, dolomite-websterite, and dolomite-wehrlite.

All published univariant curves for reaction (6) are shown in Figure 2, with a simplified projection of the reaction tie-figure illustrated in the inset (with lines to

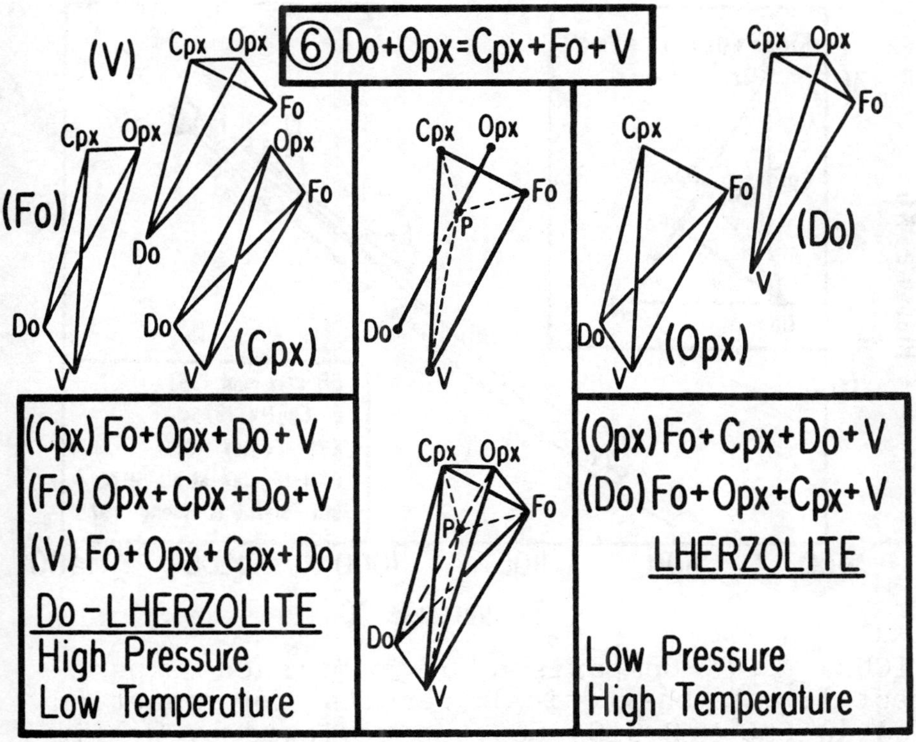

FIGURE 1. Geometry of the tie-figure for reaction (6) from within the tetrahedron CaO-MgO-SiO_2-CO_2, with the groups of four-phase tetrahedra which comprise the five-phase tie-figure on either side of the reaction. Abbreviations, see text.

CO_2 omitted for clarity). There are no experimental data below 15 kb. Wyllie et al (1984) gave reasons for selecting as the best curve the one passing through the reversed run brackets, and extrapolated down to low pressures along the EKH curve calculated by Eggler et al (1976).

Figure 2 indicates that mineral compositions in the lherzolite assemblage are forsterite,

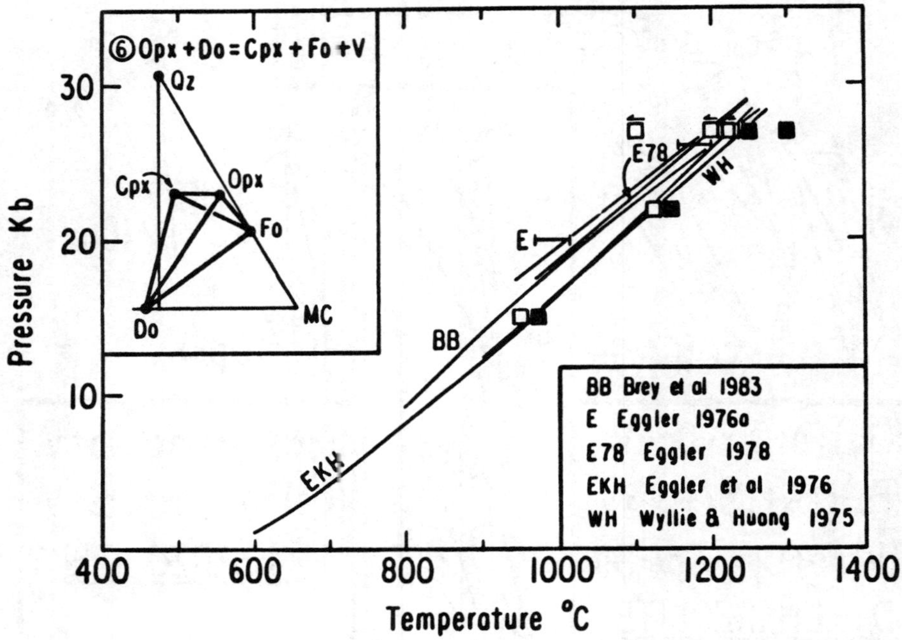

FIGURE 2. Comparison of published experimental and theoretical results for reaction (6) in $CaO-MgO-SiC_2-CO_2$, as reviewed by Wyllie et al (1984).

pyroxenes (coexisting across the solvus), and a calcic dolomite. Wyllie et al (1984) determined the compositions of some carbonates and pyroxenes in this reaction with electron microprobe and x-ray diffraction, with results at 27 kb shown in Figure 3. A trace of liquid in run 1068 is certainly responsible for the wide spread and change in composition of the carbonate compared with that in runs 1058 and 1073. At temperatures below 1100°C at 27 kb, the dolomite in the subsolidus assemblage Opx+Cpx+Do+V is slightly calcic, but it changes to magnesian calcite (Cc) with 22% $MgCO_3$ at 1220°C. At 15 kb, where the reaction temperature is below the crest of the calcite-

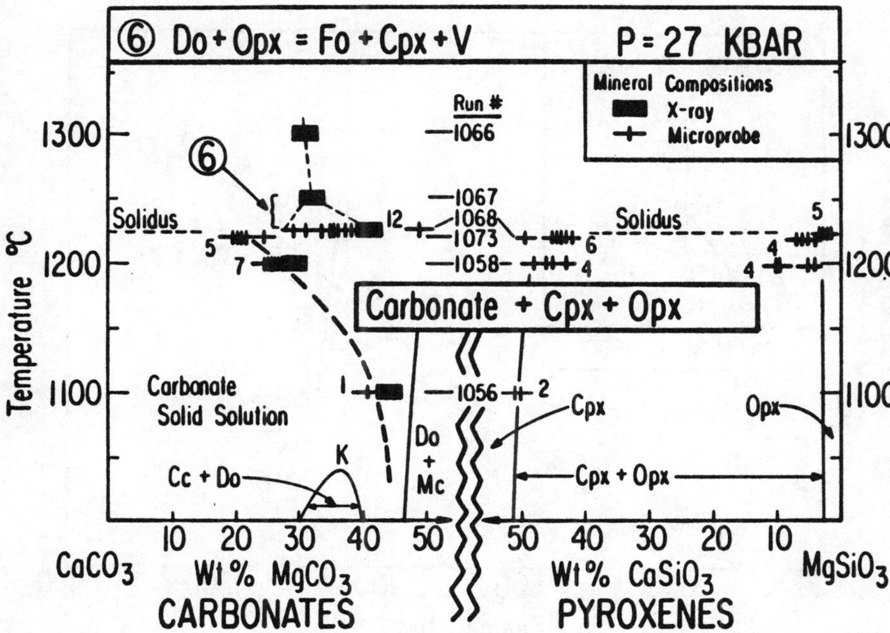

FIGURE 3. Compositions of carbonate and pyroxenes for reaction (6) and at temperatures below reaction (6), at 27 kb, compared with solvus curves for calcite-dolomite and orthopyroxene-clinopyroxene.

dolomite solvus, the carbonate composition is on the $CaCO_3$ side of the solvus, and it must be termed magnesian calcite (Cc). It has been customary to define the univariant reaction (6) at mantle pressures, as illustrated in Figures 1 and 2, in terms of a dolomite solid solution (Wyllie and Huang, 1976b; Eggler, 1978). However, at pressures and temperatures below the calcite-dolomite solvus crest, the reaction must be defined as:

$$Cc + Opx = Cpx + Fo + CO_2 \qquad (6A)$$

Käse and Metz (1980) determined reaction (9) to 10 kb, as shown in Figure 4. If extrapolated to higher pressures, the curve terminates at K_9, a critical end-point at the

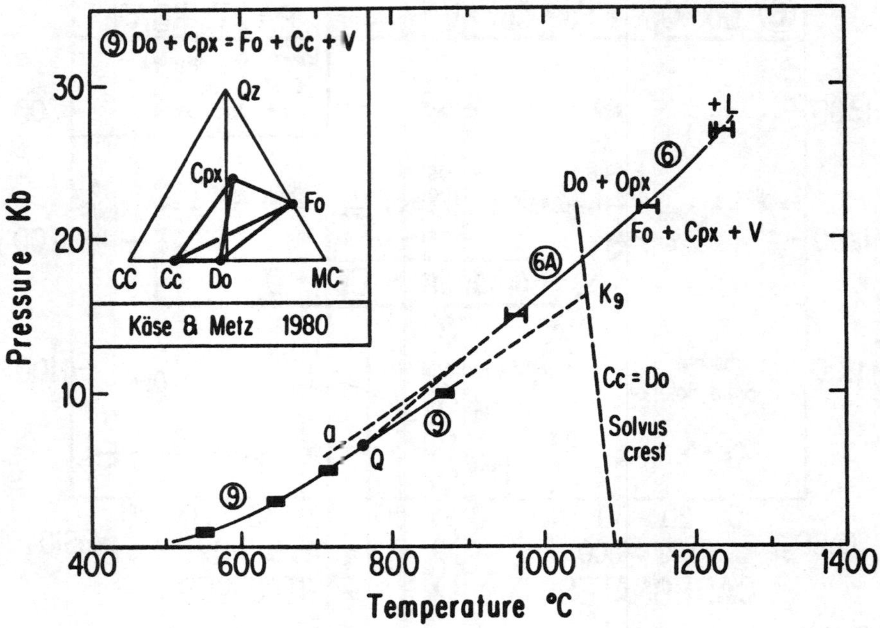

FIGURE 4. Experimental results for reaction (9) in $CaO-MgO-SiO_2-CO_2$, according to Käse and Metz (1980), based on experiments with about 10 mole per cent H_2O to enhance reaction; accuracy ±10°C. Reaction (6)-(6A) is drawn through the run brackets of Wyllie et al (1984), transferred from Figure 2. Linear extrapolation of reaction (6A) intersects reation (9) at point Q.

crest of the calcite-dolomite solvus, Cc=Do. The experimentally determined part of reaction (6) is transferred from Figure 2, and identified in Figure 4 as reaction (6A) below the crest of the calcite-dolomite solvus. Linear extrapolation of this curve produces an intersection with reaction (9) at point Q. No intersection occurs if the extrapolation is slightly curved, as shown by the line (6A)-a.

If the reaction curves (6A) and (9) do intersect, this generates an invariant point, Q, for the assemblage:

$$Cc + Do + Cpx + Opx + Fo + CO_2$$

Schreinemaker's rules provide an array of univariant reactions around this point, with theoretical arrangement of reactions (9) and (6A) differing from that indicated by the experimental data. It can also be shown that reactions (6A) and (9) cannot be in the positions given by the lines (6A)-a and (9) in Figure 4; the geometry of the reaction tie-figures makes this impossible.

The most likely interpretation for the arrangement of reaction curves for stoichiometric dolomite-lherzolite (6B) and calcite-lherzolite (6A) is that shown in Figure 7, with the reaction lines for (9) and (6A) curved to conform with the theoretical requirements, and with the invariant point Q situated somewhere between 7 kb and 13 kb. This fit can only be forced if the errors in determination of reactions (9) at 10 kb and (6A) at 15 kb are greater than those assumed by the experimenters.

CARBONATION OF MODEL HARZBURGITE

The position of this reaction was defined by Newton and Sharp (1975), as shown in Figure 5, and confirmed by the runs of Wyllie et al (1984), who also reviewed the other published curves which are shown for comparison. Addition of component CaO to assemblage MC+En+Fo+V produces a divariant assemblage Mc+Opx+Fo+V, which becomes univariant when the composition of the magnesite solid solution reaches the solvus, introducing dolomite and generating the reaction:

$$Mc + Opx = Fo + Do + CO_2 \qquad (4)$$

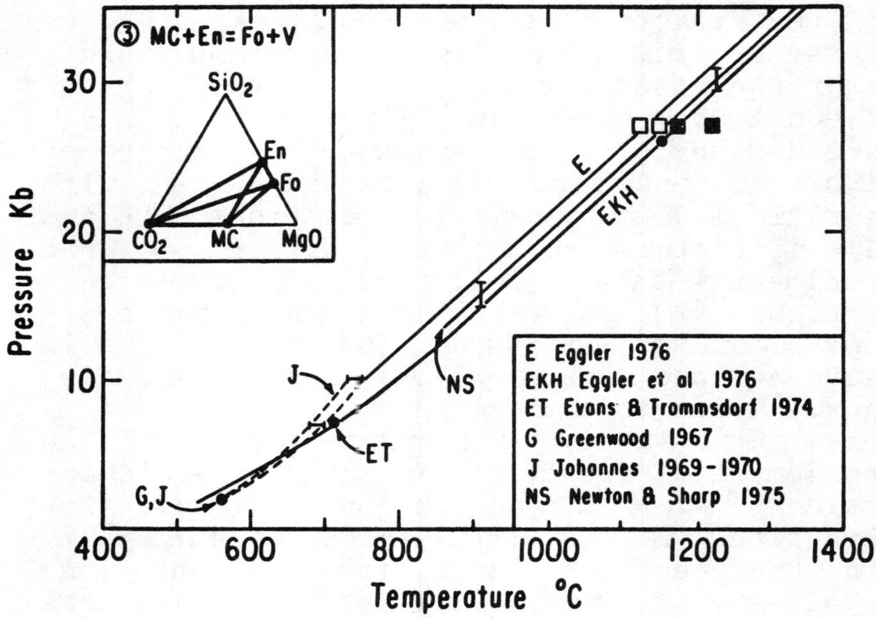

FIGURE 5. Comparison of published experimental and theoretical results for reaction (3) in MgO-SiO_2-CO_2, as reviewed by Wyllie et al (1984).

This is the quaternary equivalent of ternary reaction (3), and is situated about 5°C higher (Eggler, 1978).

CARBONATION OF MODEL WEBSTERITE

There are two vapor-absent exchange reactions involving pyroxenes and carbonates which have significance for websterites and lherzolite. Reactions (0) and (01) are degenerate. The carbonate and pyroxene compositions define a plane through the system. The degenerate reactions occurring on the plane produce mineral assemblages which may exist alone or, under suitable conditions, with forsterite, or quartz, or vapor.

Experimental and theoretical definition of these solid-solid exchange reactions is difficult, as indicated by the large differences among the five published curves for reaction (0), illustrated in Figure 6. Wyllie et al (1984) adopted the curve BB-p as the best available. Figure 6 shows the position of one calculated curve for reaction (01), but Wyllie et al (1984) concluded that reaction (01) had to be situated at pressure higher than reaction (6)-(6A). Note that reaction (01) terminates at a critical end-point (K) on the crest of the calcite-dolomite solvus (Cc-Do).

The two reactions (0) and (01) divide the PT field into three areas within which websterite coexists with different carbonates. Calcite solid solution occurs in the low pressure area below reaction (01), dolomite solid solution in the intermediate pressure interval, and magnesite solid solution at the highest pressures above reaction (0).

ARRAY OF REACTIONS IN CaO-MgO-SiO_2-CO_2

Wyllie et al (1984, Fig. 1) presented an array of reactions for the system, following a review of the available experimental and theoretical data. These reactions, with modifications, are shown in Figure 7. The positions of reactions (0), (1), (2), (5), (3) and (4) are unchanged. Determination of reaction (9) to 10 kb by Käse and Metz (1980) provides a modification of curve (6)-(6A) at crustal pressures, as discussed above. Assuming that these two curves intersect with generation of invariant point Q_2, the additional reactions plotted in Figure 7 are

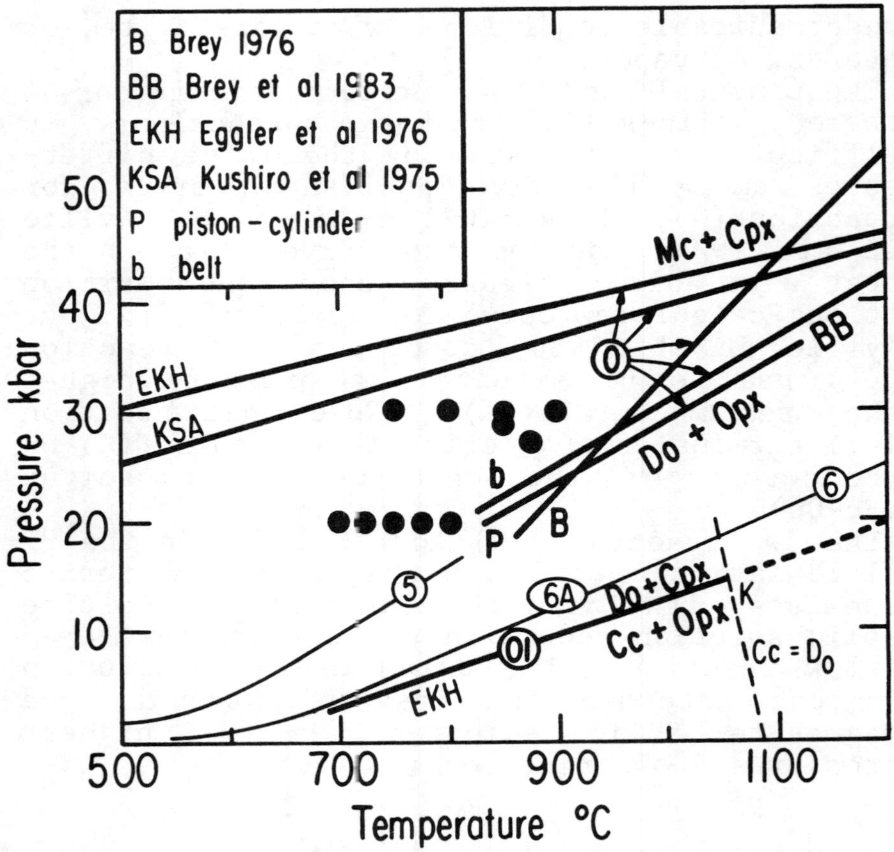

FIGURE 6. Comparison of published experimental and theoretical results for degenerate reactions (0) and (01) on the pyroxene-carbonate surface through $CaO-MgO-SiO_2-CO_2$, with closed circle runs by A. P. Byrnes, as reviewed by Wyllie et al (1984). Reaction (01) is in wrong position, see Figure 7.

required. This defines the conditions for the existence of stoichiometric dolomite-lherzolite (6B), magnesian calcite-lherzolite (6A), and high-pressure carbonate-lherzolite (6) with dolomite solid solution ranging

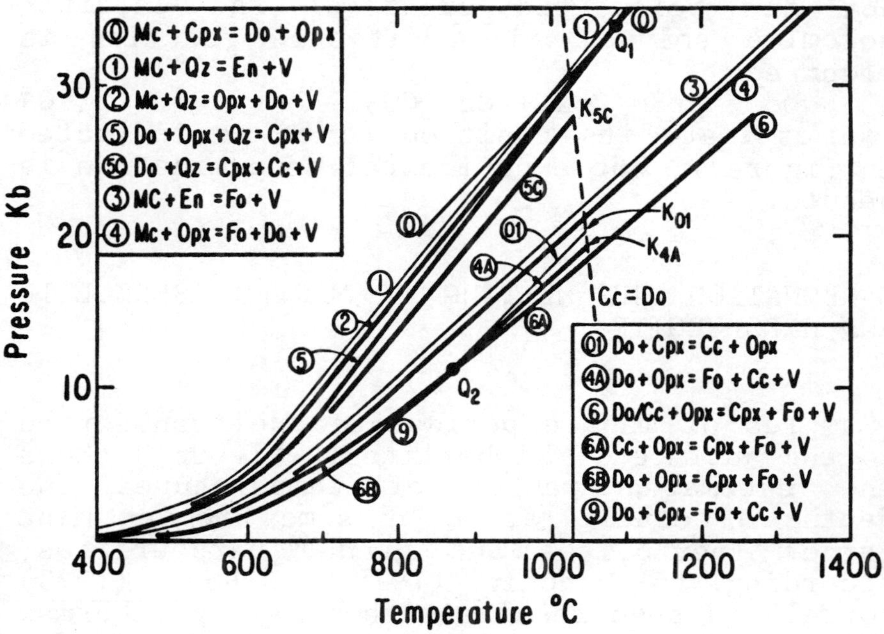

FIGURE 7. Best estimates for positions of selected reactions in the system CaO-MgO-SiO_2-CO_2, modified after Wyllie and Huang (1976a, Figure 5) and Wyllie et al (1984, Figure 1), incorporating the results for reaction (9) in Figure 4, and assuming that reactions (9) and (6A) meet at suitable angles to permit the generation of invariant point Q_2. Note the limited range of existence for reaction (01); compare Figure 6. The arrangement of reactions around Q_2 was deduced independently by E. Woermann (personal communication, 1983).

between dolomite and calcite. It limits the position and range for reaction (01), contrast Figure 6, and effectively defines the position of reaction:

$$\text{Do} + \text{Opx} = \text{Fo} + \text{Cc} + CO_2 \qquad (4A)$$

For the complete definition of progressive decarbonation at high-temperature conditions not far below the crest of the calcite-dolomite solvus, one additional reaction is required:

$$Do + Qz = Cpx + Cc + CO_2 \qquad (5C)$$

We estimate its position to be as indicated in Figure 7, but experimental determination is needed.

CARBONATION AND MELTING OF MANTLE LHERZOLITE AND HARZBURGITE

Samples of mantle peridotite are transported to the surface in kimberlites. Figure 8 shows the estimated ranges of temperatures and depths of equilibration of some Premier Mine garnet lherzolites and garnet harzburgites, according to Danchin (1979). Boyd (1973) correlated such results with fossil geotherms. Figure 8 shows phase relationships for model lherzolite and harzburgite, using the preferred reactions given in Figures 7 for the subsolidus carbonation and exchange reactions (6)-(6A), (3), and (0), respectively. These are extended to melting reactions for the model rocks in the presence of moderate amounts of CO_2. The solidus temperatures are not lowered much by CO_2, but there is a large temperature decrease for the solidus curves where the peridotites are carbonated, at pressures above the invariant points Q_6 and Q_3. The dashed lines are estimated solidus curves for carbonated peridotites with H_2O added.

These phase diagrams, and the fossil geotherms represented by the shaded areas from Premier Mine peridotites, can be used to make deductions about the petrology of the upper mantle beneath Premier at the time of eruption of the kimberlite, assuming that CO_2 and H_2O

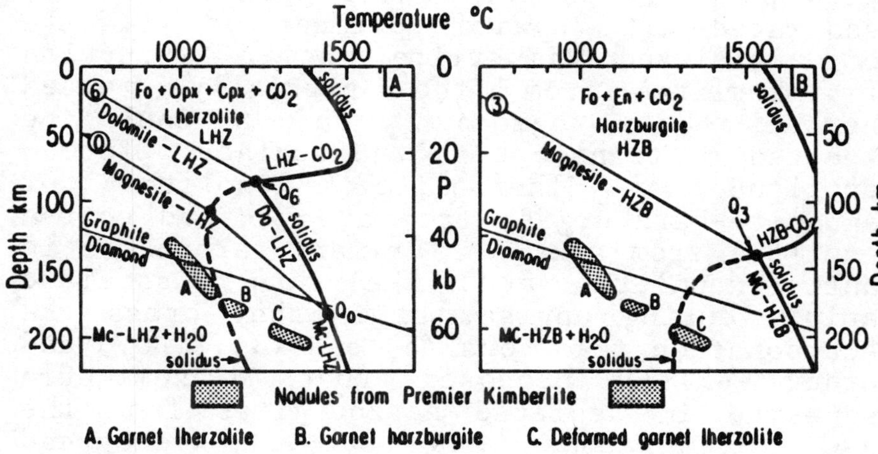

FIGURE 8. Phase diagrams for model lherzolite (LHZ) and harzburgite (HZB) with CO_2, following Wyllie (1977, Fig. 2) and Wyllie and Huang (1976c, Fig. 1), with the position of reaction (0) corrected to match BB-p in Figure 6. Dashed lines are estimated solidus curves with H_2O present as well, after Wyllie (1978, Fig. 4) and Ellis and Wyllie (1980, Fig. 9). Areas A, B, and C are estimated depths and pressures of equilibration of some mantle xenoliths from Premier Mine, South Africa (Danchin, 1979).

were both present, a condition considered to be a prerequisite for the formation of kimberlites (Wyllie, 1980). At depths greater than about 195 km, both lherzolite and harzburgite would be partially melted. Between about 195 km and 175 km, lherzolite would be partially melted, but magnesite-garnet-harzburgite, if present, would be solid. At depths shallower than 175 km, all peridotites would be solid, with or without CO_2 and H_2O.

Assume that area A consisted of coarse lherzolites and harzburgites uncontaminated by the rising mantle, which is represented by the steep portions of the fossil geotherms, B-C, associated with upward movement of volatile components and kimberlite magma. Eruption of kimberlite from the partially melted lherzolite in regions C and B could be expected to transport to the surface deformed xenoliths of lherzolite, xenoliths of magnesite-harzburgite from area B, and coarse xenoliths from area A. Primary carbonates in mantle xenoliths are likely to dissociate rapidly during uprise, after they cross the decarbonation reactions (6) and (3). Boyd and Gurney (1982) proposed that disseminated magnesite in depleted harzburgites from the base of cratons has caused widespread disruption of garnet-harzburgite nodules. The low-calcium garnets expected in garnet-harzburgites are rare in xenoliths, but common in kimberlites, and probably represent harzburgite xenoliths which exploded through magnesite dissociation during ascent. Boyd and Gurney (1982) related this to the occurrence of diamonds in the thicker cratons, where deep, depleted harzburgites were carbonated by volatiles from magmatic fluids in the asthenosphere. Figure 8 demonstrates that solid magnesite-harzburgite could coexist with partly melted lherzolite before eruption, within a specific depth interval.

ACKNOWLEDGEMENTS

This research was supported by the Earth Sciences Section of the National Science Foundation, EAR-8311758. We thank R. C. Newton for consultations, and E. Woermann for discussions.

REFERENCES

Bowen, N.L., 1940. Progressive metamorphism of siliceous limestone and dolomite. J. Geol. 48, 225-274.

Boyd, F.R., 1973. A pyroxene geotherm. Geochim. Cosmochim. Acta 37, 2533-2546.

Boyd, F.R. and Gurney, J.J., 1982. Low-calcium garnets: keys to craton structure and diamond crystallization. Carnegie Inst. Washington Year Book 81, 261-267.

Brey, G., 1976. Carbon dioxide in highly undersaturated basaltic magmas. Unpubl. Ph.D. Thesis, Australian National University, Canberra.

Brey, G., Brice, W.R., Ellis, D.J., Green, D.H., Harris, K.L. and Ryabchikov, I.D., 1983. Pyroxene-carbonate relations in the upper mantle. Earth Planet. Sci. Letters 62, 63-74.

Danchin, R.F., 1979. Mineral and bulk chemistry of garnet lherzolite and garnet harzburgite xenoliths from the Premier Mine, South Africa. In: F.R. Boyd and H.O.A. Meyer (Editors), The Mantle Sample: Inclusions in Kimberlites and Other Volcanics. Am. Geophys. Union, Washington, DC, 104-126.

Eggler, D.H., 1976. Does CO_2 cause partial melting in the low-velocity layer of the mantle? Geology 4, 69-72.

Eggler, D.H., 1978. The effect of CO_2 upon partial melting of peridotite in the system Na_2O-CaO-Al_2O_3-MgO-SiO_2CO_2 to 35 kb, with an analysis of melting in a peridotite-H_2O-CO_2 system. Am. J. Sci. 278, 305-343.

Eggler, D.H., Kushiro, I. and Holloway, J.R., 1976. Stability of carbonate minerals in a hydrous mantle. Carnegie Inst. Washington Year Book 75, 631-636.

Ellis, D. and Wyllie, P.J., 1980. Phase relations and their petrological implications in the system $MgO-SiO_2-H_2O-CO_2$ at pressures up to 100 kbar. Amer. Mineral. 65, 540-556.

Evans, B.W. and Trommsdorff, V., 1974. Stability of enstatite + talc, and CO_2-metasomatism of metaperidotite, Val d'efra, Lepontine Alps. Am. J. Sci. 274, 274-296.

Greenwood, H.J., 1967. Mineral equilibria in the system $MgO-SiO_2-CO_2-H_2O$. In: P.H. Abelson (Editor), Researches in Geochemistry. John Wiley and Sons, New York, Vol. 2, pp. 542-567.

Johannes, W., 1969. An experimental investigation of the system $MgO-SiO_2-H_2O-CO_2$. Am. J. Sci. 267, 1083-1104.

Johannes, W., 1970. Zur Entstehung von Magnesitvorkommen. N. Jb. Miner. Abh. 113, 274-325.

Käse, H.R. and Metz, P., 1980. Experimental investigation of the metamorphism of siliceous dolomites, IV. Equilibrium data for the reaction: 1diopside + 3dolomite = 2forsterite + 4calcite + $2CO_2$. Contr. Mineral. Petrol. 73, 151-159.

Kushiro, I., Satake H., and Akimoto, S., 1975. Carbonate silicate reactions at high pressures and possible presence of dolomite and magnesite in the upper mantle. Earth Planet. Sci. Letters 28, 116-120.

Newton, R.C. and Sharp, W.E., 1975. Stability of forsterite + CO_2 and its bearing on the role of CO_2 in the mantle. Earth Planet. Sci. Letters 26, 239-244.

Wyllie, P.J., 1977. Mantle fluid compositions buffered by carbonates in peridotite-CO_2-H_2O. J. Geol. 85, 187-207.

Wyllie, P.J., 1978. Mantle fluid compositions buffered in peridotite-CO_2-H_2O by carbonates, amphibole, and phlogopite. J. Geol. 86, 687-713.

Wyllie, P.J., 1980. The origin of kimberlites. J. Geophys. Res. 85, 6902-6910.

Wyllie, P.J. and Huang, W.-L., 1975. Peridotite, kimberlite, and carbonatite explained in the system $CaO-MgO-SiO_2-CO_2$. Geology 3, 621-624.

Wyllie, P.J. and Huang, W.-L., 1976a. Petrogenetic grid for siliceous dolomites extended to mantle peridotite compositions and to conditions for magma generation. Am. Mineral. 61, 691-698.

Wyllie, P.J. and Huang, W.-L., 1976b. Carbonation and melting reactions in the system $CaO-MgO-SiO_2-CO_2$ at mantle pressures with geophysical and petrological applications. Contrib. Mineral. Petrol. 54, 79-107.

Wyllie, P.J. and Huang, W.-L., 1976c. High CO_2 solubilities in mantle magmas. Geology, 4, 21-24.

Wyllie, P.J., Huang, W.-L., Otto, J. and Byrnes, A.P., 1984. Carbonation of peridotites and decarbonation of siliceous dolomites in the system $CaO-MgO-SiO_2-CO_2$ to 30 kbar. Tectonophysics, in press.

SOME GEOCHEMICAL CONSTRAINTS ON MANTLE EVOLUTION MODELS

SHEN-SU SUN

Division of Petrology and Geochemistry,
Bureau of Mineral Resources
Geology & Geophysics,
Canberra, ACT 2601
Australia

ABSTRACT

Chemical and isotopic compositions of mafic to ultramafic volcanic rocks from early Archaean to the present are used to evaluate mantle processes possibly associated with core-mantle differentiation and subsequent mantle melting. Near-chondritic ratios among refractory lithophile elements (Al, Ca, Ti, Zr, Y, Hf etc) and constant TiO_2/P_2O_5 (10 ± 1) are commonly found in fertile mantle derived magmas and nodules. They support a pyrolite type mantle source. Sm-Nd and Lu-Hf isotopic systematics also suggest a near chondritic mantle history in mantle sources of Archaean magmas. Tholeiites from Koolau Series of Oahu (Hawaiian Islands) have Nd, Sr, Hf and Pb isotopic compositions close to primitive mantle composition and He, Ar and Xe isotopic compositions of Hawaiian tholeiites suggesting close to primodial Earth's material. They have the same chemical characteristics to other mafic magmas derived from the upper mantle. Their primodial rare gas compositions and mantle convection considerations support the idea that the Hawaiian plume source is located in the lower mantle. These observations suggest that the upper and lower mantle have similar bulk chemistry and have not experienced lunar-mantle-type large scale density-controlled mineralogical-chemical differentiation during core formation.

Proceedings of the 27th International
Geological Congress, Volume 9, pp. 475–508
PETROLOGY (IGNEOUS AND METAMORPHIC ROCKS)

Several mantle evolution models are reviewed. They have much common ground but different emphasis. Since there are large uncertainties in the time related growth pattern of the continental crust, crustal recycling rate, effect of different mantle processes and convection patterns in the mantle, these models are not necessarily mutual exclusive.

Geochemical characteristics of mantle sources for basalts in mid-ocean ridges, ocean islands, island arcs and continents are described. For ocean island basalts, at least four sources are possible: 1. deep mantle plumes, 2. reactivation of the tectosphere, 3. the top of low velocity zone, 4. at various mantle depths behind island arcs associated with lithosphere subduction.

Large scale isotopic heterogeneity observed in MORB and ocean island basalts are most likely generated by different mantle processes. Combined major element (ΣFeO) and Sr, Nd, Pb and He isotope criteria indicate that at least six mantle "components" can be defined. They offer insight into geometric relationship among different types of mantle components, the timing, mechanism and nature of the processes which generate the observed chemical and isotopic variation in the mantle throughout geological time.

1. MORB source derived from the convective upper mantle depleted in incompatible elements.
2. Hawaiian plume source derived from lower mantle with close to primitive mantle composition. $^{206}Pb/^{204}Pb \sim 17.8$, $^{87}Sr/^{86}Sr \geq 0.704$, $\varepsilon Nd \sim 0$, $^{3}He/^{4}He \sim 20\text{-}50$ x air ratio.
3. Gough-Kerguelen-Tristan da Cunha type mantle possibly derived from metasomatized and reactivated or recycled tectosphere. $^{206}Pb/^{204}Pb \sim 18.4$, $^{87}Sr/^{86}Sr \sim 0.7053$ $\varepsilon Nd \sim -1$, $^{3}He/^{4}He \sim 5$ x air ratio.
4. Azores plume source derived from refractory recycled mantle material below MORB source at uncertain depths.

5. St Helena type mantle source with very high $^{206}Pb/^{204}Pb \geq 21.0$, but very low $^{87}Sr/^{86}Sr \leq 0.7029$, $\varepsilon Nd \sim +5$.
6. PLA (Planetary Lead Atlantic) type source with $^{206}Pb/^{204}Pb$ similar to St Helena type but have considerably higher $^{87}Sr/^{86}Sr > 0.704$, $\varepsilon Nd \sim 0$(?).

Mixing of different components with low velocity zone material along mid-ocean ridge axes, under oceanic islands or with the tectosphere in the continental environments is an important mechanism to generate much of the isotopic variation observed in MORB, ocean island and continental basalts. This mixing also creates heterogeneity in the upper mantle.

1. INTRODUCTION

During the last decade several conceptual models have been proposed to describe the geochemical evolution of the Earth's mantle in terms of mathematical equations with limited number of variables (e.g. Armstrong and Hein 1973; O'Nions et al 1979; Allegre et al 1980; DePaolo, 1980; Jacobsen and Wasserburg 1979). Although each model might be able to satisfy one or more sets of data and contribute to our understanding of the Earth's mantle, a unique solution which satisfies all modern and ancient geological, geophysical and geochemical boundary conditions is still lacking.

All mantle evolution models are necessarily simplified and heavily dependent on certain basic assumptions related to the thermal and tectonic evolution of the mantle. This in turn affected the Earth's melting history and mantle convection patterns which could possibly result in mineralogical and chemical layering of the mantle (e.g. Liu 1979; Anderson, 1980). The geometric relationship between different mantle reservoirs of basalts in different tectonic environment is another issue in debate.

In the paper, some geochemical boundary conditions based on studies of mantle derived mafic to ultramafic magmas of various ages and ultramafic mantle xenoliths will be used to evaluate major mantle evolution models. It is very likely that different emphasis on each boundary condition could be made by different researchers. It is hoped, however, that by defining the boundary conditions for relevant mantle processes and related consequences, it might help to improve our understanding of the evolution of the Earth's mantle.

A brief review of major mantle processes which could have taken place since core-mantle differentiation will be present to lead the discussion.

2. CORE-MANTLE DIFFERENTIATION AND SUBSEQUENT MANTLE MELTING EVENTS

Core-mantle differentiation could have started during initial accretion before the Earth reached its full size. Extensive partial melting and core formation could have both resulted from heat generated by decay of short-lived nuclei (e.g. ^{26}Al) and Earth accretion processes (e.g. Smith, 1979). The bulk of the core could have been formed within the first 100 m.y. of the Earth's history (e.g. Ringwood, 1979). Through this melting process a large proportion of siderophile elements and practically all the noble metals were segregated into the core, whereas the lithophile elements were concentrated in the silicate mantle. This core-mantle differentiation process might even lead to chemical and mineralogical stratification of the Earth's mantle (e.g. Liu, 1979, 1982; Anderson, 1980). This could then limit mantle convection to within the stratified layers which could profoundly affect mantle evolution history.

Recently Nisbet and Walker (1982) suggest that a large laterally linked Archean magma anomalies (LLLAMA) might have existed in the upper mantle at depth 150 km and below (e.g. 150-670 km?) possibly as a product of core-mantle differentiation. According to this concept the magma ocean (with MgO∿32%) could be the source for the Archaean peridotitic komatiites.

On the basis of paleomagnetic data McElhinny and Senanayake (1980) suggested that some type of plate creation and distruction has operated in the Archaean. If lithosphere subduction was involved, upper crustal material may also be involved in the recycling process to influence the chemical and isotopic evolution of the convective mantle (Armstrong, 1968, 1981).

A currently favoured model of generation of abundant Archaean tonalite and trondhjemite involves partial melting of large volumes of mafic rocks in the lower crust and/or the upper mantle (e.g. Glikson, 1972; Arth and Hanson, 1975). Seismic and gravity studies (e.g. Drummond et al., 1981) of Archaean shield areas do not indicate existence of such mafic material in the lower crust. It is very likely that the residues of partial melting have returned into the convective mantle.

Growth of the continental lithosphere (tectosphere) corresponds to the cooling history and lowering of the low velocity zone in the mantle. Much of the residual mantle peridotites from which basaltic magmas have been extracted could have been involved in the growth of the tectosphere through underplating. Since such residues are likely to be less dense than the more fertile underlying mantle (e.g. Brooks et al., 1976; Jordan, 1981). It is likely that some of the residues have returned and mixed into the convective mantle. Upward migration of melts and/or volatile rich fluid phases enriched in incompatible elements within the Archaean low velocity zone could have resulted in formation of ancient metasomatized mantle which may be preserved in the non-convective tectosphere. This part of mantle could become the source for continental basalts at a later time through reactivation of the tectosphere.

Although the growth pattern of the continental crust is not well defined, it is likely that by the end of the Archaean (∿2.5 b.y.) the volume of the continental crust might have reached some 70% or more of the present size (e.g. Taylor and McLennan, 1981). A large portion of the tectosphere was already well developed by then. It is important to recognize that lack of evidence for stable continental crust older than 3.8 b.y. does not rule out the possibility that sizeable continental crust existed in the pre Archaean (e.g. Armstrong, 1981).

Extensive meteorite bombardment during the early history of the Earth until about 3.9 b.y. ago could have catastrophic effects on the Earth's surface and induced large scale mixing of the pre-Archaean crust and mantle. It could also have profound impact on the thermal evolution of the upper mantle. It has been suggested (e.g. Chou, 1978) that this meteorite bombardment after core formation was responsible for the abundances of noble metals in the Earth's mantle.

Thermal activity in the Archaean mantle seems to end with a world-wide occurrence of mafic dykes at 2.5-2.4 b.y. A subsequent major mantle thermal event at 2.1-2.0 b.y. was represented in the form of widespread dyke swarms as observed in the Canadian shield and NW Europe and major layered complexes such as the Bushveld complex (BVSP, 1981). Continental mafic and ultramafic volcanism become prominent at 1.8-1.6 b.y. A mantle thermal event at 1.1 b.y. (Grenville) also occurred world-wide. Since then more familiar modern plate tectonic processes gradually took over (e.g. Kröner, 1982).

3. CONCEPTUAL MODELS FOR CHEMICAL EVOLUTION OF THE EARTH'S MANTLE

Three major types of mantle differentiation model will be considered:

(1) Chemical and mineralogical stratification in the mantle developed early in the Earth's history through mantle-core differentiation and associated extensive

melting, resulting in density controlled mantle-wide stratification (e.g. Liu, 1979, 1982; Smith, 1979; Anderson, 1980; Nisbet and Walker, 1982). In terms of this model, mafic and ultramafic magmas are derived from fractionated mantle layers. For example, Nisbet and Walker (1982) suggested that Archaean peridotitic komatiites were derived from LLLAMA (as defined above) whereas Anderson (1980) suggested that modern MORB were derived from an eclogite layer. This eclogite layer could have been the cooling product of LLLAMA (Nisbet and Walker, 1982).

(2) "Raisin-pudding" type of mantle generated by plate tectonics and crustal recycling (Armstrong, 1968, 1981). In this model, mantle domains of different sizes variably enriched in incompatible elements are randomly distributed in a depleted MORB type mantle. The enriched domains are the sources for ocean island volcanism. It was suggested by Armstrong that the rate of crustal growth and recycling of crustal material (including hydrothermally altered oceanic crust) through subduction were sufficient to reach a near steady state over the first eon of the Earth's history. A depleted MORB type mantle source was established in the pre-Archaean through formation of the continental crust and mantle-wide convection. Later crustal recycling might be limited to the upper mantle. Armstrong (1981) suggested that constancy of continental free board and uniformity of thickness of stable continental crust with age imply negligible crustal growth since 2.9 b.y. ago. He emphasized that available isotopic data on mantle derived rocks could be numerically satisfied by this model.

(3) Chemically layered mantle generated from an originally homogeneous pyrolite type material by continuous magma extraction and mantle convection (Fig. 1). In this model, depletion of lithophile elements in the sources of mid-ocean ridge basalts (the upper mantle) is a product of the mantle melting history. Model calculations indicate that to account for lithophile element abundances in the continental

crust only 30-50% of the mantle is of depleted (MORB source) type is sufficient (e.g. Jacobsen and Wasserbing, 1979; DePaolo, 1980). The lower mantle could either be significantly less depleted in incompatible elements than MORB type sources, or close to undepleted pyrolite material. The upper and lower mantle could have been separated by independent convection systems.

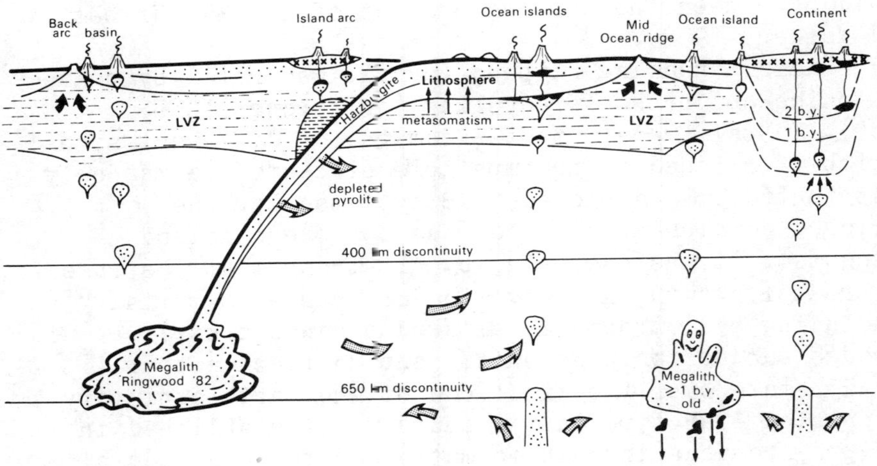

FIGURE 1. Plate tectonic related petrological model for magma generation in different tectonic environments. The geometric relationship between different magma sources is constructed based on models discussed in the text.

Various types of box model of interaction between continental crust, upper and lower mantle and core have been applied to the interpretation of isotopic data (e.g. O'Nions et al., 1979; Jacobsen and Wasserburg, 1979; DePaolo, 1980; Allègre, 1982). Among others, Allègre (1982) emphasized the importance of continental lithosphere (tectosphere) as a major potential source for the continental basalts.

Several plate-tectonics related petrological models of mantle evolution have been proposed (e.g. Hofmann and White, 1982; Ringwood, 1982). A detailed description of possible consequences of lithosphere subduction and convective recycling in the mantle has been recently given by Ringwood (1982). He suggested that the lower layer of the lithosphere consisting of relatively ductile pyrolite is stripped off during subduction and resorbed into the convecting upper mantle (Fig. 1). This material which is only depleted in higher incompatible elements (due to upward migration of small amount of melts in the upper mantle) provides a future source region of MORB on a time scale of 10^9 yrs. The slab, which subducted to 600 km at the upper-lower mantle boundary, is comprised mainly of former oceanic basaltic crust and residual harzburgite. Partial melting of this subducted oceanic crust 1-2 b.y. later due to thermal heating will contaminate the surrounding regions of former subducted harzburgite. Gravitational instability and diapirism of this newly metasomatized former harzburgite could act as the source of mantle plume activity for hot spot volcanism. Continuous underplating of these upwelling diapirs could be responsible for the growth of the continental lithosphere (tectosphere). The former subducted oceanic crust might eventually sink into the lower mantle which originally could be made of unfractionated pyrolite (i.e. primitive mantle material). In terms of this model continental crust is the product of mantle melting and differentiation processes not directly linked to the MORB sources. Periodic convective overturns in the lower mantle could result in the injection of plumes into the upper mantle and produce a different kind of plume related volcanism and mantle mixing. These periodic convective overturns could also be the ultimate cause of episodic growth of the continental crust.

4. GEOCHEMICAL CONSTRAINTS ON LAYERED MANTLE MODELS

If density controlled mineralogical and chemical stratification has developed in the mantle during mantle-core differentiation and associated extensive melting, it would be expected that the lithophile elements (e.g. Al, Ca, Ti, Zr, Hf, Y, REE, Sc etc) would have been fractionated into different layers as a result of crystal-chemical control. Consequently, chondritic ratios among these elements would not be preserved. However, in contrast to this expectation, "fertile" ultramafic mantle nodules (e.g. Jagoutz et al., 1979) commonly have chondritic ratios of these elements, except for depletion of the light REE which is most likely due to early stage extraction of small amounts of melt from "unfractionated" pyrolitic mantle (see section 3-(3). These chondritic ratios are also commonly found in Archaean non-Barberton type peridotitic komatiites and high magnesian basalts (e.g. Nesbitt et al., 1979; Arndt and Nesbitt, 1982). Since these magmas were generated by very large degrees of partial melting ($\geq 30\%$) with only olivine or olivine and orthopyroxene in the residue (e.g. Green, 1981), the observed lithophile element (except for Mg) ratios also represent those of the source. Available data on early Tertiary komatiites from the Gorgona Island also show chondritic characteristics (Echeverria, 1980; Gansser et al., 1979).

Barberton type komatiites which are commonly restricte to early Archaean (3.5-3.8 b.y.) terrains are exceptional. They have Al_2O_3/TiO_2 ratios ($\sim$10), about half the chondritic value, and are depleted in heavy REE with $(Lu/Sm)_N \leq 0.7$ (e.g. Viljoen and Viljoen, 1969; Sun and Nesbitt, 1978; Jahn et al, 1982). This deviation from chondritic ratios could either be due to a non-chondritic source composition or magma generation processes (e.g. Sun and Nesbitt, 1978; Jahn et al. 1982). It is important to note, however, that komatiites and basalts with chondritic ratios among refractory lithophile elements are also known (although less commonly) in Barberton type-locality (Viljoen and Viljoen, 1969; Sun and Nesbitt, 1978; Smith and Erlank, 1982). This also holds true for other early Archaean terrains, e.g. Pilbara Block (A.Y. Glikson, person. comm. 1983).

The possibility of a chemically layered Archaean mantle with coexisting chondritic and non-chondritic (e.g. Barberton type Lu/Sm) ratios of refractory lithophile elements in different zones could be evaluated by using $^{176}Hf/^{177}Hf$ isotopic ratios which reflect the long term ratio of Lu/Hf ($\sim$ Lu/Sm) in the system. A limited amount of Hf data on early Archaean felsic volcanics (Patchett, 1983) suggest that long term chondritic Lu/Hf ratio (i.e. flat chondritic normalized heavy REE pattern) has been maintained in their mantle sources. Additional Hf isotope studies are needed in order to evaluate this issue. However, existence of short-term chemical layering in the mantle (induced by meteorite bombardment?) may escape detection by the Hf isotope method.

It appears that available chemical and isotope data do not lend support for the model of density controlled, long lasting and large scale chemical and mineralogical stratification of the Earth's mantle. Contrasting chemical compositions of Barberton and non-Barberton type komatiites with similar MgO contents (up to 32%) would also make the LLLAMA model less attractive because it proposes a laterally linked and chemically homogenous global magma ocean. Futhermore, the LLLAMA model, as proposed by Nisbet and Walker (1982), would require extremely high temperatures ($\geq 1800^\circ C$) to exist at shallow mantle depths (~ 150 km) from 4.5 b.y. until late Archaean (2.7 b.y.). Whether or not this is consistent with the thermal history of the Archaean mantle and continental crust needs critical evaluation.

However, the book has not been closed on the possibility of layered mantle discussed in this section. Additional high quality data on lithophile element ratios in the mafic to ultramafic magmas, ultramafic mantle nodules and alpine peridotites of various ages would help to critically evaluate this issue (e.g. Palme, 1983).

5. ISOTOPIC HETEROGENEITY IN THE ARCHAEAN MANTLE AND ITS IMPLICATION FOR FORMATION OF THE PROTO-CRUST.

Existence of chemical and isotopic heterogeneity in the Archaean mantle is now well established (e.g. Sun and Nesbitt, 1978; Jahn and Nyquist, 1976; McCulloch and Compston, 1981). Variations of Sr and Nd isotope ratios observed in mantle derived volcanic rocks reflect integrated long term average Rb/Sr and Sm/Nd ratios in their source regions. For example positive εNd values (+1 to +2) for 3.5-3.8 b.y. greenstones and felsic volcanics from Isua, W. Greenland, Barberton, S. Africa, and Pilbara, W. Australia (Hamilton et al., 1981) and even higher εNd values (+3.4 to +3.7) for 2.7 b.y. greenstones from Kambalda W. Australia (McCulloch and Compston, 1981) and Vermilion, Minnesota (Ashwal et al., 1982) suggest long term depletion of light REE (and other incompatible elements) in their mantle source regions. Such depletion could have been caused by formation of the early Archaean and pre-Archaean proto-crust and/or intra-mantle differentiation such as upward migration of melts in the mantle.

McCulloch (1982) has recently suggested that highly positive εNd values ($\sim$-20 and +37) inferred for clinopyroxenes from kinberlite xenoliths yielding model age of 2.6 b.y. might represent mantle material derived from refractory cumulates precipitated from an early terrestrial magma ocean ($\sim$ LLLAMA?). It is necessary to point out, however, that kimberlite nodules are likely to come from different parts of the tectosphere and have different formation ages ranging from Archaean (e.g. Cohen et al., 1981) through late Proterozoic (Hawkesworth et al., 1983) to Recent (for the sheared nodules and megacrysts, Nixon et al., 1981). It appears that McCulloch's conclusion is not consistent with these observations. The real meaning of these Nd isotope data await further clarification.

6. GEOCHEMICAL CHARACTERISTICS OF MANTLE SOURCES FOR BASALTS IN DIFFERENT TECTONIC ENVIRONMENTS AND THEIR IMPLICATIONS FOR MANTLE EVOLUTION.

Several mantle evolution models were established on the basis of chemical and isotopic data of and geometric relationship among, mantle source regions for basalts in different tectonic environments. It is essential to describe the source characteristics before discussing an integrated mantle evolution model.

(a) mid-ocean ridge basalt (MORB)

MORB sources show depletion of highly incompatible elements (e.g. Cs, Ba, Rb, K, Th, U, Nb, Ta, light REE) which is most likely due to extraction of small amounts of melt in previous mantle melting events (e.g. Gast, 1968; Ringwood, 1982). The observed $^{176}Hf/^{177}Hf$ isotope variation (εHf +9 to +23) among typical MORB (with εNd $\sim$+11) indicates that some long term depletion of Hf (10-30%, assuming a model age for the MORB sources = 2.7 b.y.) relative to Lu existed in the MORB sources (Patchett, 1983). Since Hf behaves similarly to middle REE Sm during mantle partial melting (Patchett, 1983), it can be inferred that the degrees of Sm depletion relative to Lu in the MORB sources are also relatively small.

It is not clear, however, to what extent lithophile major elements (Al, Ca, Ti, Na) and moderately incompatible elements (Zr, Y, heavy REE, Sc) are depleted in MORB source. High magnesian MORB ($\geq$ 9% MgO) with low TiO_2 contents (0.6-0.7%) have been considered the product of large degree ($\sim$30%) of mantle melting with olivine and orthopyroxene in the residue (e.g. Sun et al., 1979). They often show chondritic ratios among Al, Ca, Ti, Zr, Y and heavy REE (Sun et al., 1979). Their chondritic normalized heavy REE patterns are quite flat. These features are the same as those observed in "fertile" alpine peridotites (Frey, in press), ultramafic nodules (Jagoutz et al., 1979) and Archaean non-Barberton type komatiites and high magnesian basalts (e.g. Sun and Nesbit, 1978). High magnesian MORB also have Ti/Zr ($\sim$110) and TiO_2/P_2O_5 ($\sim$11) ratios very similar to those estimated for the primitive mantle (110 and 10 respectively, Sun and Nesbitt, 1977; Sun, 1982). These data would suggest thatMORB sources are not significantly depleted in lithophile major elements relative to the primitive mantle composition. Further support for this conclusion comes from early Tertiary komatiites of the Gorgona Island (Echeveria, 1980). They have similar $^{143}Nd/^{144}Nd$ ratios but low La/Sm ratio (i.e. even more depleted) than the typical MORB. They also have close to chondritic ratios among Al, Ca, Ti, Y, Zr, heavy REE and Sc.

It is possible that some highly refractory harzburgite domains are heterogeneously distributed within the MORB sources but were not subjected to significant partial melting during generation of MORB thus escaping detection. In this connection, Zindler et al. (1981) suggested that $^{87}Sr/^{86}Sr$ (0.7027-0.7032) and $^{143}Nd/^{144}Nd$ (0.5130-0.5132) variations observed in seamount basalts from 9°-14°N East Pacific Rise reflect restricted melting of the more fertile domains (blobs or veins) distributed within the depleted MORB source. Therefore, small scale heterogeneity in the MORB sources can be inferred.

Systematic correlation among Pb-Sr and Nd isotopic compositions has been established for typical MORB from the Atlantic, Pacific and Indian Oceans (e.g. Cohen and O'Nions, 1982a). There is a positive correlation between $^{206}Pb/^{204}Pb$ and $^{87}Sr/^{86}Sr$ and inverse correlation between $^{87}Sr/^{86}Sr$ and $^{143}Nd/^{144}Nd$ composition, with $^{206}Pb/^{204}Pb \sim 17.6$-$18.6$, $^{87}Sr/^{86}Sr \sim 0.7022$-$0.7028$, εNd +9-+13. These data are consistent with the concept of an ocean-basin-wide mantle convection system for the MORB sources.

Plume type and transitional type MORB commonly occur close to ocean islands. Their chemical and isotopic compositions indicate mixing of mantle sources for typical MORB and ocean island basalts (e.g. Schilling 1973; Sun, 1980; Dupré and Allègre, 1983).

(b) ocean island basalts (OIB)

OIB may have different types of mantle source: i.e., (Fig. 1): 1. those derived from the mantle plumes (Hawaiian Islands, Iceland, Azores etc) have sources at various mantle depths including the lower mantle (e.g. Schilling, 1973); 2. those near continental margins (e.g. Walvis Ridge in the S. Atlantic) may have been derived from reactivation of the tectosphere (e.g. Richardson et al., 1982); 3. those associated with passive fracture zones (e.g. alkaline basalts in the Samoan Islands) could have sources in the top of the LVZ; 4. those occurring behind oceanic island arcs could have sources at various mantle depths triggered by lithospheric subduction.

Chemical composition of OIB sources may range from fertile pyrolite to refractory harzburgite. Langmuir and Hanson (1980) pointed out that some ocean island tholeiites (e.g. from Hawaii and Iceland) are more enriched in ΣFeO than the typical MORB. Either the OIB sources have greater FeO/MgO ratios (i.e. are more fertile) than the MORB sources, or melting begins at significantly greater pressure beneath ocean islands than beneath oceanic ridges. In contrast, OIB from some other ocean islands such as Azores have lower ΣFeO than typical MORB suggesting mantle sources depleted in basaltic components (i.e. are more refractory). Plume type MORB from the north Atlantic also commonly has considerably lower FeO/MgO ratios than typical MORB.

Some ocean island tholeiites have light REE depleted patterns (e.g. Iceland) or light REE enriched but concave downward REE patterns (e.g. some Hawaiian tholeiites), suggesting originally light REE depleted sources; others show light REE enriched patterns suggesting their mantle sources are enriched in incompatible elements.

Sun and Hanson (1975) suggested that alkaline basalts world-wide have sources enriched in incompatible elements with La/Ce ratio ∿1.3 times that of chondrites. This suggestion is further supported by new data $(La/Ce)_N = 1.27 \pm 0.10$), although a few exceptions do exist. It was proposed that such enrichment was caused by metasomatic processes involving addition of kimberlite-type melt to the OIB sources.

The timing of enrichment event(s) could be either ancient (e.g. precambrian) or more recent. It could even be part of the magma generation process. On the basis of Pb, Sr and Nd isotope data, an example of ancient enrichment (≥800 m.y.) was proposed by Richardson et al. (1982) for the source of Walvis Ridge basalts of S. Atlantic near Tristan daCunha. A recent enrichment event has been suggested for the source of Hawaiian post-erosional alkaline basalts (Sun, 1980; Chen and Frey, 1983). These alkaline basalts with $(La/Ce)_N \sim 1.3$ have Pb, Sr and Nd isotope data (Tatsumoto, 1978; Sun, 1980; Chen and Frey, 1983) intermediate between those for the Hawaiian plume source with $(La/Ce)_N \sim 1.0$ and the MORB source (LVZ). Since the alkaline basalts erupted after the tholeiitic plume activity the enrichment process must have taken place after initiation of the Hawaiian volcanism (i.e. recent).

Source of metasomatizing fluid in the case of Hawaiian alkaline basalts is likely to be within the LVZ (Sun, 1980, Chen and Frey, 1983) or the bottom of the oceanic lithosphere (Tatsumoto, 1978). Ringwood (1982, see section 3(3)) suggested that the metasomatizing melt and/or fluid for plumes with refractory harzbingite (e.g. Azores) could have been derived from the subducted oceanic crust. Wood et al. (1981) on the other hand, proposed that the metasomatizing fluid was introduced from a fertile plume source into a refractory harzburgite layer in the upper mantle. Alternatively the metasomatizing fluids could come from the subducted oceanic crust into the mantle wedge above. Recycling of this part of modified refractory wedge into mantle convection could become a potential source for OIB at a later time. No geochemical criteria (e.g. Ti, Nb depletion, K-Ba-Rb-Sr overabundances) have yet been used to evaluate and distinguish these possibilities.

A natural consequence of diapirism of mantle plumes into the partially molten LVZ (MORB source) is mixing of the two types of sources. Systematic Pb, Sr and Nd isotopic variations between the plume and MORB components observed in many ocean islands and nearby oceanic ridge segments clearly support this concept. (Schilling, 1973; Sun et al. 1975; Sun, 1980; Dupré and Allègre, 1983). Influx of the fertile plume components into the depleted upper mantle would also create chemical heterogeneity and affect the isotopic evolution of the upper mantle.

(c) <u>continental basalts</u>

Recent isotopic studies (e.g. Allègre et al. 1982, Menzies et al., 1983) indicate that crustal contamination is often a serious problem in study of continental basalts. According to these studies, continental basalts could have different types of mantle sources (Fig. 1): 1. those produced by reactivation of the tectosphere. Its chemical composition could be inferred from mantle nodules in kimberlites; 2. those associated with spreading center and mantle plume activity similar to that in the oceanic environment; 3. those occurred behind the destructive margins triggered by lithosphere subduction. They could have sources at various mantle depths. Mixing of different types of mantle sources could also be important in generation of some continental basalts.

Mantle metasomatism in the tectosphere could be produced by processes related to previous episodes of basaltic volcanism, or leaking of metasomatizing fluids from the downward migrating LVZ with time as the tectosphere growths thicker. The "freeze-in" time could be as old as Archaean (e.g. Menzies et al., 1983; McCulloch et al., 1983). Some parts of the tectosphere, especially that near previous continental margins, could have once been the mantle wedge above the subduction zone: e.g. Sierra-province in western U.S.A. (e.g. Menzies et al., 1983) and Papua New Guinea (Johnson et al., 1978). They may bear chemical characteristics of the island arc basalts (e.g. Nb, Ti depletion, K-Rb-Ba overabundances).

On the basis of chemical, isotopic and mineralogical data of kimberlites and their ultramafic nodules and megacrysts, Nixon et al. (1981) proposed a two stage model for kimberlite genesis: a proto-kimberlite melt derived from the deeper mantle (similar to OIB source) perhaps in connection with a rising diapir, admixed with or contaminated by a melt (enriched in incompatible elements) generated at the base of the tectosphere. Although most kimberlite samples show isotopic characteristics similar to those of IOB, Sr and Nd isotope data for some kimberlite-like potassium-rich lamproites from Western Australia indicates that they could have been derived from ancient ($\geq$1 b.y.) metasomatized sources with large negative ($<$-10) εNd values (e.g. McCulloch et al., 1983).

(d) island arc basalt (IAB)

Geochemical and isotopic data of IAB in oceanic environments suggest that their mantle sources, the wedge above the subduction zone, have been modified by fluids and/or melts derived from the subducted, altered oceanic crust and enclosed sediments (e.g. Cohen and O'Nions, 1982b). Lithophile major element composition of the wedge could be either similar to that of the MORB sources or show more refractory, residual character (e.g. Sun and Nesbitt, 1978b). The mantle wedge is an important source for the formation of tectosphere at the continental margin (Oxburgh and Parmentier, 1978). Other possible source for IAB involves incompatible element enriched domains in the raisin pudding type of mantle. Morris and Hart (1983) recently suggested that partial melting of the enriched plums ($\sim$ raisins of Armstrong's model) embedded in the depleted upper mantle matrix could generate IAB with some geochemical characteristics similar to the ocean island basalts. These plums might be the products of influx of plumes from the lower mantle.

Basalts from the back arc spreading centers also often show contributions from the subducted oceanic crust and sediments (e.g. Weaver and Tarney, 1981; Cohen and O'Nions, 1982b; Marcelot et al., 1983).

Through induced convection in the mantle wedge or reversal of the subduction polarity, the wedge material could be returned to the convecting mantle. It will bear the chemical and isotopic characteristics of IAB. It is not clear, however, how much of the modified mantle wedge and altered oceanic crust with sediment components survived the subduction zone processes and were recycled into the convective mantle and affected its chemical evolution. Repeated dehydration and partial melting of the subducted oceanic crust as it dives deeper into the mantle might create a highly refractory layer deprived of incompatible elements.

7. TOWARDS AN INTEGRATED MANTLE EVOLUTION MODEL

The issue of mantle evolution is related to questions concerning geometric relationship among different types of mantle domains, and the timing, mechanism and nature of processes which generated the observed chemical and isotopic variations in the mantle throughout geological time. Integrated isotopic studies of Sr, Nd, Pb, Hf, and He and other noble gases (Ar, Xe) on mantle derived magmas and xenoliths offer an insight into the nature of source regions and help to identify "primitive", "depleted" and "recycled" mantle material. For example, He isotopic studies of Hawaiian and Icelandic samples (e.g. Kurz et al., 1982a, 1982b; Kaneoka, 1983, Allègre et al. 1983) indicate $^3He/^4He$ ratios (20-50 x air ratio of 1.4 x 10^{-6}) considerably higher than those of typical MORB sources (8-10 x air ratio). Since high $^3He/^4He$ is a characteristic of primodial terrestrial matter, it is possible that Hawaiian and Iceland plumes were originally derived from the primitive mantle. This conclusion is supported by Xe and Ar isotopic data for Hawaiian basalts from Loihi. Allègre et al (1983) showed that they have similar $^{36}Ar/^{40}Ar$ (313-392) and $^{129}Xe/^{130}Xe$ (~6.45) ratios to those of the atmosphere (295.5, 6.48) but quite different from MORB (≳7000, ~ 6.7). They suggested that formation of the atmosphere and isolation of the Hawaiian plume source in the lower mantle was 4.4 b.y. ago. Pb isotope model calculation for time of core-mantle differentiation based on Pb isotopic data of Hawaiian tholeiites from Koolau Series (Sun, 1982) also suggested an age of 4.4 b.y..

These data support a layered mantle model with the Hawaiian and Iceland plume sources located at greater depth (i.e. the lower mantle) than depleted MORB source in the convective upper mantle with homogeneously low $^3He/^4He$. Convective overturns in the upper mantle could result in homogenization by convection and diffusion (Kurz et al. 1982a; Kaneoka, 1983). In contrast to the Hawaiian case, $^3He/^4He$ ratios for basalts from many other "plume" related regions are even lower than MORB ratio, e.g. Tristan da Cunha (5-6 x air ratio), Azore (7-8 x air ratio), FAMOUS region (8 x air ratio), suggesting they may have been derived from recycled material such as subducted oceanic crust, mantle wedge above the subduction zone or reactivated tectosphere.

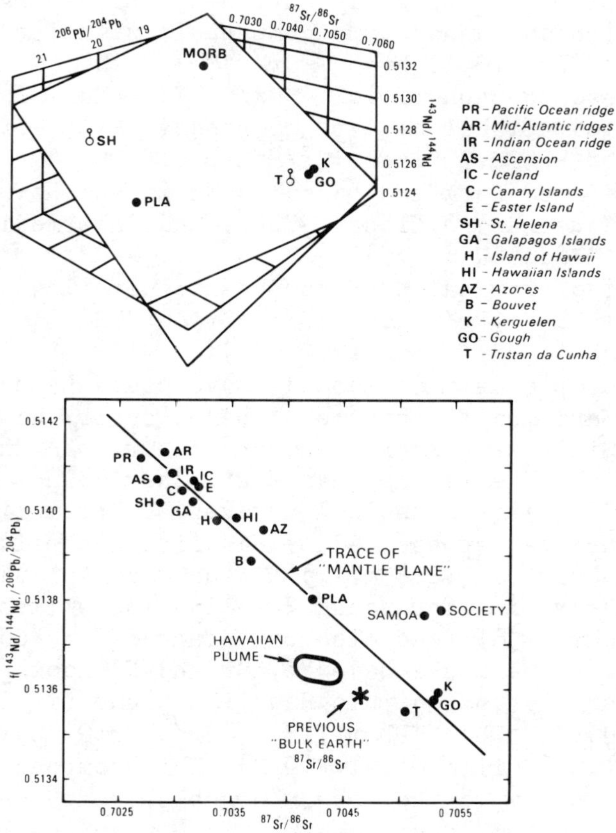

FIGURE 2 (a) Three-dimensional plot of "mantle plane" for average $^{206}Pb/^{204}Pb$, $^{87}Sr/^{86}Sr$ and $^{143}Nd/^{144}Nd$ isotope data of ocean island and ridge basalts. Positions of "end members" are shown (modified after Zindler et al., 1982). $^{87}Sr/^{86}Sr$ value of 0.7029 instead of 0.7032 was used for St. Helena. See text for explanation. (b) Two-dimensional projection taken parallel to the "mantle plane" which is normal to the paper. It is constructed after Zindler et al. (1982) with addition of data from Society and Samoan Islands (W.M. White, 1983, written comm.), PLA, and "pure" Hawaiian plume component.

1. "primitive" or "slightly enriched" Gough-Kerguelen (-Tristan da Cunha) type component: with low $^{206}Pb/^{204}Pb \sim 18.4$, high $^{87}Sr/^{86}Sr \sim 0.7053$ and close to chondritic $^{143}Nd/^{144}Nd$ ($\varepsilon Nd \sim -1$).
2. "Recycled" St Helena type component: with very high $^{206}Pb/^{204}Pb \sim 20.8$, low $^{87}Sr/^{86}Sr \sim 0.7029$ (values used here are based on White and Hofmann, 1982; Cohen and O'Nions 1982b), and intermediate $^{143}Nd/^{144}Nd$ ($\varepsilon Nd \sim +5$).
3. "depleted" typical MORB component: with low $^{206}Pb/^{204}Pb$ 18.3, low $^{87}Sr/^{86}Sr \sim 0.7027$ and high $^{143}Nd/^{144}Nd$ ($\varepsilon Nd \sim +12$).

The He isotopic data discussed above could be used to complement and modify this model based on Sr, Nd and Pb isotopes. Since samples from Gough and Tristan da Cunha have lower $^{3}He/^{4}He$ ratios than MORB, they do not represent "primitive mantle". However, a "primitive" component may be represented by Hawaiian tholeiites from Koolau series ($\sim$2.5 m.y.) of the Oahu Island, if high $^{3}He/^{4}He$ ratios for young Hawaiian tholeiites (e.g. Kaneoka, 1983) can also be inferred for Koolau tholeiites. These rocks have Pb, Sr and Nd isotopic compositions closest to pure Hawaiian plume (if it is homogeneous): $^{206}Pb/^{204}Pb \sim 17.8$, $^{87}Sr/^{86}Sr \sim 0.7041$ and $\varepsilon Nd \sim +1.0$ (Stille et al. 1982). This component (Fig. 2) has similarity to the Gough-Kerguelen component of Zindler et al (1982) but does not fall near their "mantle plane". Iceland plume component may share these Hawaiian characteristics (e.g. Sun, 1980, p.434). This component appears also to have fertile major element composition as inferred by high ΣFeO contents in tholeiites from Hawaiian Islands and Iceland (Langmuir and Hanson, 1980; see Section 6b).

Another major group of oceanic basalts which does not plot near the "mantle plane", includes samples from the Society and Samoan Islands (White and Hofmann, 1982, White, 1983 person. comm.), Sao Miguel of Azores (Hawkesworth et al. 1979). These samples have $^{206}Pb/^{204}Pb$ 18.5 - 20.0, $^{87}Sr/^{86}Sr$ 0.704 - 0.706 (high) and εNd - 1 to + 6. Hawkesworth et al. (1979) and White and Hofmann (1982) suggested that they may come from recycled oceanic crust and sediment, whereas Zindler et al. (1982) considered that their deviation from the "mantle plane" may be caused by mantle metasomatism. $^{3}He/^{4}He$ ratios for Samoan samples are high (up to 18 x air ratio) and apparently support Zindler et al's interpretation (Rison et al. 1982). Furthermore, it has been proposed that the positive correlation between Pb and Sr isotopic compositions in many north Atlantic ocean islands and plumes to normal ridge segments (e.g. Sun et al., 1975; Dupré and Allègre, 1980; Cohen and O'Nions, 1982a; Sun, 1980) could be caused by mixing of a plume component with the depleted MORB source. Extrapolation of the $^{206}Pb/^{204}Pb$ values (17.8 - 19.7) and $^{87}Sr/^{86}Sr$ (0.7023 - 0.7035) values to higher $^{206}Pb/^{204}Pb$ = 21.0, similar to the St. Helena component, will require a corresponding $^{87}Sr/^{86}Sr$ value $\sim$ 0.7042, much higher than the observed value for St. Helena component ($\sim$ 0.7029). This hypothetical component is equivalent to the Planetary Lead Atlantic (PLA) component of Dupré and Allègre (1980) and it does fall near the "mantle plane" of Zindler et al (1982).

From the above discussion, it is clear that more than six mantle components are required to satisfy available data of Pb, Sr, Nd and noble gases.

Lower $^{3}He/^{4}He$ values than MORB for some mantle components (e.g. Gough, Azores) indicate that they represent outgassed and/or recycled material. The refractory, low ΣFeO source inferred for Azores plume and plume-ridge segment in the North Atlantic (Langmuir and Hanson, 1980), mentioned in Section 6b, is consistent with the concept of recycled material. However, this does not necessarily mean that it has processed through a subduction zone (e.g. Kurz et al. 1982b).

It is important to note that mantle components, except for typical MORB and St. Helena, defined by Pb, Sr and Nd isotopic compositions for the oceanic mantle have also been observed in continental basalts, kimberlites, carbonatites and clinopyroxenes from ultramafic nodules believed to be derived from the tectosphere (e.g. Kramer, 1979; Kramer et al., 1981; Menzies et al., 1983; McCulloch et al., 1983; Grunenfelder et al., 1982).

On the basis of this general review, it appears that large scale mantle heterogeneities found for different mantle components are likely to be products of different mantle processes and derived from different depths of a layered mantle i.e., 1. MORB derived from the convective depleted upper mantle, 2. fertile Hawaiian plume derived from a primitive lower mantle, 3. refractory Azores plume derived from recycled mantle material below the MORB source at uncertain depth, 4. PLA (and St. Helena?) type plume (with high $^{206}Pb/^{204}Pb \geq 21.0$) possibly derived from mantle domains affected by core-lower mantle interaction process (Dupré and Allègre 1980); however, it might also be a product of some mineral controlled metasomatic process, 5. plume derived from recycled LVZ material (e.g. Kay, 1975;) 6. Gough, Kerguelen, Tristan da Cunha, Walvis Ridge type plume possibly derived from metasomatized, and reactivated or recycled tectosphere (e.g. Richardson et al., 1982; McKenzie and O'Nions, 1983). 7. Seamount basalts from 9-14°N East Pacific Rise derived from metasomatized domains distributed in depleted MORB source (Zindler et al. 1981), 8. Samoan alkaline basalts derived from the top of the metasomatized LVZ (Natland, 1980) which could have been swamped by the Samoan plume originated from great depth, responsible for the building of tholeiitic shield volcanoes.

In order to identify the different chemical processes and tectonic environments in which these components were developed and to work towards a unified mantle evolution model, it is necessary to integrate isotopic studies with other geochemical, geological and geophysical investigations. Establishment of trace element abundance regularity patterns (and anomalies) for tholeiites and alkaline basalts derived from different mantle components and secular variation of such regularity patterns throughout geological time could greatly complement the isotopic studies. Eu anomalies observed in fresh, modern primitive basalts could be of special interest (e.g. Sun and Hanson, 1975, Jahn et al., 1979; Hawkesworth and Vollmer, 1979). It may help to evaluate the oxygen fugacity in the environments where different tectonic and physiochemical processes took place.

ACKNOWLEDGEMENT:

Drs. Andrew Glikson and John Sheraton, both at BMR, reviewed this paper. This paper is published with the permission of the Director of Bureau of Mineral Resources, Geology and Geophysics.

REFERENCES

Allègre, C.J., (1982). Chemical geodynamics. Tectonophys. 81, 109-132.

Allègre, C.J., Brevart, O., Dupré, B., and Minster, J.F., (1980). Isotopic and chemical effects produced in a continuously differentiating convecting Earth mantle. Phil. Trans. R. Soc. London A.297, 447-477.

Allègre, C.J., Dupré, B., Richard, P., Rousseau, D., and Brooks, C., (1982). Subcontinental versus suboceanic mantle, II. Nd-Sr-Pb isotopic comparison of continental tholeiites with mid-ocean ridge tholeiites, and the structure of the continental lithosphere. Earth planet, Sci. Lett. 57, 25-34.

Allègre, C.J., Staudacher, T., Sarda, P., and Kurz, M., (1983). Constraints on evolution of Earth's mantle from rare gas systematics, Nature 303, 762-766.

Anderson, D.L., (1980). Early evolution of the mantle. Episodes, 3, 3-7.

Armstrong, R.L., (1968). A model for the evolution of strontium and lead isotopes in a dynamic earth. Rev. Geophys. 6, 175-199.

Armstrong, R.L., (1981). Radiogenic isotopes: the case for crustal recycling on a near-steady-state no-continental-growth Earth. Phil. Trans. R. Soc. London A301, 443-472.

Armstrong, R.L., and Hein, S.M., (1973). Computer simulation of Pb and Sr isotope evolution of the Earth's crust and upper mantle. Geochim. Cosmochim. Acta 37, 1-18.

Arndt, N.T., and Nesbitt, R.W., (1982). Geochemistry of Munro Township basalts. In: Arndt, N.T., Nisbett, E.G. (eds.) Komatiites. Allen and Unwin, London, 309-329.

Arth, J.G., Hanson, G.N., (1975). Geochemistry and origin of the early Precambrian crust of northeast Minnesota. Geochim. Cosmochim. Acta 39, 325-362.

Ashwal, L.D., Wooden, J.L., Morrison, D.A., Shih, C-Y, and Wiesmann, H., (1981). The Bad Vermilion Lake anorthosite Complex, Ontario: Sr and Nd isotopic evidence for depleted Archean mantle. Abst. with program, Geol. Soc. Am. 13.

Basaltic Volcanism Study Project, (1981). Basaltic volcanism on the terrestrial planets. Pergamon, New York.

Brooks, C., James, D.E., and Hart, S.R., (1976). Ancient lithosphere: Its role in young continental volcanism. Science 193, 1086-1094.

Chen, C-Y, and Frey, F.A. (1983). Origin of Hawaiian tholeiite and alkalic basalt. Nature 302, 785-789.

Chou, C.L., (1978). Fractionation of siderophile elements in the Earth's upper mantle. Proc. Lunar Sci. Conf. 9, 219-230.

Cohen, R.S., and O'Nions, R.K., (1982a). The lead, neodymium and strontium isotopic structure of ocean ridge basalts. J. Petrol. 23, 299-324.

Cohen, R.S., and O'Nions, R.K., (1982b). Identification of recycled continental material in the mantle from Sr, Nd and Pb isotope investigations. Earth planet. Sci. Lett. 61, 73-84.

Cohen, R.S., O'Nions, R.K., Dawson, J.B., (1982c). Pb, Nd and Sr isotopes in ultramafic xenoliths: Evidence for ancient subcontinental mantle? EOS 63, 460.

De Paolo, D.J., (1980). Crustal growth and mantle evolution: inferences from models of element transport and Nd and Sr isotopes. Geochim. Cosmochim. Acta 44, 1185-1196.

Drummond, B.J., Smith, R.E., and Horwitz, R.C., (1981). Crustal structure in the Pilbara and northern Yilgarn blocks from deep seismic sounding. In: Archaean Geology, Glover, J.E. and Groves, D.I. (eds). Second Intern. Symp. Perth 1980, Special pub. No. 7, Geol. Soc. Australia, 33-42.

Dupré, B., Allègre, C.J., (1980). Pb-Sr-Nd isotopic correlation and the chemistry of the North Atlantic mantle. Nature 286, 17-22.

Dupré, B., Allègre, C.J., (1983). Pb-Sr isotope variation in Indian Ocean basalts and mixing phenomena. Nature 303, 142-146.

Echeverria, L.M., (1980). Tertiary or Mesozoic komatiites from Gorgona Island, Columbia: Field relations and geochemistry. Contrib. Mineral. Petro. 73, 253-265.

Frey, F.A., (in press). Rare earth element abundances in upper mantle rocks. In: Henderson, P. (ed). Rare element geochemistry, Elsevier, Amsterdam.

Gansser, A., Dietrich, V.J., and Cameron, W.E., (1979). Paleogene komatiites from Gorgona Island. Nature 278, 545-546.

Glikson, A.Y., (1972). Early Precambrian evidence of a primitive ocean crust and island nuclei of sodic granite. Geol. Soc. Am. Bull. 83, 3323-3344.

Green, D.H., (1981). Petrogenesis of Archaean ultramafic magmas and implications for Archaean tectonics. In: Kroner, A., (ed). Precambrian plate tectonics. Elsevier, Amsterdam, 469-489.

Grunenfelder, M.H., Tilton, G.R., Bell, K. and Blenkinsop, J., (1982). Lead isotope relationships in the Oka Carbonatite Complex, Quebec. EOS, 63, 1134.

Hamilton, P.J., Evensen, N.M., O'Nions, R.K., Glikson, A.Y., and Hickman, A.H., (1981). Sm-Nd dating of the North Star Basalt, Warrawoona Group, Pilbara Block, Western Australia. In: Glover, J.E., and Groves, D.I. (eds), Archaean Geology. Second Intern, Symp. Perth. 1980, Special Pub. No. 7. Geol. Soc. Australia. 187-192.

Hawkesworth, C.J., Norry, M.J., Roddick, J.C., and Vollmer, R., (1979). $^{143}Nd/^{144}Nd$ and $^{87}Sr/^{86}Sr$ ratios from the Azores and their significance in LIL-element enriched mantle. Nature 280, 28-31.

Hawkesworth, C.J., Vollmer, R., (1979). Crustal contamination versus enriched mantle: $^{143}Nd/^{144}Nd$ and $^{87}Sr/^{86}Sr$ evidence from the Italian volcanics. Contrib. Mineral. Petro. 69, 151-165.

Hawkesworth, C.J., Erlank, A.J., Marsh, J.S., Menzies, M.A., and Van Clasteren, P., (1983). Evolution of the continental lithosphere: Evidence from volcanics and xenoliths in southern Africa. In: Hawkesworth, C.J., and Norry, M.J. (eds.) Continental basalts and mantle xenoliths. Shiva, U.K. 111-138.

Hofmann, A.W., White, W.M., (1982). Mantle plumes from ancient oceanic crust. Earth planet. Sci. Lett. 57, 421-436.

Jacobsen, S.B., and Wasserburg, G.J., (1979). The mean age of the mantle and crustal reservoirs. Jour. Geophys. Res. 84, 7411-7427.

Jagoutz, E., Palme, H., Baddenhausen, H., Blum, K., Cendales, M., Dreibus, G., Spettel, B., Lorenz, V. and Wanke, H., (1979). The abundances of major, minor and trace elements in the earth's mantle as derived from primitive ultramafic nodules. Proc. Lunar Sci. Conf. 10, 2031-2050.

Jahn, B-M., Gruau, G., and Glikson, A.Y., (1982). Komatiites of Onverwacht Group, S. Africa: REE geochemistry, Sm/Nd age and mantle evolution. Contrib. Mineral. Petro. 80, 24-40.

Jahn, B.M., Nyquist, L.E., (1976). Crustal evolution in the early earth-moon system: Constraints from Rb-Sr studies. In: Windley, B.F. (ed) Early history of the Earth. J. Wiley and Sons, 55-76.

Jahn, B-M., Sun, S.S., and Nesbitt, R.W., (1979). REE distribution and petrogenesis of the Spanish Peaks Igneous Complex, Colorado. Contrib. Mineral. Petro. 70, 281-298.

Johnson, R.W., Mackenzie, D.E., and Smith, I.E.M., (1978). Delayed partial melting of subduction-modified mantle in Papua New Guinea. Tectonophys. 46, 197-216.

Jordan, T.H., (1981). Continents as a chemical boundary layer, Phil. Trans. R. Soc. London. A 301,. 359-373.

Kaneoka, I., (1983). Noble gas constraints on the layered structure of the mantle. Nature 302, 698-700.

Kay, R., (1975). Chemical zonation of the oceanic mantle. EOS 56, 1077.

Kramer, J.D., (1979). Lead, uranium, strontium, potassium and rubidium in inclusion-bearing diamonds and mantle-derived xenoliths from southern Africa. Earth Planet. Sci. Lett. 42, 58-70.

Kramer, J.D., Smith, C.B., Lock, N.P., Harmon, R.S., and Boyd, F.R., (1981). Can kimberlites be generated from an ordinary mantle? Nature 291, 53-56.

Kröner, A., (1981). Precambrian plate tectonics. In: Kröner, A. (ed). Precambrian plate tectonics. Elsevier, Amsterdam, 57-90.

Kurz, M.D., Jenkins, W.J., and Hart, S.R., (1982a). Helium isotopic systematics of oceanic islands and mantle heterogeneity. Nature 297, 43-47.

Kurz, M.D., Jenkins, W.J., Schilling, J.G., and Hart, S.R., (1982b). Helium isotopic variations in the mantle beneath the central North Atlantic Ocean. Earth Planet. Sci. Lett. 58, 1-14.

Langmuir, C., Hanson, G.N., (1980). An evaluation of major element heterogeneity in the mantle sources of basalts. Phil. Trans. R. Soc. London A297, 383-407.

Liu, L-G., (1979). On the 650 km seismic discontinuity. Earth planet. Sci. Lett. 42, 202-208.

Liu, L-G., (1982). Chemical inhomogeneity of the mantle: geochemical consideration. Geophys. Res. Lett. 9, 124-126.

Marcelot, G., Dupuy, C., Girod, M., and Maury, R.C., (1983). Petrology of Futuna Island lavas (New Hebrides): An example of calc-alkaline magmatism associated with the initial stages of back-arc spreading. Chem. Geol. 38, 23-37.

McCulloch, M.T., (1982). Identification of Earth's earliest differentiates (Abst.), Fifth Intern. Conf. Geochron. Cosmochron. Isotope. Geol. 244-245.

McCulloch, M.T., and Compston, W., (1981). Sm-Nd age of Kambalda and Kanowna greenstones and heterogeneity in the Archaean mantle. Nature 294, 322-327.
McCulloch, M.T., Jaques, A.L., Nelson, D.R., and Lewis, J.D., (1983). Nd and Sr isotopes in kimberlites and lamproites from Western Australia: an enriched mantle origin. Nature, 302, 400-403.
McElhinny, M.W., and Senanayaka, W.E., (1980). Paleomagnetic evidence for the existence of the geomagnetic field 3.5 Ga ago. Journ. Geophy. Res. 85, 3523-3528.
McKenzie, D.P. and O'Nions, R.K. (1983). Mantle reservoirs and ocean island basalts. Nature 301, 229-231.
Menzies, M.A., Leeman, W.P., and Hawkesworth, C.J., (1983). Isotope geochemistry of Cenozoic volcanic rocks reveals mantle heterogeneity below western USA. Nature 303, 205-209.
Morris, J.D., Hart, S.R., (1983). Isotopic and incompatible element constraints on the genesis of island arc volcanics from Cold Bay and Amak Island, Aleutians, and implications for mantle structure. Geochim. Cosmochim. Acta 47, 2015-2030.
Natland, J.H., (1980). The progression of volcanism in the Samoan linear volcanic chain. Am. Jour. Sci. 280A, 709-735.
Nesbitt, R.W., Sun, S.S., and Purvis, A.C., (1979). Komatiites: Geochemistry and genesis. Can. Mineral. 17, 165-186.
Nisbet, E.G., Walker, D., (1982). Komatiites and the structure of the Archaean mantle. Earth planet. Sci. Lett. 60, 105-116.
Nixon, P.H., Rogers, N.W., Gibson, I.L., and Grey, A., (1981). Depleted and fertile mantle xenoliths from southern African kimberlites. Ann. Rev. Earth Planet. Sc. 9, 285-309.
O'Nions, R.K., Evensen, N.M., and Hamilton, P.J., (1979). Geochemical modelling of mantle differentiation and crustal growth. J. Geophys. Res. 84, 6091-6101.

Oxburgh, E.R., and Parmentier, E.M., (1978). Thermal processes in the formation of continental lithosphere. Phil. Trans. R. Soc. London A288, 415-429.

Palme, H., (1983). The abundances of refractory lithophile elements in the terrestrial mantle. Terra cognita 3, 126.

Patchett, P.J., (1983). Importance of the Lu-Hf isotopic system in studies of planetary chronology and chemical evolution. Geochim. Cosmochim. Acta 47, 81-91.

Richardson, S.H., Erlank, A.J., Duncan, A.R., and Reid, D.L., (1982). Correlated Nd, Sr and Pb isotope variation in Walvis Ridge basalts and implications for the evolution of their mantle source. Earth Planet. Sci. Lett. 59, 327-342.

Ringwood, A.E., (1979). Origin of the Earth and Moon, 295 pp. Springer-Verlag.

Ringwood, A.E., (1982). Phase transformations and differentiation in subducted lithosphere: Implications for mantle dynamics, basalt petrogenesis, and crustal evolution. Jour. Geol. 90, 611-643.

Rison, W., and Craig, H., (1982). Helium 3: Coming of age in Samoa. EOS 63, 1144.

Schilling, J-G., (1973). Iceland mantle plume: Geochemical study of the Reykjanes Ridge, Nature 242, 565-571.

Smith, H.S., and Erlank, A.J., (1982). Geochemistry and petrogenesis of komatiites from the Barberton greenstone belt, South Africa. In: Arndt, N.R., and Nisbet, E.G. (eds) Komatiites. George Allen and Unwin. 347-397.

Smith, J.V., (1979). Mineralogy of the planets: a voyage in space and time. Mineral. Mag. 43, 1-89.

Stille, P., Tatsumoto, M., and Unruh, D.M., (1982). Sr, Nd, Hf and Pb isotope geochemistry on Oahu lavas. (Abs). Fifth Intern. Conf. Geochron. Cosmochron. Isotope Geol. 354-355.

Sun, S.S., (1980). Lead isotopic study of young volcanic rocks from mid-ocean ridges, ocean islands and island arcs. Phil. Trans. R. Soc. London A 297, 409-446.

Sun, S.S., (1982). Chemical composition and origin of the earth's primitive mantle. Geochim. Cosmochim. Acta 46, 179-192.

Sun, S.S., and Hanson, G.N., (1975). Origin of Ross Island basanitoids and limitations upon the heterogeneity of mantle sources for alkali basalts and nephelinites. Contrib. Mineral. Petro. 52, 77-106.

Sun, S.S., and Nesbitt, R.W., (1977). Chemical heterogeneity of the Archaean mantle, composition of the earth and mantle evolution. Earth planet. Sci. Lett. 35, 429-448.

Sun, S.S., and Nesbitt, R.W., (1978a). Petrogenesis of Archaean ultrabasic and basic volcanics: Evidence from rare earth elements. Contrib. Mineral. petro. 65, 301-325.

Sun, S.S., and Nesbitt, R.W., (1978b). Geochemical regularities and genetic significance of ophiolitic basalts. Geology 6, 689-693.

Sun, S.S., Nesbitt, R.W., and Sharaskin, A.Y., (1979). Geochemical characteristics of mid-ocean ridge basalts. Earth planet. Sci. Lett. 44, 119-138.

Sun, S-S., Tatsumoto, M., and Schilling J-G., (1975). Mantle plume mixing along the Reykjanes Ridge axis: Lead isotopic evidence. Science 190, 143-147.

Tatsumoto, M., (1978). Isotopic composition of lead in oceanic basalts and its implication to mantle evolution. Earth planet. Sci. Lett. 38, 63-87.

Taylor, S.R., and McLennan, S.M. (1981). The composition and evolution of the continental crust: rare earth element evidence from sedimentary rocks. Phil. Trans. R. Soc. London A301, 381-399.

Viljoen, M.J., and Voljoen, R.P., (1969). Evidence for existence of a mobile extrusive peridotitic magma from the Komati Formation of the Onverwacht Group. Geol. Soc. S. Afr. Spec. Publ. 2, 87-112.

Weaver, S.D. and Tarney, J., (1982). Mesozoic-Cenozoic volcanism in the South Shetland Islands and the Antarctic Peninsula: geochemical nature and plate tectonic significance. In: Craddock, C. (ed). Antarctic Geoscience. Univ. Wisconsin press, Madison, 263-274.

White, W.M., and Hofmann, A.W., (1982). Sr and Nd isotope geochemistry of oceanic basalts and mantle evolution. Nature 296, 821-825.

Wood, D.A., Tarney, J., Weaver, B.L., (1981). Trace element variations in Atlantic ocean basalts and Proterozoic dykes from northwest Scotland: their bearing upon the nature and geochemical evolution of the upper mantle. Tectonophys. 75, 91-112.

Zindler, A., Staudigel, H., Langmuir, C., Goldstein, S., and Weaver, S., (1981). Nd and Sr isotope variations in young Pacific seamounts: Implications for the scale of mantle heterogeneity. EOS 61, 424.

Zindler, A., Jagoutz, E., and Goldstein, S., (1982). Nd, Sr and Pb isotopic systematics in a three-component mantle: a new perspective. Nature 298, 519-523.

EXPERIMENTAL STUDIES ON K-METASOMATISM OF A MODEL PYROLITE MANTLE AND THEIR BEARING ON THE GENESIS OF ULTRAPOTASSIC MAGMAS

ALAN D. EDGAR AND MAKOTO ARIMA

Department of Geology, University of Western Ontario
London, Canada N6A 5B7

ABSTRACT

Experiments at 20 and 30 kb, and 850-950°C on pyrolite with various concentrations of K_2O in aqueous solution have been used to model K-metasomatism in the upper mantle. With even small amounts of K_2O added (<0.5 wt %) phlogopite is an additional phase at 30 kb and phlogopite and calcic amphibole at 20 kb. At 30 kb, similar results were obtained from runs at 850°C and 950°C. With increasing K_2O concentration (up to 3.5 wt %), phlogopite content increases relative to garnet, olivine, orthopyroxene, and clinopyroxene, olivine content decreases relative to clinopyroxene and orthopyroxene, while orthopyroxene content remains constant relative to clinopyroxene. Once liquids appear at higher K_2O concentrations (≥3.5 wt %), phlogopite content starts to decrease. At 20 kb, increasing K_2O concentration (up to 1.8 wt %) produces an inverse relationship between amphibole and phlogopite, and decrease of olivine and orthopyroxene relative to clinopyroxene. At higher K_2O concentration (>2.0 wt %) amphibole and spinel disappear, but phlogopite content continuously increases relative to orthopyroxene, clinopyroxene and olivine with increasing K_2O concentration. These results indicate reactions for phlogopite formation involve garnet, olivine and orthopyroxene at 30 kb, and spinel, amphibole, olivine, ortho-

Proceedings of the 27th International
Geological Congress, Volume 9, pp. 509–541
PETROLOGY (IGNEOUS AND METAMORPHIC ROCKS)

pyroxene and clinopyroxene at 20 kb. In the absence of appreciable sodium, metasomatism of the upper mantle at 20 kb may result in formation of phlogopite rather than that of K-amphibole.

INTRODUCTION

The concept of heterogeneities in the mantle produced by metasomatism is now widely accepted (c.f. Bailey, 1982). Based on major, trace and isotope geochemistry (c.f. Hawkesworth and Vollmer, 1979; Vollmer and Hawkesworth, 1980; Van Kooten, 1980, 1981); on the presence of nodules containing phlogopite, amphibole and other minerals not commonly found in nodules of lherzolitic mantle composition (Lloyd and Bailey, 1975; Lloyd, 1981) and on inferences from high pressure experimental petrology (Edgar et al., 1976, 1980; Ryabchikov and Green, 1978; Wendlandt and Eggler, 1980a, b) many ultrapotassic basalts are believed to be produced by partial melting of such metasomatized mantle sources. Nodules in such basalts even from a single area range from peridotites, with incipient degrees of metasomatism as indicated by partial replacement of earlier minerals by phlogopite and amphibole, to complete replacement of residual peridotite mineralogy producing mica amphibole-clinopyroxenites which retain relicts of their original peridotitic mineralogy (Lloyd and Bailey, 1975; Bailey, 1982). Original minerals most susceptible to metasomatic alteration are garnet, spinel, orthopyroxene and olivine whereas carbonates, apatite, Fe-Ti oxides and sulphides are formed by metasomatism (Harte et al., 1975; Erlank and Rickard, 1977; Bailey, 1982). Deformation of the mica and amphibole (Erlank and Rickard, 1977; Boyd and Nixon, 1978), stabilities of these minerals relative to the mantle solidus (Bailey, 1980) and, in some instances, Sr isotope data (McCulloch et al., 1983) indicate that metasomatism must be a subsolidus process. Although the nature of the metasomatism and the source(s) of the fluids are not clear, Lloyd and Bailey (1975) and Bailey (1982) suggest that as many as eighteen elements in varying amounts may be involved. Geochemical studies of the nodules (Lloyd, 1981) and of their host basalts indicate that alkalies (particularly K), Ti, Ba, Sr, Zr, Nb and LREE are most frequently involved. These incompatible and large ion lithophile elements are also those most commonly found in products of ultrapotassic magmas suggesting that such magmas may be products of partial melting of metasomatized mantle.

Experimental modelling of mantle metasomatism is obviously very difficult. Based on the incongruent solubility of phlogopite in an aqueous vapour in the phlogopite-H_2O and phlogopite-H_2O-CO_2 joins at subsolidus temperatures and 20 kb (Yoder and Kushiro, 1969; Wendlandt, 1977), Ryabchikov and Boettcher (1980) showed that the solubility of K (as K_2O) increased from 4g K_2O per 100g H_2O at 11 kb to 25g K_2O per 100g H_2O at 30 kb at temperatures of 1050-1100°C. These results showed that fluids could easily transport K_2O under mantle conditions. Ryabchikov et al. (1982) determined experimentally that hydrous fluids coexisting with omphacitic pyroxenes leach sodium silicate with the amount leached at constant temperature increasing with decreasing pressure. These results together with those of Ryabchikov and Boettcher (1980), suggest that metasomatism involving K and Na may cause K enrichment at deeper levels in the mantle and Na-enrichment at shallower levels where amphibole might be stable. They propose that K/Na fractionation in the upper mantle may produce K-rich and Na-rich magmatism by partial melting of alkali metasomatized mantle sources at different depths.

In this paper we have extended Ryabchikov and Boettcher's (1980) and Ryabchikov et als. (1982) work to investigate the effects of K metasomatism on model mantle pyrolite (Ringwood, 1975, Table 5-2, anal. 8) at 20 and 30 kb between 850-950°C. The main objectives are 1) to determine the amounts of K_2O (as K_2CO_3 solutions) in excess of that of pyrolite necessary to produce mineralogical changes in the pyrolite composition. With K_2O as the only added component, the characteristic minerals indicative of metasomatism at depths corresponding to these pressures are phlogopite and amphibole, 2) to assess the effects of K_2O-bearing fluids of different concentrations on pyrolite at 30 kb relative to the preferential formation of phlogopite and amphibole at 20 kb where both minerals are stable, 3) to suggest mechanisms of reaction between the pyrolite mineralogy (olivine, orthopyroxene, clinopyroxene $\pm$ garnet $\pm$ spinel) and the K_2O solutions to produce phlogopite and amphibole based on the relative proportions of minerals produced in experiments with various K_2O concentrations in the fluid.

These experiments represent a great oversimplification of metasomatism in nature. Potassium is the only element in the metasomatizing fluid in our experiments and we have assumed the principal volatile is H_2O (the amounts of CO_2 relative to H_2O are negligible for the concentrations of K_2CO_3 solutions used). The assumption of H_2O being the predominant volatile at deeper levels in the upper mantle is supported by studies of Wyllie (1979) and Ryabchikov et al. (1981). In nature metasomatism is almost certainly an open system process, proceeding over very long time periods (McCulloch et al., 1983) and probably involving more than one metasomatic "event" (Dawson, 1982). In contrast our experiments were done under closed system conditions as these are the only ones feasible at the pressures involved.

Bearing in mind the limitations imposed by the simplified nature of this experimental model and the inherent experimental difficulties, we consider that these experiments can provide only an indication for the nature of K metasomatism. If such metasomatism is involved in the production of the source material from which ultrapotassic magmas are derived by partial melting then some comparisons can be made between the compositions used in our experiments with those of metasomatized nodule material believed to represent source materials from which partial melting has produced ultrapotassic liquids.

EXPERIMENTAL TECHNIQUES AND CONSIDERATIONS

Starting materials

A mixture of pyrolite composition as given by Ringwood (1975, Table 5-2, anal. 8) was prepared (Table 1) by mixing in the requisite amounts predried 'Analar' grade TiO_2, Al_2O_3, Cr_2O_3, Fe_2O_3, MnO, NiO, MgO, CaO, $Ca_5(PO_4)_3(OH)$, Na_2CO_3 and K_2CO_3. SiO_2 was added as 'pure' quartz. After crushing and mixing, the powder was heated to 900°C for 2 hours to convert the carbonates to oxides. On heating at 950°C and 30 kb this produced an assemblage of olivine + orthopyroxene + clinopyroxene + garnet without glass. At 20 kb and the same temperature spinel replaced garnet in this assemblage. This mixture was used as a representative unmetasomatized upper mantle composition.

"Metasomatizing" solutions were prepared by dissolving the appropriate amounts (by weight) of dried 'Analar' grade K_2CO_3 in 10 g of deionized H_2O to give solutions ranging in concentration from 0.1 to 5.5 g K_2O/10 g solution. In each experiment 10% by weight of these solutions were added to the pyrolite mixture.

Apparatus

All experiments were done in a 1.27 cm diameter piston-cylinder apparatus of the Boyd and England

(1960) type, using the 'hot' piston out technique. Pressures were calibrated at 22 kb and 1300°C using the kyanite-sillimanite transition (Richardson et al., 1968) and at 16.3 kb and 600°C using the albite = jadeite + quartz reaction (Johannes et al., 1971). These agreed well with accepted values. Temperatures were measured using $Pt-Pt_{90}Rh_{10}$ and $Pt-Pt_{87}Rh_{13}$ thermocouples. No frictional correction was applied to pressure and no pressure correction to the emf of the thermocouples.

Starting compositions were enclosed in Pt or $Ag_{75}Pd_{25}$ sealed capsules. Using these capsules, iron losses by alloying with the capsules are not considered to be significant at the low temperatures used (850°-950°C) and for the run times used (c.f. Stern and Wyllie, 1975).

Choice of 20 and 30 kb was based on the fact that amphibole is stable at 20 kb but not at 30 kb at the temperatures used (Gilbert et al., 1982). In contrast phlogopite is stable at both of these pressures in this temperature range (Modreski and Boettcher, 1972).

Between 20 and 30 kb, 950°C is below the solidus of pyrolite under hydrous conditions (Green, 1973) and therefore represents reasonable temperature conditions for mantle metasomatism. This temperature is also compatible with that of the continental geotherm (Clark and Ringwood, 1964). In 30 kb experiments with high concentrations of K_2O, liquid is present at 850° and 950°C.

Identification

Experimental products were too fine grained to allow microprobe determinations of their compositions. Products were identified optically and by x-ray diffraction methods. The latter was also used to estimate the proportions of phlogopite and amphibole relative to the other major phases (olivine, orthopyroxene, clinopyroxene) in runs with added K_2O solutions. In these runs only very minor amounts of garnet were observed with increasing K_2O concentrations.

The only available method to assess the degree of metasomatism as indicated by phlogopite formation was to measure ratios of phlogopite intensity on x-ray diffractograms relative to those of olivine, orthopyroxene and clinopyroxene. Similarly, the mechanisms of metasomatism could only be estimated by comparing ratios of olivine, orthopyroxene and clinopyroxene peak intensities. To avoid erroneous ratios, peaks of each mineral involved were chosen to avoid interference with other minerals; but with sufficiently high intensities to be readily measured. For the minerals involved the only suitable peaks meeting these criteria are:- phlogopite (001), olivine (130), orthopyroxene (420-222-only a single peak was observed), clinopyroxene (221) and amphibole (110).

Each run was x-rayed and the intensities for each mineral were averaged for runs with increasing and decreasing 2Θ. Intensities in the increasing 2Θ were within 2% variation of those with decreasing 2Θ and generally less. Ratios of intensities determined for different experimental runs under identical conditions varied by $\pm$ 10%.

Distinction of quench and primary phases

In experimental studies in fluid bearing compositions, crystals precipitate as quench material from the vapour phase during rapid cooling. Of the minerals involved in this study, phlogopite is very commonly formed as quench material. Distinction between primary and quench phlogopite in this study could only be made on the basis of habit. Ryabchikov and Boettcher (1980) showed that the habit of quench phlogopite varied with pressure but that primary phlogopite morphology appeared to be independent of pressure, occurring as euhedral relatively thick crystals (see Ryabchikov and

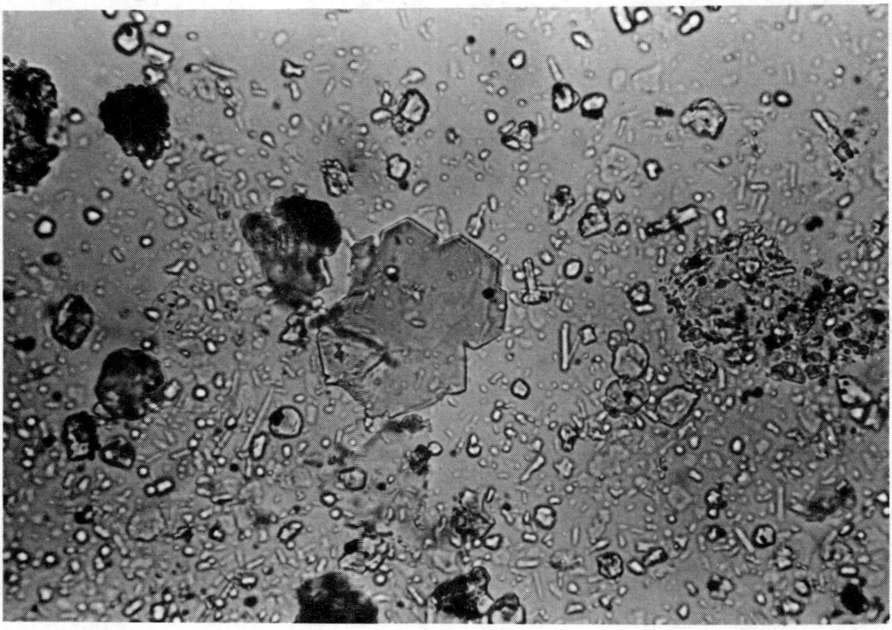

Figure 1. Quench and primary phlogopites in run at 20 kb, 950°C, 4 hours with 3.6 g K_2O/10 g solution. Large hexagonal shaped plate is primary phlogopite, fine needles of low relief throughout matrix is quench phlogopite. Plain light x 320.

Boettcher, 1980, Fig. 4). The same criteria have been used in this study (Figs. 1-3).

Some indication of the concentrations of K_2O solution where primary rather than quench phlogopite may be expected in our experiments can be estimated from Ryabchikov and Boettcher's (1980, table 1) which indicated that in the anhydrous phlogopite-H_2O system at 1100°C and 30 kb primary phlogopite was present only at compositions richer

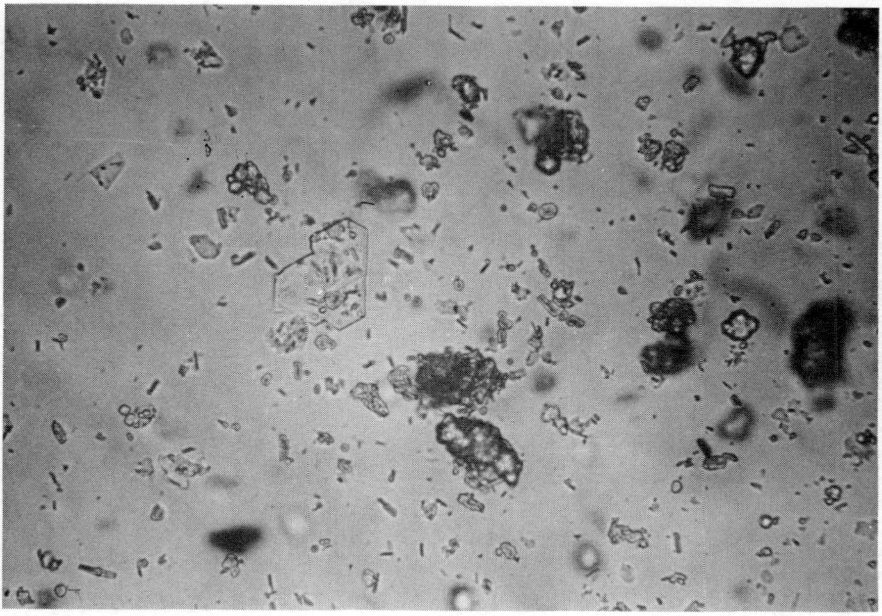

Figure 2. Quench and primary phlogopites in run at 20 kb, 950°C, 4 hours with 1.8 g K_2O/10 g solution. Hexagonal shaped crystals of primary phlogopite and tiny needles of quench phlogopite with low relief in matrix. Plain light x 320.

in anhydrous phlogopite than 65%. At 20 kb and 1100°C primary phlogopite occurred only at compositions greater than 33% anhydrous phlogopite. On Fig. 4 the compositions used in this study are recalculated in terms of pyrolite-H_2O-anhydrous phlogopite. Assuming that the solubility of MgO in the fluid coexisting with pyrolite is negligible relative to the K_2O, Al_2O_3 and SiO_2 of phlogopite (Ryabchikov and Boettcher, 1980) and that the

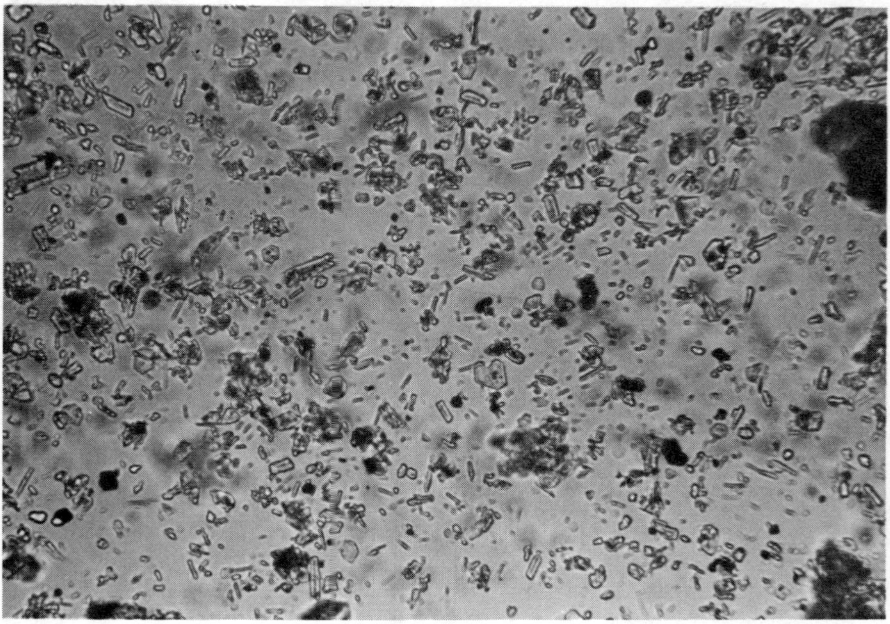

Figure 3. Quench and primary phlogopite in run at 30 kb, 950°C, 4 hours with 3.6 g K_2O/10 g solution. Small, low birefringent, hexagonal shaped plates are primary phlogopite, very tiny needles of low relief quench phlogopite. Plain light x 120.

solubility of diopside and garnet are also negligible, the compositions likely to produce primary phlogopite at 30 kb lie within the area pyrolite-anhydrous phlogopite-A (Fig. 4). At 20 kb primary phlogopites may exist within the area pyrolite-anhydrous phlogopite-B (Fig. 4). The points A and B from Ryabchikov and Boettcher (1980) are plotted on the basis of the data for 1100°C, at 30 and 20 kb respectively. In contrast, in our experiments at lower temperatures the solubility of K_2O in the aqueous phase is likely to be lower and therefore slightly less K_2O-rich fluids may produce primary phlogopite.

RESULTS

20 kb, 950°C, 4 hours with 10% solution

Intensity ratios in runs at 20 kb, 950°C for 4 hours using 10% by weight of solutions of varying K_2O concentrations are plotted on Fig. 5. Optical examination indicated that experiments with solutions 1 and 2 (Fig. 4) produced only quench phlogopite whereas both primary and quench varieties occurred in experiments with all other solutions. This is the result predicted from Ryabchikov and Boettcher's (1980) experiments extrapolated to the pyrolite-anhydrous phlogopite-H_2O composition (Fig. 4). Amphibole was detected in runs with 10% H_2O and no added K_2O and with solutions containing up to 1.8 g K_2O/10 g solution. Under comparable conditions, Green (1973) detected primary pargasitic amphibole in the (pyrolite-40% olivine) - 6% H_2O system, suggesting that amphibole in our experiments is also primary.

Fig. 5a indicates an increase in the amount of phlogopite relative to olivine, orthopyroxene and clinopyroxene with increasing concentrations of K_2O in the solutions. Phlogopite first appears in the x-ray patterns at 0.6 g K_2O/10 g solution and shows a steady increase relative to the other minerals up to about 3.5 g K_2O/10 g solution. At higher concentrations the relative increase in phlogopite is much greater. Similarities in these curves suggest that the formation of phlogopite involves reactions with olivine, orthopyroxene and clinopyroxene. Spinel is absent in all x-ray diffraction patterns but was detected optically in runs with low K_2O concentrations.

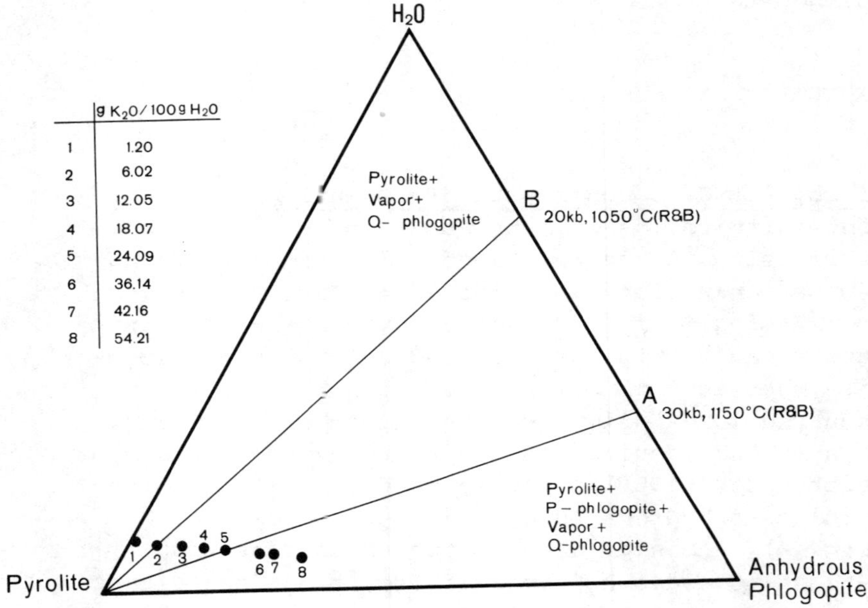

Figure 4. Postulated composition within which primary and quench phlogopite crystallize in the system pyrolite-anhydrous phlogopite-H_2O. Primary phlogopite occurs at 30 kb within the area A - pyrolite-anhydrous phlogopite and at 20 kb within the area B -pyrolite-anhydrous phlogopite. A and B based on data of Ryabchikov and Boettcher (1980) under conditions shown. Dots represent compositions used in this study as g K_2O/100 g solution.

Amphibole shows an inverse relationship to phlogopite, decreasing with increasing concentration of K_2O in the solutions (Fig. 5a). Amphibole was not observed at K_2O concentrations greater than 1.8 g K_2O/10 g solution. If this amphibole is primary its decrease with increasing concentration of K_2O in the solutions suggests it is not a K-bearing amphibole.

Fig. 5b shows the variations in the relative proportions of olivine, orthopyroxene and clinopyroxene. Although there is some scatter in this data, the variations in the olivine/orthopyroxene ratios show a positive trend and whereas both the olivine/clinopyroxene and orthopyroxene/ clinopyroxene ratios decrease with increasing concentrations of K_2O solutions up to about 2.5 g K_2O/10 g solution. The latter suggests that clinopyroxene may be involved in the formation of amphibole (and phlogopite) as well as olivine and orthopyroxene.

Variations in the amphibole contents relative to olivine, orthopyroxene and clinopyroxene (Fig. 5a) show a sharp decrease with increasing concentrations of K_2O from 0 to 1.8 g K_2O/10 g solution. The sudden decrease in amphibole in runs with zero concentration of K_2O relative to 0.1 g K_2O/10 g solution (Fig. 5a) appears to be reflected in the sharp increase in the olivine/orthopyroxene and decrease in the orthopyroxene/clinopyroxene ratios (Fig. 5b).

<u>30 kb, 950°C, 4 hours with 10% solution</u>

Results of experiments at 30 kb and the same time, temperature and concentration conditions as those at 20 kb are shown in Fig. 6. Under these conditions amphibole is absent and only quench phlogopite occurs in runs with solutions 1, 2, 3 and possibly 4 (Fig. 4). Primary phlogopite is definitely present in runs using solution 5 (Fig. 4); the latter composition lying on the line separating quench from primary phlogopite (Fig. 4).

In contrast to the 20 kb experiments, no continuous increase in the relative proportions of phlogopite with increasing concentration of K_2O in the solutions is observed. Fig. 6 shows an increase up

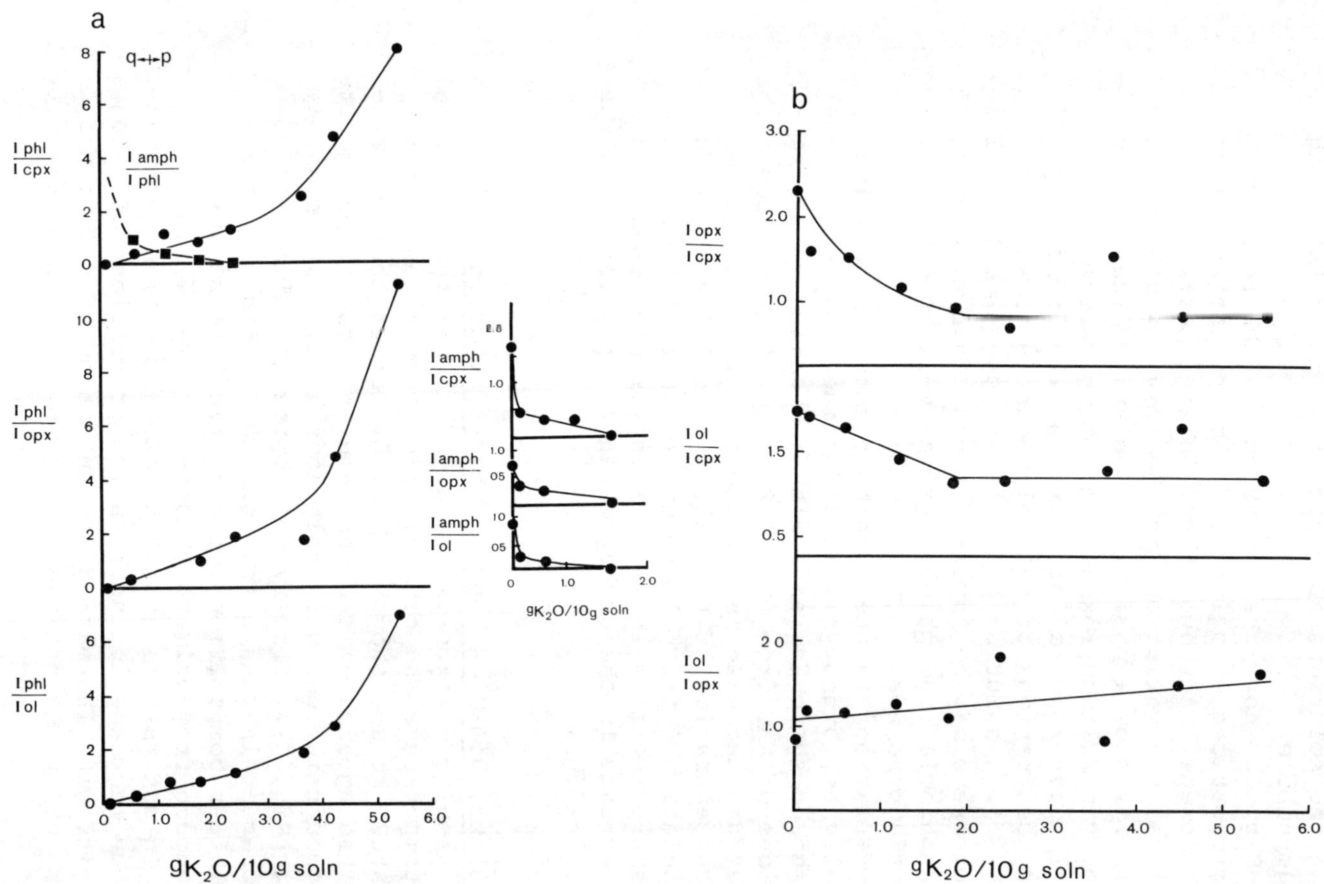

a
b
q↔p
I phl / I cpx
I amph / I phl
I phl / I opx
I phl / I ol
I amph / I cpx
I amph / I opx
I amph / I ol
I opx / I cpx
I ol / I cpx
I ol / I opx
gK₂O/10g soln
gK₂O/10g soln
gK₂O/10g soln

Figure 5a. Intensities of phlogopite and amphibole relative to olivine, orthopyroxene and clinopyroxene for experiments with various concentrations of K_2O in solution for runs at 20 kb, 950°C. q and p represent quench only and primary + quench phlogopite.

b. Relative intensities of olivine, orthopyroxene and clinopyroxene for experiments with various concentrations of K_2O in solution for runs at 20 kb, 950°C.

to 3.6 g K_2O/10 g solution followed by a systematic decrease in phlogopite relative to the other minerals using solutions with higher K_2O concentrations. In runs with 4.2 and 5.2 g K_2O/10 g solutions, glass was observed in the products suggesting that the lower relative amounts of phlogopite in these runs was due to the preferential partitioning of K into the liquid rather than into phlogopite. This implies that K_2O (in excess of about 3.0 g K_2O/10 g solution) to pyrolite lowers its solidus temperature at 30 kb relative to its value at 20 kb (Figs. 5, 6). To test this hypothesis, runs were made at 30 kb, 850°C and 7 hours.

<u>30 kb, 850°C, 7 hours with 10% solution</u>

Results of runs under these conditions are also shown in Fig. 6. Within the probable limits of error, they produce the same pattern as those at 30 kb, 950°C, 4 hours. Optical examination of runs at 850°C with more than 3.6 g K_2O/10 g solution indicated that glass was present. Thus the most likely explanation for the shape of the curves at 30 kb (Fig. 6) relative to those at 20 kb (Fig. 5a) is that K is preferentially incorporated into the liquid as the solidus is lowered by more than 100°C at the higher pressure and K_2O concentrations.

Comparison of the olivine/orthopyroxene, olivine/clinopyroxene and orthopyroxene/clinopyroxene ratios for experiments at 20 kb (Fig. 5b) and 30 kb (Fig. 6) differ, particularly in the decreasing olivine/orthopyroxene with increasing concentrations of K_2O solutions at 30 kb relative to 20 kb and for the much larger scattering in the orthopyroxene/clinopyroxene ratios at 30 kb relative to 20 kb. These variations and the presence of amphibole in some of the 20 kb runs may suggest that different processes are involved in the formation of phlogopite at both pressures. This is supported by the larger amounts of phlogopite relative to olivine at 30 kb (Fig. 6) compared to

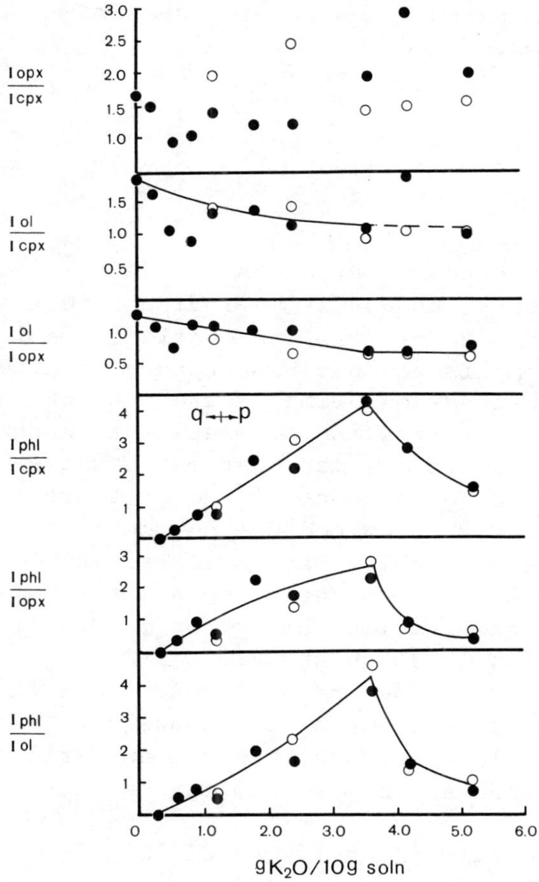

Figure 6. Intensities of phlogopite relative to olivine, orthopyroxene and clinopyroxene for experiments with various concentrations of K_2O in solution for runs at 30 kb. Solid circles - 950°C, 4 hours; open circles - 850°C, 7 hours.

that at 20 kb (Fig. 5a) for experiments with less than 3.6 g K_2O/10 g solution. The amounts of phlogopite relative to ortho and clinopyroxenes appear to be comparable under both pressure conditions (Figs. 5a, 6).

DISCUSSION

Implications from the experiments

Our experiments at 20 and 30 kb indicate that K metasomatism of a model pyrolite mantle composition is readily accomplished under laboratory conditions. Although these experiments are subject to constraints and are simplistic relative to natural conditions, it may be assumed that metasomatism at depths corresponding to those of the pressures used is a common process, provided the source(s) of K-rich fluids are available. As indicated in Figs. 5a and 6, phlogopite produced by metasomatism may occur at K_2O concentrations of approximately 1.8 g K_2O/10 g solution at 30 kb and 1.2 g K_2O/10 g solution at 20 kb. Under natural conditions in which fluid may pass through continuously in an open system for long periods, the concentrations required to metasomatise pyrolite may be quite different.

The presence of liquid as low as 850°C at 30 kb in runs under volatile present conditions with high concentrations of K_2O (Fig. 6) suggests that K-metasomatism under near solidus conditions, comparable to those suggested by the mantle geotherm (Clark and Ringwood, 1964), may be limited relative to those at shallower depths, corresponding to 20 kb, where no liquid was detected at 950°C and the amounts of phlogopite relative to other minerals increase with increasing concentrations of K_2O (Fig. 5a). Thus, in the absence of Na and other elements in the metasomatizing fluid, K metasomatism might be more a more feasible process at shallower mantle depths.

Hypothetical K-metasomatism reactions

The formation of phlogopite in our experiments, and in nature, is dependent not only on the amount of K available but also on the source and availability of Al. In normal mantle of pyrolite composition, Al occurs mainly in garnet at depths equivalent to 30 kb and in spinel at depths equivalent to 20 kb. Minor Al is also present in pyroxenes in the mantle source. Alternatively Al could occur in the metasomatizing fluid as the $KAlSiO_3(OH)_2$ molecule (Lloyd and Bailey, 1975). If Al is present only as garnet and/or spinel, the amount of phlogopite produced during metasomatism will cease when garnet and/or spinel have reacted out.

In our experiments with added K_2O neither garnet nor spinel were detected in any of the x-ray patterns. Optically both minerals decrease with increased K_2O concentrations but in 30 kb runs magnetite(?) appears to increase. In 30 kb runs, the Al for phlogopite formation comes from garnet whereas in 20 kb runs, where amphibole is present in experiments without K_2O added, the Al for phlogopite may be provided from both amphibole and spinel. The absence of garnet and spinel and presence of amphibole in metasomatized phlogopite-bearing nodules from Uganda and Eifel areas (Lloyd and Bailey, 1975) and kelyphitic rims of phlogopite on garnets in nodules in kimberlites (Schandl and Clarke, 1982) support the roles of garnet and spinel as sources of Al for the formation of phlogopite in K-metasomatic processes.

Figs. 5 and 6 indicate that reactions producing phlogopite at 30 kb involve olivine and orthopyroxene but at 20 kb clinopyroxene and amphibole are also involved. As the fine grain sizes of all minerals did not permit microprobe analyses, the compositions of these minerals and hence the exact nature of the reactions are unknown. However some possible reactions which account for the observed decrease in the relative proportions of olivine and increase in orthopyroxene at 30 kb below the solidus with increasing K_2O concentrations (Fig. 6) and the increases in olivine and clinopyroxene and decrease in orthopyroxene at 20 kb at low to moderate K_2O concentrations (Fig. 5b) may be postulated. In these reactions we assume that the minerals are pure Mg end members, that the amphibole is a pargasite, and that under subsolidus conditions all available K_2O is in phlogopite.

At 30 kb, clinopyroxene is probably not involved in the formation of phlogopite as it contains negligible CaO. The general reaction for this pressure can be written as:-

$$\text{5 olivine + orthopyroxene + garnet} \xrightarrow[H_2O]{K_2O}$$

$$\text{phlogopite + (5 - x) olivine +}$$

$$\text{(1 + y) orthopyroxene} \qquad (1)$$

where the proportions of olivine, orthopyroxene and garnet are approximately those of pyrolite-type mantle at 30 kb (Wyllie, 1975). For low concentrations of K_2O, and hence low amounts of phlogopite, the orthopyroxene coexisting with phlogopite may be aluminous (containing $MgAl_2SiO_6$) as all the available Al in the garnet will not be required for phlogopite.

A more quantitative approach is to calculate the reaction using a modified CIPW norm in which the available Na_2O is incorporated as the jadeite molecule in preference to albite and the residual Al_2O_3 is in the pyrope garnet molecule at 30 kb and spinel and amphibole molecules at 20 kb. This results in a composition comparable to that given by Ringwood (1975). Assuming that clinopyroxene is not involved at 30 kb, that only Mg end members are used and that all K_2O enters phlogopite, this gives an equation for the reaction with 1.0 g K_2O in solutions as:-

$$\underset{\text{ol.}}{4.8\ (Mg_2SiO_4)} + \underset{\text{opx.}}{(MgSiO_3)} +$$

$$\underset{\text{gt.}}{1.3\ (Mg_3Al_2Si_3O_{12})} \xrightarrow[H_2O]{1.0\ K_2O}$$

$$\underset{\text{phl.}}{(K_2Mg_6Al_2Si_6O_{20}(OH)_4)} + \underset{\text{opx.}}{3\ (MgSiO_3)} +$$

$$\underset{\text{ol.}}{2.4\ (Mg_2SiO_4)} + \underset{\text{Mg-tsch.}}{0.3\ (MgAl_2SiO_6)} \qquad (2)$$

This reaction accounts for the observed decrease in ol/opx and ol/cpx intensities (Fig. 6) for low concentrations of K_2O in solution under sub-solidus conditions, assuming that clinopyroxene is not involved. Comparison of the oxide contents between the reactants and products shows a slight deficiency in MgO and an excess in SiO_2 in the products. These discrepancies may result from this not being a mass balance calculation (sensu stricto) because FeO has been ignored, and/or because an oxide phase may be involved.

The involvement of an oxide mineral in the reaction is supported by optical observation which indicated an apparent increase in an opaque mineral with increasing K_2O concentrations. If this mineral is magnetite, chromite or magnesiochromite it could undoubtedly result in the SiO_2 discrepancy.

The reaction at 20 kb seems more complex than that at 30 kb as both amphibole and spinel are potential sources of Al_2O_3 in runs at low K_2O concentrations and only spinel is available at higher K_2O concentrations where amphibole is no longer present (Fig. 5a). Due to the low K_2O available in experiments with amphibole as a product, it is almost certainly a Ca-rich amphibole which implies that clinopyroxene must be involved in this reaction in contrast to the 30 kb experiments. A very generalized reaction analogous to that for 30 kb may be suggested for 20 kb runs in which amphibole no longer appears:-

$$\begin{aligned}&4 \text{ olivine} + \text{orthopyroxene} + \text{clinopyroxene} + \\ &\text{amphibole} + \text{spinel} \xrightarrow[H_2O]{K_2O} \text{phlogopite} + \\ &(4 + x) \text{ olivine} + (1 - y) \text{ orthopyroxene} + \\ &(1 + z) \text{ clinopyroxene} \qquad (3)\end{aligned}$$

The rates of increase and decrease in the reactants relative to the products are highly dependent on K_2O concentrations as indicated by the changes in slopes of the ratios of these minerals (Fig. 5).

The results in Fig. 5 suggest that three different reactions are involved for the ranges in K_2O concentrations used:-

(1) for the formation of amphibole with no K_2O in the solution,

(2) for the formation of amphibole and phlogopite

between 0.6 and about 2.5 g K_2O/10 g solution, and

(3) the rapid increase in phlogopite relative to other minerals above about 3.5 g K_2O/10 g solution.

The conditions of disappearance of amphibole and rapid increase in phlogopite nearly overlap. Possible reactions for each of these situations can be modelled based on the modified CIPW norm using the same method and assumptions as for the 30 kb conditions. Additional modifications are that all the Na in the jadeite molecule is used to make amphibole. The reactios are listed in the same sequence as given above. The formation of amphibole from pyrolite + 10% H_2O is:-

$$\underset{ol}{4.3(Mg_2SiO_4)} + \underset{opx}{2.5(MgSiO_3)} + \underset{cpx}{(CaMgSi_2O_6)} +$$

$$\underset{jd}{0.2(NaAlSi_2O_6)} + \underset{sp}{0.5(MgAl_2O_4)} \xrightarrow{H_2O}$$

$$\underset{amph.}{0.2(NaCa_2Mg_4Al_3Si_6O_{22}(OH)_2)} + \underset{ol.}{4.1(Mg_2SiO_4)} +$$

$$\underset{opx}{2.7(MgSiO_3)} + \underset{cpx}{0.6(CaMgSi_2O_6)} + \underset{sp}{0.3(MgAl_2O_4)} \quad (4)$$

Using these reactants, phlogopite may be formed in experiments with low K_2O concentrations in the fluid according to the equation:-

$$0.2(NaCa_2Mg_4Al_3Si_6O_{22}(OH)_2) + 4.1(Mg_2SiO_4) +$$

amph. ol.

$$2.7(MgSiO_3) + 0.6(CaMgSi_2O_6) + 0.3(MgAl_2O_4)$$

opx. cpx. sp.

$$\xrightarrow[H_2O]{xK_2O} x\ (K_2Mg_6Al_2Si_6O_{20}(OH)_4) + (4.3 + x)$$

phl.

$$(Mg_2SiO_4) + (2.5 - 7x)\ (MgSiO_3) + (CaMgSi_2O_6)$$

ol. opx. cpx.

$$+ 0.2(NaAlSi_2O_6) + (0.5 - x)\ (MgAl_2O_4)$$

jd. sp. (5)

where x K_2O represents low to moderate concentrations of K_2O in the solution (<2.0 g K_2O/10 g solution). As x must have a small value in this equation, the degree of increase in clinopyroxene will be greater than that of olivine whereas orthopyroxene will decrease. This reaction is in accord with the observed ratios of these minerals at concentrations of K_2O less than 2 g K_2O/10 g soln. (Fig. 5b). With higher concentrations of K_2O, the increased amounts of phlogopite relative to the olivine, orthopyroxene, and clinopyroxene (Fig. 5a) will be formed according to the reaction:-

$$x(K_2Mg_6Al_2Si_6O_{20}(OH)_4) + (4.3 + x)(Mg_2SiO_4)$$

phl. ol.

$$+ (2.5 - 7x)(MgSiO_3) + (CaMgSi_2O_6) +$$

opx. cpx.

$$0.2(NaAlSi_2O_6) + (0.5 - x)(MgAl_2O_4) \xrightarrow[H_2O]{yK_2O}$$

jd. sp.

$$(x + y)(K_2Mg_6Al_2Si_6O_{20}(OH)_4) + (4.3 + x + y)$$

phl.

$$(Mg_2SiO_4) + (2.5 - 7x - 7y)(MgSiO_3) +$$

ol. opx.

$$(CaMgSi_2O_6) + 0.2(NaAlSi_2O_6) + (0.5 - x - y)$$

cpx. jd.

$$(MgAl_2O_4)$$

sp. (6)

where y represents the concentrations of K_2O in solutions greater than about 2.0 g K_2O/10 g solution. This equation is in accordance with the increase in the ratios of ol/opx using these concentrations and may account for the ol/cpx and opx/cpx ratios (Fig. 5b). By this stage the amount of spinel is very low, which is in accordance with the optical observations in these experiments.

Amphibole-phlogopite relations at 20 kb

The antipathetic relationship between amphibole and phlogopite and the probable lack of a K-richteritic amphibole in experiments with pyrolite + K-bearing solutions is unexpected as K-metasomatism at depths corresponding to about 20 kb is considered likely to produce a K-rich amphibole (Ryabchikov et al., 1982). If the amphibole in our experiments is a Ca-rich species, the addition of K to pyrolite appears to suppress amphibole formation with K being preferentially incorporated into phlogopite. This result may be due to the very low Na activity in our experiments. Ryabchikov et al. (1982) have shown that the Na_2O/H_2O ratio in aqueous fluids is much higher at 20 than at 30 kb and that in metasomatism caused by fluids containing both Na and K, K will be preferentially fixed in phlogopite at the greater depth while Na would be retained in the fluid phase and not precipitate until the stability of amphibole is reached at depths corresponding to about 20 kb. This implies that in experiments with fluids containing both K_2O and Na_2O as K or Na rich amphibole may be more likely to occur. Future studies will be done to investigate this.

Metasomatized mantle as a source material for ultrapotassic magmas

The high K, Ti, LIL and other incompatible elements in ultrapotassic mafic lavas suggests they represent partial melts of a heterogeneous (metasomatized) mantle source (cf. van Kooten, 1980). Analyses of nodules in such lavas, believed to represent possible source materials, are very sparse (Table 1). The wide variations in the

TABLE 1. Compositions of Starting Material, K-Metasomatized Pyrolite and Nodules from Ultrapotassic Rocks and Kimberlites

	1.	2.	3.	4.	5.	6.	7.	8.
SiO_2	45.1	43.54	38.19	45.71	44.87	46.49	45.03	45.82
TiO_2	0.2	0.19	7.61	3.77	0.11	0.14	0.28	0.09
Al_2O_3	4.6	4.44	7.43	6.73	1.03	1.23	3.31	3.03
Cr_2O_3	0.3	0.29	-	-	0.45	0.37	0.28	0.40
Fe_2O_3	0.3	0.29	6.49	3.58	2.56	2.04	5.27	2.96
FeO	7.6	7.34	7.37	6.08	4.83	4.41	4.98	4.43
MnO	0.1	0.10	0.15	0.17	0.13	0.11	0.12	0.15
NiO	0.2	0.19	-	-	0.27	0.27	0.10	0.26
MgO	38.1	36.78	12.45	13.95	42.95	42.56	37.81	40.00
CaO	3.1	2.99	15.32	15.56	1.50	0.88	2.56	2.76
Na_2O	0.4	0.39	0.43	0.73	0.33	0.19	0.20	0.18
K_2O	0.02	3.62	3.39	3.49	0.81	0.99	0.24	0.15
P_2O_5	0.02	0.02	0.14	0.19	0.00	0.01	-	-
TOTAL	100.04	100.18	98.97	99.96	99.84	99.69	100.18	100.23
$\frac{K}{K + Na}$	0.03	0.98	0.98	0.97	0.64	0.77	0.44	0.36

1. Average mantle pyrolite (Ringwood, 1975, Table 5-2).
2. Recalculated pyrolite with 3.6 g K_2O/10 g solution added.
3. Average of 67 nodules of the Katwe-Kikarongo volcanics + 17 nodules from Bunyarunguru volcanics, Uganda (unpublished analyses from Dr. F.E. Hough)
4. Phlogopite clinopyroxenite nodule, chabachu crater, Bunyarunguru, Uganda (unpublished analyses from Dr. F.E. Hough)
5. Phlogopite bearing peridotite in kimberlite, Bulfontein, Kimberley, S. Africa (Aoki, 1975, Table 1)
6. Phlogopite bearing garnet peridotite in kimberlite. Locality and reference as in 5.

7,8. Garnet lherzolite nodules with kelyphitic runs of phlogopite or garnet, Premier Mine, S. Africa (Danchin, 1979, Table 3).

mineralogy of these nodules (Lloyd and Bailey, 1975, Fig. 1) suggests that, if they represent metasomatized mantle sources, the intensities and types of metasomatism may also have been very variable.

Both nodules (Table 1) have comparable K enrichment (as K/K + Na mol.) of about 0.97. Comparison of the compositions of the natural materials with the recalculated model metasomatized mantle composition for experiments with 3.6 g K_2O/10 g solution (Table 1, analysis 2) show good agreement in K_2O, Na_2O (and hence K/K + Na), reasonable agreement in SiO_2, FeO and MnO and poor agreement in TiO_2, Al_2O_3, Fe_2O_3, MgO and CaO values. This suggests that the mechanism we propose based on our experiments is applicable for potassium but that the compositions of the solutions are too simplistic; particularly with respect to Ti and Al. Discrepancies in MgO, TiO_2 and FeO may be related to the susceptibility of olivine to alter during metasomatism and the formation of Fe-Ti oxides during the process (Bailey, 1982). Although comparison of the compositions of the nodules with those of our metasomatized mantle pyrolite is encouraging with respect to K it is obvious that unknown processes must account for the discrepancies in other elements.

Nodules from kimberlites which may also represent metasomatized mantle sources (cf. Dawson, 1980) indicate much lower K_2O values and slightly lower K/K + Na ratios relative to the recalculated metasomatized model mantle compositions used in our experiments (Table 1). Comparison of other components in these kimberlite nodules with our model metasomatized pyrolite indicate similar SiO_2, TiO_2, MgO, FeO and CaO but different Al_2O_3 and Fe_2O_3. This may imply that the metasomatic process involved in the production of kimberlites differs from that for ultrapotassic magmas.

ACKNOWLEDGEMENTS

This study was begun at the University of Manchester where the senior author was a Royal Society-Nuffield Foundation Bursar and British Council Scholar. He would like to thank the faculty and staff of the Manchester Geology Department for their assistance. Both authors are grateful to the Natural Science and Engineering Research Council of Canada for financial support for this project.

REFERENCES

Aoki, K. (1975). Origin of phlogopite and potassic richterite bearing peridotite xenoliths from South Africa. Contrib. Mineral. Petrol., 53, 145-156.

Bailey, D.K. (1980). Volcanism, Earth degassing and replenished lithosphere mantle. Phil. Trans. R. Soc. Lond. A297. 309-322.

Bailey, D.K. (1982). Mantle metasomatism-continuing chemical change within the Earth. Nature, 296, 525-530.

Boyd, F.R. and Nixon, P.H. (1978). Ultramafic nodules from the Kimberley pipes, South Africa. Geochim. Cosmochim. Acta. 42, 1367-1382.

Boyd, F.R. and England, J.L. (1960). Apparatus for phase equilibrium measurements at pressures up to 50 kb and temperatures to 1750°C. J. Geophys. Res. 65, 741-748.

Clark, S.P., Ringwood, A.E. (1964). Density distribution and constitution of the mantle. Rev. Geophys. 2, 35-88.

Danchin, R.V. (1979). Mineral and bulk chemistry of garnet lherzolite and garnet harzburgite xenoliths from the Premier mine, South Africa. In: The Mantle Sample: Inclusions in Kimberlites and Other Volcanics. Boyd, F.R., and Meyer, H.O.A. (eds). American Geophysical Union, Washington, D.C., pp. 104-126.

Dawson, J.B. (1980). Kimberlites and Their Xenoliths. Berlin, Springer-Verlag.

Edgar, A.D., Condliffe, E., Barnett, R.L., and Shirran, R.J. (1980). An experimental study of an olivine ugandite magma and mechanisms for the formation of its K-enriched derivatives. J. Petrol. 21, 475-497.

Edgar, A.D., Green, D.H., and Hibberson, W.O. (1976). Experimental petrology of a highly potassic magma. J. Petrol. 17, 339-356.

Erlank, A.J. and Rickard, R.S. (1977). Potassic richterite bearing peridotites from kimberlite and the evidence they provide for upper mantle metasomatism. Extended Abs., Second Intl. Kimberlite Conf., Santa Fe, New Mexico, U.S.A. unpaged.

Gilbert, M.C., Helz, R.T., Popp, R.K., and Spear, F.S. (1982). Experimental studies of amphibole stability. In: Amphiboles: Petrology and Experimental Phase Relations. Veblen, D.R. and Ribbe, P.H. (eds). Mineralogical Soc. America, Washington, D.C., 229-353.

Green, D.H. (1973). Experimental melting studies on a model upper mantle composition at high pressure under water-saturated and water-undersaturated conditions. Earth Planet. Sci. Lett. 19, 37-53.

Harte, B., Cox, K.G. and Gurney, J.J. (1975). Petrography and geological history of upper mantle xenoliths from the Matsoku kimberlite pipe. Phys. Chem. Earth 9, 477-506.

Hawkesworth, C.J. and Vollmer, R. (1979). Crustal contamination versus enriched mantle: $^{143}Nd/^{144}Nd$ and $^{87}Sr/^{86}Sr$ evidence from the Italian volcanics. Contrib. Mineral. Petrol. 69, 151-165.

Johannes, W., Bell, P.M., Mao, H.K., Boettcher, A.L., Chipman, D.W., Hays, J.F., Newton, R.C. and Seifert, F. (1971). An interlaboratory comparison of piston-cylinder calibration using the albite breakdown reaction. Contrib. Mineral. Petrol. 32, 24-38.

Lloyd, F.E. (1981). Upper mantle metasomatism beneath a continental rift: clinopyroxenes in alkalic mafic lava and nodules from South West Uganda. Mineral. Mag. 44, 315-323.

Lloyd, F.E. and Bailey, D.K. (1975). Light element metasomatism of the continental mantle: the evidence and the consequences. Phys. Chem. Earth 9, 389-416.

McCulloch, M.T., Jaques, A.L., Nelson, D.R., and Lewis, J.D. (1983). Nd and Sr isotopes in kimberlites and lamproites from the West Kimberley, Western Australia: an enrihed mantle origin. Nature 302, 400-403.

Modreski, P.J. and Boettcher, A.L. (1972). The stability of phlogopite + enstatite at high pressure: a model for micas in the interior of the Earth. Am. J. Sci. 272, 852-869.

Richardson, S.W., Bell, P., and Gilbert, M.C. (1968). Kyanite-sillimanite equilibrium between 700°C and 1500°C. Am. J. Sci. 266, 513-541.

Ringwood, A.E. (1975). Composition and Petrology of the Earth's Mantle. McGraw-Hill, New York.

Ryabchikov, I.D. and Boettcher, A.L. (1980). Experimental evidence at high pressure for potassic metasomatism in the mantle of the Earth. Am. Mineral. 65, 915-919.

Ryabchikov, I.D. and Green, D.H. (1978). The role of carbon dioxide in the petrogenesis of highly potassic magmas. In: Problems of Earth's Crust and Upper Mantle, Trudy Instituta Georogii Geofiziki, So An SSR, 403, Novosibirsk, Nauka Publishers.

Ryabchikov, I.D., Green, D.H., Wall, V.J., and Brey, G. (1981). The oxidation state of carbon in the environment of the low velocity zone. Geokhimia, 221-232.

Ryabchikov, I.D., Schreyer, W., and Abraham, K. (1982). Compositions of aqueous fluids in equilibrium with pyroxenes and olivines at mantle pressures and temperatures. Contrib. Mineral. Petrol. 79, 80-84.

Scandl, E.S. and Clarke, D.B. (1982). Metasomatism in the mantle beneath pipe 200, Northern Lesotho. Terra Cognita 2, 265-266.

Stern, C.R. and Wyllie, P.J. (1975). Effect of iron absorption by noble-metal capsules on phase boundaries in rock-melting experiments at 30 kilobars. Am. Mineral. 60, 681-689.

Van Kooten, G.K. (1980). Mineralogy, petrology and geochemistry of an ultrapotassic basaltic suite, central Sierra Nevada, California, U.S.A. J. Petrol. 21, 651-684.

Van Kooten, G.K. (1981). Pb and Sr systematics of ultrapotassic and basaltic rocks from the central Sierra Nevada, California. Contrib. Mineral. 76, 378-385.

Vollmer, R. and Hawkesworth, C.J. (1980). Lead isotopic composition of the potassium rocks from Roccamontina (South Italy). Earth Planet. Sci. Lett. 47, 91-101.

Wendlandt, R.F. (1977). The system K_2O-MgO-Al_2O_3-SiO_2-H_2O-CO_2: Stability of phlogopite as a function of vapour compositions at high pressures and temperatures. Carnegie Instn. Wash. Yb. 76, 435-441.

Wendlandt, R.F. and Eggler, D.H. (1980a). The origins of potassic magmas: I. Melting relations in the systems $KAlSiO_4$-Mg_2SiO_4-SiO_2 and $KAlSiO_4$-MgO-SiO_2-CO_2 to 30 kilobars. Am. J. Sci. 280, 385-420.

Wendlandt, R.F. and Eggler, D.H. (1980b). The origins of potassic magmas: 2. Stability of phlogopite in natural spinel lherzolite and in the system $KAlSiO_4$-MgO-SiO_2-H_2O-CO_2 at high pressures and high temperatures. Am. J. Sci. 280, 421-458.

Wyllie, P.J. (1979). Magmas and volatile Components. Am. Mineral. 64, 469-500.

Yoder, H.S. and Kushiro, I. (1969). Melting of a hydrous phase: phlogopite. Am. J. Sci., Schairer Volume 267-A, 558-582.

POLYGENETIC TRONDHJEMITE

WILLIAM B. SIZE
Department of Geology
Emory University
Atlanta, Georgia 30322
U.S.A.

ABSTRACT

Rocks of trondhjemitic composition can originate from polygenetic paths related to continental margin tectonism. Plagiogranite precursors show significant differences from trondhjemite even though they outwardly resemble them and are frequently found associated with trondhjemite in allochthonous units. Trondhjemitic rocks have characteristics indicative of more primitive protoliths such as oceanic lithosphere or early crustal tonalite. Trondhjemite petrogenesis is frequently episodic during orogenesis and the rocks can be used to date the timing and emplacement of allochthonous units that they are restricted within. Trondhjemite in the Scandinavian Caledonides and the southern Appalachian orogens have many compositional and structural-setting similarities, however their petrogenesis is interpreted as resulting from anatexis of a tholeiitic-basalt parent in the former case and from a crustal tonalite in the latter case. Mechanical mixing of xenocrystic or, anatectically derived, autocrystic restite into the trondhjemitic melt can have major effects on petrogenetic interpretations.

INTRODUCTION

The general characteristics of trondhjemitic rocks are given by Barker (1979) and a discussion of the Trondheim, Norway type-area is given by Size (1979). Trondhjemitic rocks can originate from polygenetic paths with contributions coming from both chemical

Proceedings of the 27th International
Geological Congress, Volume 9, pp. 543–559
PETROLOGY (IGNEOUS AND METAMORPHIC ROCKS)

and mechanical processes. They do, however, have a common spatial relationship with basalts (low-potassium tholeiites, greenstones or amphibolites), copper-sulfide mineralization and major structural features such as nappes and overthrusts, indicative of subduction/obduction zones. Trondhjemite magmatism is also frequently related to continental margin tectonics but petrogenesis of such rocks is quite varied, including: (1) incipient anatexis (metatexis) of basaltic rocks in orogenic zones, (2) migmitization of high grade (K-depleted) gneissic and granulitic rocks, (3) partial melting of altered ophiolites, possibly depleted in potassium due to sea-water exchange, (4) partial melting of ensimatic sediments (greywacke) from island-arc and back-arc sequences, (5) remobilization and anatexis of tonalitic rocks due to renewed tectono-magmatic activity, (6) late-stage immisicible liquids from ocean ridge basalts, (7) fractional crystallization in a gabbro-tonalite-trondhjemite suite.

Trondhjemite petrogenesis is also related to several stages of tectono-magmatic activity starting with plagiogranite in ophiolite sequences. Such rocks are characterized by very low K20 content($<$0.5 wt.%), relatively low in A1203 ($<$15 wt.%) and LREE depletion with a negative Eu-anomaly. Later stage, anatectic trondhjemite, generally have a higher K20 content (ca. 1.5%), slightly higher A1203 values (14-16%) and a slightly LREE-enriched fractionated pattern with little or no Eu-anomaly. Late stage trondhjemite gneisses have a higher K20 content ($>$ 2%), but having greater variation, higher A1203 content ($>$15%) and a highly fractionated LREE enrichment pattern with a positive Eu-anomaly.

TRONDHJEMITE-PLAGIOGRANITE ASSOCIATION

Occurring within, and genetically related to, ophiolite slices and greenstones are small masses of both pre-tectonic oceanic plagiogranite and syn- to late-tectonic trondhjemite (Saunders, et al., 1979; Brown, et al., 1979). Plagiogranites and trondhjemites can be used to correlate magmatic sequences to timing of tectonic sequences such as in the Appalachian-Caledonide orogen. However, since these rock types commonly occur together and outwardly resemble one another as leucocratic, mainly quartz-plagioclase plutonic rocks, differentiating them requires closer inspection. The differences between the characteristics of oceanic plagiogranite and anatectic trondhjemite are given in Table 1. The K20-Si02 diagram in Fig. 1 also shows the compositional differences between these two rock types.

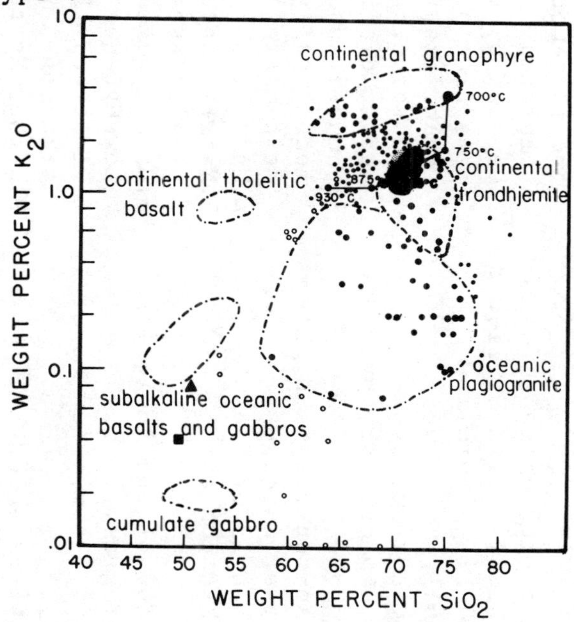

Figure 1. Plot of SiO_2 vs. K_2O wt.% for trondhjemitic and related rocks (data from Size, 1979 and Barker,1979). Compositional fields after Coleman (1977). Solid dots = trondhjemite; open circles = plagiogranite; solid triangle = Picture Gorge tholeiite starting material in partial melting experiments (Helz, 1976) from 700° C to 930° C as shown by line connecting dots.

CHARACTERISTICS OF OCEANIC PLAGIOGRANITE	CHARACTERISTICS OF ANATECTIC TRONDHJEMITE
1. Restricted to ophiolites; commonly situated in the sandwich zone above the cumulate gabbro as a layer or as small intrusions into the overlying sheeted dikes.	1. Commonly associated, but not restricted to, the upper zone of ophiolites. Also occur as small intrusions in ocean floor basalts (greenstones, amphibolites, etc) and as members of gabbro-trondhjemite complexes.
2. Develop by primary processes of fractional crystallization or liquid immiscibility during ophiolite crystallization.	2. Develop by a secondary process of partial remelting of low-K tholeiitic basalt or remobilization of other low-K protoliths in orogenic zones (i.e. greywacke or quartz diorite).
3. Compositionally heterogeneous.	3. Compositionally homogeneous, unless mechanically disturbed (i.e. crystal-melt mixing).
4. Mafic minerals commonly are pyroxene and amphibole.	4. Mafic mineral commonly is biotite.
5. Wide range of An-content in highly-zoned plagioclase. An_{10-60}	5. Narrow range of An-content in zoned plagioclase. An_{10-30}
6. Usually less than 15 wt.% Al_2O_3.	6. Usually greater than 15 wt.% Al_2O_3.
7. Extremely low K_2O content (0.01-1.0 wt.%).	7. Low K_2O content (1.0-3.0 wt.%).
8. Plot near the end of the tholeiitic trend on AFM diagram.	8. Plot near the end of the calc-alkaline trend on AFM diagram.
9. Low Rb content (5 ppm); low Rb/Sr ratio (0.015).	9. Higher Rb content (20-50 ppm); higher Rb/Sr ratio (0.015).
10. Slightly fractionated HREE enrichment pattern; negative Eu-anomaly.	10. Highly fractionated LREE enrichment pattern; possible small negative or positive Eu-anomaly.

TABLE 1. Characteristics between oceanic plagiogranite and anatectic trondhjemite.

Oceanic plagiogranites appear more related to primary crystallization in ophiolites and are commonly concentrated in a zone above the cumulate gabbro and below the, plated-downward, roof and sheeted-dikes. For example the Sarmiento ophiolite in southern Chile (Saunders, et al., 1979) includes both trondhjemite and plagiogranite which, together with granophyre, make up approximately 10% of the complex. Here plagiogranite forms an almost continuous layer just below the sheeted dike complex. The plagiogranite contains only about 0.01 wt.% K2O which is extremely low, even for plagiogranites, and may indicate that hydrothermal activity at the spreading center leached K (plus Rb and Ba) shortly after or during their formation. The plagiogranite at Sarmiento is also associated with ferrogabbros and ferrobasalts. This association supports the immiscibility experiments on primitive oceanic tholeiite from the Galapagos spreading center (Dixon and Rutherford, 1979). In the Sarmiento complex later stage trondhjemite dikes cut plagiogranite and invade the upper zone of the ophiolite sequence.

The timing and petrogenesis of trondhjemite, tonalite, plagiogranite and their volcanic correlatives, quartz keratophyres and dacites, is related to the formation of oceanic crust at spreading centers, plate convergence and ocean floor subduction and island-arc magmatism. However, in the post-Archean, they do not appear related to any ensuing ensialic, within-plate, magmatism. The radiometric age of ophiolite-bound trondhjemite serves as a lower age limit for obduction of allochthonous units onto the continental margin. Relatively younger and less deformed trondhjemite contained within the allochthon can date the upper limit of major deformation.

Discovery of ophiolite fragments intercalated with previously identified allochthonous ocean-floor tholeiites (greenstones) in the Norwegian Caledonides (Furnes, et al., 1980) has strengthened the earlier view of Gale and Roberts (1974) that these rocks represent oceanic lithosphere obducted eastward onto the Baltic continental margin during early Paleozoic time. Examples include the Karmøy ophiolite (Sturt, et al., 1979), Trondheim-Løkken ophiolite (Furnes, et al., 1980) and Leka ophiolite (Prestvik, 1972). Trondhjemite occurs in these ophiolite slices in the cumulate gabbro and sheeted dike sequences. The Karmøy ophiolite in Norway displays plagiogranite positioned above the plutonic zone, into the sheeted dike complex (Furnes, et al., 1980). It is of interest to note that plagiogranite comprises an estimated 2-3% of the Karmøy ophiolite which is similar to the amount of plagiogranite produced in the silicate liquid immiscibility experiments on oceanic lithosphere (Dixon and Rutherford, 1979). Plagiogranite and trondhjemite are apparently totally lacking in the later stage, ensialic, within plate magmatism in the Scandinavian Caledonides. A crustal origin for rocks produced in this setting is evidenced by a higher $^{87}Sr/^{86}Sr$ initial ratio, such as the granite from the West Karmøy igneous complex which has an initial ratio of 0.7210 and has xenoliths of crustal rock and a much higher K_2O content (Furnes, et al., 1980).

ANATECTIC TRONDHJEMITE

Rocks of trondhjemitic composition were important components during early stages of crustal margin accretion and irreversible differentiation. However, the environment which generated abundant Archean grey-gneiss may not have analogs in younger orogens such as the Appalachian-Caledonides due to increased thickness of continental crust (Barker, et al., 1976). In younger orogens trondhjemite occurs as small intrusive bodies contained within allochthons and also as localized differentiates in ophiolite sequences. Here low initial isotopic ratios indicate that their immediate precursors were predominantly juvenile additions to the crust and intermediate stages, if any, had short residence times.

Trondhjemites have characteristics that

collectively indicate they had a primitive source, such as oceanic lithosphere, rather than continental crustal rocks. Of importance is the low initial $^{87}Sr/^{86}Sr$ ratios for trondhjemites in the range of 0.7015-0.7039 for various early Paleozoic trondhjemites (Peterman, 1979) and, more specifically, 0.7049-0.7053 for trondhjemites from the central Norwegian Caledonides (Klingspor and Gee, 1984).

Indicative of a anatectic source for trondhjemites are the results of melting experiments on low-K_2O basalts (Helz, 1976). Figure 2 shows that a trondhjemite melt is produced at 825°C (P_{H_2O}=5Kb) from a tholeiitic basalt starting material. The first melt produced at 700°C is richer in K_2O (i.e. granitic) which decreases with higher temperature towards a melt of trondhjemitic composition. For comparison the average composition of the Follstad, Norway, type-area trondhjemite and the associated Storen gabbro are also shown. In the K_2O-SiO_2 diagram (Fig. 1) the composition melting curve for Picture Gorge tholeiite (solid triangle) used by Helz is plotted between 700-930°C.(solid line). The distribution for naturally-occurring trondhjemitic rocks (solid dots) is concentrated about the melting curve and centered in the 825°C temperature range. This is also the zone representing the average composition of the type-area trondhjemite (shown in stiple pattern). This close relationship between the compositions of natural trondhjemites and artificially produced trondhjemite melts supports the contention that low-K tholeiites are likely protoliths for trondhjemite petrogenesis by anatexis.

The liquid immisicibility experiments of Dixon and Rutherford (1979) are also of interest to trondhjemite petrogenesis. Using a mid-ocean ridge basalt (shown as a solid square in Fig. 1) the magma remaining after 95% crystallization at 1010°C split into a ferro-basalt and a trondhjemite melt. The composition of the trondhjemite melt also falls along the melting curve of Helz between 825-875° C (Fig.1).

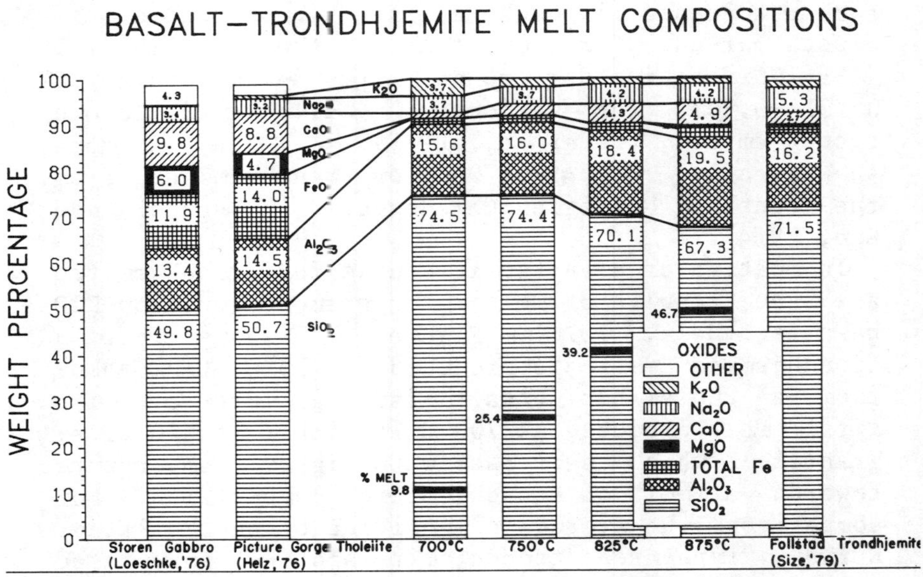

Figure 2. Graph showing partial melt compositions for Picture Gorge tholeiite between 700°C to 875°C (after Helz, 1976). Follstad, Norway type-area trondhjemite and Støren gabbro shown for comparison.

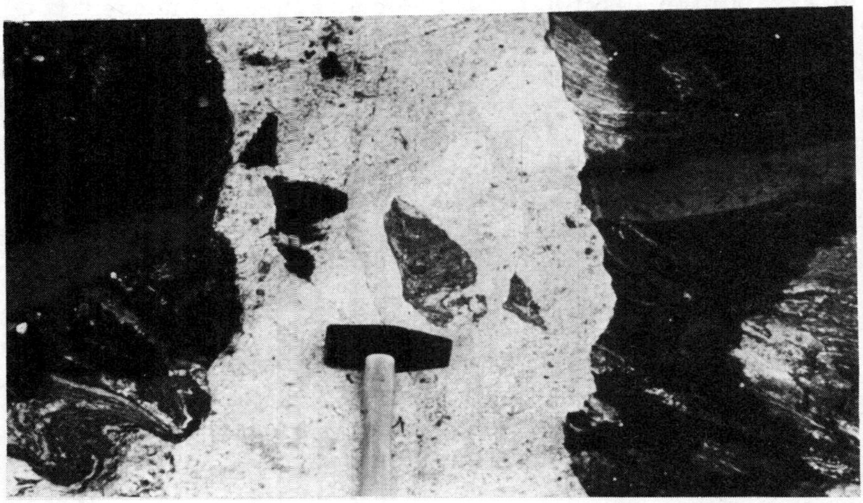

Figure 3. Trondhjemite dike from type region in the Central Norwegian Caledonides showing cross-cutting relationships to allochthonous Gula Group rocks.

For the type area trondhjemite located in the central Norwegian Caledonides (Fig. 3), the model that best fits the data is partial melting of a low-K tholeiitic basalt formed during anatexis in an orogenic zone (Size, 1979). This is based on: (1) the common association of trondhjemite with basalt, gabbro, and amphibolite of ocean floor affinities, (2) the restriction of trondhjemite to allochthonous units, (3) low initial $^{87}Sr/^{86}Sr$ ratios(i.e. close to that of a basaltic parent), (4) oxygen isotope values similar to basalt, (5) results of partial melting experiments on low-K basalts in temperature and pressure and temperature ranges associated with subduction zones and regional metamorphism, and (6) trace element and REE-analysis partition coefficients that support a partial melting model for trondhjemite.

In the central Norwegian Caledonides trondhjemite petrogenesis was episodic with three or four periods of activity identified (Size, 1979). This may directly relate to the model proposed for this region as having developed in phases of spreading followed by minor crustal thickening events from Ordovician to at least early Silurian time (Grenne and Roberts, 1980). Recent radiometric age dates for trondhjemites from the central Norwegian Caledonides (Klingspor and Gee, 1984) ranges between 509 +/- 35 Ma to 451 +/- 13 Ma. These early to middle Ordovician ages place the period of trondhjemite activity during the pre-Middle Arenig obduction of the ophiolite slices in the Storen greenstone onto the continental margin (Furnes, et al., 1980) but prior to the post-Llandovery age for final nappe translation onto western Norway(Gee, 1978).

ANATECTIC TRONDHJEMITE IN THE SOUTHERN APPALACHIAN OROGEN

The southern Appalachian orogen consists of Paleozoic sedimentary rocks contained within imbricate thrust-bounded blocks and nappes. General northwestern movement along the southeastward dipping thrust planes juxtaposed rocks of varying lithologies, deformation styles and degrees of metamorphism. The major orogenic event in the southern Appalachians is dated at post-early Pensylvanian, but was probably active episodically through much of Palaeozoic time (Thomas, et al., 1980). The thermal peak of regional metamorphism in the southern Appalachians of Alabama occurred in Lower to Middle Devonian time.

In the southern Appalachians trondhjemites and associated rocks are concentrated along a narrow, northeasterly trending tectonic unit named the Tallapoosa Block in Alabama (Fig. 4). This block is a unit of the northern Alabama Piedmont and is bounded by the Brevard Fault zone to the southeast and the Valley and Ridge province to the northwest. The northern Alabama Piedmont also contains the Coosa and Talladega blocks that are also bounded by faults that have imbricated and cut out lithologies (Tull, 1982). These blocks have been defined as suspect terranes by Williams and Hatcher (1982).

Trondhjemite and related rocks in the Tallapoosa Block include over 20 intrusions ranging in area from over 1000 km square to less than 1 km square. Most intrusions intrude the Wedowee Group rocks which are mainly graphitic metapelites (Fig. 4). In the southern portion of this block the Elkahatchee Quartz Diorite outcrops over an area 65km long by 20 km wide. It is one of the oldest and largest plutons in the southern Appalachian Piedmont. U-Pb dates from zircons yielded a date of 516 Ma and whole-rock Rb/Sr age of 490 +/- 26 Ma. The Elkahatchee Quartz Diorite has a low initial $^{87}Sr/^{86}Sr$ ratio of 0.7036 +/- 0.0004 (Russell, 1978). The modal mineral composition of this rock show a high quartz content coupled with an extremely low K-spar content and intetrmediate plagioclase composition, classifying the rock more as a tonalite or biotite-rich trondhjemite.

Further northeastward along structural trend in the Tallapoosa Block, the Elkahatchee Quartz Diorite

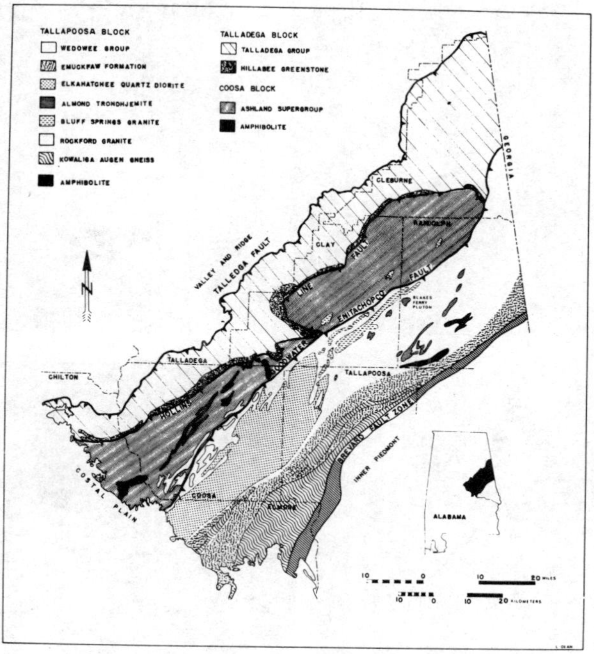

Figure 4. Diagram showing the geologic subdivisions of the southern Appalachian orogen in Alabama. Trondhjemite and related rocks are confined to the Tallapoosa block (allochthon).

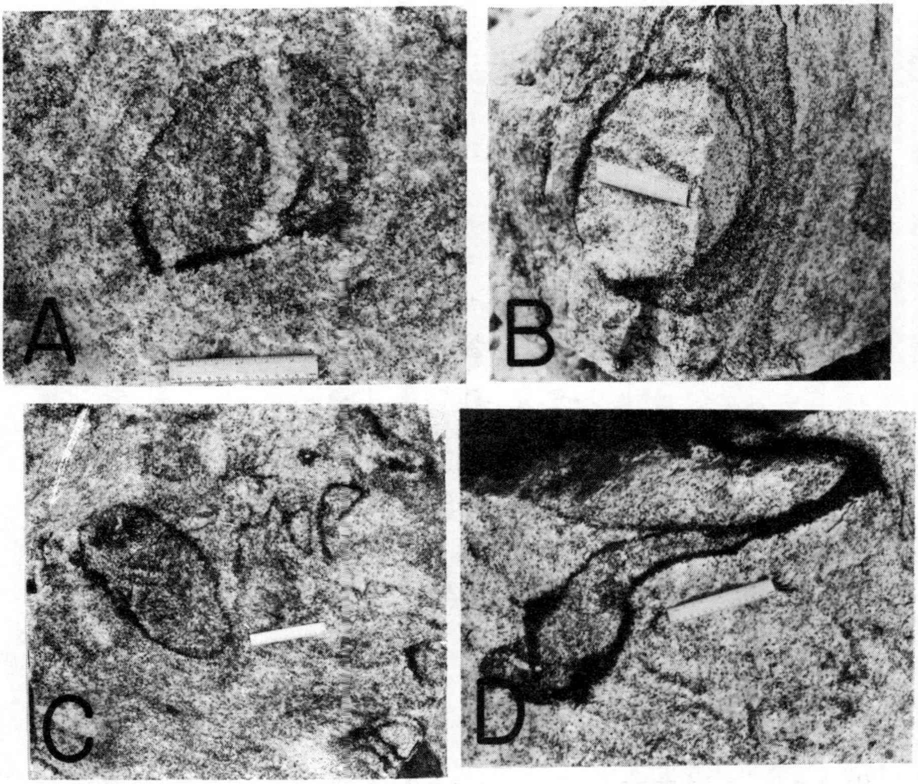

Figure 5. Photographs of the Blakes Ferry trondhjemite from the northern Alabama Piedmont showing progressively deformed tonalite xenoliths. 5A - shows a broken, but relatively undeformed tonalite xenolith with a thin biotite restite. 5B - shows xenoliths having thicker biotite rims in anatectic trondhjemite. 5C - shows dispersed biotite restite around a xenolith (note the high degree of compositional heterogeneity. 5D - shows a deformed xenolith with a thick biotite rim drawn out into schlieren. Compare with Figure 6.

contains an increasing proportion of trondhjemite dikes and isolated pods (Fig. 5). This grades into a series of over 20 intrusions collectively called the Almond trondhjemites (Fig. 4). They show varying degrees of deformation and metamorphism. A member of the Almond trondhjemites, the Blakes Ferry pluton, displays the least amount of deformation and was chosen for detailed investigation (Fig. 4).

Field and petrologic data indicate that the trondhjemite at Blakes Ferry originated by partial anatexis of a tonalitic protolith similar to the Elkahatchee tonalite. Tonalitic xenoliths are very abundant in the Blakes Ferry trondhjemite. They vary from angular shapes with little evidence of anatexis, to xenoliths with biotite-rich rims, and finally to ghost-outlined xenoliths showing biotite schlieran (Fig. 5). Diasaggregated xenoliths in the Blakes Ferry trondhjemite show that only a limited amount of plagioiclase and biotite remelting took place. Rather, these two minerals were mixed in with the anatectic melt as the xenocrysts were liberated and mechanically carried with the magma. During partial melting of a tonalitic protolith a significant proportion of the restite may accompany the anatectic melt as autoliths which may disaggregate or melt adiabatically as the melt rises. This tends to erase the effect of crystal-melt mixing. Therefore, even the most reliable group of chemical elements(i.e. Eu and Sr in plagioclase or Rb and Ba in biotite) used in the study of trondhjemite petrogenesis are of questionable usefulness if autocrysts were mechanically added or subtracted randomly in the melt.

The composition and texture of the xenoliths in the Blakes Ferry trondhjemite is similar to the Elkahatchee tonalite except for a lower anorthite content in plagioclase and less biotite, which is in accord with a partial anatexis process. The AFM diagram (Fig. 6) shows that the Elkahatchee tonalite and the xenoliths in the Blakes Ferry trondhjemite plot together in the middle of the gabbro-trondhjemite trend, while both the Blakes Ferry trondhjemite and leucocratic zones from the Elkahatchee tonalite plot together at the lower end of the trend (the type-area trondhjemite from Follstad, Norway also plots in the lower end of the diagram).

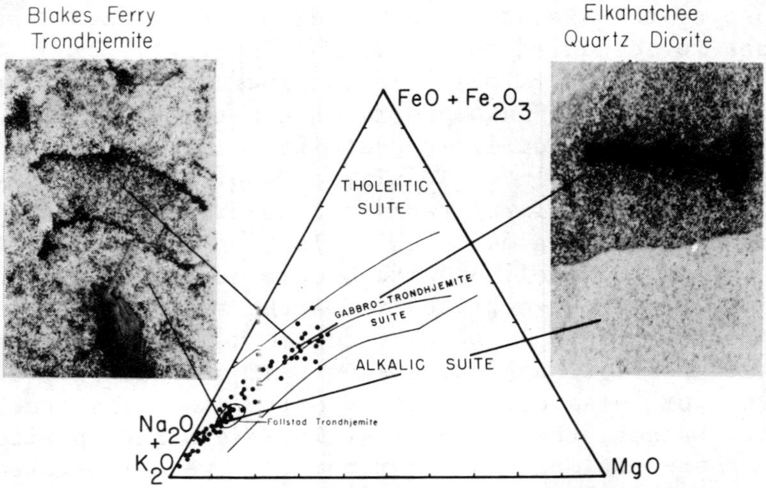

Figure 6. AFM diagram for tonalitic and trondhjemitic rocks from the Northern Alabama Piedmont. Gabbro-trondhjemite trend after Barker (1979). Xenoliths in Blakes Ferry trondhjemite plot together with suspected Elkahatchee tonalite protolith. Anatectic melt compositions in Blakes Ferry trondhjemite plot together with leucocratic anatectic zones in Elkahatchee Quartz Diorite.

SUMMARY REMARKS

The importance of trondhjemite associated with ophiolite suites is that their radiometric age serves as a lower age limit for the emplacement, onto the continental margin, of the allochthonous unit they are restricted within. Also since trondhjemite petrogenesis is episodic during orogenic phases, the younger and less-deformed trondhjemites can date the upper limit of major deformation and regional metamorphism. If present, post-tectonic, ensialic magmatic rocks within the allochthon can be used to estimate the upper age limit for emplacement of the allochthon. Even though trondhjemitic rocks can originate from polygenetic paths with both chemical and mechanical contributions during multiple periods of continental margin orogenesis, their usefulness as timing mechanisms in crustal evolution appears warranted.

REFERENCES

Barker, F., editor, 1979: trondhjemites, Dacites and related rocks. Elsevier Scientific Publishing Company, Amsterdam. 659p.

Barker, F. & Arth J.G., 1976: Generation of trondhjemitic-tonalitic liguids and Archean bimodal trondhjemite-basalt suites. Geology 4,p. 569-600.

Brown, E.H., Bradshaw, J.Y. and Mustoe, G.E., 1979: Plagiogranite and keratophyre in ophiolite on Fidalgo Island, Washington. Bull. Geol.Soc. Amer., v. 90, p. 493-507.

Coleman, R.C., 1977: Ophiolites- Ancient Oceanic Lithosphere ? Springer-Verlag, Berlin, 229 p.

Dixon, S. and Rutherford, M.J., 1979: Plagiogranite as late-stage immiscible liquids in ophiolite and mid-ocean ridge suites: an experimental study. Earth and Planetary Science Letters 45, p. 45-60.

Furnes, H., Roberts, D., Sturt B.A., Thon, A. and Gale, G.H., 1980: Ophiolite fragments in the Scandinavian Caledonides, p. 582-599. In Ophiolites - Proceedings International Ophiolite Symposium. Cyprus Geological Survey Department.

Gale, G.H. & Roberts, D.,: 1974: Trace element geochemistry of Norwegian Lower Paleozoic basic volcanics and its tectonic implications. Earth Plan. Sci. Lett. 22; p. 380-390.

Gee, D.G., 1978: Nappe displacement in the Scandinavian Caledonides. Tectonophysics 47, p. 393-419.

Grenne, T. & Roberts, D., 1980: Geochemistry and volcanic setting of the Ordovician Forbordfjell and Jonsvatn Greenstones, Trondheim region, central

Norwegian Caledonides. Contributions to Mineralogy and Petrology, v. 74, p. 375-386.
Helz, R.T., 1976: Phase relations of basalt in their melting ranges at PH2O=5Kb. Part II, Melt compositions. Jour. Petrol. 17, p. 139-193.
Klingspor, I. & Gee, D.G., 1984: The age of some Trondelag trondhjemites. In The Caledonide Orogen - Scandinavia and Related Areas. John Wiley and Sons, New York.
Peterman, Z.E., 1979: Strontium isotope geochemistry of Late Archean to Late Cretaceous tonalites and tronthjemites. p. 133-148. In Barker, F., editor, Trondhjemites, Dacites and Related Rocks. Elsevier Scientific Publishing Company, Amsterdam. 659 p.
Russell, G.S., 1978: U-Pb, Rb-Sr and K-Ar isotopic studies bearing on the tectonic development of the southernmost Apalachian Orogen, Alabama. Unpubl. PhD thesis, Florida State University, p. 197.
Saunders, A.D., Tarney, S., Stern, D.A. & Dalziel, I.W.D., 1979: Geochemistry of Mesozoic marginal basin floor igneous rocks from southern Chile. Geol. Soc. of Amer. Bull., v. 90, p. 237-258.
Size, Wm. B., 1979: Petrology, Geochemistry and Genesis of the type area trondhjemite in the Trondheim region, Central Norwegian Caledonides. Norges Geol. Under. 351, p. 51-76.
Sturt, B.A., Thon, A. & Furnes, H., 1979: The Karmoy ophiolite, southwest Norway. Geology, v. 7, p. 316-320.
Thomas, Wm. A., Tull. J.F., Bearce, D.N., Russell, G. & Odom, A.L., 1980: Geologic synthesis of the southermost Appalachians, Alabama and Georgia. p. 91-98. In Proceedings "The Caledonides in the U.S.A.". Wones, D.R., editor. Memoir no. 2. Department of Geological Sciences, Virginia Polytechnic Institute and State University, 329 p.
Tull, J.F., 1982: Stratigraphic framework of the Talledega slate belt, Alabama Appalachians. p. 3-18. In Tectonic studies in the Talledega and Carolina Slate Belts, Southern Appalachian Orogen. Bearce, D.N., Black, W.W., Kish, S.A. and Tull, J.F., editors. Geol. Soc. of Amer., Special Paper 191.,

164 p.
Williams, H. & Hatcher, R.D., Jr., 1982: Suspect Terranes and accretionary history of the Appalachian orogen. Geology, v. 10, p. 530-536.

HYBRIDIZATION OF MAGMAS ABOVE SUBDUCTED OCEANIC CRUST

TOSHIMORE SEKINE and PETER J. WYLLIE
Department of Earth Sciences, Monash University, Clayton, Victoria 3168, Australia; Division of Geological and Planetary Sciences, California Institute of Technology, Pasadena, California 91125 USA

ABSTRACT

Hydrous siliceous melts rising from subducted oceanic crust may experience hybridization with overlying peridotite. Products of hybridization have been explored in the synthetic system $KAlSiO_4$-MgO-SiO_2-H_2O. The phase diagrams were used to interpret phase relationships determined at 30 kbar for mixtures of granite, peridotite, and H_2O. Results indicate that solution of peridotite in hydrous, H_2O-undersaturated granitic liquid at 900°C, 30 kbar, causes only small changes in liquid composition, and precipitation of phlogopite-garnet-websterite. Additional experiments, set up with H_2O-undersaturated granite liquid in contact with solid peridotite at 30 kbar, 850-1050°C, produced hybrid reaction zones containing orthopyroxene and clinopyroxene in all runs, and one or more of garnet, phlogopite, and quartz in some runs. The hybridization process in subduction zones could produce discrete rock bodies dominated by pyroxenite (without olivine); there is a prospect that potassium could become concentrated into phlogopite-rich rocks. The heterogeneous rock layer produced

Proceedings of the 27th International
Geological Congress, Volume 9, pp. 561–579
PETROLOGY (IGNEOUS AND METAMORPHIC ROCKS)

above subducted crust provides aqueous fluids expelled during hybridization, which rise to generate in overlying mantle the H_2O-undersaturated basic magnesian magma which is the probable parent of the calc-alkalic rock series erupted at the volcanic front.

INTRODUCTION

It was proposed a decade ago by Nicholls and Ringwood (1973) that a critical stage in the petrogenesis of andesites and other calc-alkaline magmas involves hybridization. They suggested that hydrous siliceous melts resembling rhyodacites were generated in subducted oceanic crust, and then reacted with the overlying mantle peridotite, producing hybrid zones of olivine pyroxenite, at depths of 100-150 km. Diapirs of olivine pyroxenite would rise episodically from this zone, and subsequent partial melting of these wet, hybrid rocks would produce a wide range of calc-alkaline magmas, which may differentiate before eruption.

Much experimental effort has been directed towards determination of the products of partial fusion of the deep-seated source rocks of magmas. The melting relationships of peridotite, eclogite, amphibolite, and gabbro, as well as relevant synthetic model systems such as $CaO-MgO-Al_2O_3-SiO_2$, have been determined under various conditions. Similarly, the liquidus relationships of particular komatiites, basanites, basalts, and andesites have been studies with a view to relating them to specific source rocks, and identifying primary magmas. These forward and inverse approaches to petrogenesis were reviewed in detail by Wyllie et al. (1981). There has been very little experimental effort directed towards determination of the products of interactions among magmas, rocks, and aqueous fluids at mantle depths.

One way to investigate the products of hybridization is to model the phase relationships in synthetic silicate systems. Sekine and Wyllie (1982a) explored the data available in a system which includes representatives of (1) hydrous siliceous magma from subducted oceanic crust - the eutectic liquid in the system $KAlSi_3O_8$-SiO_2-H_2O, and (2) the overlying mantle - the assemblage forsterite + enstatite in the system Mg_2SiO_4-SiO_2-H_2O. They extended their exploration in a series of systems with one additional component, $NaAlSiO_4$, $CaMgSi_2O_6$, $CaAl_2Si_2O_8$, and Al_2O_3 (Sekine and Wyllie, 1982b).

Another approach to the problem is to take whole-rock samples as starting materials. This approach involves too many components for graphical representation in standard phase diagrams. Sekine and Wyllie (1982c) investigated the phase fields intersected by mixtures of a natural granite and a natural peridotite, in the presence of H_2O, at 30 kbar pressure.

The equilibrium phase relationships in model and mixed-rock systems provide a basis for understanding hybridization processes in the mantle, but in the dynamic, subduction-zone environment hybrid rock masses may be produced by non-equilibrium, crystal fractionation processes. Sekine and Wyllie (1983) described experiments in which granite liquid is in contact with massive peridotite, in an attempt to simulate hydrous siliceous magma rising through mantle peridotite, and to determine the crystalline products produced by reaction at the boundaries between them.

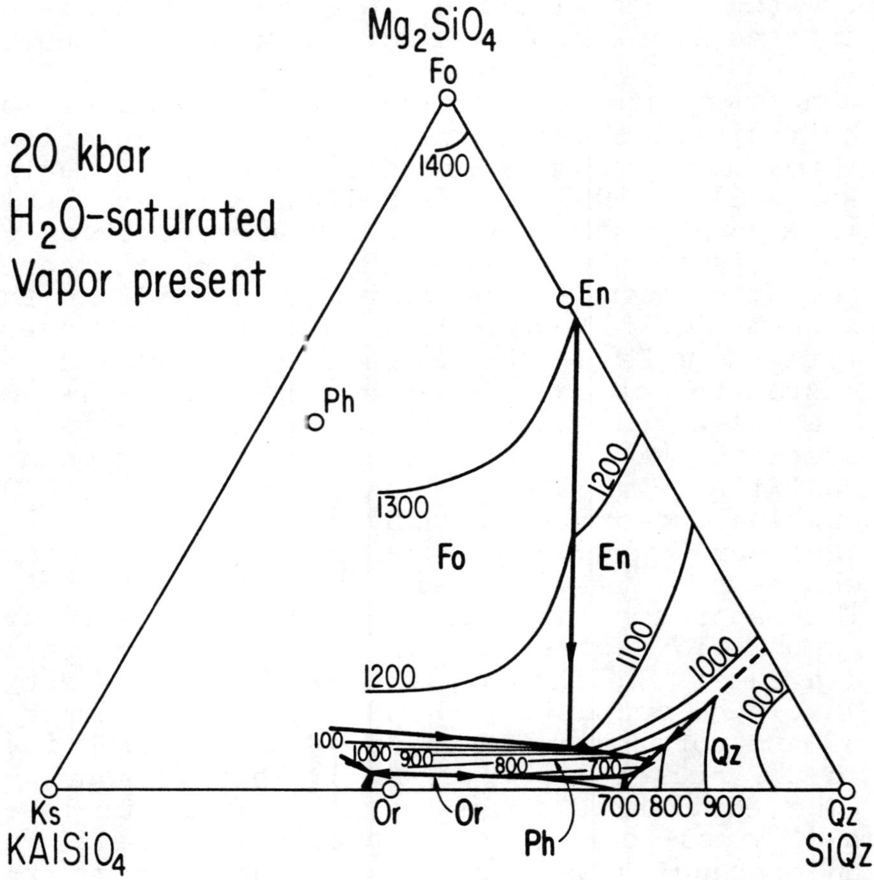

FIGURE 1. H_2O-saturated liquidus for part of the system $KAlSiO_4$-Mg_2SiO_4-SiO_2-H_2O at 20 kbar, based on available experimental data, following Yoder (1976, p. 39).

MODEL SYSTEM $KAlSiO_4$-Mg_2SiO_4-SiO_2-H_2O

The H_2O-saturated liquidus surface for the system Mg_2SiO_4-$KAlSiO_4$-SiO_2 at 20 kb is illustrated in Fig. 1. Liquidus fields for forsterite, enstatite, and quartz extend from

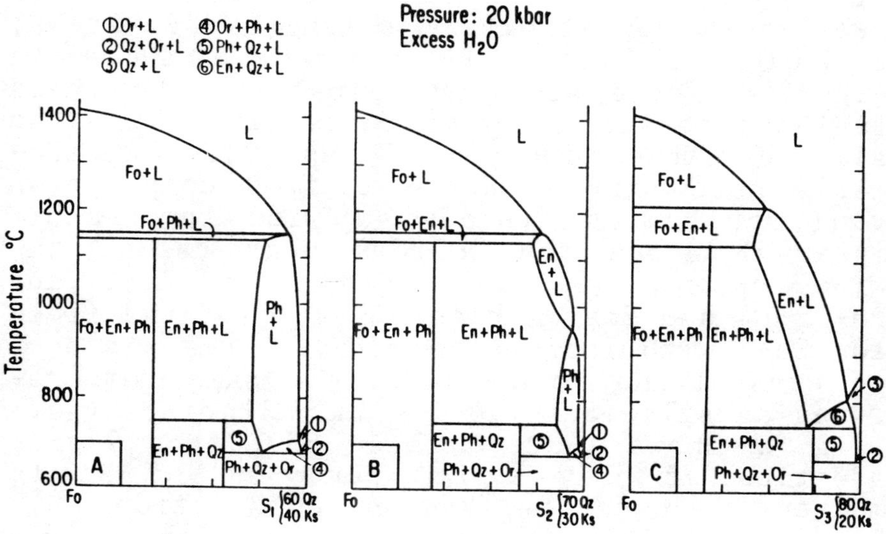

FIGURE 2. Isobaric T-X sections for mixtures of Fo with S_1, S_2 and S_3 in the system $KAlSiO_4$-Mg_2SiO_4-SiO_2-H_2O at 20 kbar constructed from Fig. 1. Points S_1, S_2, and S_3 along the join Or-Qz represent compositions of silicous melts (Sekine and Wyllie, 1982a).

the side Mg_2SiO_4-SiO_2 until they meet liquidus fields for phlogopite and sanidine. The field boundary between sanidine and quartz extending from the low-temperature eutectic on the side $KAlSiO_4$-SiO_2 terminates a very short distance into the system at the quaternary eutectic reaction at 680°C. The low-temperature liquids occupy a limited area close to the side $KAlSiO_4$-SiO_2. The liquidus surface for phlogopite rises steeply from the eutectic and the boundaries limiting the fields for sanidine and quartz. The other two isobaric invariant liquid points are peritectics, representing incongruent melting reactions at 750°C, and 1130°C.

The three reactions are represented by three subsolidus compatibility triangles, Ph-Or-Qz, En-Ph-Qz, and Fo-En-Ph which are shown in Fig. 2. These sections illustrate the phase fields intersected by mixtures of Fo with the silicic compositions S_1, S_2 and S_3. Particularly noteworthy are the limited composition range of liquids through a wide temperature interval above the solidus near Ks-Qz, the wide extent of the fields for Ph+L and Ph+Qz+L, and the increase in size of the field for En+L with increasing Qz/Or. The existence of crystalline Fo+En+Ph at the same temperatures as siliceous liquids is well displayed.

Fig. 2 shows that, between 680°C and 750°C, addition of only a few per cent of Fo to molten or partially molten compositions S_1, S_2, and S_3 is sufficient to cause precipitation of phlogopite. There is a prominent phase field for Ph+L, and the field for Ph+Qz+L increases with increasing Qz/Or. Addition of Fo to the assemblage Ph+Qz+L continues reaction until all of the liquid is used up, producing Ph+Qz.

At temperatures between 750°C and 1,130°C, the main sequences of phase assemblages produced by adding Fo to liquids with compositions S_1, S_2, and S_3 are L, En+L or Ph+L, or En+Ph+L. When the liquid is used up, leaving the assemblage En+Ph, further addition of Fo produces the assemblage Fo+En+Ph without further reaction.

Phase relationships in the model system $KAlSiO_4$-Mg_2SiO_4-H_2O described above indicate that hybridization between hydrous siliceous melts and a simplified peridotite causes solution of the peridotite components with little change in liquid composition. Phlogopite is precipitated through a wide range of pressures, temperatures, and bulk compositions. Hybridization is not a simple mixing process.

The effects of CaO, Na_2O, and Al_2O_3 were examined by reviewing available data on phase relationships with additional components $CaMgSi_2O_6$, $NaAlSiO_4$, Al_2O_3, and $CaAl_2Si_2O_8$ (Sekine and Wyllie, 1982b). Addition of Na_2O and CaO changes the hybridization reactions by reducing the amount of phlogopite and by adding jadeitic clinopyroxene to the precipitation products. Addition of CaO and Al_2O_3 includes garnet among the precipitation products. The synthetic model systems appear to represent a good basis for extrapolation to the complex natural rock system.

EQUILIBRIUM PHASE RELATIONS IN GRANITE-PERIDOTITE-H_2O

The starting materials were natural rocks with known geology, mineralogy, chemistry, and phase relationships. The crushed, mixed samples were sealed with water into gold capsules, and reacted at 30 kbar in piston-cylinder apparatus. Fig. 3 shows the estimated positions of field boundaries for the H_2O-saturated liquidus, and for the coexisting vaporus boundary.

Fig. 4 illustrates the distribution of the major phase volumes containing a small amount of vapor. Although few of the phase boundaries are closely defined by vapor-present runs, the distribution of phase volumes is unambiguous. The diagram is divided into three parts by the two vertical lines, corresponding to the boundaries separating olivine-bearing assemblages from pyroxenite, and subsolidus pyroxenite from partially melted pyroxenite. These three parts are occupied by subsolidus garnet-peridotite, garnet-pyroxenite, and quartz-garnet-pyroxenite. For each rock in the presence of H_2O there is a different solidus, with varying

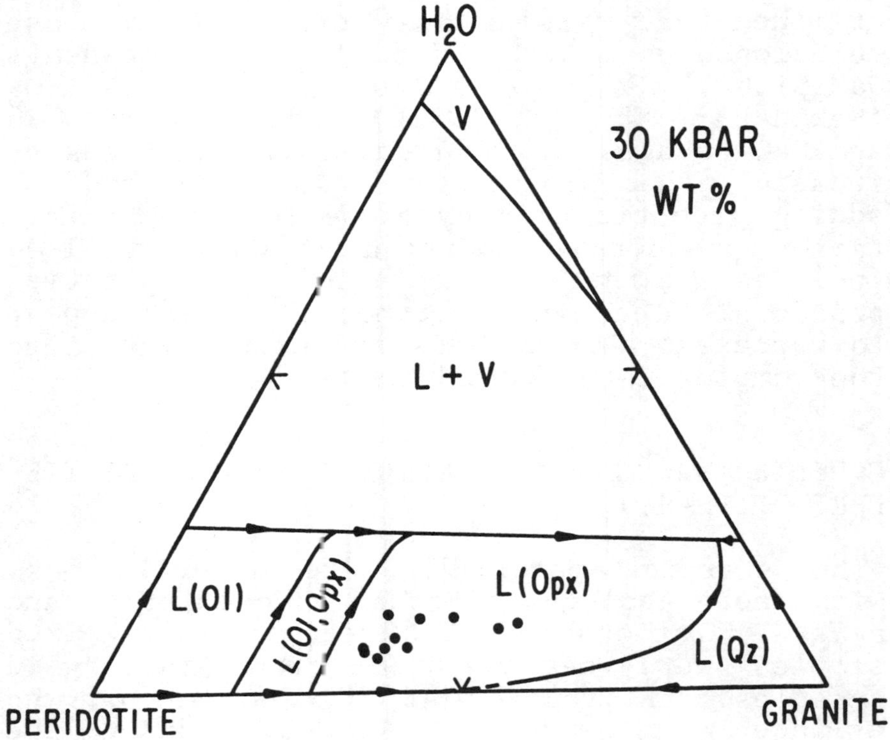

FIGURE 3. Schematic liquidus phase relationships in the system peridotite-granite-H_2O at 30 kbar. Liquidus boundaries estimated from experiments cited by Sekine and Wyllie (1982c).

temperature caused by solid solutions in the mafic minerals. There are complex adjustments near granite, but these need not concern us.

The runs demonstrate that between the field for olivine extending from peridotite-H_2O, and the field for quartz extending from granite-H_2O, the liquidus is dominated by orthopyroxene. The orthopyroxene liquidus rises steeply from the low-temperature liquidus piercing point near the granite-H_2O side.

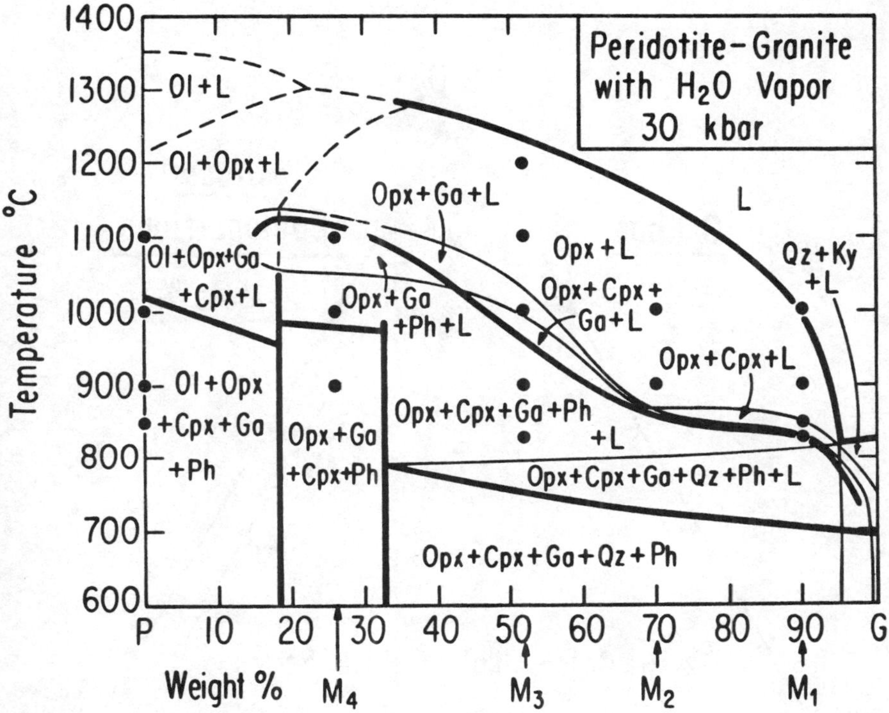

FIGURE 4. Phase fields with a small amount of vapor in system peridotite-granite-H_2O. Complex subliquidus and subsolidus phase assemblages near granite are omitted. Note the subsolidus field for phlogopite-garnet-websterite separating fields with olivine and quartz (Sekine and Wyllie, 1982c).

Glass analyses are plotted in Fig. 5, and compared with the four starting mixtures on the mixing line P-G, and with the average chemical variation of the calc-alkaline igneous series, B-G (Stern and Wyllie 1978; Sekine et al. 1981). All glasses measured in phase assemblages from mixtures M_3, M_2, and M_1 have compositions clustered towards the original

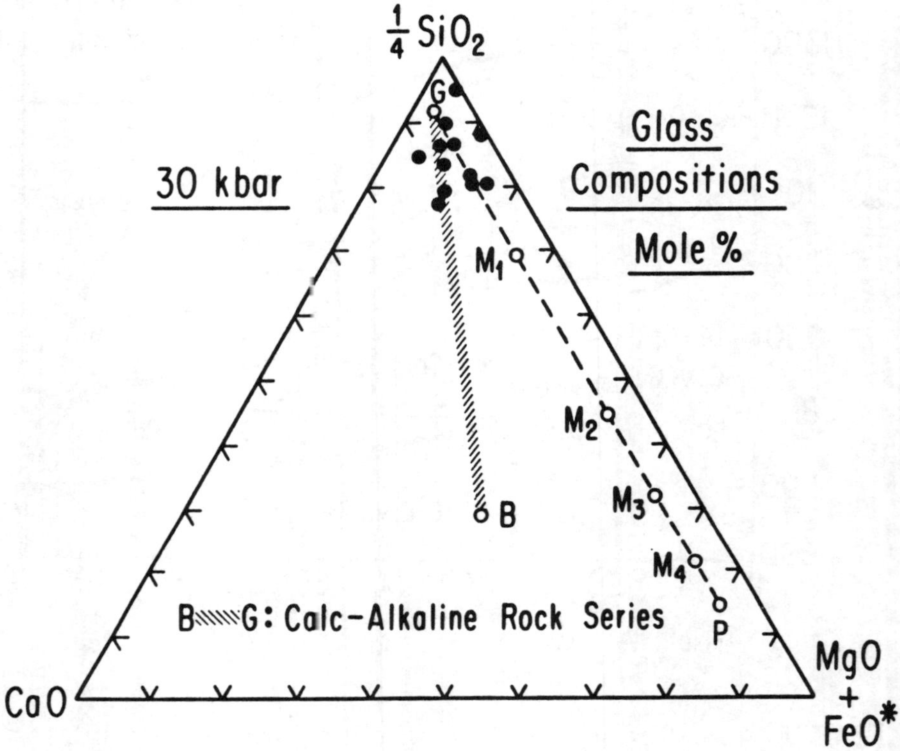

FIGURE 5. The system peridotite-granite-H_2O: compositions of glasses projected into diagram for comparison with the starting mixtures (line PG) and the average chemical variation of the calc-alkaline igneous series (BG).

granite, indicating that solution of the peridotite components in the liquid was limited.

Consider H_2O-undersaturated granite liquid L at 900°C and 30 kbar (Figs. 3 and 4). If this is emplaced within peridotite and allowed to reach equilibrium at constant pressure and temperature, the bulk composition of the mixing system would change along the line from L towards P. The liquid composition precipitates phlogopite-garnet-pyroxenite, with evolution of aqueous vapor, until the last liquid is used up at the solidus (Fig. 4). Throughout this process, the liquid composition remains rich in SiO_2 (Fig. 5). The effect of adding peridotite to the siliceous magma is to cause precipitation of cafemic minerals, with the Na_2O and K_2O from the liquid being transferred to jadeite component in clinopyroxene and to phlogopite, respectively. The final product is a body of phlogopite-garnet-websterite, with mineral compositions differing from those in the host peridotite, especially with respect to the clinopyroxene.

The mixing products in the system granite-peridotite-H_2O are directly analogous to those in the model system $KAlSiO_4$-Mg_2SiO_4-SiO_2-H_2O (Fig. 1). The join Ph-En in the model system serves the same function as a mineral-precipitation barrier between the low-temperature, hydrous siliceous liquid and the subsolidus peridotite, as does the subsolidus interval for phlogopite-garnet-websterite in the rock system (Fig. 4).

REACTION PRODUCTS BETWEEN HYDROUS GRANITE MELT AND PERIDOTITE

Sekine and Wyllie (1983) sealed crushed samples with water into gold capsules. In one arrangement, material was added to the capsule in the sequence water, granite, and peridotite, from bottom to top of the capsules, which were set vertically during a run. The peridotite was subsolidus in runs between 850°C and 1,000°C and just above the solidus at 1,050°C. The granite was completely molten in all runs. New minerals developed along the

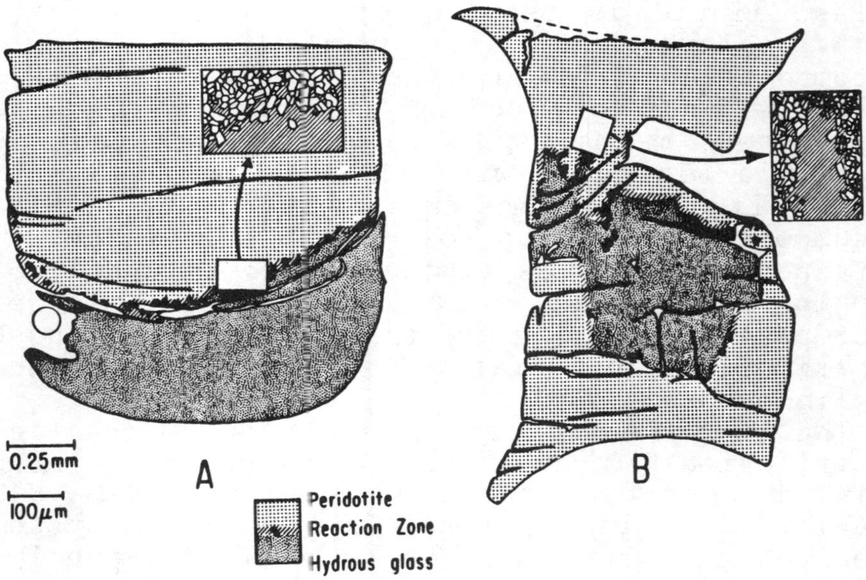

FIGURE 6. Sketches of polished sections of quenched run products for crystalline peridotite above granite melt. A. Run #30, at 1,000^{o}C for 45 hours. Minerals in the reaction zone are orthopyroxene, jadeitic clinopyroxene, and quartz. B. Run #10, at 850°C for 120 hours. Minerals in the reaction zone are orthopyroxene, jadeitic clinopyroxene, and suspected quartz. Phlogopite is distributed through the peridotite mass.

boundary between the granite liquid and the peridotite.

Fig. 6A is a sketch of the polished surface through a typical charge. The garnet peridotite consists of tightly interlocking grains, and the granite is a body of clear glass. The reaction zone between the granite glass and peridotite contained large crystals up to 50 μm diameter. In some runs, the larger minerals are isolated from the reaction zone, and completely enclosed by granite glass.

The peridotite cylinder sank into the granite liquid, as illustrated in Fig. 6A by the presence of narrow borders of granite glass occurring between peridotite and capsule walls. The reaction zone between the two materials was as well developed along the vertical boundaries as along the horizontal boundaries between peridotite and granite liquid.

Run #10 is significantly different physically from all other runs, as illustrated in Fig. 6B. The granite body has lost its regular outline. It appears to have been quenched in the process of rising through the peridotite, which now surrounds the glass completely, at least in this section. Fingers of granite glass extend into the overlying peridotite from the main granite mass.

The results are influenced by processes occurring in stages. The first stage involves the melting of granite and the reaction of peridotite with aqueous vapor. We described the product as metasomatized peridotite. During and after withdrawal of the aqueous vapor into the granite liquid, reaction between the liquid and peridotite causes precipitation of new minerals in a hybrid reaction zone. The hybridization zone remains narrow through 120 hours, the longest run completed.

The mineral assemblages produced by hybridization contain the ubiquitous products orthopyroxene and jadeitic clinopyroxene. The occurrence of additional minerals garnet, quartz, and phlogopite does not fit any obvious sequence with respect to temperature.

IMPLICATIONS FOR MAGMA GENESIS

Fig. 7 is an idealized cross-section through a subduction zone, with slab dipping at 45°, which illustrates the relative positions of source materials, dehydration and melting reactions, and identifies locations for the metasomatism and hybridization discussed below. The thermal structure corresponds to a subducted crust cooled (but not cooled to extreme conditions) by endothermic dehydration reactions (Anderson et al. 1978, 1980), and a mantle wedge warmed by induced convection (Toksöz and Hsui, 1978). With the present uncertainty about thermal structures, we have much latitude for interpretation of petrological processes. Reactions in subduction zones are commonly discussed in qualitative fashion without reference to phase boundaries with specific depth-temperature values. Possible reactions are more effectively evaluated when plotted in their correct relative positions for particular thermal structures. The locations of dehydration fronts, solidus boundaries, areas of potential melting and the variations of their positions for different thermal structures were reviewed by Wyllie (1979, 1982).

The aqueous solutions released in dehydration reactions may be transported across the solidus line for amphibolite or quartz eclogite in the subducted oceanic crust, causing partial melting represented in Fig. 7 at the location M. H_2O-rich, but H_2O-undersaturated, magma rises from the region M into the overlying peridotite. Will the liquid continue to rise, or will it hybridize and solidify?

From evaluation of the model systems, and from equilibrium mixing studies and reaction zone studies in the system peridotite-granite-H_2O, we can conclude that during hybridization

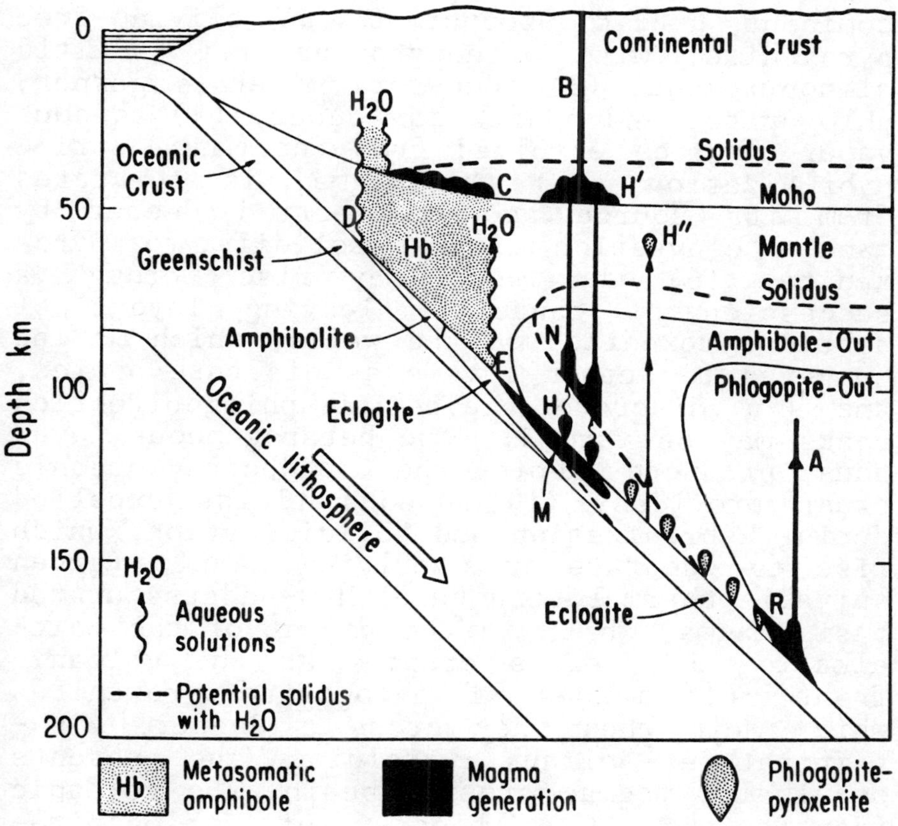

FIGURE 7. Idealized petrological cross-section through subducted oceanic lithosphere, for a specific thermal structure with isotherms dominated by cooling of oceanic crust by endothermic dehydration reactions, and warming of mantle wedge by induced convection. See Wyllie (1979, 1981, 1982) for the locations of dehydration and melting fronts for varied thermal structures.

between relatively cool hydrous siliceous magma and the hotter overlying mantle peridotite: (1) the magma experiences only minor changes in composition at temperatures below the solidus for peridotite-H_2O, (2) the dominant hybrid product is an olivine-free pyroxenite, with orthopyroxene and jadeitic clinopyroxene, (3) accessory minerals garnet, phlogopite, and quartz may occur, (4) aqueous vapor must be exsolved from the magma during hybridization, (5) the magma, if separated from its source and transported deeper by mantle convection, would solidify to pyroxenite, (6) the magma may rise through a succession of conduits, leaving layers of hybrid pyroxenites on the walls, which may be conveyed to deeper levels as discrete bodies, and (7) discrete bodies of phlogopite-rich rocks may be formed. The heterogeneous layer thus produced above the subducted oceanic crust provides: (1) aqueous fluids expelled during hybridization and solidification, which rise to generate in overlying mantle (given suitable thermal structure) H_2O-undersaturated basic magma, which is the parent of the calc-alkalic rock series erupted at the volcanic front; (2) masses of phlogopite-pyroxenites which melt when they cross a deeper, high-temperature solidus, yielding the parents of alkalic magmas erupted behind the volcanic front; and (3) blocks of phlogopite-pyroxenites which may rise diapirically for long-term residence in continental lithosphere, H", and later contribute to the potassium (and geochemically-related elements) involved in some of the continental magmatism, with geochemistry ascribed to mantle metasomatism.

If hybridization occurs in subduction zones, a great variety of possible magmatic processes is introduced, involving rocks, magmas, and aqueous fluids. Nicholls and Ringwood (1973) proposed that hybrid olivine pyroxenite generated in the mantle wedge constituted a suitable source material for the calcalkaline magma series, with suitable adjustments caused by crystal fractionation at shallower depths. We have demonstrated that the product of hybridization is an olivine-free pyroxenite, different chemically and mineralogically from other materials considered as source rocks for magmas in subduction zone environments. However, olivine would be involved for melting occurring at the boundary between peridotite and pyroxenite. The clinopyroxene is extremely enriched in jadeite compared with that in mantle peridotite, and probably richer in jadeite than the clinopyroxene in subducted oceanic crust (although uncertainties about the degree of metasomatism of oceanic crust temper this conclusion). In oceanic crust, the path of progressive fusion is controlled by clinopyroxene and garnet, whereas in a hybrid pyroxenite the liquid path would be controlled by orthopyroxene and clinopyroxene. In order to translate the results of isotope and trace element geochemistry of subduction zone igneous rocks, we need to know more not only about the compositions of melts derived by partial fusion of mantle and subducted oceanic crust, but also more about the compositions of various hybrid rocks that may be produced in the mantle wedge.

ACKNOWLEDGEMENTS

This research was supported by the Earth Sciences Section of the National Science Foundation, Grants EAR 82-06178 and EAR 83-41623.

REFERENCES

Anderson, R.N., Delong, S.E. and Schwartz, W.M. (1978). Thermal model for subduction with dehydration in the downgoing slab. J. Geol. 86, 731-739.

Anderson, R.N., Delong, S.E. and Schwartz, W.M. (1980). Dehydration, asthenospheric convection and seismicity in subduction zones. J. Geol. 88, 445-451.

Nicholls, I.A. and Ringwood, A.E. (1973). Effect of water on olivine stability in tholeiites and production of silica-saturated magmas in the island arc environment. J. Geol. 81, 285-300.

Sekine, T. and Wyllie, P.J. (1982a). Phase relationships in the system $KAlSiO_4$-Mg_2SiO_4-SiO_2-H_2O as a model for hybridization between hydrous siliceous melts and peridotite. Contr. Mineral. Petrol. 79, 368-374.

Sekine, T. and Wyllie, P.J. (1982b). Synthetic systems for modeling hybridization between hydrous siliceous magmas and peridotite in subduction zones. Jour. Geol. 90, 734-741.

Sekine, T. and Wyllie, P.J. (1982c). The system granite-peridotite-H_2O at 30 kbar, with applications to hybridization in subduction zone magmatism. Contr. Mineral. Petrol. 81, 190-202.

Sekine, T. and Wyllie, P.J. (1983). Experimental simulation of mantle hybridization in subduction zones. Jour. Geol. 91, 511-528.

Sekine, T., Wyllie, P.J. and Baker, D.R. (1981). Phase relationships at 30 kbar for quartz eclogite composition in CaO-MgO-Al_2O_3-SiO_2-H_2O with implications for subduction zone magmas. Amer. Miner. 66, 938-950.

Stern, C.R. and Wyllie, P.J. (1978). Phase compositions through crystallization intervals in basalt-andesite-H_2O at 30 kbar, with implications for subduction zone magmas. Amer. Miner. 63, 641-663.

Toksöz, M.N. and Hsui, A.T. (1978). Numerical studies of back-arc convection and the formation of marginal basins. Tectonophysics 50, 177-196.

Wyllie, P.J. (1979). Magmas and volatile components. Amer. Miner. 64, 469-500.

Wyllie, P.J. (1981). Plate tectonics and magma genesis. Geol. Rundschau 70, 128-153.

Wyllie, P.J. (1982). Subduction products according to experimental prediction. Geol. Soc. Amer. Bull. 93, 468-476.

Wyllie, P.J., Donaldson, C.H., Irving, A.J., Kesson, S.E., Merrill, R.B., Presnall, D.C., Stolper, E.M., Usselman, T.M. and Walker, D. (1981). Chapter 3, Experimental petrology of basalts and their source rocks, p. 493-630 in "Basaltic Volcanism on the Terrestrial Planets," Basaltic Volcanism Study Project, Pergamon, New York (1286 p.).

Yoder Jr., H.S. (1976). Generation of Basaltic Magma. Nat. Acad. Sci., Washington DC.

CONDITIONS FOR MELTING AND METASOMATISM IN THE EARTH'S MANTLE

PETER J. WYLLIE
Division of Geological and Planetary Sciences, California Institute of Technology, Pasadena, California 91125 USA

ABSTRACT

If we know the compositions of mantle rocks at various depths, and the geotherm in different tectonic environments, then the conditions for melting are defined by experimentally determined solidus curves. The term "metasomatism" in crustal processes is defined as reaction by solution or vapors, not by melts or magmas, and the same definition should apply to mantle processes (reaction with magmas is "hybridization"). H_2O, CO_2, or both are made available for metasomatic reactions at deep dissociation fronts, or by solidification of volatile-charged magmas. Regions eligible for metasomatism are limited by solidus curves above which the melts dissolve volatile components. Beneath the lithosphere, there can be no metasomatism between about 120 and 260 km, because melting intervenes. Solidification of kimberlitic magmas at the base of continental lithosphere is a source of metasomatic fluids. Mantle metasomatism is expected in several regions above subducted oceanic lithosphere, interspersed with magmatic events. Major differentiation of the earth is accomplished by melting, but metasomatism may cause significant redistribution of some elements.

Proceedings of the 27th International
Geological Congress, Volume 9, pp. 581–604
PETROLOGY (IGNEOUS AND METAMORPHIC ROCKS)

INTRODUCTION

The chemical differentiation of the earth into mantle and crust is accomplished by magmatism. Magmas are generated when bodies of rock are transported across the depth-temperature limits of melting curves by: (1) physical convection, (2) temperature increase arising from tectonic conditions, or (3) depression of melting boundaries to lower temperatures by influx of volatile components. The conditions for melting in the mantle would be easily defined if we knew: (1) the compositions of mantle rocks at various depths, (2) the experimentally determined solidus curves, and (3) the geotherm in different tectonic environments. However, there remain many uncertainties in the precise definition of these three items.

The abundances of major and trace elements in many magmatic rocks are difficult to explain in petrogenetic schemes involving a mantle peridotite source. The concentrations of trace elements in magmas appear to be decoupled from those of major elements. It has become common practice in recent years to alleviate this problem by appealing to metasomatism caused by unspecified fluids, to adjust the mantle composition in order to satisfy the geochemical data (Walker, 1983). There is evidence that the mantle is heterogeneous, and that fluids have migrated through mantle rocks. Some investigators describe these fluids as dense, volatile-rich solutions, and others refer to silicate melts, perhaps similar in composition to kimberlites.

It is important in petrological discussions to retain the distinction between melt (liquid), vapor (or dense gas, solution), and the descriptive term "fluid" which can refer to either phase. Liquid and dense vapor are two distinct phases with different compositions and different properties. At high pressures and temperatures, the compositions and properties of these two fluid phases approach each other, but all experimental data so far available indicate that for normal rock compositions with H_2O and CO_2 in the upper mantle, liquid and vapor maintain their separate identities. They can co-exist with each other. Liquids can exist only at temperatures above the solidus, whereas a vapor can exist through a wide range of temperatures below the solidus. Liquid may solidify and exsolve vapor. Vapor may cross a solidus boundary, and dissolve in liquid.

Metasomatism was defined originally for crustal processes, in which the compositions of rocks undergoing metamorphism were changed through the action of a mobile pore fluid, usually rich in super-critical H_2O, with CO_2 and CH_4 becoming significant in calcareous rocks. Changes in metamorphic rock compositions caused by addition of magmas, or by expulsion of melt from a rock undergoing anatexis, have not been described as metasomatic. There is a large vocabulary, including the terms migmatite or hybrid, to describe a rock formed by impregnation of magmas. If the nature of a fluid changing the composition of mantle rock can be identified as a silicate magma, then this is not a metasomatic change, unless the current definition of metasomatism in crustal rocks is extended.

Metasomatism of mantle requires the presence of volatile components. H_2O, CO_2, or both are made available for metasomatic reactions in the upper mantle by: (1) release at deep dissociation fronts, (2) solidification at depth of volatile-charged magmas, or (3) upward migration of primordial volatiles.

SOLIDUS CURVES FOR MANTLE MATERIALS

Candidates for upper mantle rocks include lherzolite, harzburgite, and eclogite. Fig. 1 illustrates solidus curves determined experimentally for synthetic pyrolite and four natural peridotites. The more refractory peridotites should have higher MgO/(MgO+FeO). The extent of depletion or enrichment of the rocks in incompatible trace elements has little effect on the phase relationships. At pressures to 30 kbar, the solidus for quartz-free eclogite is barely distinguishable (within experimental limits) from that of undepleted lherzolite; the liquidus for dry basalt is not much higher in temperature. Quartz eclogite (tholeiite) begins to melt at temperatures significantly lower. None of the solidus curves in Fig. 1 was closely defined by good experimental brackets.

Fig. 2 compares the original solidus of Kushiro et al (1968) for a spinel lherzolite (KSA in Fig. 1) with a determination by Takahashi and Kushiro (1983). In the more detailed recent study, the solidus is lower by 70-100°C. The results can be interpreted as representing three curves defining cusps at transitions from plagioclase lherzolite to aluminus-pyroxene lherzolite at about 11 kbar, and to garnet lherzolite at about 26 kbar, consistent with the results of Presnall et al (1979) for model lherzolite in $CaO-MgO-Al_2O_3-SiO_2$.

The addition of H_2O and CO_2 to peridotite introduces hydrous minerals and carbonates, and lowers the solidus. Fig. 3A is a partly schematic phase diagram for peridotite with small amounts of CO_2 (<5% by weight), H_2O (< 0.4%; > 0.02%), and with $CO_2/(H_2O+CO_2)$=0.3(mole ratio). Subsolidus

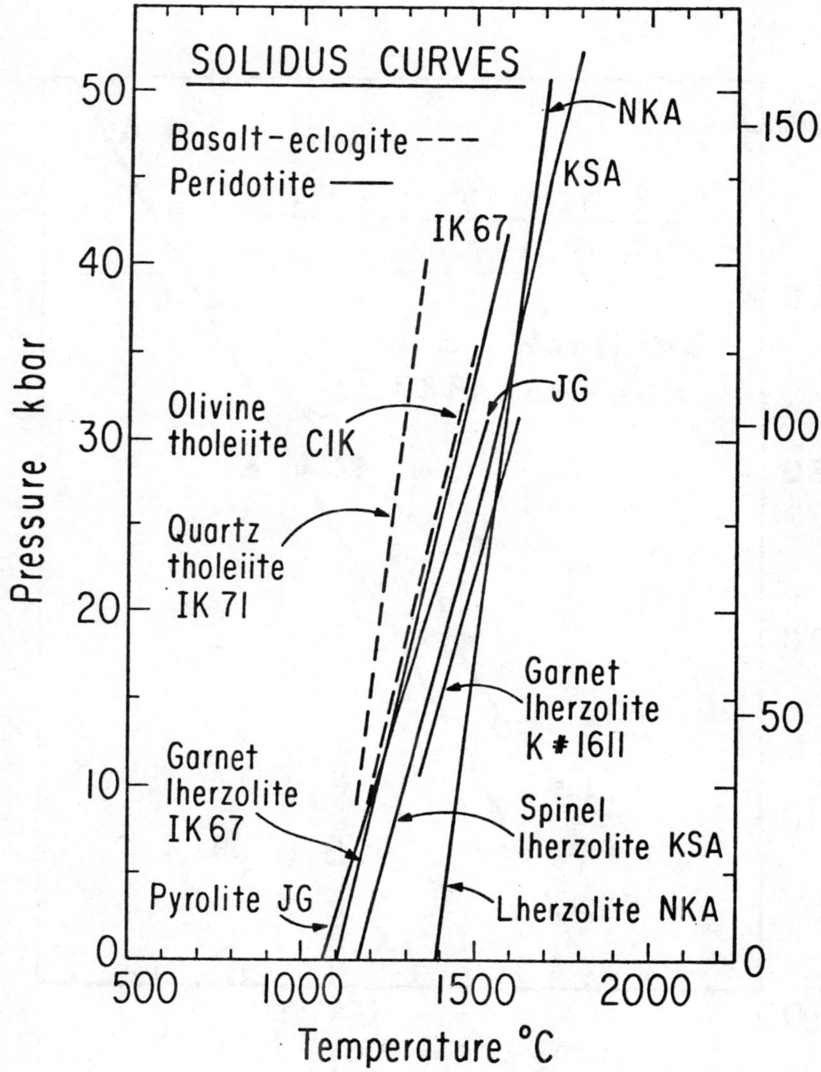

FIGURE 1. Solidus curves measured for various peridotites, for olivine tholeiite (eclogite), and for quartz tholeiite (eclogite). Abbreviations: IK, Ito and Kenedy, 1967; JG, Jaques and Green, 1980; K, Kushiro, 1973; KSA, Kushiro et al, 1968; NKA, Nishikawa et al, 1970; CIK, Cohen et al, 1967; IK, Ito and Kennedy, 1971.

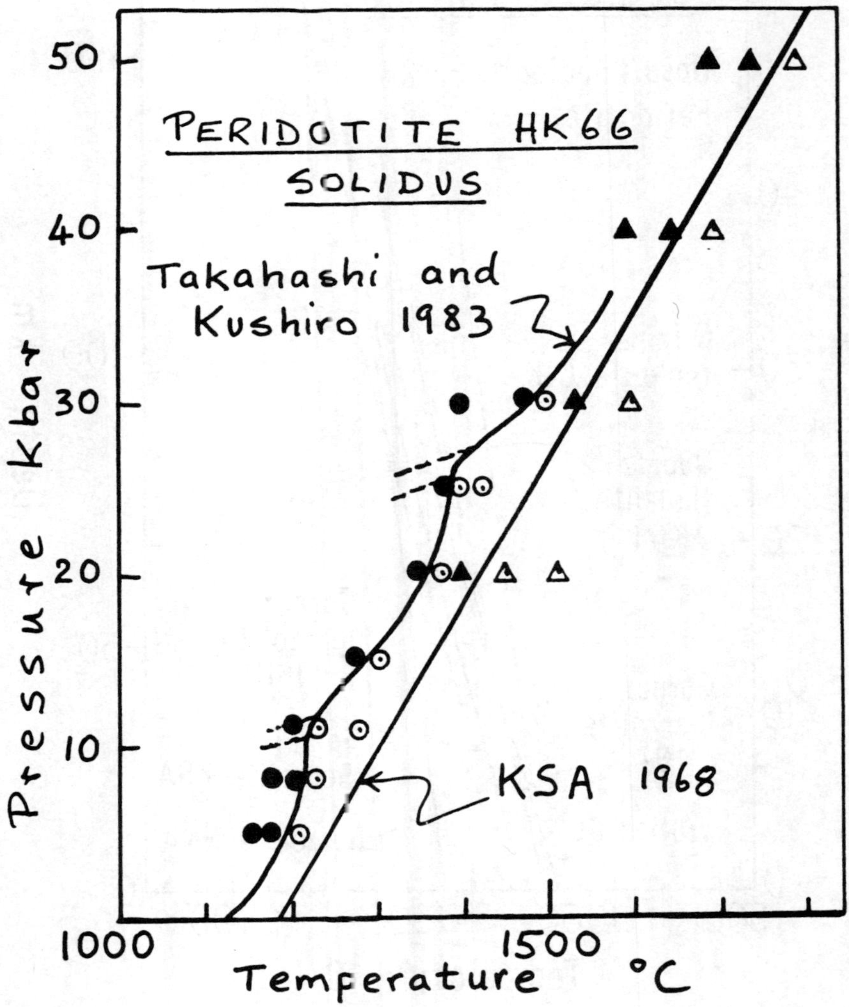

FIGURE 2. Solidus curve for dry peridotite KH66, defined by runs shown as circles, after Takahashi and Kushiro (1983). Data from previous study (Kushiro et al, 1968, triangles and straight line) correspond to KSA in Fig. 1.

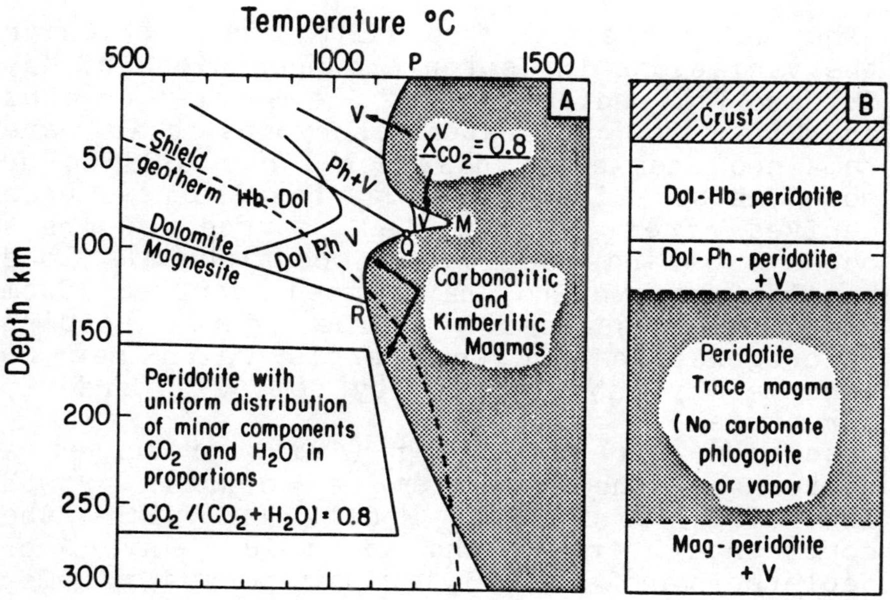

FIGURE 3. (A) The phase boundaries for peridotite-H_2O-CO_2 on a depth-temperature diagram, assuming vapor with high CO_2/H_2O. (B) The petrology of a mantle cross-section for peridotite assuming uniform distribution of minor components CO_2 and H_2O, and a shield geotherm (Wyllie, 1980).

minerals represented are phlogopite, amphibole, dolomite, and magnesite. The solidus is the line PMQR, and the shaded area shows the region of partial melting. Along QR, the vapor phase is greatly enriched in H_2O/CO_2 with increasing pressure. At depths greater than about 100 km, the solidus for peridotite-H_2O is at temperature only slightly below that for peridotite-CO_2-H_2O.

The geochemistry of basalts, in particular the variations in isotopic compositions of Nd, Sr, and Pb, indicates that the mantle contains at least two separate reservoirs that have remained chemically distinct for a billion or more years. Mid-ocean ridge basalts have been derived from a depleted source, whereas oceanic island basalts and continental flood basalts appear to have been derived from undepleted sources. The data require heterogeneities in the mantle which may be achieved by layering, or by some less orderly structure.

Jacobsen and Wasserburg (1981) presented a cartoon of the mechanism for crustal growth according to their Model I, with the continental crust and a thick layer of depleted mantle overlying undepleted mantle, from which the two overlying reservoirs had been derived by differentiation. This model can be represented in Fig. 4A by the layered mantle considered by Richter and McKenzie (1981). They assumed that the separate reservoirs, situated below the lithosphere (Layer 1), consisted of superimposed Layers 2 and 3, convecting without significant mass transfer across the boundary between them.

Fig. 4B illustrates Ringwood's (1982) modification of the pyrolite-basalt model (see Ringwood, 1975 for summary). The lithosphere is composed of harzburgite and lherzolite, depleted following extraction of basalt for oceanic crust at mid-oceanic ridges. The layered mantle structure is caused by phase transitions in pyrolite composition (undepleted). The seismic low-velocity zone is attributed to incipient melting at temperatures above the solidus, a-b, for peridotite-H_2O-CO_2 (Fig. 3). Ringwood (1982)

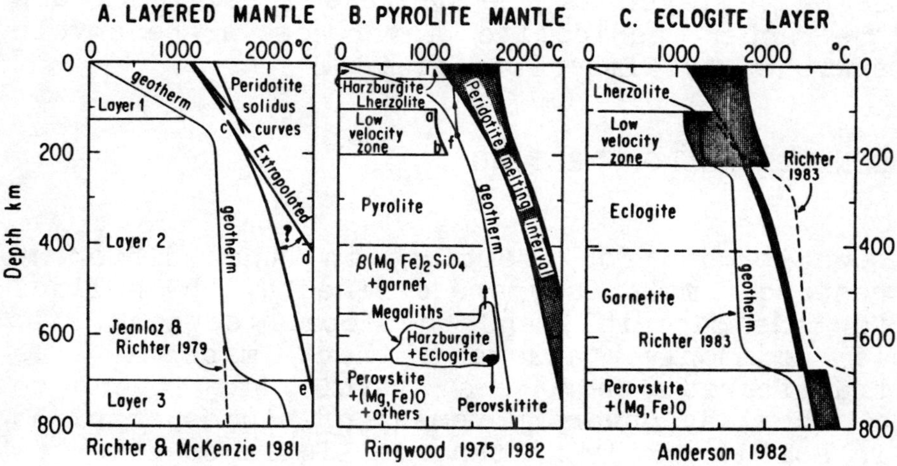

FIGURE 4. Proposed mantle structures and geotherms compared with experimentally determined and extrapolated melting conditions for peridotite and eclogite (after Wyllie, 1984). (A) Peridotite solidus curves from Fig. 1, extrapolated along either ce, or cd. (B) Estimated peridotite melting interval, after Wyllie (1981). The curve ab is the solidus QR with buffered vapor, from Fig. 3A. (C) Eclogite melting interval extrapolated from 100 km, based on solidus from Fig. 1.

identified a second mantle reservoir composed of megaliths of depleted, subducted, cool lithosphere enclosing eclogite from oceanic crust, density-balanced at about 650 km depth. Fig. 4C shows Anderson's (1982) model based on the geochemistry of basalts, kimberlites and peridotites, and consistent with seismic data. The upper mantle of peridotite composition contains a density-balanced layer of eclogite perched between 220 km and 670 km. The low-velocity zone is attributed to incipient melting, as in Fig. 4B. The eclogite layer is the depleted reservoir, and the upper peridotite layer is a reservoir enriched by kimberlite fluids.

CONDITIONS FOR MELTING

Conditions for melting in the different reservoir materials are defined by the solidus for the material, and the local geotherm. It is commonly assumed that material is transferred from one level or layer to another, by upward movement of plumes or blobs in solid, partly molten, or liquid state.

Fig. 4A shows the solidus curves for different peridotites, transferred from Fig. 1. The cusps of Fig. 2 are barely distinguishable on the scale of Fig. 4. A linear extrapolation along c-d provides very high solidus temperatures at depth. The calculated Simon-type fit along c-e (Griggs, 1972) is probably a better representation of the peridotite solidus. This is paired with an estimated liquidus curve in Fig. 4B (Wyllie, 1981). The solidus a-b is for peridotite in the presence of H_2O+CO_2. The schematic, extrapolated melting interval for peridotite in Fig. 4B is replaced between 220 km and 670 km in Fig. 4C by an estimate of the melting interval for eclogite. The extrapolation preserves the narrow melting interval known to occur through 30 kbar, and the proximity of the solidus for quartz-free eclogite with that for peridotite (Fig. 1).

Ringwood (1975) used the calculated geotherm shown in Fig. 4B, with diapiric uprise of mantle masses tracing adiabatic paths such as that rising from point *f*. Jeanloz and Richter (1979) presented an improved geotherm for a conducting lithosphere overlying a convecting mantle (Fig. 4A). Richter and McKenzie (1981) added the third Layer 3, convecting independently of Layer 2, and determined from laboratory and numerical experiments that there is a marked increase in temperature between the two convecting layers. Anderson's (1982) mantle in Fig. 4C has an additional convecting layer, compared with the model of Richter and McKenzie (1981) in Fig. 4A. This introduces another step into the geotherm. The solid line in Fig. 4C is a conservative estimate for the geotherm with this mantle model, but the dashed line is a more realistic estimate of what the additional convecting layer would do to the geotherm (Richter, 1983, personal communication).

For a mantle with separate convecting layers, as represented in Figs. 4A and 4C, the solidus curves provide constraints on the dimensions of the layers, because of the large step-like increases in temperature associated with each boundary. With the results used by Richter and McKenzie (1981) in Fig. 4A, the temperature at 700 km is not high enough to cause partial melting of Layer 3. However, they pointed out that if the bottom of Layer 2 were shallower than about 500 km, then widespread melting is suggested by their calculation. Such an arrangement would be untenable. Therefore, it is not clear how the two-layer convecting mantle could have developed into what is presently a stable condition, if Layer 2 is the depleted mantle reservoir which has grown through time at the expense of undepleted source Layer 3, as in Model I of Jacobsen and Wasserburg (1981). The problem is exacerbated if mantle temperatures were even higher in early earth history. Jacobsen and Wasserburg (1981) proposed that blobs of undepleted mantle rise from Layer 3 and intersect the solidus at shallower levels. No physical mechanism was proposed for escape of material from Layer 3.

Ringwood (1982) invoked delayed thermal buoyancy of sunken lithosphere as a mechanism for causing uprise of deep material. He suggested that the depleted lherzolite of the lower lithosphere is stripped off during subduction, and resorbed into the upper mantle providing a depleted source material. The deep, cool megaliths derived from the upper part of subducted oceanic lithosphere (Fig. 4B) reach thermal equilibrium with surrounding mantle at 650 km on a time scale of 1-2 billion years, which is followed by buoyant uprise of the harzburgite, rendered newly fertile by liquids from partial melting of entrained eclogite. The enriched harzburgite is incorporated into the lithosphere, yielding the alkaline basaltic suite by small degrees of partial melting. Parts of this model of multistage irreversible differentiation are inconsistent with the constraints. With the geotherms in Figs. 4A and 4B, the megalith at 650 km remains well below the extrapolated solidus temperature for eclogite (Fig. 4C), even in the presence of H_2O. Adiabatic uprise of buoyant megalith from the geotherm at 650 km in Fig. 4B would probably carry both harzburgite and entrained eclogite across their solidus curves in the depth interval 200-150 km. Adiabatic uprise of newly buoyant harzburgite from the geotherm in Fig. 4A at 650-700 km depth would carry it to a position below the lithosphere lid, Layer 1, without reaching the solidus curve of either harzburgite or entrained eclogite.

Anderson (1982) suggested that extensive partial melting of primitive mantle during accretion was followed by formation of a cumulate eclogite layer, as illustrated in Fig. 4C, with an overlying peridotite layer enriched by residual kimberlite fluids from below. Convection within the eclogite layer deforms the upper boundary, raising it above the solidus temperature. Partial melting produces a buoyant diapir with adiabatic ascent leading to extensive melting and eruption of mid-ocean ridge basalts. The eclogite layer is replenished by subduction of the oceanic crust derived directly from it some million years earlier. Using the more conservative of the two geotherms calculated by Richter, the temperature near the upper boundary of the eclogite layer is close to the solidus for both eclogite and peridotite. If the eclogite layer melts, one might expect the adjacent peridotite to melt as well. Clearly, we need better estimates of solidus curves for both rocks at 200 km.

The temperature of melting in the mantle is lowered by the presence of volatile components (Wyllie, 1979, for review) and potassium (Takahashi and Kushiro, 1983). If traces of these components are available, incipient melting occurs within mantle rocks in a depth interval of about 120 km to 260 km, as shown in Fig. 3B. Actual depths depend strongly on the nature of the geotherm (Fig. 4).

CONDITIONS FOR METASOMATISM

Conditions for metasomatism are limited to depth-temperature regions between the phase boundaries for decomposition of hydrous minerals or carbonates, and the solidus curves for the beginning of melting of the mantle materials in the presence of H_2O, CO_2, or other volatile components. These conditions depend on the composition of the mantle, the availability of volatile components, and the local geotherm. The geotherm depends upon the structure of the upper mantle, and the type of convection occurring. The models in Fig. 4 show a general increase in temperature as a

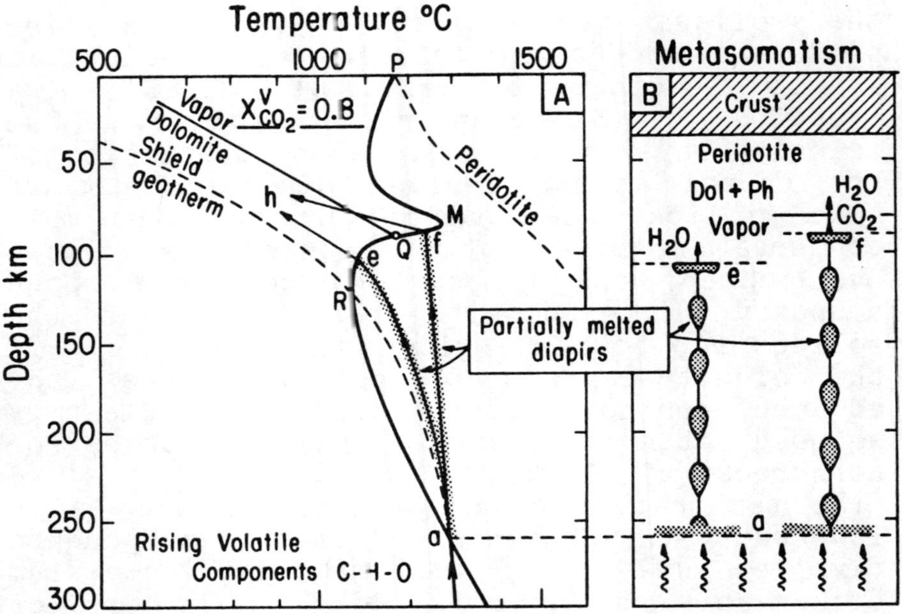

FIGURE 5. Compare Fig. 3. In the absence of volatile components, the solidus curve PMQR extrapolated to 300 km has no significance. Volatile components from greater depth rise and cause partial melting at _a_. Partly melted diapirs rise, feeding magma chambers in the lower lithosphere at _e_ and _f_, and solidification of the magmas releases volatile components for metasomatism along paths such as _e-h_ and _f-h_ (Wyllie, 1980).

function of depth. However, in other tectonic environments such as subduction, there are depth intervals within which the temperature decreases with increasing depth.

Beneath stable plates, given volatile components, Figs. 3B, 4B and 4C show that a layer of incipient melting occurs at depths between about 120 km and 260 km. Therefore, metasomatism by dense solutions or vapors is limited to upper mantle shallower than 120 km, or deeper than 260 km.

Fig. 5 illustrates the conditions for metasomatism associated with the deep fractionation of kimberlites, or other low-SiO_2, high-K_2O melts. Fig. 5A reproduces the solidus and geotherm from Fig. 3A. The solidus for volatile-free peridotite shows that the mantle cross-section in Fig. 5B remains solid unless volatile components become available. If volatile components rise along the geotherm, they induce partial melting in the peridotite as they cross the solidus at a. Partly melted diapirs may follow an adiabatic path (a-f), or a path such as a-c intermediate between the adiabat and the geotherm.

Fig. 5B includes a diagrammatic representation of kimberlite magma ponding into chambers near the base of the lithosphere, as the diapirs reach the solidus MQR (Fig. 5A). Solidification of the magmas would yield concentrated vapors. The vapors would cool along paths such as f-h or e-h, reacting with overlying peridotite to produce combinations of phlogopite, amphibole, dolomite, and other minerals. The high-temperature CO_2-rich vapors become enriched in H_2O if the cooling path crosses the dolomite phase boundary (Wyllie, 1980).

According to this scheme, metasomatism is a consequence of kimberlite magmatism, rather than its precursory cause. If the term metasomatism is extended to include composition changes caused by the introduction of silicate magmas, then the layers in Figs. 3B (incipient melting) and 5B (traversed by partly melted diapirs) may be described as metasomatized mantle. The problem then is the source of the volatile components and other trace elements in the melt which give it the capability of enriching the mantle in the required geochemical fashion. Wyllie (1980) concluded that for kimberlite magmatism the most likely source was from deeper levels within the earth, as illustrated in Fig. 5.

CONDITIONS FOR MELTING AND METASOMATISM IN SUBDUCTION ZONES

If the geotherm is lowered to temperatures below the solidus curve for peridotite-H_2O (Figs. 3A and 5A), then volatile-rich fluids, if present, can flow without the formation of silicate melt. Release of vapors from deep dissociation fronts in subducted oceanic crusts leads to migration of fluids through large volumes of upper mantle. The geometry of the arrangement of metasomatic volumes, and volumes where the fluids dissolve in magmas, is strongly dependent on the geometry of subduction, the thermal history of the plate being subducted, and the thermal structure of the subduction zone, items which have not been well defined.

The subducted oceanic crust may be relatively warm, or significantly cooled by endothermic dehydration reactions (Anderson et al, 1978, 1980). The mantle wedge may be relatively cool, chilled by the subducting slab, or it may be heated by induced convection (Toksöz and Hsui, 1978). Phase boundaries for dehydration and melting at a convergent plate boundary are shown in Fig. 6 for four thermal structures. The experimental basis for these diagrams was reviewed by Wyllie (1979, 1982), and Wyllie and Sekine (1982). The dehydration front for serpentine (D-G) approximates the greenschist facies boundary, and the dehydration front for amphibolite separates amphibole from eclogite in the subducted crust. Partial melting occurs if H_2O passes

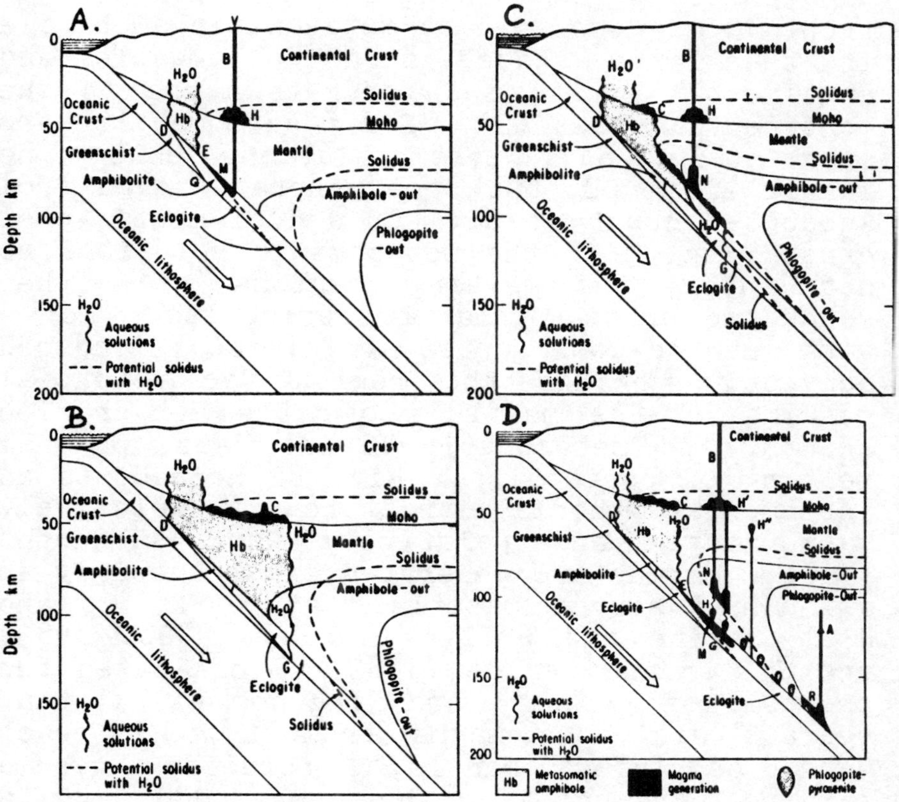

FIGURE 6. Locations of dehydration, metasomatism, and melting in subduction zones from results of experimental petrology, according to four thermal structures obtained by combinations of warm or cool (by endothermic dehydration) subducted ocean crust, with cool or warm (by induced convection) mantle. (A) Cool mantle-warm crust. (B) Cool mantle-cool crust. (C) Warm mantle-cool crust. (D) Warm mantle-warm crust (Wyllie and Sekine, 1982).

into the regions on the high-temperature sides of the dashed solidus boundaries for rock-H_2O.

Aqueous solutions rising from the dehydration front DG may precipitate metasomatic amphibole in the large shaded areas of mantle and continental crust, and generate magma in the continental crust (C). For a cool mantle-warm crust (Fig. 6A), most solutions enter the mantle along DE, but some generate magma from subducted crust at M. For a cool mantle-cool crust (Fig. 6B), the metasomatic solutions do not enter the melting regions of either subducted crust or mantle peridotite. For a warm mantle-cool crust (Fig. 6C), induced convection brings the region for potential melting of the mantle above the dehydration fronts, and metasomatic fluids cause partial melting of peridotite. Fig. 6D represents an intermediate thermal structure with relatively warm mantle and relatively warm subducted crust. Magmas are generated simultaneously in all three source materials at M, N, C, and H'. Sekine and Wyllie (1982a, 1982b, 1983) and Wyllie and Sekine (1982) investigated the proposal of Nicholls and Ringwood (1973) that hydrous siliceous magma from M would react with overlying mantle, producing hybrid olivine pyroxenite. They concluded that a series of discrete or overlapping bodies of hybrid phlogopite-pyroxenite would be transported downwards with the slab to R, where vapor-absent melting defines another possible site of magma generation. Solutions released during hybridization at H cause metasomatism, followed by partial melting in the shallower mantle at N.

COMPOSITIONS OF MAGMAS AND METASOMATIC SOLUTIONS

This topic requires the space of another paper. Recent studies of the compositions of melts from peridotite (Jaques and Green, 1980;

Takahashi and Kushiro, 1983) confirm that near-solidus melts range from basalts at moderate pressures, to alkali picrite and komatiite above 25 kbar. Wyllie (1984) reviewed available data and concluded that the generation of magmas more siliceous than basalts is limited to shallow suboceanic mantle with high H_2O contents; low-SiO_2, high-alkali magmas are generated at depths greater than 100 km by concentration of CO_2.

Wyllie and Sekine (1982) reviewed the limited experimental data on the composititions of aqueous solutions or vapors. The solubility of components is sensititve to the mineralogy of the host rocks as well as to depth and temperature. Total dissolved solute for phlogopite-bearing assemblages may reach as high as 50% (Ryabchikov and Boettcher, 1980). Mysen (1979, 1983) measured rare earth element partition coefficients between (H_2O+CO_2) vapors and silicates, and used the results to evaluate the processes of metasomatism and partial melting involved in the formation of alkali basalt from garnet lherzolite. Metasomatic vapors rising from greater to lower depths commonly follow paths of decreasing temperature (Figs. 4 and 5), but in subduction zones they may rise from lower temperature subducted oceanic slab into higher temperature mantle wedge (Fig. 6). Metasomatic vapors either scavenge peridotite for additional components, or change its composition by exchange or precipitation of components. We need much more experimental data.

PHYSICAL ASPECTS

The physical properties of rocks, rock mushes, melts, and vapors are just as important as phase diagrams and geotherms in evaluating the conditions for melting and metasomatism, but less well known. There are three mechanism for transport of magmas and metasomatic vapors through the mantle: (1) by percolation or porous flow of fluid through a crystalline matrix, (2) by the motion of rock mushes or segregated melt in diapiric form, and (3) by the propagation of elastic cracks filled with fluid. The compositions of segregated melts depend strongly on the physical conditions for the escape of melt from rock mush. Walker (1983) recently reviewed these topics. Spera (1981) presented the fluid dynamics of (H_2O+CO_2)-rich fluids rising through cracks as a first step toward a more complete theory of the physical chemistry of mantle metasomatic processes.

It is the physics of rock masses and melts and vapors which eventually controls the chemical differentiation of the earth. The major chemical differentiation is accomplished by melting and movement of magmas, but the distribution of minor and trace elements in the mantle and magmas is strongly influenced by the migration of fluids (melts or vapors) rich in volatile components.

ACKNOWLEDGEMENTS

This research was supported by the Earth Sciences Section of the National Science Foundation, Grants EAR-8311758 and EAR-8341623.

REFERENCES

Anderson, D.L., 1982. Chemical composition and evolution of the mantle. p. 301-318 in S. Akimoto and M. H. Manghani, eds., High-pressure Research in Geophysics. Reidel Publishing, Dordrecht.

Anderson, R.N., Delong, S.E. and Schwartz, W.M., 1978. Thermal model for subduction with dehydration in the downgoing slab. J. Geol. 86, 731-739.

Anderson, R.N., Delong, S.E. and Schwartz, W.M., 1980. Dehydration, asthenospheric convection and seismicity in subduction zones. J. Geol. 88, 445-451.

Cohen, L.H., Ito, K. and Kennedy, G.C., 1967. Melting and phase relations in an anhydrous basalt to 40 kilobars. Amer. Jour. Sci. 265, 475-518.

Griggs, D.T., 1972. The sinking lithosphere and the focal mechanism of deep earthquakes. p. 361-384 in E.C. Robertson, ed., The Nature of the Solid Earth, McGraw-Hill, New York, 677 p. 1972.

Ito, K. and Kennedy, G.C., 1967. Melting and phase relations in a natural peridotite to 40 kilobars. Amer. Jour. Sci. 265, 519-538.

Ito, K. and Kennedy, G.C., 1971. An experimental study of the basalt-garnet granulite-eclogite transition. p. 303-314 in J.G. Heacock, ed., The Structure and Physical Properties of the Earth's Crust. Amer. Geophys. Union Mon. 14.

Jacobsen, S.B. and Wasserburg, G.J., 1981. Transport models for crust and mantle evolution. Tectonophysics 75, 163-179.

Jaques, A.L. and Green, D.H., 1980. Anhydrous melting of peridotite at 0-15 kb pressure and the genesis of tholeiitic basalts. Contr. Mineral. Petrol. 73, 287-310.

Jeanloz, R. and Richter, F.M., 1979. Convection, composition, and the thermal state of the lower mantle. Jour. Geophys. Res. 84, 5497-5504.

Kushiro, I., 1973. Partial melting of garnet lherzolites from kimberlite at high pressures, p. 294-299 in P.H. Nixon, ed., Lesotho Kimberlites, Lesotho National Development Corporation, Maseru.

Kushiro, I., Syono, Y. and Akimoto, S., 1968. Melting of a peridotite nodule at high pressures and high water pressures. Jour. Geophys. Res. 73, 6023-6029.

Mysen, B.O., 1979. Trace element partitioning between garnet peridotite minerals and water-rich vapor: experimental data from 5 to 30 kbar. Am. Mineral. 64, 274-288.

Mysen, B.O., 1983. Rare earth element partitioning between (H_2O+CO_2) vapor and upper mantle minerals: experimental data bearing on the conditions of formation of alkali basalt and kimberlite. Neues Jahrb. Miner. Abh. 146, 41-65.

Nicholls, I.A. and Ringwood, A.E., 1973. Effect of water on olivine stability in tholeiites and production of silica-saturated magmas in the island arc environment. J. Geol. 81, 285-300.

Nishikawa, M., Kono, S. and Aramaki, S., 1970. Melting of lherzolite from Ichinomegata at high pressures. Phys. Earth Planet. Inter. 4, 138-144.

Presnall, D.C., Dixon, J.R., O'Donnell, T.H. and Dixon, S.A., 1979. Generation of mid-ocean ridge tholeiites. Jour. Petrol. 20, 3-35.

Richter, F.M. and McKenzie, D.P., 1981. On some consequences and possible causes of layered mantle convection. Jour. Geophys. Res. 86, 6133-6142.

Ringwood, A.E., 1975. Composition and petrology of the earth. 618 p. McGraw-Hill, New York.

Ringwood, A.E., 1982. Phase transformations and differentiation in subducted lithosphere: implications for mantle dynamics, basalt petrogenesis, and crustal evolution. Jour. Geol. 90, 611-643.

Ryabchikov, I.D. and Boettcher, A.L., 1980. Experimental evidence at high pressure for potassic metasomatism in the mantle of the earth. Am. Mineral. 65, 915-919.

Sekine, T. and Wyllie, P.J., 1982a. Phase relationships in the system $KAlSiO_4$-Mg_2SiO_4-SiO_2-H_2O as a model for hybridization between hydrous siliceous melts and peridotite. Contr. Mineral. Petrol. 79, 368-374.

Sekine, T. and Wyllie, P.J., 1982b. The system granite-peridotite-H_2O at 30 kbar, with applications to hybridization in subduction zone magmatism. Contr. Mineral. Petrol. 81, 190-202.

Sekine, T. and Wyllie, P.J., 1983. Experimental simulation of mantle hybridization in subduction zones. Jour. Geol. 91, 511-528.

Spera, F.J., 1981. Carbon dioxide in igneous petrogenesis: II. Fluid dynamics of mantle metasomatism. Contrib. Mineral. Petrol. 77, 56-65.

Takahashi, E. and Kushiro, I., 1983. Melting of dry peridotite at high pressures and basalt magma genesis. Amer. Miner. 68, 859-879.

Toksöz, M.N. and Hsui, A.T., 1978. Numerical studies of back-arc convection and the formation of marginal basins. Tectonophysics 50, 177-196.

Walker, D., 1983. New developments in magmatic processes. Rev. Geophys. Space Phys. 21, 1372-1384.

Wyllie, P.J., 1979. Magmas and volatile components. Am. Mineral. 64, 469-500.

Wyllie, P.J., 1980. The origin of kimberlites. Jour. Geophys. Res. 85, 6902-6910.

Wyllie, P.J., 1981. Plate tectonics and magma genesis. Geol. rundschau 70, 128-153.

Wyllie, P.J., 1982. Subduction products according to experimental prediction. Geol. Soc. Amer. Bull. 93, 468-476.

Wyllie, P.J., 1984. Constraints imposed by experimental petrology on possible and impossible magma sources and products. Phil. Trans. Roy. Soc. London, in press.

Wyllie, P.J. and Sekine, T., 1982. The formation of mantle phlogopite in subduction zone hybridization. Contr. Mineral. Petrol. 79, 375-380.